Modern Special Relativity

Johann Rafelski

Modern Special Relativity

A Student's Guide with Discussions and Examples

Springer

Johann Rafelski
Department of Physics
The University of Arizona
Tucson, AZ
USA

ISBN 978-3-030-54351-8 ISBN 978-3-030-54352-5 (eBook)
https://doi.org/10.1007/978-3-030-54352-5

This Springer imprint is published by the registered company Springer Nature Switzerland AG.
The registered company address is: Gewerbestrasse 11, 6330 Cham, Switzerland

Preface

This volume presents special relativity (SR) in a language accessible to students, while avoiding the burdens of geometry, space-time symmetries, the introduction of 4-vectors, and tensor calculus. The search for clarity in the fundamental questions about SR, the discussion of historical developments before and after 1905, the strong connection to current research topics, many solved examples and problems, and illustrations of the material in colloquial discussions are the most significant and original assets of this book. SR is presented in this volume such that nothing needs to be called paradoxical or apparent; everything is explained. The content of this volume develops and builds on the book *RELATIVITY MATTERS*,[1] including new examples, discussions, clarifications, and wording adapted for accessibility by a wider interested readership. This introduction to SR does not require 4-vector tools.

Accessible does not mean superficial: This introduction of SR offers conceptual insights that reach well beyond the usual SR contents. Readers will learn about developments ensuing upon the discovery of the theory of gravity force called general (or gravity) relativity (GR) not presented in these pages. On the other hand, the arising cosmological perspective complements the contextual setting of this book.

Some may ask why SR, after GR was created, remains a relevant and independent physics research area. In a good theoretical formulation such as is the case in GR, the dynamics of particles and of the field (in this case metric) are consistently formulated. However, SR was conceived to describe electromagnetic (EM) forces, and more than a century later we are still in search of a SR-founded formulation incorporating the EM forces in a manner consistent with EM field Maxwell equations.

The EM consistency problem, inherent to the concept of the radiation reaction force, is recognized as follows: The Maxwell equations allow all type of sources and we find that accelerated charges radiate and therefore lose energy, but the EM force equation, the Lorentz force, does not account for this loss. Accordingly, awaiting completion of the theoretical EM framework, this presentation of SR connects with research topics in particle, nuclear, and high intensity pulsed laser physics empirically using the well-established rules of SR.

[1] Johann Rafelski, *Relativity Matters*, (Springer Nature 2017), ISBN 978-3-319-51230-3.

In SR, there should be no paradoxes for well-prepared students, but mastering the fundamentals can be a challenge. Let me describe an event that illustrates this fact. In the 1980s, I tried to find a textbook in English to recommend to my students but in the process I discovered confusing presentations and misunderstandings in the English texts I was inspecting. At that point I had already published a book on the subject in German,[2] but in Cape Town where I was to teach, English was spoken. I turned to John S. Bell, of Bell inequality fame, for advice, since I believed that these SR texts would provoke his interest.

Despite our collaboration and friendship I was unaware of his preoccupation with teaching SR. Indeed, by writing to John I was adding gas to a fire. Here is his answer[3] of March 12, 1985:

> The only thing I can thoroughly recommend on relativity is my own paper. I enclose a copy. I refer there to the book of Janossy.[4] But it is very long, and insufficiently explicit that the Einstein approach is perfectly sound, and very elegant and powerful, (but pedagogically dangerous, in my opinion). What Rindler says[5] in the extract you send me, seems OK to me. There is a reference to Rindler in my paper. If you look it up, you will see that he agrees with the possibility of presenting things the way I advocate. I disagree with him only on pedagogy and history (his remark about Larmor not understanding time-dilation is astonishing to me).

A few interesting SR discussions and other SR events with J. S. Bell followed. To a large extent, this volume reflects on these intellectual adventures.

In the following decades I worked in research advancing many relativity-based topics such as Relativistic Heavy Ion Collisions, Relativistic Quark Model, and Relativistic Quantum Mechanics. Some of these projects resulted in books of their own. Finally, a decade ago, I turned to creating a new SR offering, which in 2017 concluded with the publication of *Relativity Matters*. As I was planning the updating of this volume, it became evident that inclusion of the more advanced SR material (Poincaré group, acceleration, detailed radiation (reaction) discussion, just to name a few items) necessitated considerable expansion. Accordingly, I split the contents into this SR introduction, to be followed some day in the not too distant future by a research level SR research monograph incorporating the challenges in the understanding of EM theory.

As this book progresses, the qualitative and historical introduction turns into a textbook-style presentation with many detailed results derived in an explicit manner. The reader reaching the end of this text needs knowledge of classical mechanics, a good command of elementary algebra, basic knowledge of calculus, and introductory know-how of electromagnetism. In the second volume I will

---

[2]*Spezielle Relativitätstheorie*, (Verlag Harri Deutsch, Frankfurt (1984, ISBN3-87144-711-0; 1989, ISBN 3-97171-1063-4; and 1992 ISBN 3-8171-1205-X. Published in Walter Greiner's *Theoretical Physics* series).

[3]See original letter facsimile in *Relativity Matters*, (Springer Nature 2017), ISBN 978-3-319-51230-3.

[4]L. Jánossy, *Theory of Relativity based on Physical Reality* Akadémiai Kiadó, Budapest 1971.

[5]W. Rindler, *Introduction to Special Relativity* Oxford 1982.

address SR research topics employing and exploiting the advantages that the 4-vector notation and, more generally, the covariant method offer. This second volume will reveal the unresolved questions related to relativistic charged particle dynamics and, more generally, to acceleration.

A notable part of this volume is presented in the format of discussions in which the teacher, *Professor*, and his graduate student, *Student*, are challenged by a brilliant but web-self-taught student called *Simplicius*.[6]

These conversations are representative of both the foundational concepts in SR and how students have challenged this author over the years. These conversations present the opportunity to explore what often remains unsaid when teaching SR. In these discussions I further explain how one should think about SR, addressing some frequent misunderstandings in colloquial language and commenting on renditions of the subject found for example on the WWW. These discussions also offer a vibrant opportunity to address conceptual errors seen in some "Modern Physics" texts.

[6]Simplicius of Cilicia, c.490–c.560, was a Greco-Roman mathematician and philosopher who wrote extensively on the works of Aristotle. Simplicius appears in Galileo Galilei's *Dialog Concerning the Two Chief World Systems* (Florence 1632). Simplicius Simplicissimus also appears in a 1668 book written by Hans Jakob Christoffel von Grimmelshausen.

Acknowledgments

The text as presented here was for the most part read and commented on by my school and study friend Bernd Maurer, who has also been teaching Special Relativity for many years. Bernd is a mathematician by vocation and insistent on detailed correctness as well as on presentation clarity. His questions and suggestions prompted many new threads and developments. The examples and exercises of this book were also evaluated by Martin Formanek, a graduate student in physics at that time. Martin also made many valuable suggestions about the text of this volume. Another test reader, Victoria Grossack, helped to improve the clarity of the presentation.

This manuscript was prepared in latex to publishers specifications but web publication requirements created formatting conflicts which introduced the need for extended contents recast, with associated formatting difficulties. I would like to thank Victoria, Martin, Will Price and Stefan Evans for their assistance in advancing the related pre-proof-reading process. Stefan also helped to prepare many complex technical drawings seen in the book. I thank the team for their kind support.

The scope and contents of this introduction was indirectly influenced by several readers of *RELATIVITY MATTERS*; I thank Giorgio Margaritondo (EPFL-Lausanne) for his interest and ensuing collaboration. Several comments made by Magda Ericson (Lyon and CERN), Torleif Ericson (CERN), and Harald van Lintel (EPFL-Lausanne) prompted my reexamination of the relativity work by Henri Poincaré and by Paul Langevin, who wrote on what some call the twin paradox. However, Langevin had a clear understanding of time dilation and did not consider it to be a paradox. This point, absent in *RELATIVITY MATTERS*, is made strongly in this volume. Denis Weaire FRS (Trinity College, Dublin) provided additional insights into the developments in the period between Maxwell and Einstein, including the work of the "Maxwellians": G.F. FitzGerald, Oliver Heaviside, Oliver J. Lodge, and others, which has influenced my understanding regarding the period between Maxwell and Einstein.

I thank all those involved for their kind help and interest. **I alone am responsible for any errors, omissions, and personal historical remarks and anecdotes,**

which in all cases are presented completely and to my best knowledge and memory.

The University of Arizona, Tucson
Summers 2019/20/21

Johann Rafelski

Research Profile of the Author

The era of modern physics began in the early twentieth century with the understanding of the consequences of electrodynamics and the resulting formulation of special relativity theory. Rapid development continued with quantum theory, which led to the discovery of antimatter and the realization that a transformation of energy into matter is possible, allowing us to describe the processes during the Big-Bang. Our current understanding of quantum theory shows that the structure and properties of the relativistic invariant quantum vacuum (i.e., the physical properties of matter-free space) determine the laws of physics. This understanding and the potential control of energy and matter conversion could affect the future of humanity as significantly as the discoveries in the fields of electromagnetism and relativity theory influenced our last 150 years.

Dr. Johann Rafelski, theoretical physicist, formerly a scientist at CERN and Professor at the Universities of Frankfurt and Cape Town, has been working for more than 30 years at The University of Arizona (Tucson, USA). For many years, his research focused on the areas described above. He works in several disciplines of subatomic physics to further the understanding of the nature of the quantum vacuum. Among other things, he investigates the behavior of matter under extreme temperature conditions and the effects of the strongest forces. Under such conditions, the quantum vacuum is so drastically changed that it can be investigated experimentally.

With collisions of atomic nuclei at the highest available energies, one can lay the groundwork for such investigations. Dr. Rafelski, as a scientist at CERN, has collaborated in the creation and development of this research program. The temperatures reached in the laboratory – 300 million times hotter than the surface of the Sun – can transform the normal form of nuclear matter into a new phase of hot quarks and gluons. The vacuum forces surrounding formerly confined quarks are dissolved upon reaching such high temperatures.

In his scientific work addressing the behavior of the new quark-gluon state of matter, Dr. Rafelski deals with transformation processes of energy into matter and antimatter, and especially with the production of strange quarks. These particles also allow conclusions to be drawn about the properties of quark-gluon plasma, which are the subject of experiments at CERN in Geneva and at BNL in New York. Further relevant experimental facilities are being readied: the FAIR project in Darmstadt, Germany, and NICA in Dubna near Moscow.

In investigating the behavior of individual particles under extreme conditions, Dr. Rafelski is especially interested in the effect of radiation, a vacuum frictional force caused by acceleration. Every charged accelerated body emits radiation. If the acceleration reaches a certain threshold, then this radiation generates a strong friction, which changes the particle dynamics decisively. Under these conditions, force field energy is converted into matter and antimatter. These processes help in the search for the mechanisms causing the formation of quark-gluon plasma. A related field of Dr. Rafelski's work is the radiation generated by laser pulses.

Other areas of research that interested Dr. Rafelski in the past decade include: the postulated cosmic neutrino microwave background, vacuum fluctuations caused by elemental forces and their relation to dark energy, dark matter in the form of massive compact ultra-dense objects (CUDOs), and the application of laser pulses of high intensity in nuclear fusion.

These research programs have led to a large number of scientific publications involving many students. More recently, Dr. Rafelski has become interested in teaching students the basics of special relativity theory with the perspective of historical development leading to the above new fields of research. His book, *Relativity Matters* (Springer 2017), introduces undergraduates to special relativity and instructs them on that subject through their graduate studies while also explaining the current research topics introduced above. Dr. Rafelski is also the editor of the open-access book: *Melting Hadrons, Boiling Quarks* – From Hagedorn Temperature to Ultra-Relativistic Heavy-Ion Collisions at CERN – With a Tribute to Rolf Hagedorn (Springer, 2016).

Prof. Dr. Johann Rafelski is a Fellow of the American Physical Society (APS). In 1990, he received a medal from the Collège de France for his presentation of strangeness in quark-gluon plasma, and in 2008 he was an Excellence Professor of the German Research Foundation (DFG). Dr. Rafelski is a Fulbright Fellow (2018 to 2021) and also an elected member of The University of Arizona Senate (2018 to 2022).

Contents

Part III The Lorentz Transformation

Part IV Measurement of Body Properties

Part V Space, Time, and the Doppler Shift

List of Insights

Acronyms

CMB	Cosmic Microwave Background radiation
CM-frame	Center of Momentum reference frame *This is the reference frame in which the sum of all momenta vanishes,* $\sum_i \boldsymbol{P}_i = 0$. *This concept replaces the usual center of mass, which is not a well-defined concept in Special Relativity. One often refers to CM, short for CM-frame of reference.*
EM	Electro-Magnetic fields, forces, etc.
GR	Gravity Relativity *The General Theory of Relativity addresses the force of gravity, hence in this book the acronym GR is used in this meaning.*
IO	Inertial Observer
LI	Lorentz Invariant
LT	Lorentz coordinate transformation
MM	Michelson-Morley
SR	Special Relativity (experiment and/or theory)

Part I

Space-Time, Light and the Æther

Part I is about: A general overview of SR concepts appropriate for a wide range of science-interested readers is presented. Characterization of key principles governing SR and placement of these in a greater context rather than mere mathematical formulation make this discussion a distinct part of the book. Personalities key to the development of SR are introduced following the historical path. Some of difficulties besetting the teaching of SR are introduced in discussions. A summary of the common misconceptions regarding SR phenomena closes Part I.

Introductory Remarks to Part I

Part I of this book is a general introduction explaining the background and context of the theory of special relativity (SR) at a level appropriate for a wide range of science-oriented readers. Characterization of key principles governing SR in an approachable manner, and placement of these in a greater context, make this a distinct part of the book.

Much of this discussion follows the historical path. Readers will discover how and why the legacy of Maxwell was a conundrum requiring the invention of SR: Maxwell died at the age of 48, without a chance to address the riddles his theory created. Readers of this chapter will appreciate the reasoning that led Einstein to choose the title "On the electrodynamics of moving bodies" in which he presents SR without addressing the process of dynamical change of the inertial motion of 'moving bodies'. Einstein's title characterizes the conceptual question that was created by Maxwell's electromagnetism, leading to the invention of SR theory.

We begin by introducing the principle of relativity, embodied in the name of the theory. We move on to recognize time as a coordinate necessary for characterizing events and used in our daily lives, for example, when we set meeting coordinates in time and space. This leads naturally to 1+3 dimensional Minkowski space, world lines, proper time, causality, and the process of measurement of space and time. We close Chap. 1 by answering the question – *How did relativity happen?* – both in terms of physics challenges and as a historical time line, a process that was not limited to the 40 years spanning the time between the eras of Maxwell and Einstein.

We move on in Chap. 2 to discuss how we measure space and time. We follow this with the discussion of several centuries of effort to determine the speed of light: What is the speed of light? How do we know its value? Who determined it, when and how? Maxwell's characterization of the speed of light in terms of EM phenomena is described.

The modern understanding of the æther (Ancient Greek: Pure air breathed by Gods; in the nineteenth century the medium allowing light to propagate, also called "luminiferous æther") is presented in an essay. We recall that in his 1905 publication, Albert Einstein denies the existence of an æther. We describe the evolving ideas of Langevin and Einstein which result, 15 years after SR, in the resurrection of the æther concept. In a private communication with Lorentz, Einstein explains that in his 1905 paper he should have restricted his statement to nonexistence of "æther velocity." Einstein places restrictions on æther properties: The new æther has no velocity and is non-material. We describe how this new relativistic æther entered the present day views about the quantum vacuum.

After this preparation, in Chap. 3 we address the Lorentz-FitzGerald body contraction introduced in the wake of the Michelson-Morley (MM) experiment. These changes in properties of material bodies saved the principle of relativity, made absolute motion undetectable, and thus removed need for the material æther. We discuss why the Lorentz-FitzGerald body contraction and time dilation are independent of each other and why both are real. We describe some misunderstandings surrounding these important insights, demonstrating that more than a century after the formulation of SR, there is still need for clarification and explanation of the meaning of SR phenomena.

What Is (Special) Relativity? 1

Abstract

We describe a few pivotal developments in recent centuries responsible for the creation of the relativistic understanding of our Universe. This chapter introduces the principle of relativity, Maxwell's EM theory, and time as an additional coordinate. We describe 1+3-dimensional Minkowski space, world lines, proper time, and causality. A short list of key research results forming the foundation of SR caps this chapter.

1.1 Principle of Relativity

The name of the theory of relativity derives from the title of the source book seen in Fig. 1.1: *The Principle of Relativity*. We begin our discussion by introducing this principle.

Inertial Observers (IOs)

An inertial system is understood as a reference system in which we observe that a force-free body moves in a straight line at a constant speed. In the comoving reference frame this relative motion is absent. Galileo put forward the principle that the laws of physics are the same in any inertial reference system. If true, there is no absolute motion, and thus no absolute rest and therefore no 'center' of the Universe.

This principle provided the basic framework for Newton's laws of motion, with the first law of motion presented as:[1]

[1]Original in Latin: Corpus omne perseverare in statu suo quiescendi vel movendi uniformiter in directum, nisi quatenus illud a viribus impressis cogitur statum suum mutare. It appears necessary to add the word 'relative' to the translation, yet this alters Newton's context significantly, showing the conflict of the principle of relativity with Newton's absolute (fixed star) reference frame.

J. Rafelski, *Modern Special Relativity*, https://doi.org/10.1007/978-3-030-54352-5_1

FORTSCHRITTE DER MATHEMATISCHEN WISSENSCHAFTEN IN MONOGRAPHIEN

HERAUSGEGEBEN VON OTTO BLUMENTHAL

HEFT 2

H. A. LORENTZ

A. EINSTEIN · H. MINKOWSKI

DAS RELATIVITÄTSPRINZIP

EINE SAMMLUNG VON ABHANDLUNGEN

MIT ANMERKUNGEN VON A. SOMMERFELD
UND VORWORT VON O. BLUMENTHAL

LEIPZIG UND BERLIN
DRUCK UND VERLAG VON B. G. TEUBNER
1913

INHALTSVERZEICHNIS.

Fig. 1.1 A reprint volume of essential relativity principle publications of Lorentz, Einstein and Minkowski published in 1913. Today an extended English edition (based on the 1922 German edition) also containing the later GR work of Einstein and Weyl is readily available: *The Principle of Relativity* ISBN-9789650060275 (Dover Paperback, 2008)

All force-free bodies persevere in the state of (relative, JR) rest or in the state of uniform motion.

This is sometimes presented in a more elaborate translation: Unless acted upon by an external force, an object at rest tends to stay at rest and an object in motion tends to stay in motion with the same speed and in the same direction. This principle is often called **the principle of inertial motion**, or simply **the principle of inertia**. An **inertial observer (IO)** is an observer for whom Newton's first law is true. We refer to IOs often in this book. IO implies the presence of, and thus is also used with the meaning, inertial reference frame.

Galilean Transformation

Consider the conventional nonrelativistic relation between two IOs, A and A', with A' moving at velocity $\boldsymbol{v}$ relative to A. When IO A measures the velocity $\boldsymbol{u}(t)$ of a body, this will differ at all times from the measurements made by observer A' by the relative velocity:

$$\boldsymbol{u}(t)' = \boldsymbol{u}(t) - \boldsymbol{v} . \tag{1.1}$$

Galileo Galilei ***Tuscan (Italian) Physicist, 1564–1642***

Wikimedia, CC BY-SA 3.0-License From **Galileo Galilei** portrait of 1636 by J. Sustermans, Hayden Planetarium, NY

Called the **father of modern science** by Einstein, Galileo Galilei pioneered scientific reductionism, and insisted on the use of quantitative and repeatable experiments, allowing results to be analyzed with precision.

Galileo reduced the complexity of the real world by seeking to recognize key governing factors. He knew that many sub-dominant effects had yet to be included into each and every consideration, and that imprecision of measurement also hindered experimental agreement with models considered.

His adherence to experimental results and rejection of allegiance to all other authority in matters of science ushered in the development of the modern world.

The Vatican's ban on reprinting Galileo's works was partially lifted in 1718 and in full 100 years after his death.

Since the velocity of a body is the rate of change in time of the position vector, the Galilean transformations of the coordinates of a body from A to A' consistent with Eq. (1.1) must be

$$t' = t \ ,$$
$$x' = x - v_x t \ , \quad y' = y - v_y t \ , \quad z' = z - v_z t \ . \tag{1.2}$$

Principle of Relativity Today

The **principle of relativity** requires the **physical equivalence of all IOs**: that is, two observers, who differ only in that one is moving at some fixed finite velocity relative to the other, are equivalent. This statement defines a class of IOs. From now on, an IO is any member of the class of all IOs. The laws of physics are the same for any IO.

According to the theory of the geocentric Universe, the Earth was at the center, and at rest. In order to allow the Earth to move, Galileo adopted the principle of relativity in mechanics, and in 1905 Einstein adopted the principle in electrodynamics:

> …the unsuccessful attempts to recognize any motion of the Earth relative to the 'light medium' imply the hypothesis that the concept of absolute rest, which does not apply to mechanics, is also absent from the phenomena of electrodynamics. It follows that the laws of optics and electrodynamics are valid in any chosen coordinate system to which the laws of mechanics apply, as it has already been demonstrated for leading terms. We will raise this conjecture (which in the following is called "principle of relativity") to the level of a postulate.…[2]

We recognize two important statements:

(a) The principle of relativity forbids a preferred absolute rest-frame; all places in the Universe are equivalent; there is no point of origin.
(b) Laws of physics may not refer to a particular frame of reference.

According to the principle of relativity a preferred frame of reference cannot appear in the laws of physics. Thus a method is needed in order to distinguish inertial from non-inertial, i.e. accelerated, motion. Ernst Mach[3] proposed in the context

[2]Translated by the author from the following original: "…die mißlungenen Versuche, eine Bewegung der Erde relativ zum "Lichtmedium" zu konstatieren, führen zu der Vermutung, daß dem Begriffe der absoluten Ruhe nicht nur in der Mechanik, sondern auch in der Elektrodynamik keine Eigenschaften der Erscheinungen entsprechen, sondern daß vielmehr für alle Koordinatensysteme, für welche die mechanischen Gleichungen gelten, auch die gleichen elektrodynamischen und optischen Gesetze gelten, wie dies für die Größen erster Ordnung bereits erwiesen ist. Wir wollen diese Vermutung (deren Inhalt im folgenden "Prinzip der Relativität" genannt werden wird) zur Voraussetzung erheben…"

[3]Ernst Mach (1838–1916), Professor at Graz, Salzburg, Prague (for most of his life), and Vienna; remembered for the Mach number, shock waves, and Mach's principle.

of mechanics as a preferred general reference frame the cosmic fixed star frame of the Universe,[4] a topic to which we return in Sect. 20.3. A specific choice of a preferred reference frame does not violate in any way the premises of the principle of relativity. We make such choices without thinking about it in daily life. For example it is of importance to remember that the relevant reference frame when considering the velocity of an approaching car is the personal frame of reference, in which we are at rest.

A survey of professional web pages addressing the principle of relativity reveals in general incomplete renditions. Perhaps most similar to our views are the following lines of argument,[5] but there are differences:

1. *Absolute motion cannot appear in any law of physics.* One can argue this is a duplication of #2 and #3 below, combined, since these say that experiments cannot reveal this. Since experiment takes precedent, this item can be skipped.
2. *All experiments run the same in all inertial frames of reference.* This statement is equivalent to item (b) in our discussion above.
3. *No experiment can reveal the absolute motion of the observer.* Along with Mach we extend this to say: Any experiment that explores the vastness of the Universe can reveal motion with respect to Mach's cosmic reference frame.

Body Motion

The principle of relativity implies that an IO must forever remain inertial. In 1905 Einstein presented SR and, in particular, the coordinate transformations considering the case that all forces causing acceleration of a material body vanish exactly. SR is, as Einstein put it in the title of his 1905 paper, a description of (inertially) *moving bodies*. However, the study and a consistent interpretation of many SR phenomena require that we know if and when a body is accelerated.

Since the principle of relativity does not apply to non-inertial motion, in SR we cannot consider inertial and accelerated observers as equivalent. Therefore it is not permissible to describe the properties of an accelerated body using the principle of relativity. This is true irrespective of the magnitude of the acting force. We return to this important remark in Sect. 12.2 and in Example 12.2 when we examine how this distinction works in regards to time dilation. In Chap. 20 we explain in greater detail why accelerated observers are a concept beyond the SR framework.

We note that Einstein, introducing the force of gravity, renamed his 1905 theory of relativity to be *special relativity* using the word *special* as equivalent to *no*

---

[4]Ernst Mach *Die Mechanik in Ihrer Entwicklung* (title translated according to the book contents): *The Development of a Mechanical Universe Across Centuries* 3. Auflage, F.A. Brockhaus Verlag Leipzig, (1897), reprinted by Forgotten Books Pub. ISBN 978-1-332-36427-5 (2018); English edition title: *The Science of Mechanics*.

[5]J.D. Norton at: http://www.pitt.edu/~jdnorton/teaching/HPS_0410/chapters/Special_relativity_principles/, retrieved June 2016 and December 2018.

acceleration and no forces. The general theory of relativity as we understand it today addresses the force of gravity; hence in this book the acronym GR means *gravity relativity.* The generalization to incorporate other forces, and most prominently, the electromagnetic force, remains today a preeminent research topic to which this book introduces the reader.

1.2 Time, a 4th Coordinate

Need for a Time Coordinate

We deal daily with time as a coordinate almost like any other; this is indeed how time enters relativity. Yet when seen as complementing the space coordinates, it is for most a surprise. Hence we should inspect this situation closer. Normally we use three coordinates to share information about a spatial location: *I am standing on the second story of the building located on the corner of 4th avenue and 5th street.* This is shown as a point in Fig. 1.2. The bottom pane presents, using the spatial 3-dimensional x, y, and z coordinates, a specific point in space.

Suppose that besides describing a location, I want to meet you for lunch. In addition, I say, *Let's meet at noon.* Now we have, aside from x, y, z, also the fourth coordinate time t, which we illustrate in Fig. 1.2 in the form of spaced horizontal panels drawn for three different times: origin $t = 0$ at the bottom, the half-way point at 2 min, and the meeting event after a 4 min walk.

The time coordinate is required for a complete description of any event; in this case the event is a lunch meeting. The series of values of these four coordinates describing our positions in space-time is called *a world line.* Each point, including the event where we meet in the 3+1-dimensional space-time, is called *an event.*

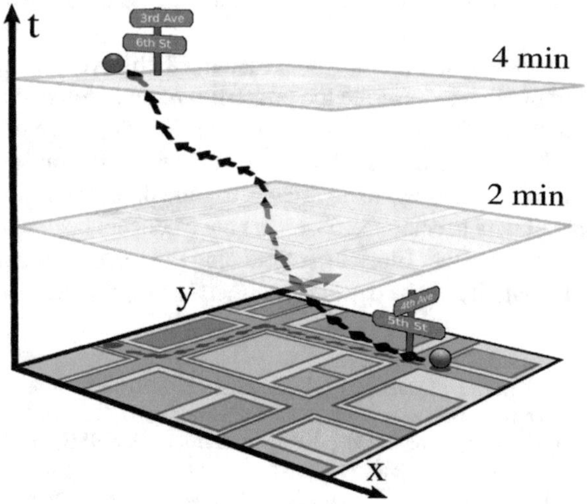

Fig. 1.2 A world line, shown here, of a four-minute walk from the corner of 4th Ave and 5th St to the corner of 3rd Ave and 6th St. Motion in the x, y, and time t dimensions is shown, with the z dimension suppressed (except for the street signs)

Albert Einstein ***German, Swiss & American Physicist, 1879–1955***

Photo: Lucien Chavan 1868–1942
Wikimedia, CC BY-SA 3.0-License
Einstein in Bern, 1904/1905

Bundesarchiv: Bild 102-10447
Albert Einstein in 1930

At the age of 15, Einstein moved to Switzerland (1901 Swiss citizen). After studying physics at ETH in Zürich, Einstein worked from 1902 to 1909 at the patent office in Bern, obtaining his Ph.D. in 1905. After professorships at the Universities in Prague and Zurich, in 1914 he moved to Berlin where he became a member of the Prussian Academy of Sciences and Director of the Kaiser Wilhelm Institute for Physics. In 1922 Einstein was awarded the Nobel Prize for physics. In 1933 he accepted an appointment at the Institute for Advanced Study in Princeton, where he worked until his death.

Much of his work while at the patent office in Bern related to questions about the transmission of electric signals and the electro-mechanical synchronization of time: problems that appear in his thought experiments. Probable contradictions in patent applications may have spurred Einstein to the formulation of SR in 1905. More than a decade later he formulated the general theory of relativity based on the equivalence principle of inertial and gravitating mass. Eddington's announcement of the observation (1919) of the predicted bending of starlight by the gravitation field of the sun galvanized Einstein's world-wide fame.

Einstein is also known for many other important contributions to physics: Brownian motion, light quantum hypothesis, lattice vibration theory of specific heat, and quantum statistics. Einstein was deeply involved in the creation of quantum mechanics, yet he could not accept the probabilistic interpretations which followed. For decades Einstein worked toward a unified field theory, searching for an understanding of how EM forces enter relativity consistently; this legacy continues today. Through his work, which revolutionized the foundations of physics, Einstein is acclaimed as the most influential physicist of the twentieth century.

Time as the 4th Coordinate of an Event

In the theory of relativity we join time and space coordinates to form the 'Minkowski Space':[6]

> The new notions about space and time, which I would like to unfold for you, are founded in experimental physics. Therein lies the argument's strength. The outcome is a new paradigm. From now on, space and time considered separately sink into obscurity; henceforth a union of the two is the new and fully adequate reality.[7]

Following Minkowski, we rely on the 1(time)+3(space) = 4-dimensional space-time framework, also referred to as 3+1-dimensional space-time or simply 4-dimensional space. The relation between time and space coordinates will be better appreciated once we consider how different observers measure the coordinates of events.

Proper Time

Before the development of SR, 'time' was assumed to be universal, advancing at the same rate for everyone, whether in London, in New York, on the Moon, on a train, in motion, or at rest in a laboratory. The pivotal feature in SR is that a clock 'ticks' differently in each moving body; each object, e.g. atom, human, man-made satellite, has its own proper time. Minkowski introduced the expression proper time, in his article *Raum und Zeit*, see Ref. [7]. A clock you carry and a clock your friend carries will show different times unless you two always stay together.

Einstein was (arguably) the first to realize that time 'belonged' to a body.[8] To human perception the difference between proper time and laboratory time is too small to notice. However, today we can observe the time dilation of an atomic clock placed in a moving car.

When setting the meeting for our lunch, one could think that we could use our watches, that is, individual proper time, to time our arrival at the meeting. However,

---

[6]Hermann Minkowski (1864–1909), German national, a mathematician well known for number theory, made decisive contributions to SR proposing unity of time with space, and introducing the concepts of proper time and world line.

[7]H. Minkowski, Opening of the address at the 80th Assembly of German Natural Scientists and Physicians (September 21, 1908, Cologne): *Physikalische Zeitschrift* **10** 104–111 (1909); and *Jahresbericht der Deutschen Mathematiker-Vereinigung* **18** 75–88 (1909), and with commentary of A. Sommerfeld reprinted in 1913, see fifth paper in the collection shown in Fig. 1.1; translated by the author from the following original: "Die Anschauungen über Raum und Zeit, die ich Ihnen entwickeln möchte, sind auf experimentell-physikalischem Boden erwachsen. Darin liegt ihre Stärke. Ihre Tendenz ist eine radikale. Von Stund' an sollen Raum für sich und Zeit für sich völlig zu Schatten herabsinken und nur noch eine Art Union der beiden soll Selbständigkeit bewahren."

[8]W. Rindler, *Einstein's Priority in Recognizing Time Dilation Physically, American Journal of Physics,* **38** (1970) 1111. However, J.S. Bell in a letter to the author, Ref. [3] in preface, suggests Larmor proposing *local time* was aware time is not universal.

this is, as we just learned, in SR not correct. Fortunately, we can make a different implicit agreement, to use the clock at the location of the meeting. This is the so-called 'laboratory' time. Every lunch participant arrives at the meeting at a different proper time; however, all meet simultaneously at the same table. We choose as the laboratory the reference frame where we perform simultaneous measurements of events at different spatial locations. We will show that for any other observer, not at rest with respect to the laboratory, this simultaneity cannot be true: Events occurring according to an observer at equal time cannot happen simultaneously for any other observer not at relative rest with the laboratory, see Chap. 8.

Causality

In SR causality means (see Chaps. 11 and 12) that the *sequence* of events in time cannot change. Keeping causality of events is a concern, since the coordinate transformation of SR involves the transformation of time. We will find that, within SR, we cannot go back, e.g., to the time of our birth. The one-dimensional, unidirectional nature of time prevents us from looping motion in time. On the other hand, it is common to revisit the same location in space. Therefore time, which is for many purposes just another coordinate tacked onto space, turns out to be, after all, profoundly different from space.

However, time ticks differently for each of us. Individuals from the past could therefore emerge from an alien spaceship to tell us how different the world used to be. And, for the same reason, if you know a friendly and technologically advanced alien, you could visit the future a thousand years from now. But once you reach the future, there is no way back; for each of us the time arrow points forward. The sequence of events cannot change. The adherence to the arrow of time for all bodies is referred to as the principle of causality.

Is There a Deeper Understanding of Time?

What is time? is perhaps the most important and yet least understood of the fundamental questions. Moreover, we now have recognized a few different ways that we can consider and measure time:

1. The laboratory time
2. The proper time, i.e. the time measured by a proper body clock
3. The cosmological time, i.e. the proper time (also known as age) of the Universe

For time measurements within SR, only the choice of the reference system matters. When studying the dynamical motion of a particle, we will consider proper time and laboratory time in parallel in this book, and make a clear distinction. However, it is worth knowing that proper time does not occur in relativistic quantum mechanics.

The cosmological time is the time measured by a clock at rest in the cosmological reference frame of the Universe. In comparison, all other time measurements are dilated; i.e. all other clocks in the Universe are slower. The expansion and aging of the Universe today are the subject of ongoing experimental and theoretical research. Theoretical models, among other insights, verify our understanding of cosmological time.

1.3 Path Toward Lorentz Coordinate Transformations

With the advent of Maxwell's electromagnetism, a way was sought to transcribe Maxwell's equations from one reference frame to another. The Michelson-Morley experiment, Sect. 3.1, added to the challenge by suggesting the unobservability and thus nonexistence of Newton's absolute motion. The question how that could happen was debated and an explanation was proposed by FitzGerald[9] and independently by Lorentz,[10] who each realized that a material body in motion contracts. This adds to the conundrum: How could such an effect be consistent with the generalized coordinate transformation?

A similar situation is known in classical mechanics. The Galilean transformation of coordinates allows us to infer the values of momentum and velocity a body has with reference to an observer in motion. There is a consistency between the coordinate transformation and the measurement of momentum and velocity of the body.

In fact the relativistic modification of the properties of a body in motion did not end with the Lorentz-FitzGerald body contraction. For the MM interferometer to produce a null result there had to be another effect, the moving body time dilation. However, these two new body effects are grossly inconsistent with the Galilean coordinate transformation. Namely, the measurements of e.g. body length made by two different Galilean observers will always produce the same result. This is inconsistent with the Lorentz-FitzGerald body contraction. Moreover, there is no time dilation, since for Galilean transformations the time-scale does not change.

We realize on account of time dilation that the new set of coordinate transformations must include, aside from the three space coordinates, also time. This means that we are looking for a transformation of the coordinates of an event t, x, y, and z observed in a frame of reference S into the coordinates t', x', y', and z', of the same event observed in the frame S' moving relative to S at a velocity $\boldsymbol{v}$.

---

[9] George Francis FitzGerald, "The Ether and the Earth's Atmosphere," *Science*, **13** 390 (1889). A historical account of complex research carried out prior to this short published note is described by B.J. Hunt, *The Maxwellians*, Cornell University Press (1991), pp. 185–197.

[10] H.A. Lorentz, §89–92 extracted from *Versuch einer Theorie der Elektrischen und Optischen Erscheinungen in Bewegten Körpern* (Leiden 1895) republished in original German, and in 1923 translated under title *Michelson's Interference Experiment*, as the first paper in the collection shown in Fig. 1.1.

Hendrik Antoon Lorentz ***Dutch physicist, 1853–1928***

Lorentz took up the chair in theoretical physics three years after completing his Ph.D. at the University of Leiden in 1878 and he remained there the rest of his life. He was awarded half of the second Nobel Prize in physics in 1902 for the elucidation of the Zeeman effect.

Lorentz was primarily interested in Maxwell's theory of electromagnetism. The force an electron experiences in an electromagnetic field, the Lorentz force, is named after him. In 1892 he explained the Michelson-Morley experiment, proposing that moving bodies contract in the direction of motion; the Lorentz-FitzGerald body contraction is named after him. In 1899 and again in 1904, Lorentz searched for the relativistic coordinate transformations; the Lorentz transformation is also named after him.[11] Lorentz's contributions to the development of SR are often seen to be on par with Einstein's.

The transformation we seek is called *the Lorentz coordinate transformation* (LT). Indeed, in several attempts Lorentz sought a coordinate transformation consistent with Maxwell's equations; the last related publication[12] of 1904 does not present LT in recognizable format. Lorentz does not use in his effort what Sir Joseph Larmor[13] published four years earlier.[14] The hidden key element is that spatial coordinates must combine with time. Already in 1899/1900 Larmor calls the transformation he

[11]The Lorenz gauge condition is not named after Lorentz.

[12]H.A. Lorentz, *Electromagnetic Phenomena in a System Moving with any Velocity Less than that of Light, Proceedings of the Royal Netherlands Academy of Arts and Sciences,* **6**, pp. 809–831 (1904), the second article in the collection shown in Fig. 1.1.

[13]Sir Joseph Larmor (1857–1942) Irish-British physicist, Lucasian Professor of Mathematics at Cambridge, credited with discovery of the Lorentz transformation, known for Larmor radiation formula, adhered to material æther.

[14]J. Larmor, *On a dynamical theory of the electric and luminiferous medium, Phil. Trans. Roy. Soc.* **190**, 205 (1897); J. Larmor, *Aether and Matter*, Cambridge University Press (1900).

found after Lorentz, given that he 'simply' solves the problem Lorentz posed in his body of work. A modern evaluation of Larmor's work[15] argues that the credit for the first presentation of the LT, including the crucial time dilation, belongs to Larmor. This is also what John Bell wrote to this author, see Ref. [3] in the preface, nearly coincident with this article.

Henri Poincaré ***French mathematician & theoretical physicist, 1854–1912***

Wikimedia, CC BY-SA 3.0-License **Henri Poincaré in 1887** *by Eugène Pirou (1841–1909)*

Jules Henri Poincaré, a polymath who excelled in all branches of exact sciences of his time, graduated from École Polytechnique in 1875, taught as of December 1879 in Caen and as of October 1881 at Sorbonne in Paris. He was elected February 1, 1887 to the Académie de Sciences and on March 5, 1908 member of the French Academy. He was also a cousin of a French statesman, Raymond Poincaré.

Poincaré's research work is characterized by high diversity and originality; a preeminent mathematician, he ventured into several fields of physics. Poincaré discovered in the study of the three-body problem the deterministic chaotic behavior, the beginning of chaos theory. In SR context he recognized the key relevance of the principle of relativity for the new mechanics. He is remembered for the (Poincaré) group properties of space-time transformations.

The naming of the transformation after Lorentz is set in stone by Henri Poincaré. He offers a short written report of his oral presentation made to the French Academy[16] describing Lorentz's latest work. We see in this work the modern

[15]M. N. Macrossan, *A note on relativity before Einstein, The British Journal for the Philosophy of Science* **37**, 232 (1986).

[16]H. Poincaré, *Sur la dynamique de l'électron,* (On the Dynamics of the Electron), *Comptes rendus de l'Académie des Sciences,* **140**, pp. 1504–1508 (1905), footnoted to be the written account of a lecture presented at an Academy session on June 5, 1905.

symmetric form of coordinate transformation, Poincaré claims that he is adding little if anything new; he checked the results of Lorentz and insignificantly improved their presentation. Poincaré shows no detail or derivation in this presentation of the textbook form of Lorentz transformation, but he remarks that the form follows from the mathematical group properties of the transformation, a remark also seen in the work of Einstein. This author is still searching for a copy of "Comptes Rendus" showing distribution on about June 9, 1905 with the SR contents. Without this evidence, it is more natural to expect that the written account reporting the Poincaré lecture of June 5, 1905 must have been composed after his Academy lecture presentation was scheduled for delivery, and before the time his long SR work[17] was completed in mid July 1905 (received at the publishers July 23, 1905).

The full and comprehensive formulation of SR is seen for the first time in the much longer, and scientifically (i.e. Physics) revolutionary and complete work of Einstein[18] written in German. *Annalen der Physik* (Berlin) dates Einstein's paper as received on June 30, 1905. This Poincaré-Einstein often discussed relativistic coordinate transformation priority question is without merit as the actual credit for the first publication of the LT is due to Sir Joseph Larmor, see Ref. [15], while the two works of Einstein and of Poincaré make complementary contributions in regard to properties of these transformations. Einstein presented the required and sufficient physical principles and explored their consequences, while Poincaré consolidated the mathematical context foundations.

Einstein holds priority in presenting a derivation of the relativistic coordinate transformation from first principles, opening with his work the door to a new domain of physics, SR. Einstein's complete characterization of SR, see Ref. [18], which cites no other paper, was written in his capacity as a junior patent clerk without a doctoral degree; that is, in present day language by a self-supporting graduate student, was published in arguably the most prestigious physics research journal of the epoch by the editor Prof. Max Planck.[19] Within a few months Einstein recognizes the pivotal equation of the twentieth century, publishing the $E = mc^2$ relation.[20]

For clarification of the historical time line: Einstein was awarded a doctorate by the University of Zurich in January 1906 and in April 1906 was promoted to be a second-class technical expert at the Swiss Patent Office; in January 1908 he obtained *venia docendi* (habilitation, the right to teach) at University of Bern, a procedure

[17] H. Poincaré, *Sur la dynamique de l'électron, Rendiconti del Circolo Matematico di Palermo,* **21**, pp. 129–175 (December 1906), https://doi.org/10.1007/BF03013466.

[18] A. Einstein, *Zur Elektrodynamik bewegter Körper,* (translated: *On the electrodynamics of moving bodies,*) *Annalen der Physik* **17** 891 (1905).

[19] Max Planck (1858–1947), a renowned German theoretical physicist, after whom the Planck constant $\hbar$ (quantum of action) is named for which he was awarded the 1918 Nobel prize. Planck recognized and advanced the work of Einstein from the beginning.

[20] A. Einstein, *Ist die Trägheit eines Körpers von seinem Energieinhalt abhängig?* (translated: *Does the inertia of a body depend upon its energy content?*), Annalen der Physik **187**, 639 (1905); received by publishers on September 27, 1905.

delayed for a few months by rules and regulations governing the process. Even so at the time of his habilitation, Einstein was only 28 years old.

Einstein obtained the LT from the following general principles, see Chap. 6:

1. The isotropy of space and the homogeneity of space-time. This is rarely discussed in elementary SR presentations.
2. The principle of relativity, meaning here the equivalence of all IOs.
3. A universal speed of light c.

No further tacit or contextual assumptions are required. This means in particular that the medium in which Maxwell's waves propagate is undetectable by an *inertial* observer, a point made by Einstein in his 1905 work.

In 1911 Paul Langevin provides a more general hypothesis concerning the æther, recognizing that bodies can interact with the æther when their motion is non-inertial. Upon completion of the theory of gravity Einstein returned to the æther challenge, seeing the need for further understanding to characterize forces governing electromagnetism. In 1919/1920 Einstein introduced the non-ponderable (we would say today, non-material) æther. We describe these developments in Sect. 2.3.

1.4 Highlights: How Did Relativity Happen?

Let us briefly summarize several developments that led to Einstein's formulation of SR:

1. Within Maxwell's theory EM waves propagate at the same speed as light. Light is an electromagnetic wave, taking part in electric and magnetic dynamics.
2. The speed of light in matter-free space is universal and independent of the motion of the emitter or the observer; it is the same when measured (in a vacuum) in a laboratory on Earth, or in interstellar space, and the same at any wavelength.
3. Since wave velocity is universal, Maxwell's theory cannot be invariant under Galilean transformations.
4. The Michelson-Morley experiment shows that absolute motion is undetectable; its result is interpreted in terms of the Lorentz-FitzGerald body contraction hypothesis. Larmor recognizes time dilation.
5. Einstein demonstrates that Galileo's principle of relativity can be reconciled with all of these new revolutionary insights: the speed of light being independent of motion of the source or observer, the non-universality of body size, and time not being universal.

We list now in *chronological* order a subjectively selected list of scientific works that are most relevant to these key developments.

1863 Mach (1838–1916)	***The Development of Mechanics: A Critical and Historical Perspective*** Mach, Ernst, *Die Mechanik in ihrer Entwicklung. Historisch–kritisch dargestellt.* (Leipzig: F.A. Brockhaus, first edition 1863 – sixth edition 1908) reprinted: (Forgotten Books, ISBN-10: 1332364276, 2018): (English book translation title *The Science of Mechanics*).
1856–1873 Maxwell (1831–1879)	**A Dynamical Theory of the Electromagnetic Field** Maxwell, James Clerk, *Philosophical Transactions of the Royal Society of London* **155** (1865) 459–512. This article accompanied a December 8, 1864 presentation by Maxwell to the Royal Society. To be remembered together with his two publications, of 1856 and 1861, and the 1873 *A Treatise on Electricity and Magnetism.*
1887 Michelson (1852–1931) & Morley (1838–1923)	**On the Relative Motion of the Earth and the Luminiferous Aether** Michelson, Albert, and Morley, Edward, *The American Journal of Science and Arts* **34** 333–345 (1887).
1889 FitzGerald (1851–1901)	**(1) The Ether and the Earth's Atmosphere; (2) See: *George Francis FitzGerald*** (1) FitzGerald, George Francis, *Science* **13** 390 (1889); (2) Dennis Weaire, Editor (Living Edition 2009).
1897–1900 Larmor (1857–1942)	**(1) On a Dynamical Theory of the Electric and Luminiferous Medium; (2) *Aether and Matter*** Larmor, Sir Joseph, (1) *Phil. Trans. Roy. Soc.* **190** 205–300 (1897); (2) *Aether and Matter* (Cambridge, 1900).
1904 Lorentz (1853–1928)	**Electrodynamic Phenomena in a System Moving with any Velocity Smaller than that of Light** Lorentz, Hendrik A., *Proceedings de l'Académie d'Amsterdam (KNAW)* **6** 809–831 (1904).
1905 Einstein (1879–1955)	**(1) On the Electrodynamics of Moving Bodies; (2) Does the Inertia of a Body Depend on Energy Content?** Einstein, Albert, (1) *Zur Elektrodynamik bewegter Körper, Annalen der Physik* **17** 891–921 (June 1905); (2) *Ist die Trägheit eines Körpers von seinem Energieinhalt abhängig? Annalen der Physik* **18** 639–641 (September 1905).
1905/1906 Poincaré (1854–1912)	**(1) Dynamics of an Electron; (2) On the Dynamics of the Electron** Poincaré, Henri, (1) *La Dynamique de l'Électron, Comptes Rendue de L'Académie de Sciences* **140** 1504–1508 (June 1905); (2) *Sur la dynamique de l'électron, Rendiconti del Circolo Matematico di Palermo,* **21** 129–175 (December 1906).
1907 von Laue (1879–1960)	**The Drag of Light by Moving Bodies According to the Principle of Relativity** von Laue, Max, *Die Mitführung des Lichtes durch bewegte Körper nach dem Relativitätsprinzip, Annalen der Physik* **23** 989–990 (1907).
1909 Minkowski (1864–1909)	**Lectures on Space and Time held 21 September 1908** Minkowski, Hermann, *Raum und Zeit* see article 5 in *Principle of Relativity* as shown in Fig. 1.1; and *Jahresbericht der Deutschen Mathematiker-Vereinigung* **18** 75–88 (1909).
1911 Langevin (1872–1946)	**The Evolution of Space and Time** Langevin, Paul, *L'Évolution de L'Espace et du Temps, Scientia* **X** 31–54 (1911); and English translation by J.B. Sykes, *Scientia* **108** 285–300 (1973).

Light and the Æther

2

Abstract

The process of measurement of space and time requires the understanding of the speed of light. Several centuries of effort to determine the speed of light are described, addressing propagation on Earth and between stars. Maxwell recognized the speed of light as the speed of EM waves. In an essay we describe the evolving understanding of the luminiferous æther.

2.1 Measuring Space and Time: SI-Unit System

The question 'what is time?' relates directly to the question 'where are we?' To answer both questions we must first establish how each is measured. We adhere in this book to the International System of Units (SI) as this is the system students know. We will discuss further the wisdom of SI-unit choices, introducing additional elementary energy units, see Insight 16.1 on *Elementary energy units*. Similarly, we describe choices available in electromagnetic theory in the Insight 20.1 on *SI and Gauss EM units*.

SI-Unit System

The time-unit is the same in scientific and in everyday life. The measurement of time has been directed for thousands of years by the wealth of astronomical observations and by easily transportable clocks and watches. This helped create a universal unit, the second, as the fraction $1/86{,}400 = 1/24 \times 1/60 \times 1/60$ of the average day length. Since 1967, the 'scientific' second is the period equal by definition to 9,192,631,770 cycles of the atomic clock unit frequency. This originates in the elementary atomic transition between two hyperfine interaction split levels of the atomic ground state of Cesium-133. This definition makes the cesium oscillator, referred to as an atomic

J. Rafelski, *Modern Special Relativity*, https://doi.org/10.1007/978-3-030-54352-5_2

clock, the primary standard for time measurement. Note that we employ as time unit the material clock *proper time*.

At the beginning of the creation of the SI-unit system, the meter, the unit of material size, was chosen arbitrarily. A consequence of this choice was the value of the speed of light[1] c, which provides the connection between these two arbitrarily chosen units of body size and proper body time. Recognition of a universal value of the speed of light created a different definition opportunity implemented in 1983. In the SI-unit system:

1. The speed of light c has a defined value and is not measured anymore

$$c \equiv 299{,}792{,}458\ \frac{\text{m}}{\text{s}}\ . \tag{2.1}$$

2. The meter, as a unit, has been redefined as the distance traveled by light in vacuum within a given fraction of a second

$$1\,\text{m} \equiv c\,\frac{1\,\text{s}}{299{,}792{,}458}\ . \tag{2.2}$$

By defining the meter in this way, the accuracy of spatial distance measurement is limited only by time-keeping, which is very precise. However, this definition shifts the unit of length to be a property of space-time and not that of a body, while the unit of time is based on the proper time of a material clock.

In astronomy, it is common to refer to distance based on how far light travels in a year; that is, the light-year (ly). Defined by the International Astronomical Union (IAU), 1 ly is the distance that light travels in a vacuum in one 'Julian' year (Jyr)

---

[1]The path to the letter c becoming the symbol for the speed of light is described in: K.S. Mendelson, *The story of c*, *Am. J. Phys.* **74**, 995 (2006). A few key points from this article: Einstein switches from V to c three years after inventing relativity. Why c? The German nineteenth century *Constante*, instead of 'Konstante' today, seen in writing by W. Weber who interprets magnetic forces in the pre-Maxwell world; and the Latin word for speed *celeritas,* though this interpretation appears explicitly first in 1959 in *The Magazine of Fantasy and Science Fiction* writings of I. Asimov "c for celeritas." The general use of c follows M. Abraham 1903 "Prinzipien der Dynamik des Elektrons, (Principles of Electron Dynamics)," *Ann. Phys.* **10**, 105 (1903), and his influential 1904 textbook: M. Abraham, and A. Föppl *Theorie der Elektrizität: Einführung in die Maxwellsche Theorie der Elektrizität (Theory of Electricity: Introduction to Maxwell's Theory)* (Leipzig, Teubner 1904). It is important to appreciate that Abraham, according to the educational norms of his time could not avoid being fluent in Latin and thus certainly knew that speed = *celeritas*. Similarly we note that Larmor in his year 1900 book *Æther and Matter,* used (capitalized) C.

where Jyr = $3.15576 \cdot 10^7$s; and a Julian light-year (ly) by definition

$$1 \text{ ly} = 9.4607304725808 \cdot 10^{15}\,\text{m}\,. \tag{2.3}$$

A related natural unit of laboratory distance is the light-nanosecond, 10^{-9} fraction of a light-second $\equiv 1\,c\,\text{ns} \equiv 0.2997\,\text{m}$, which corresponds nearly to an American unit of length, the foot, ($1\text{ ft} \simeq 0.3048$ m). It can be useful to remember this coincidence: Light takes a bit more than a nanosecond to travel one foot.[2]

Natural Units

The vast majority of theoretical physicists working in the fields such as relativity, astrophysics, particle physics and related areas use a natural system of units in which, as just discussed, time can be used to measure distance. To see how this works, imagine a doctor's visit where the nurse determines the patient's height by triggering a laser pulse at the patient's feet, bouncing the pulse from a mirror at the top of the patient's head. A good clock measuring the time interval will for a 6 foot (1.83 m) tall person score (double) light flight time distance of 12 lns ('l' for light). In the final step we multiply with c to get a precise body height in usual length unit, meter or feet.

This measurement shows clearly that an independent unit of length as such does not exist, and the last step seems superfluous, since the nurse can record the height in units of time. However, this means that we redefined the meaning of c

$$\text{speed of light} \equiv c = \frac{\text{distance traveled}}{\text{time}} = \frac{1\text{ light}\cdot\text{s}}{1\,\text{s}} = 1\ (\text{light} = \text{no units}). \tag{2.4}$$

In unifying the measurement of time and space:

(a) We abandon the distinction between unit of time and unit of length.
(b) We eliminate from use the arbitrary factor 299,792,458, Eq. (2.2), connecting the unit of time to an arbitrary unit of distance, arising solely from historical context.
(c) We can either: (a) use the time duration of light travel to determine distance (akin to light year), or (b) replace the unit of time by specifying the distance with the light travel time (akin to a mirrored light-pulse clock).

[2]This author believes that $1\ c$ ns $\simeq$ 0.9833 ft $\equiv$ 1 'lns' (light nano second) by lucky chance a good scale both at the human and in fundamental considerations could be adopted as the SI-unit of length. Several nuisances would disappear from the SI-unit tables; for example, Eq. (2.2) simplifies since $c = 10^9$ lns/s. Even today a unit of length based on 1 lns could result in a more widely accepted unit system, perhaps finding worldwide acceptance, as it would replace the arbitrary and different numerical values in use today.

We now see that the principle that the speed of light is universal, and the unification of space and time into space-time in SR allows a universal unit relation between space and time. This means that in this book we practically always will see next to t a c, implying that a unit of distance ct traveled by light is used as a measure of time t. Agreeing to this procedure we render the presence of the symbol c superfluous; there is no need to write c into the equations. When this is done the unit system is called *natural*. The reader will therefore see in many other books the same equations as in this book without an explicit c. However, in this book we keep writing c for didactic reasons.

2.2 Speed of Light

Many different precise measurements of the speed of light c were necessary to recognize the universal nature of c. We will describe how we learned that the speed of light is observed to be the same in the vast Universe and in the Earthbound laboratory, and that c is found to be independent of the state of motion of the source or of the observer. Arguably these experimental results, obtained over the span of about two centuries, led to the development of SR.

Astronomy and the Speed of Light

The speed of light has fascinated scientists for millennia, but only as recently as the seventeenth century did the invention of the telescope allow the speed of light to enter the realm of measurable quantities. Around 1679, Danish astronomer and later statesman Ole Roemer, working at the Paris Observatory, was puzzled by the orbital period of Jupiter's moon Io varying slightly as measurements were carried out during the year, as it was understood at that time that a moon's orbital period must be constant. Roemer believed that this effect had its origin in a finite value of the speed of light c.

The effect was further studied by Christiaan Huygens with whom Roemer corresponded. Both Roemer and Huygens considered the time light takes to travel from Io to an Earth observer. Light must travel an additional distance when Jupiter and Earth move apart during one Io orbit, which takes 42 h. This extra distance that light must travel is added inadvertently to the raw orbital period. Similarly, the light travel distance is shorter when Jupiter and Earth approach each other, in which case the observed orbit time of Io appears shorter. The effect accumulates with each Io orbit. The effect of Earth motion dominates the motion of Jupiter: Jupiter is both slower than Earth by a factor 2.3, and its larger orbit takes 12 Earth years to complete.

To gauge the magnitude of the effect of Earth orbital motion, we recall that light takes about 1000 s to cross the diameter D of the Earth orbit around the Sun. It follows that as the Earth moves through half of its orbit, corresponding in time to about a hundred 42 h Io orbits, the magnitude by which the Io orbital appearance could vary is 16 min; that is, the time light needs to cross the Earth orbit diameter around the Sun. In an analysis of the Roemer results, Huygens obtained $c \simeq 220{,}000$ km/s (note that SI-units were not yet defined; the value reported here has been converted). The difference from the true value is attributed today to the quality of available astronomical data and not to omissions in the theoretical analysis.

The speed of light was first precisely obtained by James Bradley,[3] who worked with Samuel Molyneux[4] until Molyneux's death in 1728. They intended to measure the parallax of Gamma Draconis, or 33 Draconis, a star in the zenith of London. The zenith is an imaginary point in the sky directly opposite the direction of the vertical defined by the direction of the apparent gravitational force at that location. The idea of exploring Gamma Draconis goes back to an experiment carried out in 1669 by Hooke,[5] who in 1674 claimed to have detected a Gamma Draconis parallax, but his result was not confirmed by work that followed. One can say:[6] "Hooke apparently saw what he wanted to see amidst a storm of instrumental error."

Parallax is the difference in the apparent position of an object viewed along two different lines of sight, i.e. from two different positions by an observer at relative rest. The parallax method is widely used in determining distance. In astronomy today distance is defined in terms of half of the angular displacement that a star appears to move relative to the celestial sphere as Earth orbits the Sun. The unit of distance parsec (pc) is invented to correspond to arcsecond sized parallax; thus $1\,\text{pc} \equiv 3.2615638\,\text{lyr} \equiv 1''$. The measurement of the Gamma Draconis parallax of $(21.2 \pm 0.1) \times 10^{-3}\,''$ implies a star distance $R = 154.5 \pm 0.7$ lyr.

After three years of work Molyneux and Bradley discovered, however, another effect, the aberration of light, which allowed the measurement of the speed of light with unprecedented precision. We will address (stellar) aberration in technical detail in Sect. 13.3. The observed Gamma Draconis 'movement in the sky' consisted of nearly circular motion of Gamma Draconis around the vertical defined on Earth. This motion was large and not consistent in its annual periodicity features with the expected parallax effect. The interpretation of this effect required a new idea.

---

[3] James Bradley (1693–1762), English astronomer, Astronomer Royal from 1742. Appointed to the Savilian chair of astronomy at Oxford in 1721, Bradley's work provided the first direct evidence for the movement of the Earth around the Sun as well as a precise measurement of the speed of light based on the newly discovered aberration effect.

[4] Samuel Molyneux (1689–1728), member of the British parliament and an amateur astronomer, Fellow of the Royal Society in 1712. Molyneux commissioned precise telescopes and engaged James Bradley.

[5] Robert Hooke (1635–1703), Oxford natural philosopher and polymath, opponent of Newton; Hooke's law is named after him.

[6] Todd K. Timberlake, "Seeing Earth's Orbit in the Stars: Parallax and Aberration," *Phys. Teach.* **51**, 478 (2013).

The existence of an aberration of the line of sight by relative motion can be recognized by considering as an example the observer driving in a car through a snowstorm. Driving into the falling snowflakes, the angle of impact onto windshield tilts apparently to be more horizontal. When the car speed is high compared to the falling snowflakes, the snowflakes hit from the front, instead of falling from the top; the aberration due to motion makes them appear to travel from the front. Clearly we can measure the speed of snowflakes using the car speedometer. The aberration of line of sight of Gamma Draconis is thus due to the variation in the relative speed of the observer on Earth with respect to the observed star and in analogy to above we can evaluate the speed of light. The variation of observer's speed is bounded by the already known magnitude of the orbital velocity vector of the Earth, $\simeq \pm 30\,\mathrm{km/s}$.

Astronomers recognize a fundamental difference between the parallax effect and the aberration effect, even though on first sight the observational outcome appears the same, a circular movement of a star in the sky. Parallax displacement of the location of a star in the sky is a function of stellar distance as compared to the distance between two observation points maximized by the Earth orbit diameter. On the other hand, the aberration effect depends on relative motion.

It was fortuitous that the Gamma Draconis location is nearly in the plane in which Earth moves around the Sun. As result, Gamma Draconis was in apparent motion synchronized in a specific way with Earth's orbital motion around the Sun, and this new 'aberration' effect had to be due to the motion of the Earth around the Sun, as Bradley realized shortly after Molyneux's death in 1728.

In the context of relativity theory it is important to recognize that the effect of light aberration, though recorded by an observer without any additional knowledge of the condition of the source, depends on the *relative* velocity between this observer and the source and the velocity vector relation to the apparent line of sight to the light source. For this reason it is possible to measure the light speed using the zenith star observation alone. How this is possible is explained in the more technical parts of this book in Sect. 13.3. The understanding of the light aberration effect relies on the application of principles of SR. Without proper in-depth study in the context of SR, the aberration of the line of sight dependence on relative velocity can seem very mysterious and has often been misunderstood.

Bradley determined the speed of light from the aberration angle measurement using a nonrelativistic version of the method described in Sect. 13.3. Bradley's result $c = 301{,}000\,\mathrm{km/s}$ remained the best value for 130 years until the terrestrial measurement of mid nineteenth century by Fizeau and Foucault, which we discuss in the following section. The importance of Bradley's determination of the speed of light is that a value derived from the far reaches of the Universe paved the way to recognizing the universality of the speed of light once it was later measured to be the same on Earth.

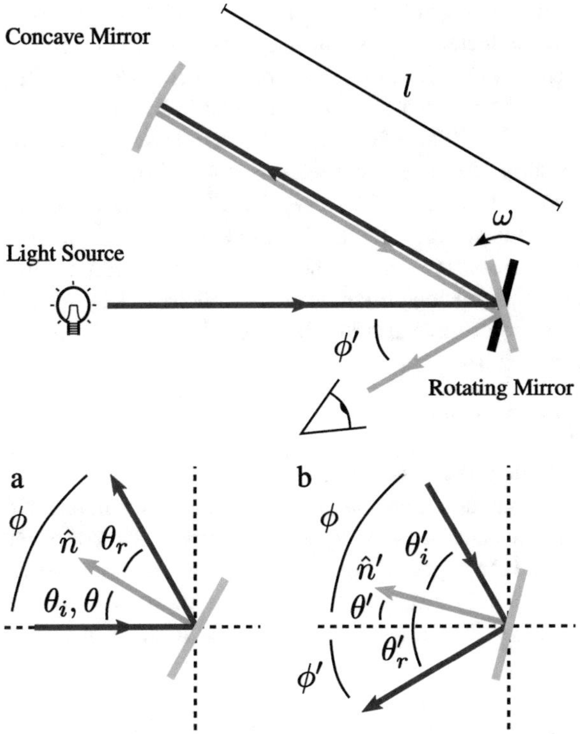

Fig. 2.1 The principle of the Fizeau-Foucault apparatus to measure the speed of light. Top: a beam reflects off the rotating mirror and travels a distance l to a concave mirror, where it is reflected back traveling along nearly the same path to the rotating mirror. The light makes an angle ϕ' with the original beam after reflecting back off the slightly rotated mirror. Bottom: two stages (**a**) and (**b**) of the rotating mirror geometry, see Example 2.1

Terrestrial Measurement of the Speed of Light

The first measurement of the speed of light near the surface of the Earth was made by Louis Fizeau[7] using two fixed mirrors, with one partially obscured by a rotating cogwheel. Fizeau's 1849 value for the speed of light was 5% too high. A year later Fizeau and Foucault[8] used a much improved apparatus based on light reflecting off a rotating mirror toward a stationary concave mirror about 35 kilometers away (see Fig. 2.1). The mirror, rotating at an angular velocity ω, moves slightly in the time it

[7](Armand-Hippolyte-) Louis Fizeau (1819–1896), French physicist. In 1849 he published the first 'terrestrial' measurement of the speed of light, and in 1850 with Foucault, this measurement was considerably refined. In 1851 he demonstrated the Fresnel drag (see Example 7.8).

[8](Jean Bernard) Léon Foucault (1819–1868), French physicist. Foucault made a precise terrestrial measurement of the speed of light together with Louis Fizeau. He is world renowned for experiments with the Foucault pendulum carried out in 1851.

takes for the light to bounce off and return from the stationary mirror. The returning light is therefore deflected by a small angle ϕ' from the original beam. A more detailed discussion of the Fizeau-Foucault experiment is found in Example 2.1.

By 1862 Foucault achieved a precision similar to Bradley's aberration measurement, reporting $c = 298{,}000 \pm 500$ km/s. This result was close enough to Bradley's to nurture the notion of the universal nature of speed of light. An editorial review in 1886 by *Nature* entitled *The Velocity of Light*[9] presents the perspective on many efforts to measure the speed of light. The article is written in the historical context of a period-contemporary material æther, showing the many efforts to understand speed of light in the nineteenth century. We conclude: Light travels at the same speed on the surface of the Earth and between the stars. Moreover, unlike sound, light travels through airless space.

Example 2.1

The Fizeau-Foucault Experiment

Obtain the deflection angle, ϕ', in the Fizeau-Foucault experiment (Fig. 2.1), as a function of the angular velocity ω and the distance l to the concave reflection mirror. Calculate the expected deflection ϕ' for a separation between the rotating mirror and the concave mirror of $l = 35$ km and an angular velocity of $\omega =$ 10 revolutions/s $= 20\pi$ rad/s. Discuss sources of error in this measurement of light speed. ◀

▶ **Solution** The incident light initially bounces off the rotating mirror when the mirror's normal unit vector $\hat{n}$ makes an angle $\theta_i = \theta$ with the horizontal, as shown in bottom detail (a) of Fig. 2.1. By the law of reflection

$$\mathbf{1} \quad \theta_r = \theta_i = \theta \ .$$

Therefore, the angle ϕ of the reflected beam is

$$\mathbf{2} \quad \phi = 2\theta \ .$$

When the light hits the concave mirror, it reflects directly back at the same angle ϕ to the horizontal. The time Δt it takes the light to travel from the rotating mirror to the concave mirror and back is

$$\mathbf{3} \quad \Delta t = \frac{2l}{c} \ .$$

[9]*Nature* pp. 29–32, May 13, 1886.

In this time, the mirror has rotated by an angle $\Delta\theta$ to have a new angle θ' with the horizontal, as seen in bottom detail (b) of Fig. 2.1. We obtain $\Delta\theta$ in terms of the angular velocity ω of the mirror

4 $$\Delta\theta = \omega\Delta t = \frac{2l\omega}{c}, \qquad \theta' = \theta - \Delta\theta \ .$$

From the geometry detail (b) in Fig. 2.1 we see that the incident angle of the returning light is

5 $$\theta_i' = \phi - \theta' \ .$$

We now obtain ϕ', the final deflection of the light from the horizontal using again the law of reflection, $\theta_i' = \theta_r'$ evaluating $\Delta\theta$

6 $$\phi' = \theta_r' - \theta' = \phi - 2\theta' = 2\theta - 2\theta' = 2\Delta\theta = \frac{4l\omega}{c} \ .$$

With $l = 35$ km and $\omega = 20\pi$ rad/s we expect to observe a deflection of magnitude

7 $$\phi' \approx \frac{\pi}{100} \text{ rads} = 1.7^\circ \ .$$

Conversely, a precise measurement of ϕ', for a prescribed distance l and a rotation frequency ω, yields the speed of light

8 $$c = \frac{4l\omega}{\phi'} \ ,$$

with an error δc, where

9 $$\delta c \equiv \delta\phi' \frac{\partial c}{\partial \phi'} = -\delta\phi' \frac{4l\omega}{\phi'^2} \ .$$

This can be simplified to read

10 $$\frac{\delta c}{c} = -\frac{\delta\phi'}{\phi'} \ .$$

Beside the measurement error in ϕ', the experimental set-up requires a precise determination of the distance l, and the control of the rotation frequency ω. These normally can be done to a much greater precision compared to the measurement error $\delta\phi'/\phi'$. ◀

Speed of EM Waves

More than one hundred and fifty years ago, the scientific context was shaped for what we now call SR. To be specific about the birth of SR, in a letter dated 5th of

January 1865, James Clerk Maxwell wrote to his cousin about his latest scientific work:[10]

> I have also a paper afloat containing an electromagnetic theory of light, which, till I am convinced to the contrary, I hold to be great guns.

James Clerk Maxwell ***British-Scottish physicist, 1831–1879***

Wikimedia, CC BY-SA 3.0-License Stipple engraving by G. J. Stodart from a Lithography by Fergus of Greenack

Maxwell began his studies at the age of 16 at the University of Edinburgh. He moved to Cambridge in 1850 and graduated from Trinity College in 1854 with a degree in mathematics.

In 1855 Maxwell was made a fellow of Trinity College. He returned to Scotland a year later as professor of natural philosophy at Marischal College, Aberdeen. He was laid off following an academic reorganization and moved in 1860 to King's College, London. In 1871 he moved to the newly endowed Cavendish Laboratory at Cambridge.

> From a long view of the history of mankind–seen from, say, ten thousand years from now–there can be little doubt that the most significant event of the nineteenth century will be judged as Maxwell's discovery of the laws of electrodynamics.

(Richard P. Feynman, *Lectures on Physics,* Volume II)

Introducing for purely theoretical reasons the missing term into the system of EM equations developed before,[11] Maxwell saw the presence of electromagnetic wave solutions within the system of equations we now name after him. Maxwell showed that the wave speed can be expressed in terms of the ratio of the strengths of the electric and magnetic fields (at the time called forces).

[10] F. Everitt, *James Clerk Maxwell: a force for physics, Physics World* **19** 32 (December 2006).

[11] J. Clerk Maxwell, "A Dynamical Theory of the Electromagnetic Field," *Phil. Trans. R. Soc. Lond.* **155** 459–512 (1 January 1865).

Maxwell found the (theoretical) velocity of the EM-waves, within the measurement error, to be close to the speed of light $c_{\text{EM}} = 310{,}740{,}000\,\text{m/s}$. He wrote,

> light and magnetism are affections of the same substance, and light is an electromagnetic disturbance propagated through the field according to electromagnetic laws

Maxwell comments as follows on the measurement of light speed within his theoretical approach:

> The only use made of light in the experiment [to measure the speed of light, JR] was to see the instruments.[12]

In 1886/1887 Heinrich Hertz[13] confirmed Maxwell's waves. By the time of Einstein's SR, Maxwell's wave velocity had been measured[14] to be $c_{\text{EM}} = 299{,}710 \pm 30\,\text{km/s}$. These EM measurements would lead to the speed of EM wave propagation in air: the value we can compute today using air index of refraction $n_{\text{air}} = 1.00027$ is $c_{\text{air}} = c/n_{\text{EM air}} = 299{,}712\,\text{km/s}$ in precise agreement with EM wave speed of 1905/1906. This means that Maxwell's hypothesis about light being an EM wave was precisely confirmed, a fact well known by Einstein and his contemporaries.

Light, Particles, Æther

Within the corpuscular view of light, a moving emitter is 'throwing' the light-particles. Therefore, the Galilean view of light velocity follows from Eq. (1.1); the source and light velocities add vectorially. On the other hand, we know that the speed of sound is a property of the medium in which sound propagates (air, water, etc.), and is not dependent on the motion of the emitter. However, the motion of material, such as air, can change the speed of sound:

$$c_s' \stackrel{?}{=} c_s + v, \tag{2.5}$$

where v is the 'wind' speed, and not the velocity of the source.

Even though the form of Maxwell's equations made c independent of the velocity of the wave source, and independent of the wavelength of the wave, the spreading of Maxwell waves representing light was at first understood in analogy to the spreading of sound. Maxwell considered a medium, a ponderable and transparent to matter æther, necessary for his waves to propagate. Since the speed of light was the property of the material medium, only a modification of the state of the material æther, and in particular, æther wind, could modify the observed velocity of light.

---

[12]See page 499 in Ref. [11].

[13]Heinrich Rudolf Hertz, (1857–1894), German physicist disavowed by Nazis, proved the existence and studied the propagation of electromagnetic waves. The frequency unit 'Hz' is named after him. He presented the Maxwell's equations in the format we learn today. Thus in literature from before 1933 these equations are often called *Maxwell-Hertz* equations, and that is how Einstein referred to them in 1905.

[14]E.B. Rosa, and N.E. Dorsey, *The Ratio of the Electromagnetic and Electrostatic Units, Bulletin of the Bureau of Standards,* **3** (6) 433 (1907) and *Phys. Rev. (Series I),* **22**, 367 (1906).

The presence of a material æther is in principle not compatible with the principle of relativity for the EM phenomena. The results of the Michelson and Morley experiments would ultimately restore in the young student Albert Einstein the trust in the Galilean principle of relativity and his analytically-deductive study of this question created SR where Einstein postulated that for all IOs

$$c' = c \,. \tag{2.6}$$

Equation (2.6), the universality of the speed of light, is arguably the key input into Einstein's formulation of SR. Paraphrasing Einstein's words, there cannot be an æther velocity; see essay *Æther and Special Relativity*, Sect. 2.3.

Let us restate the above: Maxwell completed a theory that unified what were, at the time, three fundamental phenomena in physics: electricity, magnetism, and light. The unified theory of Maxwell necessitated SR because it is inconsistent with Galilean coordinate transformations. After 40 years of confusion Einstein found the solution of how to extend the principle of relativity to the new EM phenomena. He added another principle, that the speed of light in the vacuum is universal. He formulated a new fundamental theory framework, later called Special Relativity.

Discussion 2.1 Speed of light

About the topic: We kick off the conversations with two students and the professor choosing the speed of light as our topic: What is special about the speed of light? Is the speed of light really universal, and why? This and many other conversations that follow actually took place, but we generally present extended and/or dramatized versions. Here Student is a more seasoned helper to the Professor, and Simplicius is a well-read novice.

- *Simplicius:* Professor, why can't we exceed the speed of light? We break the sound barrier all the time. Why not the light barrier?
- *Professor:* Let us begin by looking at the differences between the propagation of light and sound.
- *Simplicius:* Is there a difference? Light and sound are both waves.
- *Professor:* They are both waves but there is more to it. Although they both travel as waves, light travels as a transverse wave, whereas sound travels as a longitudinal wave.
- *Simplicius:* So, what?
- *Student:* Sound waves are created by the compression of air molecules. The material properties of air make this possible. The wave oscillates in the direction of motion.

(continued)

- *Professor:* However, Maxwell recognized that his electromagnetic waves, which of course include light, oscillate transversely to the direction of propagation of the wave, instead of longitudinally. For sound this type of behavior can be attributed to propagation in an incompressible solid medium.
- *Student:* This also reminds us that the properties of the medium affect the speed of sound. The speed of sound in air decreases with increased elevation or a change into a denser medium, for instance. The speed of sound in steel, say, is much faster than the speed of sound in air.
- *Professor:* Right, the speed of sound varies, and just for this reason is there no indication that the speed of sound could not be exceeded. This is not true of light. That there is a difference you note by remembering that in empty space for all conditions we have been able to establish that the speed of light is universal.
- *Student:* In the SR class I learned that Einstein used the principle of the universal speed of light to derive the Lorentz transformation without needing to look at the properties of the newly formulated Maxwell equations, a difficulty that derailed earlier efforts.
- *Simplicius:* Postulating that the speed of light is universal does not explain why the speed of light is universal.
- *Student:* Remember, it is the speed of light in space free of matter, the vacuum, that is universal and always the same.
- *Professor:* That is the key: No ponderable matter whatsoever, and yet light can propagate and the value is always the same. When you 'throw' light quanta, no matter how fast the source moves, the speed of light is the property of space that is uniform and homogeneous, and everywhere we can go the same. One could think that there is something else and non-material that supports light propagation. After 1919/1920 Einstein called that stuff relativistically invariant æther. We discuss this topic next, see Discussion 2.2.
- *Simplicius:* I heard that material bodies can move almost with the speed of light but can never quite reach it.
- *Professor:* Right, but light is massless and therefore must travel at the maximum velocity.
- *Student:* That of course is true in the vacuum. However, elementary particles, when shot into material bodies at ultra-relativistic speed, achieved with the help of an accelerator, show radiation features that resemble supersonic phenomena for sound.
- *Professor:* Indeed, the motion of material bodies through matter can generate for both sound and light a number of similar phenomena. However, these waves are very different: In space empty of ponderable matter, the vacuum, sound does not propagate while light does.

(continued)

- *Student:* For me that is the biggest difference between sound and light.
- *Professor:* Let me recapitulate: The speed of light in the vacuum is an upper universal limit. A body that has a rest mass must move slower than light in the vacuum, and the same is true for the propagation of light in materials. Light reaches maximum speed in the vacuum and no material body can catch up!
- *Simplicius:* What happens to light in material bodies?
- *Student:* The speed of light is only universal in the vacuum. In materials the speed of light is slowed down: The way Maxwell set up his equations guaranteed that there would never be a material that would have a propagating speed faster than light.
- *Simplicius:* Maybe we just need to discover a material in which the speed of light is faster?
- *Student:* I believe material bodies could only get in the way of light.
- *Professor:* Aside of the argument that space-time is homogenous and uniform, there is another way to argue the universality of c: Do you remember $E = mc^2$? If the speed of light were not universal, we would need to reformulate the equivalence between energy and mass.
- *Simplicius:* Can I say that all material bodies are made of light? That seems to me to be the contents of Einstein's $E = mc^2$.
- *Professor:* I myself say this too. And Einstein was explicit that he saw mass in this way. Early in his life Einstein laid the foundation for his understanding of rest mass and light. He realized that photons, the quanta of light, carry energy and thus mass from their source and can transfer this mass increment, moving the energy to a destination where they are absorbed. This is the photoelectric effect he published in 1905. From this insight ultimately the mass-energy equivalence emerged. Interestingly, the only known massless particle is the photon, moving with speed c and connecting different material bodies while transferring a mass increment $\delta m = E_{\mathrm{photon}}/c^2$.

Discussion 2.2 Is there an æther?

About the topic: Very few books today address the æther. Even if this idea was wrong, we should learn why. And if it was not quite wrong, maybe we should look again at this context.

- *Professor:* Maxwell, Lorentz and just about every scientist before 1905 believed that the reason light can propagate is due to the presence of

(continued)

an undiscovered light carrier substance, the luminiferous (light carrying) æther.

- *Simplicius:* Since Einstein's 1905 paper, everybody knows there is no æther.
- *Professor:* Surely they mean no ponderable material æther as was discussed by Maxwell and many followers.
- *Student:* In any case, 'what everybody says,' is not a scientific argument. By the way, who is everybody?
- *Simplicius:* My high school teacher. A popular book I have read. Wikipedia: "Aether fell to Occam's Razor[15] ...".
- *Professor:* By the year 1919 Einstein, Langevin, Lorentz and many other physics giants saw this differently. In fact Einstein in explicit words regretted his 1905 words. These were pronounced according to Occam's Razor: something that is not needed in SR does not exist. After the formulation of GR his SR argument was incorrect. GR needs an (imponderable) æther.
- *Student:* To illustrate the æther problem, let me ask some provocative questions. (Turning to Simplicius:) Do you think the Earth is flat?
- *Simplicius:* What?
- *Student:* We remember, once upon a time, practically 'everybody' believed this truth proved by Occam's Razor! (Turning to Simplicius:) Do you believe that the Earth moves around the Sun? Occam's Razor would claim sitting still is simpler.
- *Simplicius:* But Earth motion was evident to Copernicus, Galileo...
- *Student:* ...Copernicus did not dare to speak (but his associate did after his death). And, Galileo, well, you trust this condemned scientist rather than a well established philosophical principle according to which Earth is the center of the Universe in which it is at rest?
- *Simplicius:* Religious courts cannot judge scientific truth, therefore I trust Galileo; I trust Einstein.
- *Student:* Which Einstein, the one at age 26 (in 1905), or the other, at age 40 (in 1919)?
- *Simplicius:* I would think the older Einstein knows more.
- *Student:* But did you know that by the time Einstein turned 55, he was a refugee? Can you really trust a homeless scientist?
- *Simplicius:* How could the greatest living scientist of the epoch become homeless?
- *Professor:* Here we are, let me exaggerate – it all started with the æther!
- *Simplicius:* How?

(continued)

[15]William of Ockham's, also written as "Occam" (c. 1288–1347) razor is a principle rooted in an antique philosophy, seeking the simplest hypothesis possible when explaining phenomena.

- *Professor:* This is my personal view: Philipp Lenard, a 'very German' colleague of Einstein's, rejected relativity, and sought to reinterpret the relevant phenomena, such as the MM-experiment, or the perihelion rotation of Mercury, within his own ponderable æther context. He was not alone, and he was able to present supporting experimental results. Einstein refuted these claims, pointing out experimental mistakes. Lenard enlisted with the Nazis and introduced the concept of German science. In his view and that of all Nazis, Einstein and his æther was on all counts wrong. Lenard's argument was a Nazi version of Occam's Razor: Einstein, being a Jew, had to be wrong. When the Nazis seized power in Germany in early 1933, Einstein was by chance abroad. After exchanging a few letters with the Prussian Academy, he resigned his academic appointments and while doing this he became a stateless and homeless refugee.
- *Simplicius:* I don't believe that the Nazis came to power over the question of æther.
- *Professor:* In a highly polarized German society of the epoch the æther controversy divided scientists – to this day discussion of the æther is practically a taboo in Germany.
- *Simplicius:* If æther was so important once, I am at a loss to see why I hear so little about it today.
- *Professor:* The subtle nuance – there is æther but it is not made of usual matter – was very hard to grasp in the 1920s as is evidenced by the Einstein-Lenard controversy, and was too hot to handle just after WWII. Instead everyone focused on the simple and uncontroversial 'no æther' perspective presented by Einstein in the 1905 SR paper.
- *Student:* This was a step back into Einstein's footsteps he made as a Ph.D. student at the age 26. Even if this naive view allows us to teach SR effectively, this deprives us of the important insights he presented at the age of 40, which as I believe, are very interesting in the present day context.
- *Professor:* It is important to recall here the related arguments of Langevin invoking observability of absolute acceleration, see Sect. 2.3. And, before Einstein and Langevin era arrived, Ernst Mach (see Ref. [4] in Chap. 1) recognized the need to distinguish inertial from accelerated reference frames. This insight is of such an importance that we call it Mach's principle, and argue in depth upon several versions (not in this book). In my view Einstein's non-ponderable æther of 1920 is needed to determine locally if a body is accelerated.
- *Simplicius:* Can you explain why several generations later I only read about the 'young' Einstein point of view?
- *Professor:* In part this is so since today the concept of imponderable æther has been replaced by 'quantum vacuum'.

(continued)

- *Simplicius:* What a funny combination of words: 'a quantum of nothing'!
- *Student:* Checking Wikipedia on this general topic I read: "...the Higgs[16] field, arising from spontaneous symmetry breaking, is the mechanism by which the other fields in the theory acquire mass."
- *Simplicius:* Can you explain?
- *Professor:* The Higgs field provides the measuring stick for the masses of particles. That is just like Einstein's æther and his 1920 conceptual writings.
- *Student:* In 1919/1920 Einstein wrote that æther is "...space is endowed with physical qualities." Now the Higgs field endows particles everywhere with a part if not all of their inertial mass.
- *Simplicius:* So the Higgs field brings back the æther...
- *Professor:* ...in its quantum vacuum dynamical reincarnation. The properties of the quantum æther are just what Einstein saw as being the permitted properties of the æther before quantum mechanics was invented. Quantum mechanics brings in additional dynamical features absent in the classical æther theory.
- *Simplicius:* Why in the end is Wikipedia's remark about Occam's Razor evidence against æther wrong?
- *Professor:* Existence of æther is a scientific question. Occam's Razor is, however, a logical argument. Occam's Razor advises you, if you do not know better, to take the simplest point of view. The adjective 'simple' is used here synonymously with 'known'. Occam's Razor argument suggests avoiding the invention of new theories. However, if the known 'simple' has contradictions, we must propose new ideas which need time to find their simple form. Often these new ideas appear at first considerably more complex compared to the simple and known. Clearly we know much more today about the nonponderable æther and the quantum vacuum properties.
- *Simplicius:* Can I learn by example how new and complicated becomes simple and known?
- *Student:* Have you tried reading Newton's *Principia* or Maxwell's *Electromagnetism*? Good Luck! According to Occam's Razor, you would be convinced they must be wrong.
- *Simplicius:* I have read and I could understand Einstein's 1905 *Relativity*.
- *Professor:* This is an exemplary case of an exception that confirms the rule: We recall that Einstein did not cite anyone in that revolutionary work. Being a graduate student entirely on his own, with no position, no grants, no reputation (as yet), he could ignore the complexities of all

(continued)

[16]Peter W. Higgs (1929–), British theoretical physicist who developed quantum vacuum structure ideas allowing decisive characterization of the mass of elementary particles. Nobel Prize 2013.

past misunderstandings and follow his own path, writing in part in a new manner we since adopted and are familiar with today.

- *Simplicius:* In his 1905 work Einstein disposes of the æther.
- *Professor:* Using Occam's Razor he pushes it aside as unobservable. In 1919/1920 Einstein explains that he went too far with his 1905 existential argument. In the intervening 15 years he understood that to assure the consistency of relativity there could not be an æther wind. An immovable imponderable æther of 1920 creates consistency with the general theory of relativity. Today we need the æther also in the context of quantum theory, where we have introduced Einstein's non-material æther, renaming it the quantum vacuum.

2.3 *Essay:* Æther and Special Relativity

We look here at rapidly evolving views about the æther at the time relativity theory was created and developed. Our discussion centers around Einstein's elaborate relativistic æther that we call today the structured quantum vacuum. It is important that all readers become aware in what format Einstein accepted and defended the concept of a relativistic æther:[17]

> (To summarize:) We can say that according to the general theory of relativity, space is endowed with physical qualities; in this sense the æther exists. According to the general theory of relativity, space without æther cannot be considered. Without æther light not only could not propagate, but also there could be no measuring rods and clocks, resulting in nonexistence of space-time distance as a physical concept. On the other hand, this æther cannot be thought to possess properties characteristic of ponderable matter, such as having parts trackable in time. Motion cannot be inherent to the æther.

These words of 1919/1920 echo some of the insights presented by Paul Langevin[18] in 1911. However, the ideas of Einstein are rooted in his GR and address dynamical constraints and properties of the æther.

[17]A. Einstein, *Äther und Relativiäts-Theorie (Æther and the Theory of Relativity)* (Verlag Julius Springer, Berlin 1920), reprinted in *The Berlin Years Writings 1918–1921*, M. Janssen, R. Schulmann, J. Illy, Ch. Lehner, and D. K. Buchwald, Eds; see pp. 305–309; and p. 321, Document 38 in *Collected Papers of Albert Einstein*, (Princeton University Press, 2002), translated by the author from the original: "(Zusammenfassend können wir sagen:) Nach der allgemeinen Relativitätstheorie ist der Raum mit physikalischen Qualitäten ausgestattet; es existiert also in diesem Sinne ein Äther. Gemäß der allgemeinen Relativitätstheorie ist ein Raum ohne Äther undenkbar, denn in einem solchen (Raum ohne Äther) gäbe es nicht nur keine Lichtfortpflanzung, sondern auch keine Existenzmöglichkeit von Maßstäben und Uhren, also auch keine räumlich-zeitlichen Entfernungen im Sinne der Physik. Dieser Äther darf aber nicht mit der für ponderable Medien charakteristischen Eigenschaft ausgestattet gedacht werden, aus durch die Zeit verfolgbaren Teilen zu bestehen; der Bewegungsbegriff darf auf ihn nicht angewendet werden."

[18]Paul Langevin, *L'Évolution de l'espace et du temps Scientia* **X** 31–54 (1911) (Proceedings of "Conférence au Congrès de Philosophie de Bologna"); and translation by J.B. Sykes, *The Evolution of Space and Time, Scientia*, **108** 285–300 (1973).

The topic of æther deserves consideration in the context of SR. This is so since practically all readers think that æther does not exist: Einstein in his 1905 manuscript proved that it is a spurious concept. However, Einstein elaborated on this in more detail in 1919/1920. He was not the first to realize our views about the æther needed to be revisited. We will show that already in 1911 Paul Langevin, *loc.cit.*, points out that only inertial motion cannot be perceived. It follows that IOs can be equivalent only if during the relevant time period they were not subject to an acceleration.

A few years after Langevin and under the influence of insights that the understanding of gravity force generated, Einstein takes Langevin's insights one step further, giving the æther the properties which assure that Langevin's argument is valid. Therefore, as soon as we use acceleration to transit an observer from one state of inertial motion to another, this observer is different from observers who always remain inertial. Practically all paradoxes attributed by other authors to SR originate in the misunderstanding of this one pivotal point. In this essay we therefore explain Einstein's views about æther as formulated in 1919/1920. Before, however, we need to dedicate a few words to the prehistory of the topic.

Maxwellians and the Æther

For the 40 years between the eras of Maxwell and Einstein, the greatest minds were unable to reconcile classical dynamics with electromagnetism. Arguably this was so since the contemporary way of thinking was distracted by the imposition of a material *luminiferous* æther: the adjective *luminiferous* (meaning light-bearing) reflects on the common speed of both light and Maxwell's waves. The luminiferous æther was like air is for sound, the medium capable of vibrations.

To appreciate this difficult context in full we need to step back to the times of Maxwell's unified theory of electricity, magnetism and light. Previous theories had assumed that the energy was located at, or on, magnets and/or electrically charged bodies. In his work Maxwell recognized the EM field as partaking in energy. He thus answered the question where the electromagnetic energy resides. Maxwell further searched for the answer to how light propagates in space, devoting a great amount of effort to the understanding of EM wave medium. Seeing that his waves oscillated transverse to the direction of propagation, Maxwell developed models based on material incompressible æther, not allowing density compression waves.

Æther became an obsession of the scientific elite. By the mid-1890s, EM was regarded as one of the most fundamental and fruitful of all physical theories. Much of ongoing research work was focused on the understanding of how EM waves and light propagate in the otherwise invisible æther. This prompted exploration of many hypothetical properties of this new form of matter – a situation not different from present day where both 'dark matter' and 'dark energy' play comparably enigmatic roles. A reminder of this material æther research effort is the volume edited by Sir Joseph Larmor, (Ref. [13] in Chap. 1) in memory of FitzGerald in 1902; see Fig. 2.2. We see here titles of FitzGerald's works demonstrating the prevailing æther research context in this time period.

Despite his original contributions to the development of LT and radiation emission by accelerated charged particles, Larmor's own book of 1900 made, if anything, the situation worse. It is sufficient to look at Fig. 2.3 to realize the cause. One of his colleagues characterized the book as, "All Æther, no Matter," in consideration of Larmor's view that all matter is created by æther density

THE

SCIENTIFIC WRITINGS

OF THE LATE

GEORGE FRANCIS FITZGERALD

Sc.D., F.R.S., Hon. F.R.S.E.

Fellow of Trinity College, and Erasmus Smith's Professor of Natural and Experimental Philosophy in the University of Dublin

COLLECTED AND EDITED WITH A HISTORICAL INTRODUCTION

BY JOSEPH LARMOR, Sec. R.S.

FELLOW OF ST JOHN'S COLLEGE CAMBRIDGE

DUBLIN: HODGES, FIGGIS, & CO., Ltd., GRAFTON STREET
LONDON: LONGMANS, GREEN, & CO., PATERNOSTER ROW

1902

Fig. 2.2 The title page of the FitzGerald memorial volume edited by Sir J. Larmor with a few titles discussing the æther

AETHER AND MATTER

A DEVELOPMENT OF THE DYNAMICAL RELATIONS OF THE AETHER TO MATHERLAL SYSTEMS

ON THE BASIS OF THE

ATOMIC CONSTITUTION OF MATTER

INCLUDING A DISCUSSION OF THE INFLUENCE OF THE EARTH'S MOTION ON OPTICAL PHENOMENA

BEING AN ADAMS PRIZE ESSAY IN THE UNIVERSITY OF CAMBRIDGE

BY

JOSEPH LARMOR, M.A., F.R.S.

FELLOW OF ST JOHN'S COLLEGE, CAMBRIDGE

CAMBRIDGE
AT THE UNIVERSITY PRESS
1900

Fig. 2.3 The title page of the year 1900 flagship Larmor volume: *Æther and Matter: A Development of the dynamical relations of the æther to material systems on the basis of the atomic constitution of matter, including a discussion of the influence of the Earth's motion on optical phenomena,* by Joseph Larmor, (Cambridge University Press 1900) compression. Clearly, the question, "What is the æther?" remained unresolved and was perhaps the paramount challenge facing Einstein.

In his 1905 work Einstein, a graduate student at the time, dispensed with æther, thus dismissing work by top scientists of the preceeding period including but not limited to: FitzGerald, Heaviside, Hertz, Larmor, Lodge, Lorentz, and Maxwell. Einstein considered in 1905 æther in the new relativistic world of inertially moving bodies as unobservable, and thus not needed. This is in general what a physics student learns today. However, this was merely Einstein's first encounter with the æther. Perhaps we should also teach these students that more than a century ago a self-supporting graduate student could publish prominently (Ref. [19] in Chap. 1) a foundational SR research paper that invalidated – I guess here – 30% of fundamental physics research of preceeding decades.[19]

Paul Langevin ***French physicist, 1872–1946***

Albert Einstein & Paul Langevin **'Twins' without a paradox** *Wikimedia, CC BY-SA 3.0-License* Segment from group photo 1911 Solvay Conference, Brussels

Langevin studied physics in Paris and Cambridge, UK under the guidance of Pierre Curie, Joseph John Thomson, and Gabriel Lippmann, Ph.D. in 1902. He became the professor of physics at Collège de France in 1909, and was elected to the Académie des Sciences in 1934.

Langevin profoundly impacted modern physics directly and indirectly; among his students: Louis de Broglie, Léon Brillouin, and Irène Joliot-Curie. Langevin is credited for the description of the "Twin Paradox" – for Langevin (as for Einstein) a paradox did not exist when in 1911 he interpreted SR, and went beyond Einstein in regard to the conceptual meaning of acceleration and the æther. His deep understanding of time dilation's relation to the presence of non-inertial motion was presented in elaborate French, and was not translated for many decades. This could be the origin of the common misunderstanding of this pivotal idea-based contribution to the theory of relativity.

[19]This author recollects vividly a conversation with Jack Sandweiss, a prominent journal editor, who described internal editorial discussion acknowledging that none of Einstein's path-breaking 1905 publications could have passed the present day peer review process.

Paul Langevin and the Æther

In 1911–1912, while Einstein was writing a long technical review of EM and SR he would never publish,[20] Paul Langevin composed a long foundational paper, see Ref. [18], developing further the principles of SR. Langevin clarifies what Einstein achieved about the understanding of the æther, saying:

> ...a uniform translation motion in the æther is not experimentally detectable. ...From this it should not be concluded, as has sometimes happened prematurely, that the æther must be abandoned having no physical reality since it cannot be experimentally probed. Only the uniform velocity relative to the æther cannot be detected, any change of velocity, that is, any acceleration, has an absolute meaning.[21]

Occasionally the last words from the above citation 'toute accélération a un sens absolu' are interpreted to be a name of an 'Absolute Acceleration' theory. However, such a theory was not transmitted to the present time. Grammatically this is also an incorrect translation. The meaning of these words is 'acceleration cannot be transformed away with coordinate transformations, and in this sense it can be measured universally (absolute) by any IO'. This interpretation is also supported by Langevin's arguments, that an accelerated motion cannot be subject to the relativity principle.

A few lines after the æther paragraphs, Langevin considers time dilation[22] and makes an unequivocal statement:

> Concluding, we can say it is sufficient to be set in motion, to experience acceleration in order to age less quickly.

Langevin is clarifying that in his view acceleration is an inseparable part of the time dilation effects. Langevin's insight is transmitted to the present day in a fragmentary manner and is frequently misunderstood. One sees in books and web pages confusing and even outright wrong accounts, introducing the 'twin paradox'.

---

[20]The manuscript is published as a facsimile of the original handwritten document, with original German, English translation, and historical introduction: A. Einstein and H. Gutfreund, *Einstein's 1912 Manuscript on the Special Theory of Relativity,* ISBN 0807615323 (George Braziller 2004); the publication of this manuscript was delayed by the outbreak of WWI. After WWI Einstein disavowed a late publication of his pre-WWI 1912 special relativity review since the scientific context had evolved far beyond the SR of pre-WWI. Today, of course, it has evolved even further.

[21]Translated by the author from the following French original, top p. 47 *Scientia* X (1911): "...une translation uniforme dans l'éther n'a pas de sens expérimental. Mais il ne faut pas conclure pour cela, comme on l'a fait parfois prématurément, que la notion d'éther doit être abandonnée, que l'éther est inexistant, inaccessible à l'expérience. Seule une vitesse uniforme par rapport à lui ne peut être décelée, mais tout changement de vitesse, toute accélération a un sens absolu."

[22]French original p. 49, second paragraph *Scientia* X (1911) "On peut dire encore qu'il suffit de s'agiter, de subir des accélérations pour vieillir moins vites."

Einstein and the Æther

After the formulation of the general theory of relativity, Einstein revised his position concerning the æther. In a letter to H.A. Lorentz of November 15, 1919, he says[23]:

> It would have been more correct if I had limited myself, in my earlier publications, to emphasizing only the non-existence of an æther velocity, instead of arguing the total non-existence of the æther, for I can see that with the word æther we say nothing else than that space has to be viewed as a carrier of physical qualities.

Einstein refines his better understanding of æther in Spring 1920 in an essay based on a lecture he is scheduled to present in Leiden, see Ref. [17], in honor of Lorentz. Einstein begins by explaining why in 1905 he disposed of the æther, saying

> ...considered alone in the context of special theory of relativity [used in the meaning of 'theory of electromagnetism', JR], the æther hypothesis is a hypothesis devoid of content. In the field equations of electromagnetism, in addition to charge densities, only the electric and magnetic field strengths enter. It appears that in a vacuum, these equations consistently determine the outcome of all electromagnetic processes uninfluenced by any other physical quantities. The electromagnetic fields appear as the ultimate, irreducible realities. Therefore, the consideration of electromagnetic fields as states of the hypothetical homogeneous, isotropic æther-medium appears, on a first inspection, superfluous.[24]

Einstein proceeds to explain carefully the new developments that causes his retreat from his earlier criticism of the æther. Einstein argues in detail and presents his reasoning that æther is a necessary ingredient. The reality of the æther is due to GR. He makes it clear that the new æther differs decisively from proposals made by others, including Lorentz. His æther is a nonmaterial, nonponderable substance that can satisfy the principles of SR. He contrasts the situation with that of space free of matter in EM theory. In Einstein's words:

> The principal difference between the æther of general theory of relativity and the case of the Lorentzian æther is that in the first case the state of the æther at any point is determined by the presence of matter in this neighborhood. This is so since laws of physics have the form of differential equations. On the other hand, the state of the Lorentzian æther in the absence of electromagnetic fields is determined by nothing other than itself, and the state is everywhere the same.[25]

---

[23]See page 2 in *Einstein and the Æther*, L. Kostro, Apeiron, Montreal (2000).

[24]Translated by the author from the original German in Ref. [17]. The original text conveys the exact meaning intended by the author. The original paragraph reads: "Allerdings erscheint die Ätherhypothese vom Standpunkte der speziellen Relativitätstheorie zunächst als eine leere Hypothese. In den elektromagnetischen Feldgleichungen treten außer den elektrischen Ladungsdichten nur die Feldstärken auf. Der Ablauf der elektromagnetischen Vorgänge im Vakuum scheint durch jenes innere Gesetz völlig bestimmt zu sein, unbeeinflußt durch andere physikalische Größen. Die elektromagnetischen Felder erscheinen als letzte, nicht weiter zurückführbare Realitäten, und es erscheint zunächst überflüssig, ein homogenes, intropes Äthermedium zu postulieren, als dessen Zustände jene Felder aufzufassen wären."

[25]Translated by the author from the original in Ref. [17]: "Das prinzipiell Neuartige des Äthers der allgemeinen Relativitätstheorie gegenüber dem Lorentzschen Äther besteht darin, daß der Zustand

Einstein emphasizes that the field equations of GR allow us to evaluate everywhere the response of the space-time metric to the presence of matter, a feature absent in SR context. By implication in GR the concept of æther is meaningful. He argues that Lorentz's æther cannot be quantified without the presence of EM fields. Considering Einstein's ensuing effort it appears he knew that the framework of GR does not solve the problem of EM forces and he was uncertain how and if the presence of EM fields influence the metric and could be integrated into GR.

We have seen now how Einstein evolved along the path of Langevin, and how in the GR context he adds significantly by discussing the properties of the æther.

Æther, the Structured Quantum Vacuum, and the Origin of Mass

Einstein's reestablishment of the relativistic æther was intentionally timed to occur in the context of his scheduled visit at Leiden in order to clarify how and why he approached the views of his æther-loving host, Lorentz. Einstein's new conceptual framework became well known and was much debated as the author experienced in conversations with Victor Weisskopf.[26] Weisskopf arrived at the physics scene a decade after Einstein's æther essay. He shared his thoughts about "Einstein relativistically invariant æther" (Weisskopf's words) with the author in the summer of 1979. Our conversation was centered on how Einstein's views about æther paved the path for the development of the structured quantum vacuum. Weisskopf should know this as he wrote a pivotal article about the response of the quantum vacuum to applied EM fields.[27]

In our conversation Weisskopf recognized the path of Einstein's evolution from his initial rejection of æther, by way of acceptance of non-material, non-divisible, relativistically invariant æther, to the formulation of its properties rooted in GR frame in a form compatible to our understanding of the structured quantum vacuum. We discussed how the pre-quantum physics æther ideas of Einstein have found their way into the quantum-field theory structured quantum vacuum, with measurable properties related, for example, to quark confinement, the research topic that overshadowed our 1979 conversations.

The structured quantum vacuum reaches beyond Einstein's æther. In particular, the phenomenon of quark confinement originates in the response of the quantum

---

des ersteren an jeder Stelle bestimmt ist durch gesetzliche Zusammenhänge mit der Materie und mit den Ätherzuständen in benachbarten Stellen in Gestalt von Differentialgleichungen, während der Zustand des Lorentzschen Äthers bei Abwesenheit von elektromagnetischen Feldern durch nichts außer ihm bedingt und überall der gleiche ist."

[26]Victor Weisskopf (1908–2002), an Austrian-born American theoretical physicist, and Director General of CERN 1961–1966. A student of Max Born, he earned his PhD at the age of 23; collaborator of Werner Heisenberg, Erwin Schrodinger, Wolfgang Pauli and Niels Bohr.

[27]V. Weisskopf, "Über die Elektrodynamik des Vakuums auf grund der Quantentheorie des Elektrons (On electromagnetic quantum vacuum structure induced by quantum electron fluctuations)," *Kongelige Danske Videnskabernes Selskab, Matematisk-fysiske meddelelser* **XIV**, No 6 (1936).

structure to the presence of quarks: the quantum vacuum (æther) abhors free quarks.[28] However, confinement can melt, liberating quarks within the localized volume of a quark-gluon plasma at temperatures achieved in relativistic heavy ion (nuclear) collision experiments. We return to heavy ion scattering in Sect. 3.3. To understand the mass of strongly interacting composite particles, for example of a proton, we need to account for both the material contribution of quarks and the dominant contribution stored in the local modification of the space-time domain confining the quarks. This quantum æther deformation thus determines the origin of the mass of all matter surrounding us.

The present day understanding of the origin of mass of nucleons and other related composite particles as originating in the quantum structured vacuum deformation aligns well with Maxwell's ideas. In Maxwell's theory, electromagnetic energy could be stored in the deformation described by a 'field' of the carrier of the waves, the luminiferous æther. Einstein argues along the same line, noting that elementary particles are condensations of field energy:[29]

> Considering that the elementary constituents of matter are in their nature nothing else but condensations of the electromagnetic fields ....

In context of his æther views Einstein interpreted the mass of particles as a condensed field energy.

Accordingly, the fundamental understanding of the origin of the mass of elementary particles does not conflict with Einstein's deductive arguments about the æther either. For Einstein the mass was frozen in EM field energy, in the standard model of particle physics, the material elementary particle mass is a condensation of the so called Higgs field. In the standard model of particle physics this mass is generated in the interaction with the Higgs field, see Discussion 2.2. This model offers a concrete realization of Einstein's comments about the æther providing the space and time measurement units and thus also units of the mass and energy of elementary particles.

We conclude: In the century that elapsed since Langevin and Einstein we discovered quantum physics, and have developed the standard model of elementary particle physics. In the context of these discoveries we developed the new picture of the quantum vacuum, and with it, in an increasingly quantitative manner, we improved our understanding of the origin of mass rooted in the quantum vacuum, the æther as we understand it today.

End of *Essay:* Æther and Special Relativity

---

[28] See for example: J. Rafelski, *Melting hadrons, boiling quarks, Eur. Phys. J. A* **51** 114 (2015).

[29] Translated by the author from original in Ref. [17]: "Da nach unseren heutigen Auffassungen auch die Elementarteilchen der Materie ihrem Wesen nach nichts anderes sind als Verdichtungen des elektromagnetischen Feldes,. . . ."

Discussion 2.3 Acceleration and the æther

About the topic: A relativistically invariant æther helps justify absolute acceleration – 'absolute' meaning that all inertial observers agree on the presence and strength of acceleration the observed body experiences. We discuss how in SR, where the word 'relative' enters prominently, acceleration can have an absolute, i.e. non-relative dynamical value. We ask how the aether can become visible. This leads to a survey of the understanding of the Universe, addressing some of the most surprising results in the field of cosmology, including dark energy.

- *Simplicius:* Can you remind us what an inertial frame is?
- *Student:* An inertial reference frame is one that has no net external forces acting on it.
- *Professor:* A body retains its inertial motion forever if no forces are acting upon it. Only relative motion and relative velocity can be determined.
- *Simplicius:* Why does acceleration have an absolute meaning?
- *Professor:* Consider the following situation: If I was driving down the road and you were walking beside me and my car, and if we were both at the same speed, neither one of us can say which of us is moving. However, if I slam the gas or the brake, I could not claim you were slowing down or speeding up. It would not make sense as no force is being applied on you; yet my brakes or the car's engine are applying a force on me. Force creates acceleration and it is specific to the body to which it applies.
- *Student:* I remember that no matter which frame of reference you decide to observe from, you cannot just make gravitational attraction; it is a distortion of space-time regardless of which reference frame it is viewed from.
- *Professor:* That argument works nicely for gravity force, embedded into General Relativity (GR), which in my view would be better called Gravity Relativity. We must remember that one of the problems with SR is that despite a long effort by Einstein and many others, a comparable to GR foundational level of understanding how the other forces work has not been achieved. This remark is true in particular for the electromagnetic forces, the Lorentz force.
- *Simplicius:* That seems straight forward enough; if acceleration is absolute, what is the best frame to consider acceleration from then?
- *Professor:* Any inertial frame is good enough – there is no need to use a special frame such as the cosmological reference frame. However, there is the relativistically invariant Einstein's æther of 1919/1920, which can give acceleration objective meaning, 'how' was first described by Langevin as you just saw in the Essay: *Æther and Special Relativity*.

(continued)

- *Simplicius:* Was not the æther a theory debunked by the Michelson-Morley experiment?
- *Professor:* The Maxwell-FitzGerald-Larmor æther theory was debunked. After the formulation of GR Einstein recognized that a new non-ponderable æther was necessary.
- *Student:* This æther has some interesting properties; in Einstein's address at Leiden, see Ref. [17], he notes that the æther had imponderable properties; that is, no kinematic or mechanical properties.
- *Simplicius:* How can we even be sure that something we cannot measure exists?
- *Professor:* Such æther allows a body subject to acceleration to know it is accelerating without having to ask a distant inertial observer. In Einstein's GR this is the space-time metric. I believe that seeking to understand the meaning of the space-time metric Einstein revised his views about the æther. And, last but not least, æther can have the property of energy density, an energy density we call dark energy, that would be present all over the Universe.
- *Simplicius:* What significance does energy density have?
- *Professor:* It has tremendous importance in cosmology where all forms of energy – usual matter, dark matter, and this new form, dark energy – combine and fix the value of the Hubble parameter[30] which describes how fast the Universe is expanding. We believe today that just a few percent of energy inventory today is due to the visible matter in different forms we are familiar with. This has changed profoundly our understanding of the Universe. For example, we now believe that the age of the Universe is 13.8×10^9 years.
- *Simplicius:* How can you distinguish energy X from energy Y?
- *Student:* Dark energy is an energy not associated with matter, a measurable property of the æther. This feature can be represented in Einstein's GR field equations introducing the cosmological constant. It influences the structure of cosmological equations in a way different from energy that is inherent in normal matter.
- *Simplicius:* This sounds complicated and, I heard, Einstein did not like his constant ....
- *Student:* It is. And the 'biggest blunder', as Einstein for some time referred to his cosmological constant, was in the end another stroke of genius.
- *Simplicius:* And how can I distinguish normal matter from dark matter?

(continued)

[30] H is named after Edwin Hubble (1889–1953), an American astronomer who recognized cosmological redshift, see Insight 13.1.

- *Student:* In the old days, we literally counted stars we could see with the biggest telescopes. Then we compared our count to the speed of Universe expansion today and said: 'lots of stuff seems invisible'.
- *Professor:* Since then we have become much more sophisticated. We found that our normal 'baryonic matter' is just 4.84% fraction in the energy inventory, where the number refers to the total energy content, and mass is converted to energy using $E = mc^2$.
- *Simplicius:* How can you tell this to three digits?
- *Student:* We are looking at matter fluctuations, temperature fluctuations, computing how the early Universe made light elements, ...
- *Professor:* ...experimental cosmology is today a vast research field. To conclude: There remains much to be discovered about material body acceleration and the relativistically invariant æther. Among important insights to retain from this discussion is that there is a profound significance to absolute acceleration. However, its understanding requires extension of Special Relativity as a dynamical theory at least to include EM interactions. In this book we rely on the recognition of the difference between inertial and accelerated bodies. This allows us to clarify the physical meaning of some of the principles of SR. We also keep in mind that in the context of SR there are only inertial observers. This is also true in GR in a generalized sense. Literature introducing accelerated observers must therefore be studied with extreme caution.

Material Bodies in SR

3

Abstract

The Michelson-Morley experiment is described. Time dilation and the Lorentz-FitzGerald body contraction are two properties of a material body required to assure unobservability of absolute body speed. We show how these effects enter the daily life of physicists in the context of experimental particle and nuclear physics. We explain why the body contraction is as real as is time dilation. We list and discuss diverse SR misunderstandings.

3.1 The Michelson-Morley Experiment

Earth's Motion and the æther

The discovery of EM waves paved the way for the search for how these propagate, and the introduction of material æther. Since the material æther was believed to be at rest in some specific reference frame, one could measure the Earth's absolute velocity vector. Furthermore, observation of changes in the speed of light could tell us about the properties of the material æther. Such experiments were naturally of great interest.

In 1881 Michelson conducted an experiment in an attempt to measure (a) the movement of the Earth relative to the material æther, and (b) the effect of the movement of the Earth on the speed of light arising from the æther wind. In 1887 an improved experimental effort by Michelson with Morley[1] followed.

[1]Edward Williams Morley (1838–1923), Professor of Chemistry from 1869 to 1906 at what is now Case Western Reserve University.

J. Rafelski, *Modern Special Relativity*, https://doi.org/10.1007/978-3-030-54352-5_3

The Michelson-Morley (MM) apparatus consists of a two-armed interferometer with the two light paths as depicted in Fig. 3.1. We note the light source Q, and a silver coated glass surface P which both transmits and reflects, splitting the beam towards mirrors M_1 and M_2. A piece of glass G assures that the path across glass material is the same for both light waves.

Albert Abraham Michelson ***American physicist, 1852–1931***

Picture: Smithsonian Institution Dibner Library Wikimedia, CC BY-SA 3.0-License

Michelson arrived as a child in the United States in 1855. From 1869 to 1881 he served in the marines, eventually as an instructor in physics. He undertook his postgraduate studies in Berlin and Paris.

Starting in 1880, and collaborating later with E.W. Morley, Michelson developed precise experiments to investigate the effect of the Earth's velocity on the speed of light. In 1893 Michelson joined the University of Chicago, creating its acclaimed Department of Physics. In 1907 he was the first American to receive the Nobel prize for physics "…for his optical precision instruments and the spectroscopic and metrological investigations carried out with their aid."

In Fig. 3.1 we assume an initial geometry such that one of the arms of the MM apparatus is aligned parallel, and another perpendicular, to the relative velocity $\boldsymbol{v}$ of the Earth with respect to the light-carrying æther. While the light wave is traveling the mirror M_1 is either approaching or receding from the light source. Similarly, the light path from P to M_2 and back is triangular as P moves. Motion with respect to light-carrying æther clearly influences the length of both optical paths. Therefore when this apparatus is made to rotate around an axis we expect to see in the detector T a time dependent interference fringe-shift.

Michelson's objective was to push the precision so that he could measure the variation of v caused by Earth's orbital speed $\delta v = \pm 30$ km/s. This variation perturbs the Earth peculiar speed with respect to the Cosmic Microwave Background

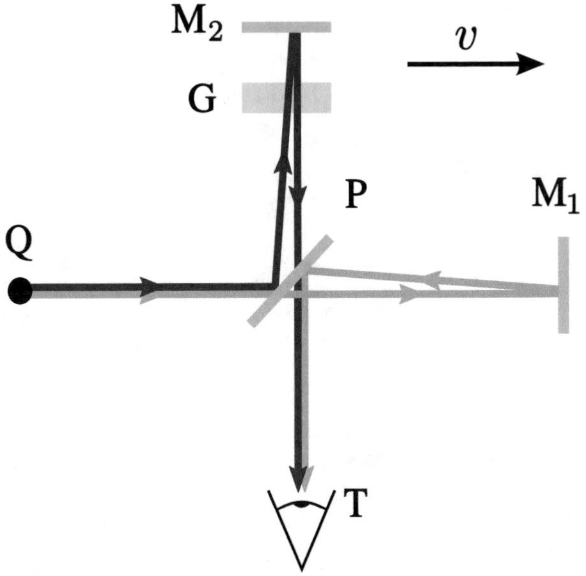

Fig. 3.1 The Michelson-Morley Interferometer. The light follows two partially overlapping paths: one QPM_2T (shown in blue) and the other QM_1PT (shown in green). The glass plate G seen in the former light path compensates the greater path length of the QM_1PT beam which penetrates through the silver coating of the upper side of the mirror P. The two beams could produce variable interference in the observer's T-detector device depending on the instantaneous orientation with respect to velocity $\boldsymbol{v}$ of the Earth with respect to the light-carrier

(CMB), which as we know today is about 12 times greater than the orbital velocity. The three main components of Earth's velocity vector $\boldsymbol{v}$ with respect to the CMB are illustrated qualitatively in Fig. 3.2: the smallest is the orbital speed around the Sun; the orbital speed in the Galaxy is 10 times larger; and the velocity of our Galaxy with respect to the CMB is yet about twice larger. The net velocity with respect to the CMB[2] is 370 km/s.

The detailed mathematical description of the light and mirror motion inherent to the MM experiment is presented in our study of the light-clock in Sect. 4.2. We defer a detailed description of the MM optical paths to this discussion. Here we note that the optical paths were defined by mirrors attached to a common rigid material body. Any changes of this body as it travels through the æther thus influences the outcome of the experiment, in addition to the expected modification of the propagation of light dependent on the relative motion of the MM apparatus with respect to the æther. The opinion in the late nineteenth century was that Maxwell's equations were valid only with respect to the æther at rest. Given the large magnitude of the speed of light, it was thought that it would take elaborate experimentation to discover

[2]P.A. Zyla et al. (Particle Data Group), *Prog. Theor. Exp. Phys.* 2020, 083C01 (2020) and 2021 update at https://pdg.lbl.gov/; see Table 2: Astrophysical Constants and Parameters.

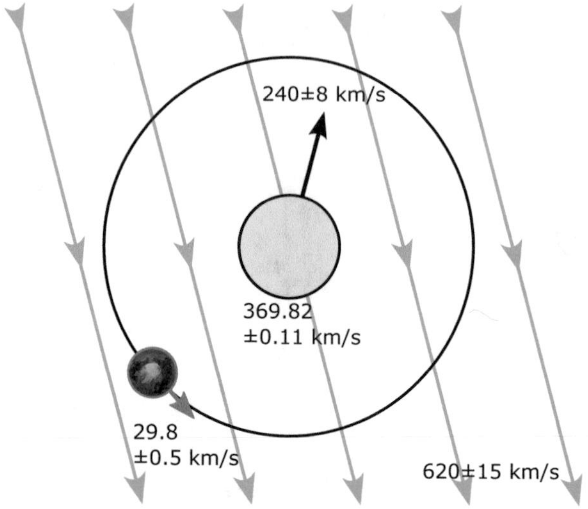

Fig. 3.2 The Earth's net motion against the CMB includes an orbital speed of 29.8 ± 0.5 km/s around the Sun, which moves with $\boldsymbol{v} = 369.82 \pm 0.11$ km/s (status 2021, Ref.[2]) against CBM. This value comprises: The Sun orbital speed within the Milky Way 240 ± 8 km/s; Milky Way galaxy group speed 620 ± 15 km/s

the limits of the validity of Maxwell's equations inherent in the local motion of the æther. Thus an objective of the MM experiment was to find the possible æther "wind".

Michelson and Morley's experimental null result, a result of unprecedented precision at that time, at the level of $2.5 \cdot 10^{-5}$ of the speed of light, was a sensation. Neither the motion of the apparatus nor any influence on the light speed was detected at the upper bound of an 8 km/s speed differential relative to a stationary material æther in which light propagates at the speed $c \simeq 300{,}000$ km/s. The outcome of this pivotal experiment was a paradigm shift to the belief that the state of motion of an IO is not observable by a local in space and time experiment. But how was that possible?

How this unobservability is possible was first explained in terms of body contraction in a terse and definitive way by George F. FitzGerald:[3]

> I have read with much interest Messrs. Michelson and Morley's wonderfully delicate experiment attempting to decide the important question as to how far the ether is carried along by the earth. Their result seems opposed to other experiments showing that the ether in the air can be carried along only to an inappreciable extent. I would suggest that almost the only hypothesis that can reconcile this opposition is that the length of material bodies changes, according as they are moving through the ether or across it, by an amount depending on the square of the ratio of their velocity to that of light. We know that electric forces (EM fields in today's language, JR) are affected by the motion of the electrified bodies relative to the ether, and it seems a not improbable supposition that the molecular forces are affected by the motion, and that the size of a body alters consequently....

[3]George Francis FitzGerald (1851–1901), Irish physicist, worked at Trinity College, Dublin; see Ref. [9] in Chap. 1.

The quoted text constitutes 2/3 of the one paragraph published note, see Ref. [9] in Chap. 1. The last cited phrase shows that FitzGerald recognized the body contraction as paralleling the already observed EM field changes for different observers. This is our current interpretation of the body contraction phenomenon: electron orbitals contract due to the motion of the source of the EM force, the atomic nuclei.

3.2 Body Contraction and Time Dilation

Body Contraction

A body traveling through a resisting material medium such as material æther could be subject to deformation. When the MM experimental news hit the scientific world in the æther era, this explanation was recognized as necessary. It was thus not surprising that the insight of FitzGerald, the shortening of the physical length of the interferometer along the direction of motion such that

$$L(v) = L_0\sqrt{1 - v^2/c^2}\,, \tag{3.1}$$

which was hidden in a letter to the editor of an American journal practically invisible to the European scientific community, was soon rediscovered by Hendrik A. Lorentz.[4] Note that in this relation Eq. (3.1), the proper length L_0 is on the right hand side and the laboratory observed length $L(v)$ is on the left. Body length in directions orthogonal with respect to the direction of motion remains unchanged.

After the introduction of the Lorentz-FitzGerald body contraction hypothesis to explain the MM experiment, it was widely believed that a smaller MM interference effect should still arise. Namely, if the presence of the material æther leads to body contraction, then conversely, æther wind must be created by the passage of the body through the æther; i.e., the body drag of the æther. The æther might be dragged along by the Earth, just like water is dragged along by a ship, or air is dragged along by a moving car.

Given that the EM Maxwell waves are transverse to the direction of motion, unlike the density waves of sound in air, æther should be difficult to compress. Therefore an Earth-dragged æther wind would have a velocity smaller than that of the Earth. For this reason the Lorentz-FitzGerald body contraction makes a more precise MM type experiment necessary. The study of the possible æther-drag continues today! The current best experimental limits, see Fig. 19.1, reached a

[4] H. A. Lorentz, "The relative motion of the Earth and the æther," *Versl. Kon. Akad. Wetensch.*, **1**, 74 (1892). Having learned of the earlier work by FitzGerald, Lorentz spoke of FitzGerald contraction.

precision at the relative 10^{-18} level, and demonstrate an upper speed limit of æther wind of 3Å/s (Å= 10^{-10}m).

Time Dilation

Lorentz (Ref. [4]) recognized his body contraction result when seeking to transcribe the Maxwell equations from one reference frame to another. Larmor (Ref. [15], Chap. 1), carrying on this work, realized that in this transformation time could not remain unchanged. The physical reason for this is that the synchronization of clocks located at different locations takes 'time' to complete. Einstein (Ref. [18], Chap. 1) realized that each body has its own 'body time'; later Minkowski (Ref. [7], Chap. 1) called this body time 'proper time'.

To understand the importance of this new proper time concept, let us compare the coordinate time, t, to the clock time on a train departing from a station and thus measuring train proper time τ of which an increment is

$$d\tau = dt\,\sqrt{1 - v^2/c^2}\,. \tag{3.2}$$

We note that in this relation the placement of proper time is on the left and laboratory time on the right, conceptually opposite to what we saw in Eq. (3.1).

Since $d\tau < dt$, we speak of time dilation. Furthermore, after the train returns to the station, the cumulative time dilation effect is shown by the train clock; based on this simple experiment the effect of time dilation is hard to misinterpret. Time dilation is as inherent to point-sized elementary particles as it is to large structured bodies. The first to recognize time dilation was (arguably) Sir Joseph Larmor, see Ref. [15] in Chap. 1. We return to discuss Larmor's work in Sect. 6.2, see page 110. In the last sentence of the letter of John Bell on page vi, we see that he believes that Larmor was fully aware his transformations implied time dilation.

Absolute Inertial Motion Is Unobservable

When considering the light paths of the MM experiment in Sect. 4.2 we will learn to appreciate better the two relativistic properties of material bodies, body contraction and time dilation. Their presence assures consistency of the relativistic behavior of material bodies with the generalized relativistic coordinate transformations, the Lorentz transformations. Specifically: In the MM experiment the Lorentz-FitzGerald contraction affects the laboratory table, a material body. The result is an optical light path length independent of interferometer orientation with respect to the direction of motion.

However, in terms of the laboratory time, the optical light path takes longer to complete compared to a reference experiment that remains at rest in the laboratory. The observer in motion cannot notice that his proper time is subject to time dilation Eq. (3.2), as the unit of time of the moving clock is lengthened (dilated) in the same way. This explains how it is possible that a velocity of the MM interferometer is undetectable by a comoving observer traveling along with the MM interferometer. We conclude: When both effects, body contraction and time dilation, are accounted for, there is no residual effect in the MM experiment at all; the optical paths are not dependent on the motion of the MM interferometer; absolute uniform motion is undetectable with the MM interferometer.

Since an absolute motion is not perceptible, it follows that we need to reconsider the exact meaning of the speed v entering the body contraction, Eq. (3.1), and time dilation, Eq. (3.2). Presenting v we always mean a velocity acquired when the body is set in motion (accelerated) with reference to a laboratory observer. We do not need to be present when the motion begins and develops generating the effects that formulas Eqs. (3.1) and (3.2) describe. Thus v can have any value that we find convenient to consider. This speed of a material body can in particular arise when a body is set in motion with reference to the inertial laboratory. In this case v has a prescribed and often time dependent value.

These considerations rely on the velocity of light c being a universal constant which is the key postulate that Einstein needed in order to create the SR framework. This idea was not seen at all in earlier work; indeed the MM experiment was in many ways an attempt to measure the variation of c arising from Earth motion with respect to the light-carrier, the luminiferous æther. The present day MM experiments demonstrate this postulate of Einstein to a very high precision. We return to this matter in Chap. 19.

Concluding Remarks

Body contraction and the properties of Maxwell fields invoked in this context by FitzGerald are in glaring conflict with the Galilean coordinate transformation, which needs to be adapted to this new situation. In Chap. 6 we therefore introduce the Lorentz transformations which resolve the inconsistency of the coordinate transformation with the effect of the Lorentz-FitzGerald body contraction and time dilation.

The MM experiment was and is today a test of the principle of relativity and a device to set a limit on the variation of the speed of light c, just as discussed in Einstein's 1905 manuscript. One should not interpret this experiment as a demonstration or even a measurement of the body properties such as the Lorentz-FitzGerald body contraction or time dilation of the comoving observer.

3.3 Is the Lorentz-FitzGerald Body Contraction Measurable?

The effects a body experiences have to be consistent with the coordinate transformation of relativity, the Lorentz coordinate transformations (LT), which we will study in Chap. 6. This new transformation will allow us to interpret consistently the same experiment in different inertial frames. However, using LT we can 'undo' the body contraction by choosing to perform a measurement in a frame of reference in relative rest. Is the body contraction real then, or merely a mathematical trick?

Lorentz always had the explicit opinion that the Lorentz-FitzGerald body contraction is "real." The views of Einstein were less explicit and more subtle, to the point of a misunderstanding he needed to clear up in 1911. The key statement reads:[5]

> The claim by the author (Variĉak) of a difference between Lorentz's view and that of mine with regard to the physical properties (of the Lorentz-FitzGerald contraction) is not correct. The question as to whether the Lorentz-FitzGerald contraction is a physical phenomenon or not can lead to a misunderstanding. For a comoving observer it is not present and as such it is not observable; however, it is real and in principle observable by physical means by any non-comoving observer.

In this paragraph Einstein confirms Lorentz's view that a body is truly contracted. He makes sure the reader understands that while this cannot be observed by anyone comoving with the body, the body contraction is observable from outside. This is just like the case of a passenger in a car being unable to measure car momentum, which a pedestrian standing outside the vehicle would notice. It is worth remembering that 'inertial navigation systems' can guide a flying body by integrating the effect of acceleration. In the same sense one can imagine building an instrument that could, from within a body, evaluate the instantaneous body contraction with reference to the body size measured in the laboratory.

We prove an effect is real by considering an experiment where the outcome demonstrates the phenomenon. For example, kinetic energy is 'real' even if we can identify a frame of reference wherein it vanishes. In the same way, the Lorentz-FitzGerald body contraction is a real phenomenon if we can present an experimental set-up where the outcome depends on the change in body length. The existence of a reference frame, or better said, of an inertial observer (IO) who cannot measure body contraction, is of no consequence in determining if body contraction is real.

[5] A. Einstein, "Zum Ehrenfestschen Paradoxon. Eine Bemerkung zu V. Variĉaks Aufsatz," (On Ehrenfest's Paradox. A remark regarding V. Variĉaks contribution) *Physikalische Zeitschrift* **12**, pp. 509–510 (1911); Original: "Der Verfasser hat mit Unrecht einen Unterschied der Lorentzschen Auffassung von der meinigen mit Bezug auf die physikalischen Tatsachen statuiert. Die Frage, ob die Lorentz-Verkürzung wirklich besteht oder nicht, ist irreführend. Sie besteht nämlich nicht "wirklich", insofern sie für einen mitbewegten Beobachter nicht existiert; sie besteht aber "wirklich", d. h. in solcher Weise, daß sie prinzipiell durch physikalische Mittel nachgewiesen werden könnte, für einen nicht mitbewegten Beobachter." *The author of this book translated this paragraph, taking into account the shift in usage of the scientific language in the past 100 years.*

Detecting the reality of relativistic physical phenomena depends on the availability of instruments that can measure the effects. The Lorentz-FitzGerald body contraction will be a recurrent topic in this book. The reality of the body contraction effect, i.e. the ability to measure it, will be addressed in depth. John S. Bell, known for his interest in the clarification of the difference between classical and quantum reality, further developed Lorentz's and Einstein's remarks on the reality of the Lorentz-FitzGerald body contraction. He refined a thought (Gedanken) experiment that allows an observer to record the effect of the Lorentz-FitzGerald contraction.

The Bell instrument compares the length scale of a body, measured prior to acceleration being applied, with a length scale in the body that is now speeded up. We return to this topic in Chap. 10. We will show in Sect. 10.3 that it is, in principle, possible to build a 'length clock' that both measures instantaneous contraction by comparing the laboratory body length to the instantaneous proper length, and retains the information about the accumulated effect of the Lorentz-FitzGerald contraction. Note that an ordinary clock measures only proper time and at return to base we know the cumulative effect of time dilation. An instrument that scores both proper time and laboratory time in the moving body frame would also need a new design akin to our discussion in Sect. 10.3. Such measurements are only possible if we start a body in a laboratory and allow it to acquire relative speed.

Let us consider as an example a 'train', departing a station at the foot of a mountain, gently accelerated, entering after a long time a mountain tunnel at relativistic speed. An observer could at that time be taking a movie and with luck a frame will show only the mountain, with the train being entirely inside the tunnel. This can happen due to the Lorentz-FitzGerald body contraction, even if the train is much longer than the tunnel, provided the train is sufficiently fast.

To create a good experiment an observer should be at the same distance from both ends of the observed body, so that the light from each body end takes the same time. The choice about the simultaneity of measurements of the body length matters in a normal nonrelativistic situation, and certainly, it is of great importance in SR where time and space combine to Minkowski 4-space. We return to discuss this in Chap. 9. Therefore, the measurement of the moving train (now in the tunnel) by an observer remaining at rest with respect to the point of origin (the station) must be arranged with care so that a blink of light originating at the same time at both train ends can hit the camera simultaneously.

The observation of a relativistic train is not a practical experiment for demonstrating the Lorentz-FitzGerald body contraction. A possibly feasible experimental arrangement for measuring the Lorentz-FitzGerald body contraction has been proposed.[6] This proposal involves a rapidly accelerated material foil fragment, accelerated to a high speed over a short distance in a very short time with a high intensity pulsed laser. From the other side another laser probes the flying foil fragment, evaluating the Lorentz-FitzGerald body contraction of the flying fragment compared to the remainder of the material foil still at rest in the laboratory.

[6]J. Rafelski, "Measurement of the Lorentz-FitzGerald body contraction," *Letter Eur. Phys. J.* **A 54** 29 (2018).

Discussion 3.1 Train in a tunnel and the principle of relativity

About the topic: We discuss why a train accelerated to a finite speed is contracted.

- *Simplicius:* There cannot be any reality to the Lorentz-FitzGerald body contraction, right? Given the principle of relativity, all measurement does is compare lengths and give the relative change; we do not know if the train or the tunnel is shorter.
- *Professor:* To decide this matter we set up a thought experiment. To begin with, the train is at rest with respect to the station and the mountain, and we establish that it is longer than the tunnel by a measurement in the common rest-frame. This can be done e.g. by taking a photo of the train parked at the station in front of the mountain. After that measurement, we ask the train driver to gently, very gently accelerate so that any effect that acceleration may have, would seem negligible. Once the train is fast enough, we let it coast inertially into the tunnel. We take a movie of this with a very fast camera. Based on SR we predict that by inspecting the frames we will establish that the fast train can fit into the tunnel. To finish the experiment, we ask the driver to slow and stop the train, which we observe to have again its original length.
- *Simplicius:* As long as nobody on the train has any memory of being shorter, I could claim it was the mountain that was shorter, or longer, or whatever. As long as there is no possibility of measurement we can claim anything, right?
- *Student:* I disagree: Professor already noted we did nothing to the mountain, so for any measurement we perform it was always there and of the same size. Second, we started with a train that was at rest and later was accelerated, and finally decelerated to come back. I believe any effect of motion would have to be with the train, not the mountain.
- *Professor:* Expanding on this argument: Only the train can be shorter as compared to the tunnel. The tunnel could not be longer just because nearby some train was set in motion. Moreover, even if passengers will not remember having contracted, we can observe this effect with a device to keep score of both momentary relative size, and the accumulated contraction effect, based on ideas made popular by John S. Bell, see Chap. 10. With Bell's device we can measure the momentary length of the train from within the train with reference to the length standard from prior to the acceleration process.
- *Simplicius:* However, by virtue of the principle of relativity I could place Bell's device in the mountain and claim that the mountain is being accelerated towards the train. I would expect to demonstrate that it is the tunnel and not the train that became shorter.

(continued)

- *Professor:* This argument is wrong and misleading, inflicts again and again much grief! We must never apply the principle of relativity to accelerated, non-IOs. Here we are accelerating the train, not the mountain. Acceleration of the train makes it different and, more to the point of your remark, it makes Bell's device report a change in body size of the train. The second device left with the mountain will report no change in body length properties.
- *Simplicius:* I have one more question: If only the train and the tunnel are present in the Universe, how can you tell which is subject to acceleration?
- *Professor:* I believe that Einstein reintroduced relativistically invariant æther in order to be able to recognize what is accelerated locally.
- *Simplicius:* But how can experiment tell which of the two bodies is accelerated? How and why can we ignore inertial forces?
- *Professor:* A well-known manner in which acceleration manifests itself is by emission of radiation: A body accelerated with reference to Einstein's æther (in general) emits either gravitational and/or electromagnetic radiation.

3.4 Experiments Requiring Body Contraction

After this in-depth discussion of a not as yet directly measured Lorentz-FitzGerald body contraction, one can wonder if this is a footnote without any physics consequences worth remembering. It turns out that this effect plays a very important role in experiments involving the collision of two relativistic heavy nuclei, called 'heavy ions'. The experimental program has been ongoing for a few decades now and continues at several accelerator facilities. These experiments are conducted for the purpose of forming the quark-gluon plasma state of matter. This form of matter filled the Universe when it was less than about 25 μs old. All matter around us originates in this primordial form of matter.

The colliding heavy nuclei are, viewed from any reference frame, subject to Lorentz-FitzGerald body contraction. It is best to consider collisions of ultra relativistic heavy ions coming from opposite directions at the same speed; the situation is analogous to two Lorentz-FitzGerald body contracted 'trains' colliding, each compressed by a very large factor in the range ten to ten thousand, depending on how fast the collision is. Given the original ball-like geometry of atomic nuclei, an observer in the laboratory will see two thin pancakes colliding as shown on the left in Fig. 3.3. Under the current experimental conditions the pancakes are flattened to one-tenth, or even down to one ten-thousandth, compared to the nuclear diameter, and their density in the CM system is increased by the same factor. In the middle of Fig. 3.3 we see the new quark-gluon plasma phase. After the collision, on the right in Fig. 3.3 we typically see a lot of newly created matter emerge. Observation

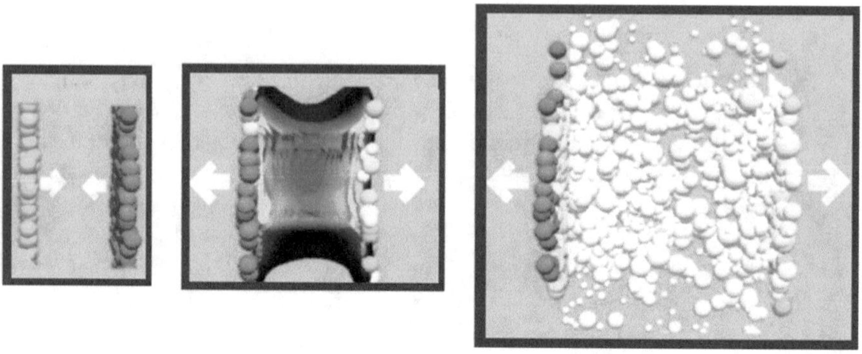

Fig. 3.3 On the left two Lorentz-FitzGerald body contracted nuclei approach each other; in the middle these 'heavy ions' crossed and there is a new phase of quark-gluon matter forming between them; to the right the remnants separate, with many new matter particles produced

of these particles helps us understand the processes that transpire, and helps us to unravel the secrets of the early Universe.

Aside of collider experiments represented in Fig. 3.3 there is an extensive body of experimental data for a relativistic heavy ion beam hitting a laboratory fixed target. These results are typically considered with the help of a suitable LT in the center of momentum frame of reference, as we discuss at the end of Chap. 18. In heavy ion experiments the duration of the collision process and the number of collisions are both of physical relevance. For the nuclei to cross each other they need to travel across the contracted atomic nucleus. We recall that contracted nuclei have a higher density. This means that the effects of thickness and density could cancel, but if QGP is formed in the process, this may not be the case.

Another related and important issue is: Can we treat the colliding nuclei as a rigid body, or is it a heap of independent particles? One can argue based on quantum physics that at sufficiently low energy the nuclei act as a strongly bound rigid body. On the other hand one can also argue that at a very high energy the nuclei could appear as an unbound cloud-like group of particles. To read more about this see Discussion 10.1. The relevant question is if an atomic nucleus retains its composite structure during the acceleration process, and this is the case. Therefore as the individual nuclei approach each other they are, as we stipulated before, Lorentz-FitzGerald body contracted. A concrete experiment has been proposed to measure the contraction effect.[7]

The consideration of relativistic heavy ion collisions we presented is an example where many of the SR results we discussed were put to use. These were considered so far only in a qualitative manner. In the following chapters they will be derived with mathematical tools and precision. Beyond this example one can say that in

[7] A. Bialas, A. Bzdak and V. Koch, "Femtoscopy of stopped protons," *Phys. Rev. C* **99** 034906 (2019).

many contexts of modern physics the proper understanding of SR is a necessary prerequisite.

3.5 Resolving Misunderstandings of SR

From the very beginning SR suffered from misunderstandings. The problem originated without doubt in the insistence of the Maxwellians in the four decades between Maxwell and Einstein's work of mid-1905, Ref. [18] in Chap. 1, on the presence of material luminiferous æther, the carrier of light waves. Many to this day still relevant research results, such as Larmor's radiation formula were contaminated by the material æther context. The continuing reverberations of these partially false works are the probable cause of recurring false interpretations of relativistic phenomena: It is hard to draw a distinct line between the relativistic æther of Langevin and Einstein and the material views of Maxwell, Larmor and many others, see Sect. 2.3.

Many students find SR of sufficient importance to seek a better understanding on their own. They procure popular relativity books and visit open web sources, but these may not always improve their comprehension. SR is a topic that attracts a crowd of contributors who often have skill with words but lack science preparation to discern right from wrong sources of information. The current pages of Wikipedia in my opinion add fuel to the common confusion about many SR topics. I have tried to make some necessary corrections but these were promptly removed by guardians of the Wikipedia pages. The procedure required to make corrections is not practical; scientific accuracy is not decided by a majority vote in a debating club dominated by self-appointed Wiki guards.

In the following I describe a few problems that are often seen on-line and in popular presentations of relativity. Before, a cautionary remark:

- In a world without forces, and thus without the related acceleration, knowledge of coordinate transformations suffices for understanding the behavior of point particles always remaining in inertial motion with respect to each other. This is how Einstein's 1905 SR is formulated. To be able to address the real world we must apply SR principles in a world in which material bodies have a finite size and acceleration allows for a physical transfer of an observer from one state of inertial motion to another.

Misunderstanding 1: Space Is Contracted

When posing the question, *Is the Lorentz contraction that of space or of a body?* I am told frequently that space is contracted. However, the Lorentz-Fitzgerald body contraction cannot be a contraction of space, for the simple reason that SR does not address the properties of the space-time in which we live. A later theory, Einstein's GR of 1916, looks at this question in order to create a relativistic form of Newton's law of gravity. That is why in this book I always speak of "body contraction," rather than simply "contraction." It is important that the reader always remembers: space and time are not impacted in any way in SR; in particular, they are not impacted by the inertial motion of particles or extended material bodies. The fact that one IO measures event coordinates that are different from those measured by another IO does not mean that there is a change to the space-time manifold.

Misunderstanding 2: The Lorentz-FitzGerald Body Contraction and Time Dilation Confirm Each Other

In SR both the Lorentz-FitzGerald body contraction and time dilation are unrelated body property phenomena (unlike energy and momentum which are related). That they are unrelated is easily recognized by remembering that an elementary practically point particle can experience large time dilation but cannot experience a Lorentz-FitzGerald body contraction of comparable absolute magnitude.

An unstable particle experiences time dilation irrespective of another material body being present, introduced to facilitate concurrent observation of an unrelated Lorentz-FitzGerald body contraction. Therefore, the additional claim that the two effects, body contraction and time dilation, confirm each other, is not a logically correct line of argument: This claim depends on a material body that is not required in the study (of e.g. the flight distance) of the unstable particle.

Misunderstanding 3: The Lorentz-FitzGerald Body Contraction Is Not Real

The fact that the Lorentz-FitzGerald body contraction and time dilation are largely independent phenomena contributes to claims that body contraction is not real, i.e. observable, while time dilation is real and measurable by the common clock. A Lorentz-FitzGerald body contraction 'clocking' instrument does not exist today. However, in Chap. 10 we will describe an instrument sensitive to the Lorentz-FitzGerald body contraction. This assures that, like time dilation, the body contraction is real and can be measured, see further discussion in Ref. [6].

To avoid the need for a decision on the matter of the reality of the Lorentz-FitzGerald body contraction some books speak of distance contraction, generally without clarifying what this expression means, and never mention body contraction. In Chap. 9 we will introduce the measurement process that creates a contracted space-time event separation which is an interpretation of the meaning of distance contraction. We present this measurement procedure to show the consistency of the LT with the Lorentz-FitzGerald body contraction.

Misunderstanding 4: Small Acceleration Is Always Irrelevant

The principle of relativity refers exclusively to IOs. Einstein's 1905 SR work did not mention acceleration. Einstein's effort to understand acceleration and forces in SR resulted in understanding the gravity force in the GR framework: Relying on the principle of equivalence of gravitational and inertial mass GR generalization of SR is specific to gravity force alone.

Accelerated observers are never equivalent to IOs for the simple reason that no matter how small the acceleration is, we can tell it is present. Langevin makes the point, see Ref. [22] in Chap. 2, that the clock that travels away and back, and thus is accelerated, will always score a shorter time compared to another clock placed in an inertial laboratory. Parallel to this argument, Bell argues, see Chap. 10, that no matter how small is the acceleration which in this example propels two rockets away, only a material body connecting them will be contracted, and not the spatial separation between these rockets. What matters in these examples is the presence of acceleration, no matter how small it is.

Misunderstanding 5: Time Dilation Is Observer-Reversible = 'Twin Paradox'

A returning space traveler will always be younger compared to his twin on Earth. The twin paradox is created by invoking the relativity principle and claiming that exchange of the twins is possible, thus it should be the laboratory twin that is younger. However, such exchange is not possible since only the laboratory twin was inertial, not the traveler. Moreover, the measurement process must include a definition of how both space and time are measured. This definition must be maintained as one compares measurements made by different observers, or else contradictory results may be reported: We are specifically not allowed to exchange time measurements between two bodies without consideration of the space measurement.

This is so since only so called Lorentz invariant quantities, such as the proper time of a body, are measured to be the same by all IOs. Therefore, for each body only its proper time is a meaningful measure of time flow; that is the time measured by a clock at relative rest with that body. Focusing on proper time and remembering that bodies subject to an acceleration are not equivalent to inertial bodies fixes the conceptual challenges associated with twins. We will discuss this at length in Chap. 8.

Misunderstanding 6: Relativistic Doppler Effect

Many SR textbooks claim that the relativistic Doppler effect is created in two distinct phenomena: (1) by time dilation at the source, in the instant when radiation is formed and (2) during propagation process, in this part the argument goes just like for sound in air. This author strongly disagrees with these two points:[8]

1. Time dilation of the source cannot be part of the Doppler effect since the relative speed with respect to the undetermined observer is not known at the time of light emission. Moreover, different time dilation effects would be needed, depending on the motion state of different observers of the same light emission process.
2. Since the luminiferous æther is non-material, there is no change of wavelength during propagation, contrary to the behavior of sound in air.

Einstein concluded that the light wave must carry to the observer the information about the source, allowing later decoding of the relative motion and thus the determination of the relative shift in frequency and wavelength carried out at the actual observation of the light signal. We conclude: the SR Doppler shift is created in the process of observation of the incoming light. For more complete discussion, see Chap. 13. Note: The cosmological redshift (not the Doppler effect) impacts light while traveling from a source to an observer, see Insight 13.1.

Misunderstanding 7: Extended Bodies Have No Place in SR

Not true! In SR we strive to comprehend what happens to extended material bodies. It is in this context that the Lorentz-FitzGerald body contraction emerges as a pivotal concept. A cohesive extended body is naturally different from a cloud of non-interacting particles. Since space does not contract, a particle cloud does not either (assuming a density well below some interaction range). All cohesive material bodies are contracted.

Between a non-interacting cloud and a rigid stick are many other complicated structures as will be discussed in Discussion 10.1, but not further explored in this book. This does not mean that SR is somehow not applicable to such objects or that it could not with success be used in their study.

Conclusions

The situation with SR is rarely immediately clear; misconceptions can be deeply hidden. We saw in Ref. [5] that Einstein needed to clarify how he should be understood when responding to a well-known scientist of the epoch. We read in

[8] G. Margaritondo and J. Rafelski, "The relativistic foundations of synchrotron radiation," *J. Synchrotron Rad.* **24** 898–901 (2017).

Ref. [3] in the preface a letter from John Bell telling the author to beware of Einstein pedagogy and ensuing misunderstandings.

These problems are compounded by the increasing presence of unvetted open internet based presentations containing SR misunderstandings. Equipped with the list of the few classic misunderstandings which were listed here the reader can recognize more easily when SR contradicting arguments are presented.

Part II

Time Dilation and the Lorentz-Fitzgerald Body Contraction

Part II is about: Two key physical body properties are presented: (1) time dilation and (2) the Lorentz-FitzGerald body contraction. Historically the understanding of these body effects preceded the development of the Lorentz coordinate transformation which we study later in Part III. We study the light-clock with mirrors mounted on a single solid body. We find that only in the presence of both time dilation and body contraction this clock ticks the same way irrespective of clock motion and orientation in space.

Introductory Remarks to Part II

The reconciliation of the understanding of all physics laws of his epoch with the principle of relativity was achieved by Einstein in 1905. In this new SR framework he introduced the (proper, but not yet called in this way) body time; that is, a clock ticking differently in each and every moving body. Given the universality of the speed of light, a convenient starting point in the conceptual development of SR, we develop the (proper) body time measurement by a light-clock with time unit created by a light beam bouncing forth and back between two body mounted mirrors.

The study of the light-clock in motion will lead us to the quantitative recognition of the two independent effects that accompany relativistic motion of a material body: (a) time dilation, and (b) the Lorentz-FitzGerald body contraction. The understanding of the light-clock is equivalent to the understanding of why the MM interferometer cannot detect body motion.

As a body is taken from a laboratory reference frame to a relative velocity, its body clock measures its proper time, which is different from the laboratory time. This time dilation effect is moreover accumulated by any clock set in motion. Thus the elapsed time shown by a clock depends on the history of the motion with reference to the laboratory observer. Clock comparison after the clocks are reunited is referred to as the twin experiment.

The effect of time dilation can be observed with precision clocks placed on planes and in satellites, and in the study of the lifespan of moving, naturally decaying elementary particles, such as muons. All unstable elementary particles have a mean proper lifespan. When we observe the formation and decay events of a moving particle, we find a longer lifespan as compared to a measurement of the lifespan carried out at rest in the observer frame. We return to this topic repeatedly in the book.

The relativistic Lorentz-FitzGerald body contraction is recognized, demanding that a light-clock time not depend on the orientation of the clock mirrors with respect to the direction of motion. This finding is akin to the arguments advanced by FitzGerald and Lorentz in the wake of the MM experiment. For the light-clock we need that the travel time between mirrors be orientation independent, while in the MM interferometer the optical path should remain the same while the interferometer is rotated.

In Einstein's approach the principle of relativity was the paramount reason why the MM interferometer motion is unobservable. On the other hand one could also argue that the orientation independence of the light-clock measured proper time created the necessity to introduce both time dilation and the Lorentz-FitzGerald body contraction. In this way the Lorentz-FitzGerald body contraction is recognized to be as real as is time dilation. Consideration of possible experiments demonstrating the body contraction make this argument stronger.

Time Dilation

4

Abstract

We explore the properties of a relativistic clock consisting of a light pulse traveling between two body mounted mirrors. As our first example we choose the orientation of the optical path normal to the direction of motion of the clock with respect to the laboratory observer. The moving clock requires a longer optical path and ticks fewer times compared to the laboratory clock. This is time dilation. We discuss how time dilation can be observed by clocks placed on planes and satellites. Time dilation of a fast unstable particle, the muon, is introduced and muon travel range discussed.

4.1 Proper Time of a Traveler

We agree to compare time always at the common spatial location of two clocks. By making this agreement we do not mix the question of measuring time with the question of how to measure time that passes when we exchange information between two different space locations. This agreement means that in order to compare clocks both have to be co-located at the same spatial point at least at two different instances.

Applying the principle of relativity in terms of relative velocity, it is impossible to distinguish the two clocks. However, at least one of these two clocks must be subject to acceleration, or else we could not compare their time as discussed above. Acceleration distinguishes these clocks; a traveler cannot claim that it is the inertial laboratory that accelerates to approach her.

By determining which of the clocks is subject to an acceleration we have defined a measurement prescription. The clock of the accelerated body remains always at the same location with reference to the body, while the measurement in the laboratory must take into account the fact that the body clock is set in motion and thus each time-tick occurs at a different location.

J. Rafelski, *Modern Special Relativity*, https://doi.org/10.1007/978-3-030-54352-5_4

To be specific, we consider $d\tau$ to be the increment of the body proper time, dt that of the laboratory, and $\boldsymbol{v} = d\boldsymbol{x}/dt$ and $|\boldsymbol{v}| < c$, the velocity of the moving body observed from the laboratory. The time dilation Eq. (3.2) can be stated in the format

$$(d\tau)^2 = (dt)^2 - (d\boldsymbol{x}/c)^2 \,, (dt)^2 = (d\boldsymbol{x}/c)^2 + (d\tau)^2 \quad \rightarrow$$

$$(dt)^2 \geq (d\tau)^2 \,. \tag{4.1}$$

We see that the increment of proper time $(d\tau)^2$ is always smaller compared to the increment of laboratory time $(dt)^2$; both are equal only if the traveler is at rest, $d\boldsymbol{x} = 0$, as observed in the laboratory.

This view of the proper time measurement resolves the argument that once the relative motion is established (that is, after the acceleration phase is over), both the moving clock observer and the laboratory observer could claim to be time dilated; the difference that remains is in how the time is measured, i.e. which of the two observers measures at the location of the clock.

We presented here a measurement that assures absence of reciprocity in the time dilation, since accelerated observers are not part of SR. Remembering this procedure we can obtain this time dilation result using the relativistic Lorentz coordinate transformations, see Part IV.

However, in Chap. 13 we will see the reciprocity of the Doppler wavelength (or frequency) shift: two inertial observers looking at each other find reciprocal Doppler effects. The Doppler effect originates in the observation of the traveling light wave, propagating independently of the state of the light source – there are two different traveling waves that are inspected by the two observers. This also means that, contrary to popular view, the relativistic Doppler effect has nothing to do with time dilation, see also Misunderstanding 6 at the end of Chap. 3.

Integrating Eq. (3.2) we obtain a relation between the proper time of the traveler with her world line (see $\boldsymbol{x}(t)$ Sect. 1.2) defined in the laboratory frame

$$\tau_2 - \tau_1 = \int_1^2 dt \sqrt{1 - (d\boldsymbol{x}/cdt)^2} \,, \tag{4.2}$$

and we see again that always $\tau_2 - \tau_1 \leq t_2 - t_1$. The time recorded by a clock attached to a body set in motion is always less compared to the time recorded by the clock that remains at rest in the laboratory which observes the body in motion.

Along with an in-depth discussion of the time dilation considerations, Langevin also presented in his 1911 lecture (see Ref. [18], Chap. 2) an academic example of time dilation, in order to describe the enormous magnitude of effects that Eq. (4.2) makes possible. His study case is today common at particle accelerators where such huge time dilation effects occur for short-lived, unstable particles. Even so, in the following Example 4.1 we consider a small time dilation effect we can experience daily and which require sensitive clocks to observe.

Specifically, in Example 4.1 we explore time dilation that an intercontinental traveler experiences. We already know, also in view of Eq. (4.1) that a traveler on a the plane (and the Earth satellite) will be 'younger' compared to the laboratory 'twin'. Experiments have confirmed the phenomenon of clock time dilation to high precision, we return to this topic in Sect. 19.4. The challenge in observing the time dilation effect occurring due to motion near Earth surface is in separation of the motion related SR time dilation effect from the also present time dilation due to gravity force, the GR effect. Both SR and GR time dilation effects can be comparable in size.

Example 4.1

Time dilation in airplanes and satellites

Compare the time measured by a clock at rest on Earth with that measured in a passenger plane (traveling at 1000 km/h), see Fig. 4.1. Compare to the effect at a near-Earth orbiting satellite, evaluating for a 90 min orbit. Is the case of a GPS satellite the same? ◀

▶ **Solution** We compare here clocks of which one is on Earth scoring the time t and another flies on either a plane (or a satellite) scoring proper time τ. The clocks are synchronized before the plane or satellite takes off, $\tau_1 = t_1 = 0$, and we ignore the short take-off time with variable speed in our evaluation. A large passenger plane flies at a typical cruising speed near 1000 km/h. This corresponds to $v/c = 1/(300 \times 3600) = 0.926 \times 10^{-6}$.

From Eq. (4.2) we find

1 $$\frac{t_2}{\tau_2} = \frac{1}{\sqrt{1-(v/c)^2}} \approx 1 + \frac{1}{2}\left(\frac{v}{c}\right)^2 ,$$

and with this

2 $$\frac{t_2}{\tau_2} \approx 1 + \frac{1}{2}(0.926)^2 \times 10^{-12} .$$

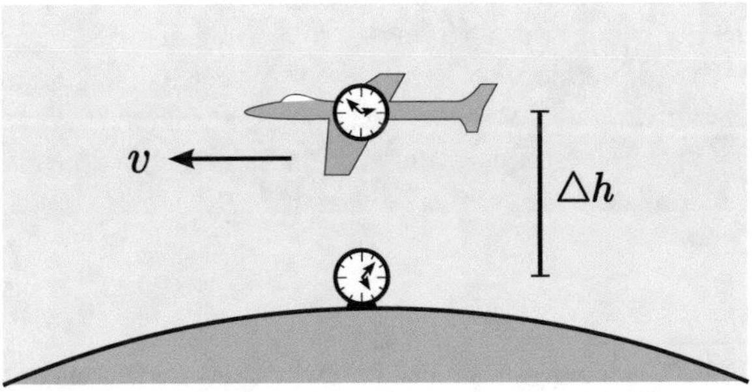

Fig. 4.1 Clock comparison between an Earth stationary clock with a clock on an airplane

We see that after 7 h of flight between Europe and New York one can expect a mismatch of clocks of $3600 \times 7/2 \times (0.926)^2 \times 10^{-12}$s $= 10.8$ ns. This effect is not difficult to measure in the twenty-first century.

A near-Earth satellite orbits the Earth 27 times faster compared to the passenger plane as it travels about 41,000 km in about 1.5 h. Accordingly, the effect of time dilation would be $27^2 \simeq 730$ greater. Therefore the reference clock on such a near orbit satellite is out of synchronization with the Earth clock by a factor that is about 730 times greater compared to our final result; that is 3.1×10^{-10} fraction of the elapsed time. Thus after each 1.5 h. satellite orbit the orbiting clock time would be short by 1.7 μs due to motion compared to Earth clock.

For a GPS satellite the numbers are a bit different: These satellites are in an orbit about 20,000 km above surface of the Earth moving with a speed of 14,000 km/h with an orbital period of about 12 h. Considering the difference in speed the effect is 14^2 greater than that on an airplane, the SR effect per orbit is about 3.6 μs. Actually, as this qualitative consideration shows, the GPS navigation system clocks must be continuously adjusted to comply with relativity, both SR and GR. The GPS signal adjustment procedure by itself provides a sensitive test of SR;[1] we return to this matter in Sect. 19.4. ◀

Discussion 4.1 Clock on a (relativistic) train

About the topic: We discuss time dilation observed on a moving (relativistic) train.

- *Student:* Did you check the clock time on the train and at the station before and after the trip?
- *Simplicius:* Indeed, I see the train-bound clock is slow.
- *Student:* That is the effect of time dilation. The traveler on the train aged less than the observer at the station did.
- *Simplicius:* How can you be sure that it is the train clock that is slow, and not the station clock that is fast? With 'relativity', how can you know that?
- *Student:* We measure time at 'equal space'; that is, the clock in the train is always at the same spot in that train-frame of reference, and the watch on your hand is wherever you are. Therefore, to compare time, we refer to time measurements performed at the same location.
- *Simplicius:* How does this explain that the train clock is slow?
- *Student:* The train was first at rest with respect to the station reference frame; we synchronized the clocks at that time. After, the train-bound clock moved away; later it returned with the train; it measured time moving along

(continued)

[1] P. Wolf and G. Petit, "Satellite Test of Special Relativity Using the Global Positioning System," *Phys. Rev. A* **56**, 4405 (1997).

with the train. A simple equation, Eq. (4.1), shows why taking into account accelerated train motion the clock located on the train is slow.

- *Simplicius:* Are you sure? I sat in the train and could see, along with many other passengers who do this daily, how you went away and later came back. From this perspective I expect the opposite result; the station clock should remain younger. How can this ever be resolved?
- *Student:* The point is that for two clocks moving apart and coming together later, at least one of the clocks needed to undergo acceleration, changing direction and speed of motion. In our example it is the train clock that accelerates, turns around, and slows down to stop at the station again.
- *Professor:* Indeed, the relativity principle does not extend to an accelerated observer. The on-board observer is non-inertial, is set in relative motion and as the train accelerates away all its clocks measuring proper time slow down.
- *Simplicius:* What if there were two different trains?
- *Professor:* The fun part is to be able to tell the outcome when all clocks were (gently) accelerated while their carriers were exploring the Universe. Consider several travelers; the one who travels farthest will come back youngest. We will return to this point again, see Examples 12.2 and 23.2.

4.2 Relativistic Light-Clock

Our next objective is to understand better the process of measuring time dilation. To this end we consider a generic light-clock comprised of two mirrors firmly attached to a solid body. The tick of time is the reflection of returning light in a mirror. This assures that, using our device, we study time ticking within this body, but we insulate the time tick from the specific material character of the body.

Our objective is to compare the proper body time between two, or more, light-clocks. For the purpose of our following discussion, a gentle acceleration is acting on one of the two light-clocks. It is as small as needed so that we can argue the case without developing a new motion model adapted to non-inertial motion. We only require that the clocks differ in that one is subject to some (small) acceleration.

In a first step, we set up the light-clock such that the effect of the Lorentz-FitzGerald body contraction is not present. To achieve this we let light travel between mirrors normal to the direction of motion. In a second step to which we return in Sect. 5.1 we show that, as long as we allow for body contraction, the outcome of time measurement is the same for a clock where the light travels in the direction of motion. We return in Sect. 5.3 to show that irrespective of the orientation of the light path with respect to the motion of the light-clock, the clock scores the same time, see Example 5.3 demonstrating that the light-clock is an acceptable time measuring device.

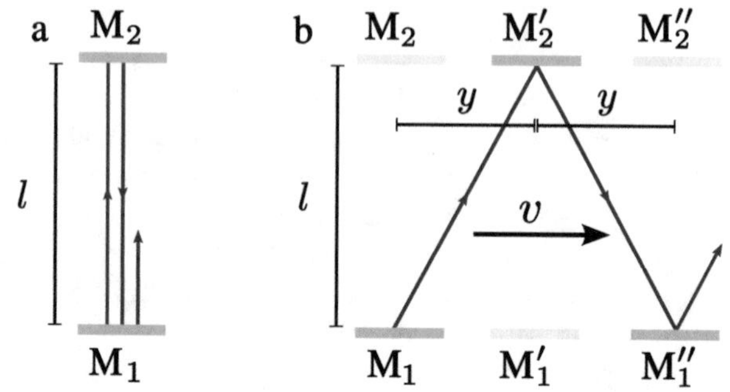

Fig. 4.2 (**a**) Stationary with respect to laboratory light-clock; (**b**) light-clock set in motion with relative velocity $\boldsymbol{v}$ in direction nearly transverse to the optical paths

Two light-clocks are shown in Fig. 4.2. In Fig. 4.2a we see the light-clock at rest with respect to the laboratory observer, and in Fig. 4.2b the clock is in motion with speed v normal to the original light path. In Fig. 4.2a an observer is in the same inertial frame as the clock. Therefore this clock measures the proper time of this observer. We see that in this measurement light travels between two mirrors, M_1 and M_2, firmly attached to one solid material body. The proper time measured in this way is a property of this material body.

The two mirrors in Fig. 4.2a are separated by the body distance l. The measurement of the material body distance l between both mirrors is made at the same time in the rest-frame of the body. For example, we carve into the supporting rigid body the units of a meter-stick and we read off this body-proper rigid stick the mirror separation l.

The unit of body-clock proper time as measured in Fig. 4.2a is given by the time that light, traveling in space, takes for the return trip between the mirrors

$$I_{(a)} = \frac{2l}{c}. \tag{4.3}$$

The time τ elapsing in the body is measured as number of units $I_{(a)}$. Choosing $l = 14.896229$ cm we have $I_{(a)} = 1$ ns.

Normal to Motion Optical Path

Now we apply a very gentle acceleration to the body, achieving a state of body motion with reference to the laboratory observer. In the laboratory frame the traveler's clock Fig. 4.2b has a non-vanishing velocity $\boldsymbol{v}$. For simplicity, we first assume that our light-clock moves at a constant velocity $\boldsymbol{v}$ relative to the original mirror normals, as is depicted in Fig. 4.2.

The time that the light needs to travel from M_1 to M_2' and back to M_1'' is

$$I_{(b)} = \frac{2\sqrt{l^2 + y^2}}{c} . \tag{4.4a}$$

We can also write this time in terms of how long it takes the mirror to move the distance y

$$I_{(b)} = \frac{2y}{v} . \tag{4.4b}$$

Solving these two equations for y yields

$$y = l \frac{v/c}{\sqrt{1 - (v/c)^2}} . \tag{4.4c}$$

The period of the moving clock is found using Eq. (4.4c) in Eq. (4.4b)

$$I_{(b)} = \frac{2y}{v} = \frac{2l}{c} \frac{1}{\sqrt{1 - (v/c)^2}} , \quad \Rightarrow \quad I_{(b)} = I_{(a)} \frac{1}{\sqrt{1 - (v/c)^2}} . \tag{4.5}$$

This result is easily understood by considering the longer optical path seen in Fig. 4.2b as compared to Fig. 4.2a, which requires $I_{(b)} > I_{(a)}$. For the laboratory observer the moving clock has a longer optical path compared to the identical clock at rest in the lab.

To understand who stays 'younger', imagine that the speed v achieved is so close to the speed of light c that the traveling clock only has a chance to tick a few '$I_{(b)}$-times' before reaching its destination. Clearly the traveler is in that case much younger when compared to his twin for whom the laboratory clock at rest ticks many '$I_{(a)}$-times'. As we increase the speed v the number of '$I_{(b)}$-times' decreases and in the limit of $v \to c$ the traveler will not age covering any finite distance. The traveler catches up with the flow of time.

To summarize: The light pulse ticking the time in a light-clock takes a longer light path when the clock is set in motion in comparison to a clock at rest. This effect is called time dilation; the twin set in motion remains younger.

Further reading: The effect of time dilation was experimentally observed for the first time in 1941 by Rossi and Hall.[2] Rossi and Hall found that the metastable particle, the muon, observed in cosmic radiation, see Fig. 4.3, takes longer to decay when it moves faster. This is described below in the Insight on "Elementary Particles and the Muon", the following Example 4.2, and in discussions in Sect. 4.3. For discussion of time dilation tests of SR see Sect. 19.1.

[2] B. Rossi and D.B. Hall, "Variation of the Rate of Decay of Mesotrons with Momentum," *Phys. Rev.* **59**, 223 (1941) (note: mesotrons = muons today).

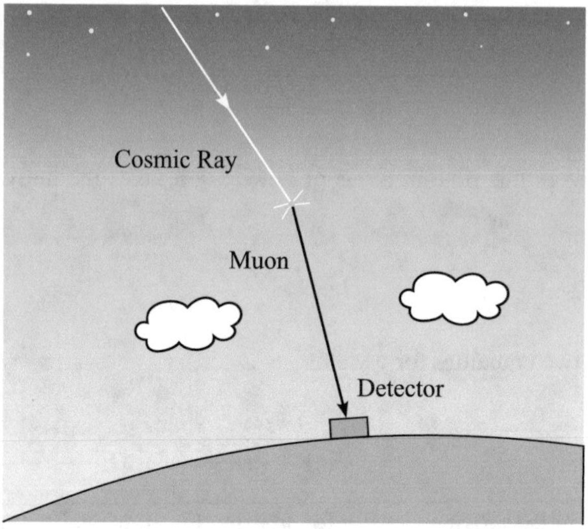

Fig. 4.3 A high energy cosmic ray (typically a proton) collides with an atomic nucleus creating a pion decaying into a muon high in Earth's atmosphere. After traveling several km this muon can later be detected in an Earth surface mounted detector

▶ **Insight: 4.1 Elementary Particles and the Muon** All matter around us is made of electrons e and nucleons (protons p and neutrons n). Electrons are bound by the EM force in large atomic orbits to atomic nuclei. These are much more localized on account of the stronger mutual interactions between nucleons, and $\simeq$ 1840 times higher mass, comparing a nucleon to the electron. Exploration of nucleon properties demonstrates that these particles have further constituents, the lighter and fractionally charged u (up, $Q = +2/3|e|$) and d (down, $Q = -1/3|e|$) quarks. These two quark types along with the electron and its weakly interacting partner, the neutrino ν_e, round off the 'family' of 4 distinct stable elementary particles, the building blocks of matter.

However, we have further discovered two repetitions of these 4 fundamental elementary particles, for a total of three matter particle 'families'. Among the 8 additional elementary matter particles is the muon. It is a 206.77 times heavier sibling of the electron. Both electrons and muons are subject to exactly the same electromagnetic and weak interactions. However, it is the heavier muon, which by emitting neutrinos can decay to the lighter electron by a weak-interaction process. The proper mean lifespan of the muon is $\tau_\mu = 2197\,\text{ns}$, $c\tau_\mu = 658.654\,\text{m}$. We also note that the half-life of a muon; that is, the time it takes for half of the muons to decay, is $\tau_{\mu\ 1/2} = \tau_\mu \ln 2 = 1523\,\text{ns}$.

Cosmic muons are produced predominantly in the decay of charged pions, themselves produced in the upper atmosphere by the impact of high energy cosmic protons. Electrically charged pions π^-, π^+, have a lifespan $\tau_\pi = 26\,\text{ns}$, $c\tau_\pi = 7.805\,\text{m}$; therefore they decay after a flight of a few dozens of meters, creating a significant flux of muons μ^-, and, respectively, anti-muons μ^+, in the upper atmosphere. A good fraction of these muons reaches the surface of the Earth, see Fig. 4.3.

Example 4.2

Travel range of muons before decay

We introduce particles called muons in the Insight 4.1: *Elementary Particles and the Muon*. Muons observed on Earth's surface originate in the upper layers of Earth's atmosphere from collisions of high energy cosmic ray particles with the atomic nuclei of Nitrogen and Oxygen as is illustrated in Fig. 4.3. These muons move with a high speed v. How great must v of a muon be for it to travel a distance of 13,000 m before it decays? ◀

▶ **Solution** Muons have a finite lifespan; thus their creation (in a cosmic pion decay) and subsequent decay are distinct events. Considering their natural decay lifespan, each muon has to be viewed as equipped with a built-in light-clock that measures the muon's proper time.

According to Eq. (4.2), in the Earth's rest-frame 'E', the lifespan of a muon is measured to be longer than the proper lifespan,

1 $$\tau_\text{E} = \tau_\mu \frac{1}{\sqrt{1-(v/c)^2}}\,.$$

In this dilated lifetime τ_E a muon travels the mean distance $S_\text{E} = v\tau_\text{E}$. Therefore,

2 $$\tau_\text{E} = \frac{S_\text{E}}{v}\,.$$

We combine both results and obtain

3 $$S_\text{E} = \frac{\tau_\mu v}{\sqrt{1-(v/c)^2}}\,.$$

Note that this result allows us to measure the lifespan of a particle by evaluating the distance traveled in a detector at a measured velocity v.

Without SR know-how and working with the approximation that muons move near the speed of light, their travel range is (see Insight 4.1: *Elementary Particles and the Muon*)

4 $$S_\mu \equiv c\tau_\mu = 658.654\,\text{m}\,.$$

This value is often found in particle data tables as it expresses the muon lifespan in units of length. Combining last two results

5 $$S_\text{E} = S_\mu \frac{v/c}{\sqrt{1-(v/c)^2}}\,.$$

We solve for v and obtain

6 $$v = \frac{c}{\sqrt{1 + S_\mu^2/S_\mathrm{E}^2}} \,.$$

Then using binomial approximation, it follows that

7 $$c - v = c\frac{S_\mu^2}{2S_\mathrm{E}^2} \,.$$

Substituting in $S_\mathrm{E} = 13\,\mathrm{km}$, as well as $c = 2.998 \times 10^5\,\mathrm{km/s}$, yields using two last results

8 $$v = 0.9987\,c = c - 385\,\mathrm{km/s}\,.$$

Thus a fast moving muon traveling a distance $S_\mathrm{E} = 13\,\mathrm{km}$ that is 19.7 times further than the naive expectation of the range. Muon speed is just 385 km/s shy of the speed of light. Such a muon can reach the Earth's surface due to the large time dilation effect.

This time dilation effect is often characterized by defining the Lorentz factor γ

9 $$\gamma \equiv \frac{1}{\sqrt{1 - (v/c)^2}} \,.$$

Eliminating v in favor of the path we find

10 $$\gamma = \sqrt{1 + \frac{S_\mathrm{E}^2}{S_\mu^2}} \,.$$

In this example we require the value $\gamma = 19.74$ to travel the distance $S_\mathrm{E} = 13\,\mathrm{km}$ given S_μ.

The reader should notice that in this example we did not allow for a gradual decay of the muon according to the exponential decay law. This approach suffices since the objective here was to understand that the muon has the capability to survive the trip to the Earth's surface because of the effect of time dilation.

The effect of time dilation is responsible for the extension of the travel distance of the muon. This is distinctly different from any other SR effect and in particular time dilation may never be confounded with the Lorentz-FitzGerald body contraction. Though we were discussing flight distance as occurring in the atmosphere, the presence of the atmosphere was without any relevance once the muon had been created. We could have imagined these experiments on the Moon, using an orbiting material target for cosmic rays and a surface muon detector. Clearly, there is no material rigid body between an orbit around the Moon and a surface detector and thus no body contraction. Certainly, there is no space contraction in SR.

We stress that all results of SR, and this includes the muon travel range, must be compatible with the relativistic coordinate transformations. Conversely, once we are familiar with these we can also obtain the travel range using this method. We explore this matter in the following pages, discussing this topic in Discussion 6.1, and once the Lorentz coordinate transformations are in our hands, a rather straightforward solution of this problem is offered in Example 7.2. ◀

4.3 Talking About Time (Dilation)

In the following three discussions we explore in depth the questions surrounding the time dilation effect. We discuss first a simple verification of time dilation when flying an airplane before we return to the question why the muon reaches the surface of the Earth. We then consider the question why the speed of light is special.

Discussion 4.2 How to measure time dilation

About the topic: We discuss here how best to design a flight experiment to measure the special relativity time dilation effect.

- *Simplicius:* Having read these pages I want to perform my own experiment to measure time dilation. I see that all I need is a precise clock, a plane ticket, and we are done.
- *Student:* Not really! You must have heard that light is deflected by gravity.
- *Simplicius:* What does this have to do with my experiment?
- *Student:* The just considered light-clock depended on the light traveling straight between the mirrors. If the path is curved by gravity, the path changes and time dilation changes.
- *Simplicius:* That cannot be a significant effect on Earth; maybe you are thinking of a black hole.
- *Student:* Not true: the escape speed from the Earth is a qualitative measure of the strength of Earth's gravity. Since the speed we can employ flying a plane is significantly smaller than the escape speed, the effect of gravity could be bigger.
- *Simplicius:* So how can we measure the motion related SR time dilation effect on Earth?
- *Student:* People came up with a clever idea. Note that when the effects of gravity and special relativity are small the time dilation effect is the sum of two small numbers; that is, the effects of GR and SR are additive. In order to eliminate the GR effect we form a difference of two time dilations obtained such that the velocity of motion is different, but the effect of GR is the same. This was accomplished by sending two clocks, one west around the Earth and the other east around the Earth.
- *Simplicius:* How can it matter which direction around the Earth the planes flew?
- *Student:* Assume for simplicity that the planes fly around the equator. Since the Earth rotates pretty fast, the total speed of the two planes as compared

(continued)

to an inertial observer floating in space is not at all the same. The total velocity of each plane that went on the trip is $v_\pm = v_r \pm v_p$, where v_r is the Earth's rotation speed $v_r = 40{,}000$ km/24h, and v_p the flight speed against the ground. Both planes have otherwise the same history, of take-off and landing, of flight height, the same as much as possible.

- *Simplicius:* That's neat; the Earth rotation speed is $v_r \simeq 1700$ km/h and that is faster than the speed of the planes. Thus the SR difference effect containing $v_+^2 - v_-^2 = 4v_r v_p$ is enhanced by the speed of the Earth's rotation.
- *Student:* Moreover, in the difference, the effects of gravity and acceleration drop out.
- *Simplicius:* Did it work?
- *Student:* Yes, this measurement confirmed special relativity. In fact the effects of both special relativity and gravity were measured. The GPS signal also allows a test along this line, so we do not need more plane tickets, see Sect. 19.4.

Discussion 4.3 Why muons reach the surface of the Earth

About the topic: All agree that muons can cross tens of kilometers, the distance from their origin in the upper atmosphere to the Earth's surface, see the last Example 4.2. However, even a light ray could only travel 660 meters during the muon mean lifespan $\tau_\mu = 2.2\,\mu$s. This situation is arguably the most misunderstood of SR contents; this conversation aims to clarify the circumstance. As we discuss diverse claims we discover that the errors can be deeply hidden.

- *Simplicius:* Was the two-plane experiment the first measurement of time dilation?
- *Student:* We saw that the effect of time dilation was recognized to govern particle decay and their ability to travel far. This is in general seen as the first experimental evidence of time dilation, see Ref. [2].
- *Simplicius:* I learned from you about muons living longer due to time dilation so they can travel further. However, in another book I saw that this is due to 'distance or space contraction'.
- *Student:* Within the scope of special relativity, space cannot be affected by the motion of a body. Was this a science fiction book?
- *Simplicius:* No, it was a very expensive book on Modern Physics; all physics students have to work through it. At first glance, the explanation sounds good; the book even claims that the ability to explain muon impact

(continued)

on Earth either as time dilation, or the Lorentz 'space contraction', proving the 'consistency' of special relativity. To help 'prove' this claim they place a tall mountain in the book figure that will also contract in presence of a moving muon.

- *Student:* I am confused by this argument. The words and pictures you report indicate the book author believes that the body contraction follows from space contraction which happens also to contract the mountain, aside of the distance the muon travels.
- *Simplicius:* Yes, this is my impression, and it looks odd to me now. I am asking myself what happens if I have two muons traveling together but at slightly different speeds: Is the space contracted differently for each?
- *Student:* Certainly not, and this shows there is no such a thing as 'space contraction'.
- *Professor:* Let us remember that we are trying to figure out if muons travel far because of time dilation or because of something else. To see the issues more clearly, let us talk next about an observer in the train that is going through the mountain tunnel, of our train and tunnel conversations, Discussion 4.1, and check the travel time.
- *Simplicius:* As I recall, the train clock was synchronized before the train left the station.
- *Student:* Recalling the airplane experiment we know the train clock will show an earlier time once the train returns to station after traveling through the tunnel.
- *Simplicius:* Using that other book I can say that the tunnel was contracted and I needed less time to travel through it. However, I am worried: Now the train does not fit into the tunnel. And I see another inconsistency: How is it possible that the train clock is showing an earlier time compared to station clock? If this space contraction causing tunnel contraction argument were true I would just return from the trip at an earlier time. I see now that this argument cannot be correct.
- *Professor:* For someone distorting the SR framework in this way it is easy to continue and look at the length of the tunnel using the contracted meter stick found on the train. Since the tunnel is longer than before one could instead argue it takes more time to cross the tunnel.
- *Student:* Using words one can claim whatever comes to mind in SR. There is no end of such simple arguments and the misunderstandings can be very well hidden, here in the contraction of the measurement scale.
- *Professor:* Body contraction and time dilation effects occur together and relate to the body that is making the trip. However, it is frequent to see body contraction and time dilation claims stated without a thought given to how one measures these effects. This is the most frequent source of misunderstanding. The correct arguments follow after we consider: (1)

(continued)

Which is the moving body with its proper time ticking at rest at the same location in that body? (2) Which observer is measuring in his reference frame at equal time the size of a moving body? After answering these two questions, we set the context for the presence of SR body effects in a consistent manner. And while the body contraction assures that the clock time does not depend on clock orientation, (we work on this in full generality in Example 5.3); this in no way means that the Lorentz-FitzGerald body contraction explains time dilation, or vice versa. These two body effects cannot substitute for each other.

- *Student:* A muon is our traveling body. A laboratory observer sees the newly formed muon travels at a speed v. The muon's proper time is always shorter compared to a laboratory observer clock. Since the muon is point-like and can travel in matter free space, in this consideration only time dilation enters. Had the muon a finite size it would contract in direction of motion.
- *Simplicius:* But hypothetically, could we not argue that as seen by an observer comoving with the muon, the Earth has a speed v and is instead contracted?
- *Student:* What your comoving observer sees has nothing to do with travel range of muons; the muon could travel between two (point) satellites. Satellites float at fixed distances of, say 10 km, from each other. In this case there is no contraction of anything.

Related reading in this book: For self-learners of SR many of the topics we raised in this conversation are well-known sources of misunderstanding. Therefore, after we study Lorentz coordinate transformations we will return a second time to the challenge posed by the muon traveling in space. We re-discuss the topic in Discussion 6.1. How muon time dilation works is presented in Sect. 7.1. We look at a similar problem in Example 7.2 where we obtain in another way the results we used in this discussion.

Discussion 4.4 The speed of light is very special

About the topic: What is special about the speed of light?

- *Simplicius:* Why is there such a difference between sound and light? Both propagate with velocities that are properties of the 'medium', yet I never heard that I needed to learn a new transformation to understand sound.
- *Student:* The first difference you note is that all matter around us can move faster than sound. Second, we can take air along for the ride.

(continued)

- *Simplicius:* Can you not pack the æther and run?
- *Student:* No, that is the point Einstein explained in his æther article of 1920. If the speed of light cannot be changed, then you cannot talk about taking æther along as then the the speed of light could be changed. Conversely, for the speed of light to be universal the æther must be indivisible, and the concept of velocity cannot apply to æther.
- *Simplicius:* But why do I need a new relativistic form of coordinate transformation?
- *Student:* Because we must assure that $c' = c$; in other words, the speed of light is not dependent on the state of motion of anything.
- *Simplicius:* OK, could I say that this new 'Lorentz' coordinate transformation explains time dilation and the Lorentz-FitzGerald body contraction?
- *Student:* This new transformation has to be combined with a measurement method which we must carefully define, then it is consistent with time dilation and the Lorentz-FitzGerald body contraction. This transformation also allows us to recognize that the speed of light is always the upper speed limit of all matter from which you can make mirrors.
- *Simplicius:* How does this last point arise?
- *Student:* If the mirrors of the light-clock were to move with $v > c$, they could outrun the light in the clock. That means time could stand still or even tick backwards.
- *Simplicius:* OK, I see, for the mirror not to outrun the light, c must be the upper speed limit of all matter that can interact with light.
- *Student:* All matter that interacts with light has yet more to do with light; just recall the most famous formula $E_0 = mc^2$. Note c^2 here.
- *Simplicius:* Since matter has locked rest energy which is the mass at rest multiplied with the square of speed of light, in some sense matter is made of light.
- *Student:* We therefore call it 'visible' matter.
- *Simplicius:* And 'dark' matter?
- *Professor:* That is the big conceptual question. We know by means of observing the effects of gravity that a lot of invisible, thus 'dark' matter, is around. The question, could 'dark' be gray, is crucial. 'Gray' means visible, but with more effort. Then, the gray (not dark) matter must have the same mass-energy relation $E_0 = mc^2$, the same limiting speed c. Truly dark but gravitating means dark matter does not know directly there is light. If truly dark matter exists this author considers this a very interesting new physics situation: How could such dark stuff know about the speed of light, hence that it should not move faster than c? Interesting physics may be lurking in the Universe and we, literally, need to learn to see in the dark.

Alternate reading: W.J. Swiatecki[3] (a highly regarded nuclear theorist from Berkeley, CA, USA) argues that SR should be presented as arising from the (symmetry) properties of Minkowski space alone, and that the connection with Einstein's principle of the universal light speed is superfluous – the units of length and time are the same, as are the units of mass and energy; the quantity c does not need to appear. However, by removing physics relevance from the speed of light removes light (and along with it, all EM waves) from the dynamical context of SR or any other theoretical framework. Therefore this space-time geometry based argument contradicts the SR presentation developed in this book which relies on propagation of light waves, and completly contradicts the arguments made in this last discussion, hence I mention this work here.

Swiatecki's argument is not entirely new; it reverberates in less dogmatic configuration in many space-time geometry based GR books. In the study of GR it is wise to remember that SR and GR address distinct domains of physics, despite their historical connection: GR provides in terms of the space-time geometry an interpretation of the gravity force, while SR provides the global contextual framework imposing the principle of relativity and the maximum universal speed of light on all matter and all its interactions.

However, massless photons play an important role in the understanding of space-time in GR: the path that a massless particle follows is called null geodesic. This name is signaling that: (a) the "distance" traveled in spacetime is equal to zero, and (b) the photon does not have a finite proper time associated with it; it does not age.

[3]W.J. Swiatecki, "Relativity and the Speed of Light: a 100 year old Misunderstanding," *Int. J. Mod. Phys. E* **15**, 275 (2006).

The Lorentz-FitzGerald Body Contraction

5

Abstract

We compare a light-clock with an optical path parallel to the direction of motion with a light-clock operating with an optical path normal to moving mirrors. Both time measurements agree only if the material body is subject to a Lorentz-FitzGerald body contraction in the direction of motion. We extend this to an arbitrary orientation and find always the same time dilation. Since time dilation is a 'real' effect we now conclude the Lorentz-FitzGerald body contraction is 'real' as well. We discuss the naturalness of body contraction. In a discussion we introduce tachyons, hypothetical superluminal particles.

5.1 Light-Clock Moving Parallel to Light Path

Universality of Time Measurement

In Sect. 4.1 we considered the case of a light-clock with mirrors mounted orthogonal to the direction of the relative reference clock velocity vector $\boldsymbol{v}$. Thus the optical path orientation was also orthogonal with respect to the direction of relative motion. We derived the time dilation of the time unit using only the principle that light propagates independent of the speed of motion of the body mounted mirrors.

Since we presume that space is homogeneous and isotropic, time measurement cannot depend on a particular orientation of the clock. Two inertial observers at rest with respect to each other must measure the same proper time, irrespective of the orientation of the mirror clock each is using. Should this not be the case we could see in the way the clocks score time their motion $\boldsymbol{v}$ with respect to a reference clock. Extending this argument we recognize that any two light-clocks that are not parallel

J. Rafelski, *Modern Special Relativity*, https://doi.org/10.1007/978-3-030-54352-5_5

in their light paths would coincide in their time measurement only in the absolute rest-frame. Determination of such a reference frame contradicts the principle of relativity. Thus we conclude that the light-clocks must score time independent of their orientation.

Thus these clocks must remain synchronized irrespective of their orientation. In order to achieve this result, it is necessary to introduce the Lorentz-FitzGerald body contraction, which as shown here, was probably discovered by FitzGerald when searching for an explanation of why the Michelson-Morley interferometer did not reveal the motion of the Earth. Both the light-clock and the null outcome of the Michelson-Morley experiment rely on the requirement that the optical path length is independent of the orientation of the mirrors.

Optical Path Parallel to Motion

We consider a light-clock moving at constant velocity relative to the observer, but this time with the motion in the same direction as the light beam, as depicted in Fig. 5.1.

We now determine the time t_{m1} light needs to travel from the left mirror M_1 to the right mirror, which is at a new location denoted M_2' in Fig. 5.1. We see that by the time the light reaches the right mirror, it is an extra distance x further away

$$t_{m1} = \frac{l' + x}{c} ,$$

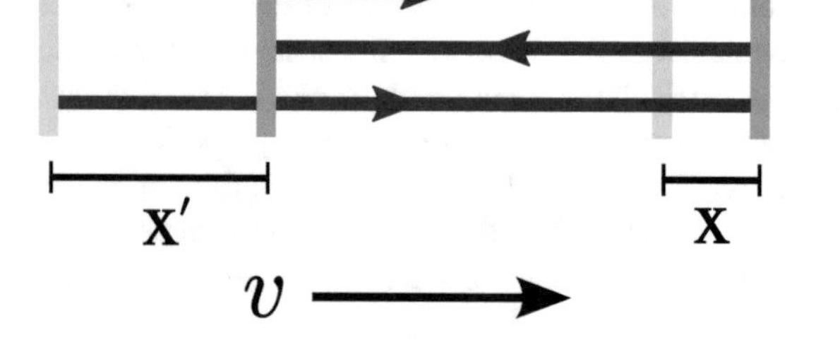

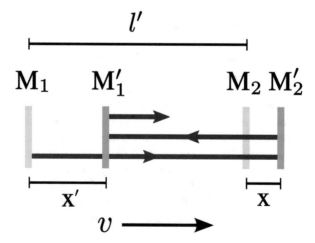

Fig. 5.1 The light-clock moving parallel to the direction of the pulse

where we use $l \to l'$ anticipating the contraction of the body. Considering the speed v of the body of the clock, the travel time of the mirror M_2 to the new position M'_2 is

$$t_{m1} = \frac{x}{v} .$$

Setting both expressions for the two flight times equal to each other

$$\frac{x}{v} = \frac{l' + x}{c} .$$

This allows us to determine x

$$x = \frac{l' \, v/c}{1 - v/c} . \tag{5.1}$$

On the return path of light, the left mirror moves to the right by x', reducing the amount of time needed by light to reach the mirror, thus respectively,

$$t_{m2} = \frac{x'}{v} , \qquad t_{m2} = \frac{l' - x'}{c} ,$$

and from these relations we obtain

$$x' = \frac{l' \, v/c}{1 + v/c} . \tag{5.2}$$

The entire distance covered by the light pulse ticking one time increment is

$$S = (l' + x) + (l' - x') .$$

Substituting x and x' from equations 5.1 and 5.2

$$S = 2l' + \frac{l' \, v/c}{1 - v/c} - \frac{l' \, v/c}{1 + v/c} .$$

After rearranging to give a common denominator we find

$$S = \frac{2l'}{1 - (v/c)^2} . \tag{5.3}$$

The period of the clock oriented in parallel to the direction of motion is then:

$$I^{\|}_{(b)} = \frac{S}{c} = \frac{2l'}{c} \frac{1}{1 - (v/c)^2} . \tag{5.4}$$

We recall for comparison the result for the rotated clock, Eq. (4.5)

$$I^{\perp}_{(b)} = \frac{2l}{c} \frac{1}{\sqrt{1-(v/c)^2}} \,. \tag{5.5}$$

We now have two different expressions for the clock period, and if $l = l'$ it is impossible that in general Eqs. (5.4) and (5.5) are equal – other than for $c \to \infty$, or $v \to 0$.

Convention:
Henceforth we write $l \to l_0$ for the not contracted body size, and $l' \to l$ for the contracted body size.

5.2 Body Contraction

It is natural to require that our light-clock should measure time independent of the orientation of its mirrors. If this were not the case we could find a reference system at absolute rest by stipulating that the time measured by differently oriented clocks be equal. Therefore we now require that the clock time be the same for the two considered cases.

The condition that the two clocks score the same time is obtained by setting Eq. (5.4) equal to Eq. (5.5); $I^{\|}_{(b)} = I^{\perp}_{(b)}$

$$\frac{2l}{c} \frac{1}{1-(v/c)^2} = \frac{2l_0}{c} \frac{1}{\sqrt{1-(v/c)^2}} \,. \tag{5.6}$$

We find that to maintain the same period, the length $l \neq l_0$. The change in length is a function of the velocity

$$l = l_0\sqrt{1-(v/c)^2} \,. \tag{5.7}$$

This equation says that the body on which the mirrors are mounted undergoes compression in the direction of motion. FitzGerald in 1889, Ref. [3] in Chap. 3, and Lorentz in 1892, Ref. [4] in Chap. 3, independently proposed the hypothesis that a body would be contracted in the direction of its motion. This effect is known today as the Lorentz-FitzGerald body contraction. We have determined the compression factor to be exactly $\sqrt{1-(v/c)^2}$ from consideration of mirror motion and the principle that the light velocity is a constant.

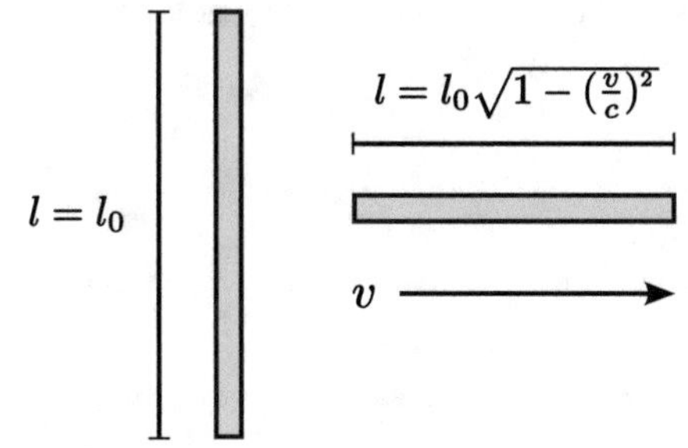

Fig. 5.2 A metal rod of length l_0 perpendicular, and with length l parallel to velocity $\boldsymbol{v}$

We observe that once Eq. (5.6) is made consistent by use of Eq. (5.7) there remains a difference in time as we already determined before in Sect. 4.2; this is time dilation. The Lorentz-FitzGerald body contraction idea preceded Larmor's and Einstein's concept of time dilation by more than a decade. This shows that a body's shape changes as it travels across space was easier for us to recognize, while the idea that we age at different, personal, rates was more difficult to perceive and accept.

Compression of a body traveling through æther in the direction of motion seems to be a natural hypothesis for reconciling the two clock time expressions seen Eq. (5.6). This hypothesis fits the view of æther action on material bodies. Given the general contraction rule Eq. (5.7) one usually expects a more complete theory to follow. SR is such a theory specifically created to bring the EM phenomena into unity with classical principles. In SR the Lorentz-FitzGerald body contraction is necessary; it is required by virtue of both the universality of the speed of light and the principle of relativity. Today we can expand these considerations to all known forces and recognize that the Lorentz-FitzGerald body contraction is a universal phenomenon applying to all visible material bodies.

Note that the comparison we have performed between the parallel and orthogonal to motion orientations of the light-clock is just the same as the comparison of the optical paths for the two arms of the Michelson-Morley interferometer, see Fig. 3.1. Thus the result that the clock will function independent of its orientation is equivalent to the explanation of why the Michelson-Morley experiment will not show a fringe shift when the interferometer is rotated.

In Fig. 5.2 the Lorentz-Fitzgerald body contraction of a metal rod is shown. When it is oriented perpendicular to its direction of relative motion, it has the length l_0; when it is turned so that it is parallel to its direction of motion, it will have the length l

$$l = l_0\sqrt{1-\beta^2}\,. \tag{5.8}$$

As we just used, the following notation appears often in these pages: the dimensionless velocity

$$\beta = \frac{v}{c}\,, \tag{5.9}$$

and the Lorentz factor

$$\gamma = \frac{1}{\sqrt{1-(v/c)^2}} = \frac{1}{\sqrt{1-\beta^2}} \geq 1\,. \tag{5.10}$$

The Lorentz-FitzGerald body contraction can be written

$$l = \frac{l_0}{\gamma}\,, \tag{5.11}$$

and time dilation becomes

$$t = \gamma t_0\,, \qquad \tau = \sqrt{1-\beta^2}\,t\,. \tag{5.12}$$

One often replaces t_0 by the proper time symbol τ.

It is the principle of relativity that assures that no physical observable can depend on the value of v. For example, a meter stick attached to the body also contracts, so measurement of the contraction effect by the comoving observer is impossible; it would amount to measurement of velocity, violating equivalence of all inertial observers and allowing for the existence of some absolute reference body. However, we should be able to measure the effect of the Lorentz-FitzGerald body contraction if we set up an experimental circumstance in which we change the body velocity and keep a record of it. We will return in Chap. 10 to discuss this type of experiment and the reality of the Lorentz-FitzGerald body contraction.

Example 5.1

Example of the magnitude of a body contraction

Determine the speed v of a metal rod of length ℓ_0 if it is found to be contracted to 3/5 of its original length in the direction of motion. ◄

► **Solution** The Lorentz-FitzGerald body contraction of the rod is set in the problem to

1 $\ell = \ell_0 \, \frac{3}{5}$.

We use

2 $\ell = \ell_0 \sqrt{1 - v^2/c^2}$.

Solving for v yields

3 $v = c\sqrt{1 - \ell^2/\ell_0^2} = \dfrac{4}{5}c$. ◄

Discussion 5.1 Relativistic train entering a tunnel

About the topic: We study the Lorentz-FitzGerald body contraction by considering the example of a train moving at relativistic speed into a tunnel too short to hide this train when at rest. This leads us to the question, how can we tell that it was the train and not the mountain that was set in motion?

- *Professor:* We need to perform an initial measurement to determine that the tunnel is indeed shorter than the train waiting for departure at the station. This must be done in the frame of reference for which the mountain and the train are at rest, e.g. the train station. After this, we start the train on the trip into the tunnel.
- *Simplicius:* I have taken many train rides and never have noticed that there was any train contraction.
- *Student:* It is hard to find relativistic trains today, so please consider this a Gedankenexperiment. However, if the train should be contracted, your measuring rod will be also contracted. Using a contracted rod, you cannot realize the train has become shorter.
- *Simplicius:* How can I measure differently?
- *Professor:* Had you missed the train departure you measure in the station's reference frame the moving train differently compared to when you are riding the train. This is so because you always measure the size of an object at equal time in your proper frame in which you are at rest. These two

(continued)

measurements (station and train) are different since what is simultaneous to one inertial observer, is not to another who is moving with a relative velocity.

- *Simplicius:* Please confirm: Two different inertial observers do not agree on what is simultaneous?
- *Professor:* Indeed this is a new insight original to SR: We have learned that as a consequence of a finite value of the speed of light, time ticks slower in the reference frame of a moving clock compared to my own 'at rest' reference clock. The relativistic coordinate transformation that also demonstrates this, predicts that two events I observe at the same time occur to any other moving observer at two different times.
- *Simplicius:* Oh, if so, then in any case I am the center of the Universe, 'my' rest-frame must be special: All experiments, measurements, and observations should refer to my frame of reference.
- *Student:* And I am another center of the Universe. Just as everybody you have your proper frame of reference, you have your proper time, your proper event simultaneity and also your proper standard of length.
- *Simplicius:* You tell me that in my proper frame I will always measure the same proper train length. How can one ever notice a body contraction?
- *Professor:* Your question is perhaps better posed as follows: How can you, riding the train, explain why someone from outside sees the longer train vanish in the shorter tunnel? Imagine the tunnel has two entry and exit switches which report the presence/absence of the moving train. These switches will report simultaneously in the mountain frame of reference that the train is inside the tunnel. However, for you riding on the train, these two signals in the front and in the back occur at two different times. This is so since what is simultaneous for a mountain is not simultaneous in the moving train frame of reference. The difference in when the time clicks of entry and exit occur is explaining why a passenger on the train can explain to his neighbor that the moving train does disappear from sight of the station observer without need to invoke body contraction.
- *Simplicius:* That is complicated. Is there a simpler explanation?
- *Student:* There are two easy ways to consider this situation: (a) An observer at the point of origin of the the train trip concludes the moving train is Lorentz-FitzGerald body contracted; (b) An observer on the train does not realize a body contraction but notes that the two signal gates, one at the front of the tunnel, the other in the back, click at different train times. Naturally, there are observers who are in some other inertial frame of reference, such as an observer riding another faster or slower train. Such observers combine and interpolate these two effect, and do not add new insights.

(continued)

- *Professor:* By realizing that events simultaneous to one inertial observer cannot be simultaneous to another (unless they have zero relative velocity), Einstein resolved observational contradictions accompanying the Lorentz-FitzGerald body contraction.
- *Student:* That is so since a body measurement requires, as we just have learned, a complete evaluation of the two space-time measurement events, including in particular how time measurement is made.
- *Professor:* An important outcome of this discussion is that the Lorentz-FitzGerald body contraction is not reversible due to the method of measurement. An observer riding a train claiming to see a contracted mountain changed the measurement's 'proper' frame. We must remember we began this conversation by comparing the lengths of the train and the mountain at the station where the train ride started. Therefore the only comparative body length measurement involving a moving train that makes physical sense is the one carried out from the station reference frame. This is the insight that connects the Lorentz-Bell and Einstein pedagogy and clarifies their equivalence. We return to these questions with due technical detail in Chap. 9.

5.3 Arbitrary Orientation of the Light-Clock

Our examination of the light-clock oriented perpendicular to its motion revealed time dilation, see Sect. 4.2. Because light travels at the same speed for any observer, we were obliged to invoke the Lorentz-FitzGerald body contraction hypothesis, see Sect. 5.2. This explains how a light-clock oriented parallel to its motion could experience the same time dilation as the one oriented perpendicular.

In Example 5.3 we now demonstrate for arbitrary orientation that any moving body subject to Lorentz-FitzGerald body contraction can provide the material body for a light-clock that registers time dilation. This further demonstrates that time dilation and the Lorentz-FitzGerald body contraction are two different phenomena, which, when combined together, lead to a consistent theoretical framework.

Note that only if the arbitrarily oriented light-clock works consistently do the two optical paths of the Michelson-Morley interferometer seen in Fig. 3.1 remain unchanged when this instrument is rotated. It follows that as long as the speed of light c is a constant we cannot observe an effect of body motion.

We explore in Example 5.2 how the Lorentz-FitzGerald body contraction alters the orientation of a material rod in motion, and how the rod length changes. The material body length is found to be orientation dependent. On the other hand, when we look at the case of arbitrarily oriented light-clock in Example 5.3 we find that the light path is orientation independent. In the following we discuss the naturalness of the Lorentz-FitzGerald body contraction and introduce tachyons, hypothetical particles moving faster than light.

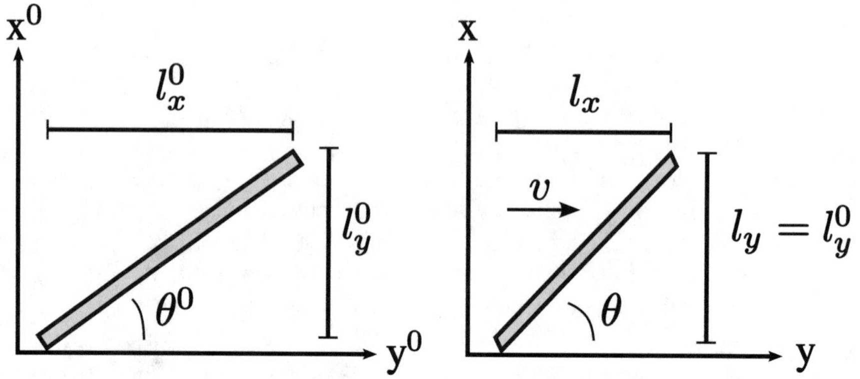

Fig. 5.3 Rod at rest (left) and in motion (right), see Example 5.2

Example 5.2

Modification of body orientation due to motion

A rod shown in Fig. 5.3 is placed at an angle θ^0 with respect to the future horizontal direction of motion. Obtain the angle θ of a rod moving with the relative speed v. Obtain the magnitude of the Lorentz-FitzGerald body contraction for this rod. ◀

▶ **Solution** In its rest-frame, the rod has length parallel to its motion of l_x^0 and perpendicular length l_y^0, related to the angle θ^0

1 $\tan\theta^0 = \dfrac{l_y^0}{l_x^0}$.

Using Eq. (5.8), we find that in the laboratory frame the component of length parallel to the motion of the rod is contracted by

2 $l_x = l_x^0\sqrt{1-(v/c)^2}$,

while the component perpendicular to the motion is unchanged:

3 $l_y = l_y^0$.

We now relate these quantities to the angle θ that the rod makes with the horizontal in the lab frame

4 $\tan\theta = \dfrac{l_y}{l_x} = \dfrac{l_y^0}{l_x^0\sqrt{1-(v/c)^2}}$.

We obtain

5 $\tan\theta = \dfrac{\tan\theta^0}{\sqrt{1-(v/c)^2}}\,, \qquad \theta = \tan^{-1}\left(\gamma \tan\theta^0\right)\,.$

This also means that the rod length is dependent on the orientation of the rod

6 $l \equiv \sqrt{l_x^2 + l_y^2} = \sqrt{l_x^{0\,2}\left(1-\dfrac{v^2}{c^2}\right) + l_y^{0\,2}} = \sqrt{l_x^{0\,2} + l_y^{0\,2} - \dfrac{v^2}{c^2} l_x^{0\,2}}$

$= l^0 \sqrt{1 - \dfrac{v^2}{c^2}\cos^2\theta^0}\,.$

We can now invert this relation combining both results shown above

7 $l^0 = \gamma\, l \sqrt{1 - \dfrac{v^2}{c^2}\sin^2\theta}\,.$

For both $\theta = 0°$ and $\theta = 90°$, i.e. for motion parallel and perpendicular to the orientation of the rod, the limits agree with our expectations. ◀

Example 5.3

Light-clock in arbitrary orientation

Assume: (i) Lorentz-FitzGerald body contraction, and (ii) a universal speed of light. Show that the light-clock time dilation is independent of the clock orientation with respect to the direction of motion. ◀

▶ **Solution** In its rest-frame the light-clock has between the mirrors a body distance a in the direction parallel to (future) motion and b in the direction perpendicular, as shown in Fig. 5.4. The distance between the mirrors in the body rest-frame is independent of any change in orientation

1 $a^2 + b^2 = l_0^2\,.$

According to Eq. (4.3) the period of this clock in the rest-frame is

2 $I_{(a)} = \dfrac{2l_0}{c}\,.$

Now we must consider its period $I_{(b)}$ as observed when in motion at velocity $\boldsymbol{v}$, as shown in Fig. 5.4. The total period is given by the time t_1 taken to travel along S_1 from M_1 to M_2', and the time t_2 taken to travel along S_2 from M_2' to M_1''

3 $I_{(b)} = t_1 + t_2\,.$

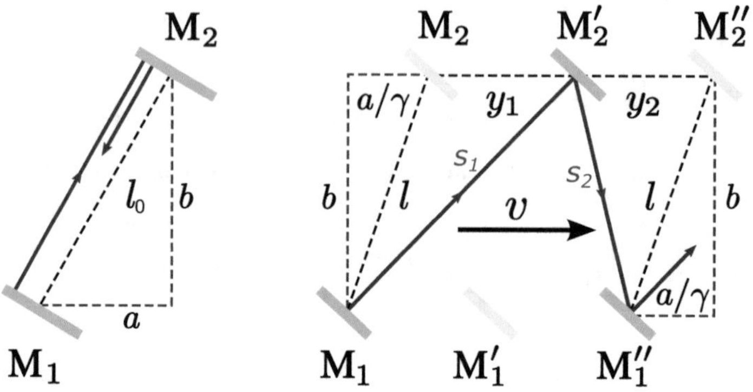

Fig. 5.4 An arbitrarily oriented light-clock: on left at rest with respect to observer, and on right in motion at velocity $\boldsymbol{v}$. As in the Fizeau-Foucault-Experiment Fig. 2.1 one uses concave mirrors to assure that a large fraction of the light pulse reaches the other mirror. See Example 5.3

Analogous to our earlier discussion of the light-clocks, we find for the time of travel at constant speed c along respective section of the path

4 $t_i = \dfrac{S_i}{c}\,, \qquad i = 1, 2\,,$

while due to body motion we also have

5 $t_i = \dfrac{y_i}{v} = \dfrac{y_i}{c\beta}\,.$

Setting these quantities equal we obtain

6 $y_i = \beta S_i\,.$

To find the path length S_1 followed by the light, we must consider also that the light-clock is contracted in the direction of its motion; the displacement between the mirrors in the direction of the motion becomes $\dfrac{a}{\gamma}$. From the geometry of the problem, see Fig. 5.4, we obtain

7 $S_1^2 = \left(y_1 + \dfrac{a}{\gamma}\right)^2 + b^2\,, \qquad S_2^2 = \left(y_2 - \dfrac{a}{\gamma}\right)^2 + b^2\,.$

We obtain

8 $y_1^2 = \beta^2\left(\dfrac{a^2}{\gamma^2} + \dfrac{2ay_1}{\gamma} + y_1^2 + b^2\right)\,, \qquad y_2^2 = \beta^2\left(\dfrac{a^2}{\gamma^2} - \dfrac{2ay_2}{\gamma} + y_2^2 + b^2\right)\,.$

Rearranging terms and recognizing that $1 - \beta^2 = \gamma^{-2}$ yields

9
$$\gamma^{-2} y_1^2 - \frac{2a\beta^2}{\gamma} y_1 - \left(\frac{a^2}{\gamma^2} + b^2\right)\beta^2 = 0\,,$$
$$\gamma^{-2} y_2^2 + \frac{2a\beta^2}{\gamma} y_2 - \left(\frac{a^2}{\gamma^2} + b^2\right)\beta^2 = 0\,.$$

We now solve this quadratic equation for y_1

10 $$y_1 = \frac{1}{2}\gamma^2 \left(\frac{2a\beta^2}{\gamma} \pm \sqrt{\frac{4a^2\beta^4}{\gamma^2} + \frac{4\beta^2}{\gamma^2}\left(\frac{a^2}{\gamma^2} + b^2\right)}\right) = a\beta^2\gamma \pm \beta\gamma\, l_0\,,$$

where we used Eq. 1 in the last term. To assure that $y_1 > 0$ at $b = 0$ we must choose the positive sign and obtain

11 $$y_1 = \gamma\beta\,(l_0 + a\beta)\ .$$

We can apply the same method to obtain y_2 resulting in replacement $a \to -a$, see Eq. 7

12 $$y_2 = \gamma\beta\,(l_0 - a\beta)\ .$$

The total time traveled by the light is given by

13 $$I_{(b)} = t_1 + t_2 = \frac{y_1}{c\beta} + \frac{y_2}{c\beta}\ .$$

Substituting our results for y_i we obtain

14 $$I_{(b)} = \gamma\,\frac{2l_0}{c} = \gamma\, I_{(a)}\ .$$

This shows that the time the light takes for a return trip between mirrors is independent of the magnitudes of a and b and thus independent of the orientation of the clock. The clock period only depends on velocity (contained in γ) and the mirror separation in rest-frame l_0. The time dilation effect is independent of the orientation of the clock. ◀

Discussion 5.2 Is the Lorentz-FitzGerald body contraction 'natural'?
About the topic: The recognition of the Lorentz-FitzGerald body contraction was a pivotal step on the winding path to relativity theory. We discuss how and why, given the MM experimental result, this step had to happen. The

(continued)

topic expands as we try to resolve many questions the reader will have at this time.

- *Simplicius:* With Maxwell's recognition that electromagnetic waves and light propagate at the same speed, even I could argue that, just like sound in air, light and Maxwell waves propagate in some new medium, the æther.
- *Student:* I agree, given what was known at that time that had to be the 'Maxwellian' point of view. For Maxwell, there had to be an æther filling the Universe. The MM interferometry experiment was devised to observe how the æther responds to the motion of the Earth orbiting the Sun moving against the stationary æther. The outcome of the experiment was a big surprise. Einstein describes in 1919/1920 a non-material æther determining the speed of of light c.
- *Simplicius:* All this I have learned here but I fail to see why the MM experiment was so surprising.
- *Student:* If there is a material æther, why is matter not hindered in motion and yet Maxwell's waves are transverse vibrations with respect to the direction of wave propagation? Maxwell understood that transverse waves arise in a hard and rigid medium. Sound in air is a longitudinal compression wave; in general this means that the medium is relatively easy to penetrate, and indeed air can be tossed around. Thus it was not understood how material bodies could move at all in the æther without an observed resistance. According to MM experiment we could not see an effect on the orbiting of the Earth around the Sun.
- *Simplicius:* I see, that is why no effect of motion reported by MM added to the enigma surrounding the æther.
- *Professor:* In this context the body contraction hypothesis seemed to offer an escape; instead of being inhibited in motion through friction, material bodies in their motion through the æther experienced a compression.
- *Simplicius:* However, there are other interpretations. I could argue that the null result of the MM experiment and the stability of Earth orbit simply mean that the Earth and the æther are comoving together around the Sun.
- *Student:* Local motion of the carrier of light would affect the path of starlight, alter the path of light and the value of the speed of light. The agreement of terrestrial and astrophysical methods of light speed measurements (see Sect. 2.2) exclude this interpretation.
- *Simplicius:* I see, but what you say about these experiments seems to be excluding any chance of any MM experiment observing an effect.
- *Student:* Foremost, that comment clearly relates to experimental precision. A big æther drag effect on light can be excluded. But some small drag seemed possible and the MM experiment was designed to be most sensitive

(continued)

to a directional difference in the speed of light and less so to the absolute modification of the speed of light.

- *Simplicius:* How could all this lead to a body contraction hypothesis? Let me take a ruler and rotate it in space. While I do this, according to our study of the light-clock, my ruler is compressed and decompressed and yet I do not notice much happening to it.
- *Professor:* The magnitude of the effect expected by Lorentz, and independently by FitzGerald, was 5×10^{-7} percent, so the body change was undetectable in the experiment you propose, but fixed the problem of the absence of any MM interference fringe shift.
- *Student:* This is much smaller than the typical thermal expansion of the ruler when temperature changes by one degree, an effect that is present when you pick up anything with your hands.
- *Professor:* This discussion also shows the need for precise control of temperature and many other environmental conditions in order to achieve the required experimental precision in modern MM experiments (see Sect. 19.2).
- *Simplicius:* I would like to understand better the idea that the body contraction is the result of æther resistance to motion. How is this possible?
- *Student:* We will return to discuss this topic; see Discussion 10.1. We will recognize that the electromagnetic force field emanating from the moving charged nuclei compresses atoms, and hence a chain of cohesively bound atoms, and more generally, any rigid material body, producing the Lorentz-FitzGerald body contraction. Thus it is not the direct reaction from the æther, but the effect of Maxwell's equations that cause body contraction.
- *Simplicius:* You are now explaining body contraction as being due to the nature of electromagnetic forces, yet we found all these effects also without any reference to forces, but rather from general considerations.
- *Professor:* Understanding body contraction is similar to understanding energy conservation. First we learn this as a principle, and later we study the outcome in terms of microscopic interactions that must be consistent with this principle.
- *Simplicius:* Except that there seems to be nothing conserved regarding the body size. We turn a ruler into the direction of motion and it contracts, ever so slightly.
- *Professor:* This works because all laws of physics respect the rules of Lorentz coordinate transformations. This assures that we can look at the effects of special relativity as being either an outcome of dynamics; i.e., forces, or a consequence related to space-time Lorentz coordinate transformations.
- *Simplicius:* This means that many physical properties of a body depend on its velocity – we have discussed energy, momentum, now size – is there anything that does not change?

(continued)

- *Student:* Yes, there are invariant properties of a physical body which all observers will measure to be the same, for example the inertial mass m. But kinetic energy or momentum are clearly not among these invariant properties.
- *Professor:* Mass is related to the proper energy of the body, the energy that is frozen inside the body when at rest; remember $E_0 = mc^2$. This is an important topic in this book.
- *Simplicius:* People also speak of proper time. Is this also an invariant?
- *Professor:* I agree, indeed as an example, any unstable particle has a proper time lifespan that is invariant. The proper time is the time measured by the clock attached to the body. Extending this argument, we recognize that any extended body such as an atomic nucleus, an atom, a long rod, any material body also has a proper size. This is how in Ref. [3] in Chap. 4 the SR without involving 'light' is presented.

Discussion 5.3 Riding the photon wave, seeking tachyons

About the topic: We review and deepen the insights and the consequences of the light-clock results.

- *Simplicius:* This root $\sqrt{1-(v/c)^2}$ in Eq. (4.5) troubles me. What if $v > c$?
- *Student:* First things first: Note that when $v \to c$ the light-clock stops ticking.
- *Simplicius:* Stops?
- *Student:* Yes, as $v \to c$ one tick of the moving clock is eternity elsewhere. The clock can go anywhere in the Universe and not even tick once…
- *Simplicius:* …running with the light wave.
- *Student:* Yes, that is an important insight. A light-clock that moves almost as fast as the speed of light will hardly make a tick, yet it can go anywhere.
- *Simplicius:* I see! Finally I understand – riding a photon the rider does not age and goes all over the Universe.
- *Student:* Yes, I would love to catch and ride the light wave!
- *Simplicius:* So again, what about $v > c$?
- *Student:* Nothing much. The mirrors would outrun the light, and the light-clock makes no sense to me.
- *Simplicius:* You seem to imply that no matter could be moving faster than light just because your clock will not work?
- *Student:* Well, the light-clock is just one example how and why matter that can interact with light must move with a speed that is limited by the speed of light c. That is what we clearly have assumed when considering the light-clock.

(continued)

- *Simplicius:* However, I did come across tachyons;[1] someone created such particles moving faster than light.
- *Student:* You mean as a Meta-Relativity idea[2] using a pen to write some words on paper?
- *Professor:* Let us not reject new ideas without proper scrutiny: We see tachyons in the scientific literature of mid-twentieth century[3] invented to help handle in quantum field theory the situation in which the sign of the mass-squared needed to be changed. However, a very different idea had at the time already resolved the problem;[4] the recently discovered Higgs particle was predicted by virtue of these considerations.
- *Simplicius:* Thank you! So what are tachyons today?
- *Student:* A word in science fiction shows and movies characterizing material bodies that can be faster than light.
- *Professor:* Even if I fully agree with this characterization, we must check to see if such particles somehow could make sense. Tachyons are beastly things that also are getting faster while losing energy. Anything that could move faster than the speed of light, and accelerate while losing energy will always run far and away.
- *Simplicius:* But we could make such particles in laboratory, and if so, they could be made by cosmic rays elsewhere and come our way as a cosmic ray.
- *Professor:* The search is on for new cosmic particles, including tachyons and many others. So far nothing that moves faster than light has been found. A discovery of a new particle, even one that is not really new in the sense that a tachyon would be, is a ticket to fame.
- *Student:* We already noted that matter and light are related, given $E_0 = mc^2$. So how is it possible to look for matter with $v > c$?
- *Professor:* Actually, these peculiar particles were invented to be dependent on, and related to, the speed of light, and thus, importantly in our present conversation, quite visible. Only their physical properties, energy and momentum, varied oddly as a function of their speed. That is why in the laboratory, our experimental colleagues would not fail to discover them, given their odd properties.

(continued)

[1]Greek: $\tau\alpha\chi\upsilon\varsigma$ (tachys), meaning "swift, quick, fast, rapid".

[2]O. M. P Bilaniuk, V. K. Deshpande and E. C. G. Sudarshan,"'Meta' Relativity," *Am. J. Phys.* **30**, 718 (1962).

[3]G. Feinberg, "Possibility of Faster-Than–Light Particles," *Physical Review*, **159** 1089 (1967).

[4]P. W. Higgs, "Broken symmetries and the masses of gauge bosons," *Phys. Rev. Lett.* **13**, 508 (1964).

- *Simplicius:* How would they appear to us?
- *Student:* We know when supersonic planes cross the sound barrier there is a sound Mach cone shock wave. Similarly, faster than light but interacting with light means Mach cone shock waves of light. We see this happen when ultrarelativistic particles travel in materials. Speed of light in matter is reduced, thus we have an effective model of tachyonic motion with all the unusual radiative properties. I am sure it would be hard not to notice production of tachyons.
- *Professor:* I agree. To my profound regret tachyons are today little more than words on paper. This also agrees with the findings of a few good people who question this theoretical idea in terms of mathematical consistency. But we should always remain open minded, even though we have not seen anything tachyonic.

Further reading: This topic is often heard about in science fiction context, thus we will return to deepen the subject in Discussion 11.1. An early and readable professional review written by Bilaniuk and Sudarshan[5] can be found on-line.

[5]O.M. P. Bilaniuk, E. C. G. Sudarshan, "Particles Beyond the Light Barrier," *Physics Today*, 43 (5) (1969).

Part III
The Lorentz Transformation

Part III is about: The relativistic Lorentz coordinate transformation is derived following the approach presented by Einstein in his 1905 work. The nonrelativistic limit, the inverse transformation, and the Lorentz invariance of proper time are among several consequences of the Lorentz coordinate transformation considered here. In the context of multiple sequential coordinate transformations we study the addition theorem of velocities and achieve a conceptual simplification by introducing rapidity.

Introductory Remarks to Part III

In Part III our objective is to understand and characterize a change in reference frame from one inertial observer to another inertial observer in the context of SR. The relativistic coordinate transformation was named by Larmor and Poincaré after Lorentz, who was the first to systematically attempt to resolve the incompatibility of Maxwell's electromagnetism with the Galilean transformations. However, these transformations were derived by Einstein in the Spring and Summer of 1905 using first principles, and, in a different manner, by Poincaré nearly at the same time. In this book we follow the presentation made by Einstein.

The Lorentz transformation would better be called the Larmor-Lorentz-Einstein-Poincaré **LLEP-coordinate transformation**, to honor also: (a) the actual first presentation of the correct Lorentz coordinate transformation by Sir Joseph Larmor; (b) Lorentz who paved the way for Larmor; (c) the SR inventor Einstein; and (d) Poincaré, who studied the group properties of relativistic coordinate transformations, see Sect. 1.3.

This accurate naming could have avoided the confounding name-connection between the relativistic Lorentz coordinate transformation and the Lorentz-Fitz-Gerald body contraction – which many simply call the Lorentz contraction. The name similarity links by association the space-time coordinate transformation to the body contraction property without a necessary discussion of the time measurement.

We begin by clarifying that Lorentz coordinate transformations are, in general, passive. The body is still in the same state of motion as before; it is the observer who is changing her frame of reference. We show how the form of Lorentz coordinate transformation is determined by these three physics inputs: the isotropy and homogeneity of space, the principle of relativity discussed in Sect. 1.1, and the universality of the speed of light, see Sect. 2.2.

We introduce the Larmor form of the Lorentz transformations, which is particularly convenient in the study of the time dilation. The consistency of Einstein's approach is explored in several examples. The nonrelativistic Galileo transformations are shown to be a limiting case; the conditions allowing study of this limit are described.

Classic results of relativity following from the Lorentz coordinate transformations are derived. The invariance of proper time under Lorentz coordinate transformations is demonstrated twice, using two different methods. The addition theorem of velocities is presented for the simple collinear motion case and the general case of two arbitrary velocities. Rapidity, replacing body speed, is introduced. The merits of rapidity are demonstrated by showing the additivity property for collinear motion. Among physics examples that have impacted the development of SR, the Fresnel light drag by a fluid in motion is shown to emerge from the velocity addition theorem.

Relativistic Coordinate Transformation 6

Abstract

Lorentz coordinate transformations describe the change of event coordinates for different inertial observers; the body is still in the same state of motion as before; it is the observer who is changing her frame of reference. We show how three physics inputs: (i) the isotropy and homogeneity of space; (ii) the principle of relativity; and (iii) the universality of the speed of light; allow us to determine the form of the Lorentz coordinate transformation. The Larmor form of the transformation is presented. The consistency of the formulation is explored in examples.

6.1 Derivation of the Lorentz Coordinate Transformation

Passive and Active Coordinate Transformations

There are, in principle, two ways to address coordinate transformations:

1. We can transform actively; that is, we can move by the transformation the coordinates of the world line traced by a particle and retain the coordinate system in place.
2. We can transform passively; that is, we change the coordinate system and we leave the world line unchanged.

In this book we obtain the Lorentz coordinate transformation as was done by Einstein, that is in a study of the passive transformation. We evaluate how the coordinates of a body change when comparing different inertial observers. It is inappropriate to consider the body comoving observer since at the instant of the change of the body state of motion this observer is non-inertial. Only in the case of

J. Rafelski, *Modern Special Relativity*, https://doi.org/10.1007/978-3-030-54352-5_6

inertial motion of a body can we consider this active transformation in the context of SR.

Only in the case of inertial motion can active and passive transformations be equivalent. As an example let us consider the twin who left home and was accelerated. Her clock is time dilated, unlike the clock that stayed at home. Forgetting the acceleration phase of the traveling clock/twin and reversing from passive to an active transformation for the traveling but at the instance of comparison, inertial twin, the so called 'twin paradox' is created. This is so, since such a reversal of the reference system is disregarding that a traveler underwent acceleration in reference to the observer. This violates the principles of the theory of relativity and this procedure is only allowed if clock synchronization is erased and both twins are in inertial motion.

The situation concerning passive and active transformations is illustrated in Fig. 6.1 where we see two observers S and S' recording the motion of the body K with velocity $\boldsymbol{u}$ on the left and $\boldsymbol{u}'$ on the right. There are two coordinate systems x, y, z, t and x', y', z', t' attached to and at rest with respect to these two S and S' observers; in Fig. 6.1 we only show x, y and x', y' coordinates.

In the left frame we see observer S at rest and observer S' in motion to the right with a speed v, while in the right frame we see observer S in motion to the left also with a speed v while S' is at rest. Thus the relative velocity of the two observers is the same in both cases, while the velocity of the body as measured by either of the two observers is different in both cases.

Under the Galilean coordinate transformation we read off Fig. 6.1 the location of the body; with time the origin of S' approaches the body so we must have, compare Eq. (1.1)

$$\boldsymbol{x}' = \boldsymbol{x} - t\boldsymbol{v} , \tag{6.1a}$$

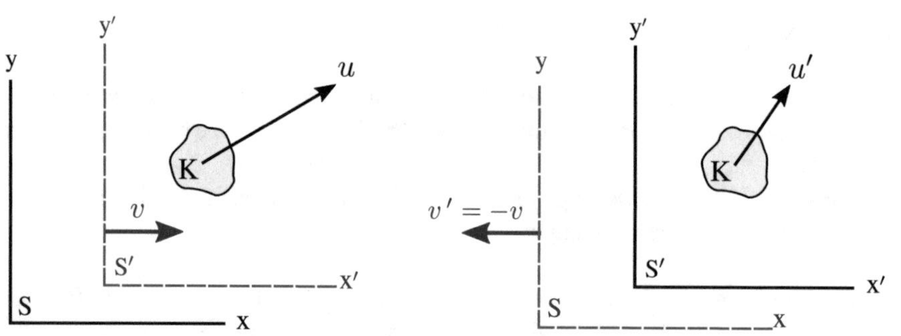

Fig. 6.1 On left: a body K is moving with velocity $\boldsymbol{u}$ with respect to the observer S, while another observer S' is in motion with velocity $\boldsymbol{v}$. On right: the same body K is moving with velocity $\boldsymbol{u}'$ with respect to the observer S', while the observer S is in motion with $-\boldsymbol{v}$. See text for further discussion

which when differentiated with respect to time produces the velocity addition theorem Eq. (1.2)

$$\boldsymbol{u}' = \boldsymbol{u} - \boldsymbol{v} \,. \tag{6.1b}$$

Exchanging the role of S with that of S' we obtain the inverse transformation

$$\boldsymbol{x} = \boldsymbol{x}' - t\boldsymbol{v}' = \boldsymbol{x}' + t\boldsymbol{v} \,, \tag{6.1c}$$

and hence

$$\boldsymbol{u} = \boldsymbol{u}' - \boldsymbol{v}' = \boldsymbol{u}' + \boldsymbol{v} \,. \tag{6.1d}$$

These relations address passive transformations; that is, of the coordinate system and not the body.

Anyone who mastered the detail about active and passive transformation in the Galilean world can verify SR results by looking at the nonrelativistic limit $v/c \to 0$ in SR. This shows that to change from passive to active transformation we substitute $\boldsymbol{v} = v\hat{i} \;\to\; \boldsymbol{v} = -v\hat{i}$.

Using the Isotropy and Homogeneity of Space

The isotropy of space means that there is no preferred direction. Assuming isotropy, it suffices to consider one arbitrary direction of motion and the results we find are valid generally. Thus we can choose the relative velocity $\boldsymbol{v}$ to be parallel to one of the coordinate axes, say the x-axis of some coordinate system S.

We consider coordinate transformation such that the transformed coordinate system S' has its x'-axis parallel to the x-axis. By making these two axes x and x' parallel, we exclude from current consideration rotation transformations superposed on top of the Lorentz coordinate transformation. The resultant transformations are called *special Lorentz transformations* or in short *(Lorentz) boost(s)*; in this book we speak of a boost.

The homogeneity of space means that the two observers S and S', with their coordinate systems x, y, z, t and x', y', z', t', respectively, both recognize the forceless motion of a body as being linear and uniform. We have called this type of body motion 'inertial'. Thus, if a body K in system S moves inertially with a constant velocity $\boldsymbol{u}$, its velocity $\boldsymbol{u}'$ in system S' should likewise be constant.

These requirements can only be fulfilled when the coordinate transformation involving both time and space is a linear transformation

$$x' = a_{11}x + a_{12}t + k_1 \,, \tag{6.2a}$$

$$t' = a_{21}x + a_{22}t + k_2 \,, \tag{6.2b}$$

$$y' = y \, , \qquad z' = z \, . \tag{6.2c}$$

We now choose coordinate transformations between S and S' such that at a prescribed instant in time $t'_0 = t_0 = 0$ we also have $x'_0 = x_0 = 0$. We must have

$$k_1 = k_2 = 0 \, . \tag{6.3}$$

This means that for both coordinate systems the origins coincide. By making this choice we do not consider here transformations which are called translation in space and/or translation in time.

As shown in Fig. 6.1 for the more general case, system S' moves relative to S with velocity $\boldsymbol{v} = v\hat{i}$. We want to consider the movement in S of the coordinate origin $x' = 0$ of S'. For the origin of S' moving with velocity v we obtain the equation

$$a_{11}x + a_{12}t = 0 \, . \tag{6.4}$$

For a small increment in space dx and time dt we thus have

$$-\frac{a_{12}}{a_{11}} = \frac{dx}{dt} \equiv v \, . \tag{6.5}$$

The last equality is the definition of v given that $x' = 0$ and our choice to consider motion along the x axis only. We use this result in Eq. (6.2a) and obtain

$$x' = a_{11}(v)(x - vt) \, . \tag{6.6}$$

Here the notation $a_{11}(v)$ reminds us that the coefficient a_{11} is in general a function of v, as is inherent in Eq. (6.5).

Using the Principle of Relativity

As we have already presented in Fig. 6.1 for the more general case, we can consider the motion in frame S'; then S moves relative to S' at $\boldsymbol{v}' = -\boldsymbol{v}$. By the principle of relativity, both points of view are equally valid. The transformation from S' to S must then take on the same form as Eq. (6.6)

$$x = a_{11}(v')(x' - v't') \, . \tag{6.7a}$$

The velocity v' in Eq. (6.7a) is the velocity of S relative to S', whereas v in Eq. (6.6) is the velocity of S' relative to S, therefore $v' = -v$. We thus have

$$x = a_{11}(-v)(x' + vt') \, . \tag{6.7b}$$

We could have chosen the coordinates such that both the x- and x'-axes pointed in opposite directions. In this case we modify relations Eqs. (6.6) and (6.7b) by transformations $x \to -x$, $x' \to -x'$ and $v \to -v$. After imposing this transformation, Eqs. (6.6) and (6.7b) read

$$x' = a_{11}(-v)(x - vt) , \qquad x = a_{11}(v)(x' + vt') . \tag{6.8}$$

To assure that the transformation we are seeking is not dependent on the choice of the direction of the coordinate axis, the coefficient a_{11} must not be dependent on the sign of the velocity; that is, on the direction of $\boldsymbol{v}$. Therefore $a_{11}(v) = a_{11}(-v)$ must be true.

We implement this by writing $a_{11}(v^2)$. We then have, restating Eq. (6.8):

$$x' = a_{11}(v^2)(x - vt) , \qquad x = a_{11}(v^2)(x' + vt') . \tag{6.9}$$

Using the Universality of Speed of Light

In order to determine the coefficient $a_{11}(v^2)$, we consider the coordinates of a light ray. At time $t = t' = 0$ the origins of the two systems coincide. At this point in time a flash of light is sent out from the origin. In system S at time t the flash of light reaches the position

$$x = ct , \tag{6.10}$$

shown in Fig. 6.2. Similarly in S'

$$x' = c't' . \tag{6.11}$$

The speed of light should be the same for all inertial observers; this is a direct experimental input into the reasoning about the new coordinate transformation we are seeking. To justify this result recall that in Sect. 2.2 we found that c was the same if measured on Earth or in interstellar space, and there we also described the result that Maxwell's equations provide for waves that propagate with light velocity.

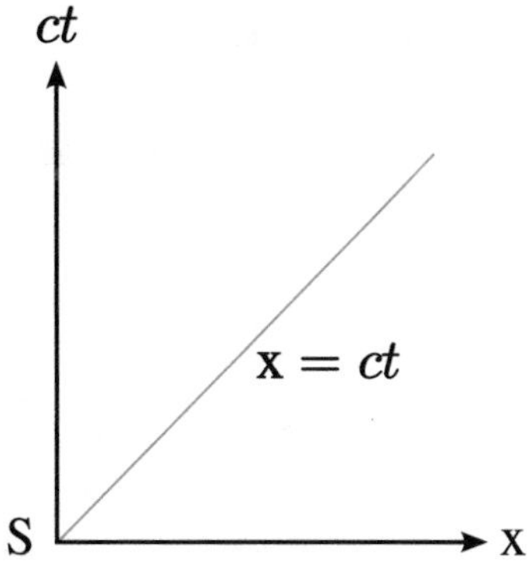

Fig. 6.2 Space-time path of a flash of light in system S

We thus demand that the speed of light is observed to be the same in all inertial systems in any state of motion, see Eq. (2.6)

$$c' = c\,. \tag{6.12}$$

Therefore in system S' at time t' the light cone reaches the position

$$x' = c't' = ct'\,. \tag{6.13}$$

We substitute these two conditions, i.e. $t = x/c$ and $t' = x'/c$, into Eq. (6.9) and obtain two transformation constraints

$$x' = a_{11}(v^2)x\left(1 - \frac{v}{c}\right)\,, \qquad x = a_{11}(v^2)x'\left(1 + \frac{v}{c}\right)\,, \tag{6.14}$$

which apply only to the light cone coordinates, but which still must be consistent. Multiplying the left-hand side and right-hand side of these two equations with each other and canceling the common factor $x'x$, we obtain

$$1 = a_{11}(v^2)a_{11}(v^2)\left(1 - \frac{v}{c}\right)\left(1 + \frac{v}{c}\right)\,, \tag{6.15}$$

and thus

$$a_{11}^2(v^2) = \frac{1}{1-(v/c)^2}\,. \tag{6.16}$$

In consideration of the limiting case $v/c \to 0$ which must yield the Galilean coordinate transformation, we choose the positive root and obtain for the Lorentz transformation (LT):

$$a_{11} = \frac{1}{\sqrt{1-(v/c)^2}} \equiv \gamma\,, \tag{6.17}$$

where we introduced the Lorentz factor γ. With Eq. (6.5) we have also

$$a_{12} = \frac{-v}{\sqrt{1-(v/c)^2}} \equiv -\beta c\gamma\,, \qquad \beta \equiv \frac{v}{c}\,. \tag{6.18}$$

Digression: The type of LT which reverses the direction of time (time reversal) and/or of space (parity) requires the negative root in Eq. (6.17). These are called "improper LT". Such transformations are part of the Poincaré group of all space-time transformations.

6.2 Explicit Form of the Lorentz Transformation

Einstein's form of the Lorentz Coordinate Transformation

With Eqs. (6.17) and (6.18) we can write for the spatial part of the coordinate transformation:

$$x' = \frac{x - vt}{\sqrt{1-(v/c)^2}}, \quad \text{or} \quad x' = \gamma(x - \beta ct)\ . \tag{6.19}$$

As always: $\beta = v/c$, and Lorentz factor $\gamma = 1/\sqrt{1-\beta^2}$.

Next we determine the coefficients a_{21} and a_{22} required in the transformation equation for the time coordinate t', Eq. (6.2b). t' is also seen in Eq. (6.9) on the right-hand side. We solve Eq. (6.9) for t'

$$t' = \frac{x}{va_{11}} - \frac{x'}{v}\,,$$

where a_{11} is given in Eq. (6.17) and x' in Eq. (6.19)

$$t' = \frac{x}{v\gamma} - \frac{\gamma(x-vt)}{v} = \gamma\left(t - x\left(\frac{1}{v} - \frac{1}{v\gamma^2}\right)\right)\,.$$

We obtain

$$t' = \gamma(t - (v/c^2)x)\ . \tag{6.20}$$

We thus have

$$t' = \frac{t - (v/c^2)x}{\sqrt{1-(v/c)^2}}, \quad \text{or} \quad ct' = \gamma(ct - \beta x)\ . \tag{6.21}$$

We can now read off Eq. (6.21) the coefficients of the transformation Eq. (6.2b)

$$a_{21} = -\frac{v}{c^2}\frac{1}{\sqrt{1-(v/c)^2}} = -\frac{1}{c}\beta\gamma\,, \qquad a_{22} = \frac{1}{\sqrt{1-(v/c)^2}} = \gamma\,. \tag{6.22}$$

The two transverse coordinates are not modified

$$y' = y, \qquad z' = z\,. \tag{6.23}$$

The relations Eqs. (6.19), (6.21), (6.23) are the set of equations forming the Lorentz coordinate transformation in the x-direction or "x-boost" in more colloquial language. We recall that spatial isotropy assures that our considerations apply to boosts along any direction in space. Therefore, the other two orthogonal boosts in y, and, respectively, z, directions follow by renaming $x \leftrightarrow y$, and respectively, $x \leftrightarrow z$.

Larmor's form of the Lorentz Transformation

We multiply Eq. (6.21) with v and form the sum with Eq. (6.19) to obtain the mixed transformation of the spatial coordinate

$$x' = x\sqrt{1-(v/c)^2} - t'v\,. \tag{6.24}$$

The spatial coordinates normal to the direction of motion remain unchanged, see Eq. (6.23). This form of transformation shows that $x' \leq x$ for $t' = 0$. We will show in Sect. 9.2 how setting $t' = 0$, Eq. (6.24), allows us to recognize the consistency of the Lorentz coordinate transformation with the Lorentz-FitzGerald body contraction.

Larmor introduced the 'local time' (see Ref. [13] in Chap. 1) which was the body proper time. This originates in the time transformation in form analogous to Eq. (6.24). We multiply Eq. (6.19) with v/c^2 and form the sum with Eq. (6.21) to obtain

$$t' = t\sqrt{1-(v/c)^2} - \frac{x'v}{c^2}\,. \tag{6.25}$$

Note that for $x' = 0$, Eq. (6.25) describes time dilation since t' is the proper time (called by Larmor local time) of a body at rest in S' at $x' = 0$, see Sect. 3.2 and Eq. (3.2), as well as Sect. 4.1 and Eq. (4.2).

As our derivation shows, the transformed coordinates x' and t' in Eqs. (6.25) and (6.24) are equivalent with those of Einstein's Eqs. (6.19) and (6.21). We accordingly call Eqs. (6.25) and (6.24) Larmor's form of coordinate transformation (see Ref. [14] in Chap. 1).

Larmor in his work assumed the presence of material æther, thus his insights were cast aside after the creation of SR, even though his coordinate transformation was accurate. Larmor obtained his result by following the lead of Lorentz in his study of the invariance transformations of Maxwell equations. Larmor claims that his result is accurate "to second order in v/c" and calls his coordinate transformations 'Lorentz coordinate transformations', a name that is repeated in Poincaré's work that followed years later, and remains in the literature.

Example 6.1

LT invariance of speed of light

Consider light traveling in frame S parallel to the x-axis with a wavefront at position $x = ct$. Confirm that in frame S', moving at an arbitrary velocity v relative to S, the speed of light remains unchanged. Since we used $c = c'$ in order to find the Lorentz coordinate transformation, this example is a cross-check of our work. ◀

▶ **Solution** In the frame of reference system S, if $x = 0$ at $t = 0$, the light expands along the light cone such that

1 $$x = ct\ .$$

In the moving frame of reference S', where the speed of light is c', we have analogously

2 $$x' = c't'\ ,$$

and with this and the transformation equations Eqs. (6.19) and (6.21),

3 $$c' = \frac{x'}{t'} = \frac{x - vt}{t - (v/c^2)x}\ .$$

Inserting the light cone condition, we find as expected

4 $$c' = \frac{c - v}{1 - v/c} = c\ .$$

Thus we have checked that the Lorentz coordinate transformation indeed contains the condition that the light velocity is the same for all observers. ◀

Example 6.2

What causes time transformation?

The derivation of the Lorentz coordinate transformation has one characteristic hypothesis that makes no sense in consideration of the Galilean transformation, namely $c' = c$. We ask: What happens to the derivation of the Lorentz coordinate transformation for $c' = c - v$? ◀

▶ **Solution** As noted we allow for two different velocities c and c'

1 $c' = c - v\,,$

a relation one could write for a 'tennis ball' corpuscular light theory. The minus sign in Eq. 1 is appropriate for the passive transformation. Note that it is the coordinate origin of S' that moves with velocity v and hence a tennis ball light has a smaller velocity in the x'-direction in S'.

Considering Eq. (6.14) but allowing for c' we find

2
$$x' = a_{11}(v^2)x\left(1 - \frac{v}{c}\right)\,,$$
$$x = a_{11}(v^2)x'\left(1 + \frac{v}{c'}\right) = a_{11}(v^2)x'\left(1 + \frac{v}{c-v}\right) = a_{11}(v^2)\frac{x'}{1-\dfrac{v}{c}}\,,$$

which makes sense only if

3 $a_{11}^2(v^2) = 1 \quad \to \quad a_{11}(v^2) = +1\,.$

Regarding the choice of sign above: We could in principle also consider $a_{11}(v^2) = -1$. However, in this book we will not explore time reversal transformations (see also end of Sect. 6.1).

We now seek to determine the coefficients a_{21} and a_{22} required in the transformation equation for the time coordinate t', Eq. (6.2b). We solve for t' seen in Eq. (6.9):

4 $t' = \dfrac{x}{va_{11}} - \dfrac{x'}{v}\,.$

Combining last two equations we obtain in analogy to the derivation of Eq. (6.19)

5 $t' = \dfrac{x}{v} - \dfrac{(x - vt)}{v} = \left(t - x\left(\dfrac{1}{v} - \dfrac{1}{v}\right)\right)\,,$

and obtain

6 $t' = t\,.$

Hence

7 $a_{21} = 0\,, \qquad a_{22} = 1\,.$

Inspecting Eq. (6.9) we recognize that we have demonstrated the Galilean transformation. This shows that we can derive the Galilean transformation of coordinates assuming the isotropy and homogeneity of space-time. To obtain the Galilean transformation we further used our opening assumption $c' = c - v$, while we obtained the Lorentz transformation using $c' = c$. **Einstein's postulate** that the speed of light is a universal constant in vacuum, $c' = c$, is the key hypothesis that changes everything. ◀

Example 6.3

Lorentz coordinate transformation of dx, dt

We consider the transformation of coordinate differences $dx = x_2 - x_1$, $dt = t_2 - t_1$ to those observed in another coordinate system S' moving with velocity v along the x-axis. ◀

▶ **Solution** We will study the Lorentz coordinate transformation of two different events $(t_1, \boldsymbol{x}_1)$ and $(t_2, \boldsymbol{x}_2)$

1
$$\begin{aligned} x_1' &= \gamma(x_1 - \beta c t_1), & c t_1' &= \gamma(c t_1 - \beta x_1)\,, \\ x_2' &= \gamma(x_2 - \beta c t_2), & c t_2' &= \gamma(c t_2 - \beta x_2)\,. \end{aligned}$$

The difference between the coordinates of these two events and their respective Lorentz coordinate transformations thus are

2 $$x_2' - x_1' = \gamma(x_2 - x_1 - \beta c(t_2 - t_1))\,, \qquad c t_2' - c t_1' = \gamma(c t_2 - c t_1 - \beta(x_2 - x_1))\,.$$

The transverse directions are not transformed

3 $$y_2' - y_1' = y_2 - y_1, \qquad z_2' - z_1' = z_2 - z_1\,.$$

This shows that event differences are subject to the same Lorentz coordinate transformation as any single event. Taking the difference between events to be small we find as a corollary the often used form

4 $$dx' = \gamma(dx - \beta c dt), \quad dy' = dy, \quad dz' = dz\,, \quad c dt' = \gamma(c dt - \beta dx)\,.$$

Since the Lorentz coordinate transformation is linear it applies to linear combinations of coordinates in the same way as it does to one coordinate. ◀

Discussion 6.1 Why muons reach the surface of the Earth II

About the topic: We return to discussing the trip that a muon makes from the upper atmosphere to Earth's surface this time using the Lorentz coordinate transformation as a means of observing the circumstance in Earth's rest-frame.

- *Simplicius:* I have learned that the Lorentz coordinate transformation is not the same as the Lorentz-FitzGerald body contraction.
- *Student:* True, by using the Lorentz coordinate transformation we evaluate how event coordinates change when observed by different inertial observers (IO).
- *Simplicius:* Is the Lorentz-FitzGerald contraction a part of the Lorentz coordinate transformation?
- *Professor:* Not directly. Using the Lorentz coordinate transformation we compute what a set of coordinate events in one frame of reference means to another IO. The Lorentz-FitzGerald contraction refers to the contraction in size of a moving material body. However, if we put our minds to it, in principle we can find the contraction using the coordinate transformation approach. This is so since the Lorentz coordinate transformation must be consistent with the Lorentz-FitzGerald body contraction.
- *Simplicius:* In another physics book I read that a muon produced at an altitude of 10km by (secondary) cosmic particles reaches the surface of the Earth because space contracts. Since I learned from you that space does not contract, how do you explain that the muon reaches the surface of the Earth? In its lifespan of $\tau_\mu = 2.2\,\mu\text{s}$, the muon can only travel 660 meters at the speed of light. That is about 1/15 of the required distance.
- *Student:* In the muon rest-frame – or, more accurately, in the frame of a muon-comoving observer – the time between two events, here the birth and the death of the muon, is $\Delta t = t_2 - t_1 \equiv \tau_\mu$, i.e. the muon lifespan. The distance traveled during that time is $\Delta x = x_2 - x_1 = 0$, i.e. for the comoving observer, the muon does not change position and lives its lifespan.
- *Simplicius:* I see, in the comoving frame of reference a muon lives its lifespan and does not move at all. But how long does it live and how far does it travel when observed from the surface of the Earth?
- *Student:* Let me answer by employing the Lorentz coordinate transformation. For an IO in any other reference frame who sees the muon at a velocity $v = \Delta x'/\Delta t'$, where the prime coordinates are those of the other (e.g. an observer on Earth), these coordinates become, see Example 6.3, according to the Lorentz coordinate transformation

$$\Delta t' = \gamma(\tau_\mu - (\Delta x = 0)v/c^2)\,, \qquad \Delta x' = \gamma((\Delta x = 0) - v\tau_\mu)\,.$$

(continued)

We indeed find that in this other reference frame the muon lives $\Delta t' = \gamma \tau_\mu$ and during that time it travels the distance $\Delta x' = -\gamma v \tau_\mu$. Note also that $\Delta x'/\Delta t' = -v$.

- *Simplicius:* I see. For an observer on Earth the muon lifespan is stretched by γ. We have derived time dilation by using the Lorentz coordinate transformation. This is the other explanation I read about, that the muon reaches the Earth since it lives effectively longer when observed by an Earthbound observer.
- *Professor:* The time dilation effect is introduced here by employing the Lorentz coordinate transformation *and* by implementing a measurement prescription, that the proper lifespan of the muon τ_μ is measured at $\Delta x = 0$. To find the time dilation we compare the clock time measured at rest, $\Delta x = 0$, with respect to the muon with a clock that is manifestly different: an inertial clock for which both t', x' are changing.
- *Simplicius:* How about the argument that space contraction confirms relativity since it provides an alternate explanation of muon time dilation and explains how the muon reaches the Earth?
- *Professor:* There is no 'space contraction'; an explanation in terms of a contracted spatial distance traveled by the muon is an incorrect application of the principles of SR.
- *Simplicius:* However, I do not like the time dilation explanation. The reason is that in the theory of relativity I was told I can reverse the point of view: I can choose to be the observer approaching the muon with velocity vector $-\boldsymbol{v}$. How does your explanation do?
- *Professor:* The situation is that any IO moving with the same speed v will, within muon proper time τ_μ, travel as far as the muon. Relativity means that it does not matter if it is the Earth or the muon that moves.
- *Simplicius:* OK, I see that my argument was circular; we already computed $\Delta x'$ as measured by the observer. Still, let me explain better the reason I do not like the time dilation explanation. Once I reverse the point of view, and I look at the clock attached to the observer moving towards the muon, the time dilation reverses, right? In this case the observer would see a muon decay in his time τ_μ/γ and thus the distance traveled would not be 10 km but (660 m/15) = 44 m.
- *Professor:* The key insight solving this riddle is that there is only one frame of reference in which the muon is at rest and that is the frame in which the muon lives 2.2 μs, which is the reference frame in which its proper time lifespan is measured by a clock that is at rest. For all other IOs the muon clock moves; the observed lifespan of the muon is therefore longer as the equation given above by the student shows using $\Delta x \neq 0$. Any observer who moves with respect to the muon can tell the time it takes in her frame for the muon to decay. This will be considered further in Sect. 7.1.

6.3 The Nonrelativistic Galilean Limit

A wealth of daily experience shows that the Galilean coordinate transformation (GT) is correct in the nonrelativistic limit in which the speed of light is so large that it plays no physical role; that is, $c \to \infty$. Any coordinate transformation replacing the GT must also agree with this experience, and thus must contain the GT in the nonrelativistic limit.

To obtain the nonrelativistic limit of the Lorentz coordinate transformation, we expand relations of interest to us by choosing as the small expansion parameter $v/c \ll 1$. We obtain

$$\gamma = 1 + \frac{1}{2}\left(\frac{v}{c}\right)^2 + \frac{3}{8}\left(\frac{v}{c}\right)^4 + \frac{5}{16}\left(\frac{v}{c}\right)^6 + \dots . \tag{6.26}$$

Using this expansion in Eq. (6.19) yields

$$x' = x - vt\left(1 + \frac{v^2/2 - vx/2t}{c^2} + \dots\right) . \tag{6.27}$$

For $|v^2/2 - vx/2t| \ll c^2$ we see the Galilean limit $x' = x - vt$.

For Eq. (6.21) we obtain

$$t' = t\left(1 + \frac{v^2/2 - vx/t}{c^2} + \dots\right) . \tag{6.28}$$

For $|v^2/2 - vx/t| \ll c^2$ we see the Galilean limit $t' = t$.

The correction terms in both Eqs. (6.27) and (6.28) have a similar format, with a term quadratic in v/c and another one that is effectively quadratic in v/c considering that in general $x \propto vt$. This means that even for body speeds that reach 10% of light speed, the relativistic corrections can often be ignored. While in daily life such speeds are rarely attained, they do occur frequently in atomic physics, and dominate particle and nuclear physics. This also means that in order to observe experimentally the effect of relativity, we must study body speeds that are closer to c.

6.4 The Inverse Lorentz Coordinate Transformation

The Lorentz coordinate transformation allows an observer to determine the coordinates of an event, if these are known to another inertial observer. The coordinates (t, x, y, z) of an event in S can be transformed into (t', x', y', z') in S'. With the velocity v of S' relative to S pointing by our choice in the x direction we have,

$$\begin{aligned} ct' &= \frac{ct - (v/c)x}{\sqrt{1-(v/c)^2}} = \gamma(ct - \beta x) , \\ x' &= \frac{x - vt}{\sqrt{1-(v/c)^2}} = \gamma(x - \beta ct) , \\ y' &= y , \quad z' = z . \end{aligned} \tag{6.29}$$

Seen from S' the motion is in the negative x' direction and thus the inverse Lorentz coordinate transformation must be:

$$\begin{aligned} ct &= \frac{ct' + (v/c)x'}{\sqrt{1-(v/c)^2}} = \gamma(ct' + \beta x') , \\ x &= \frac{x' + vt'}{\sqrt{1-(v/c)^2}} = \gamma(x' + \beta ct') , \\ y &= y' , \quad z = z' . \end{aligned} \tag{6.30}$$

The four equations of 6.30 are the inverse Lorentz coordinate transformation. We have obtained them from Eq. (6.29) through the application of the principle of relativity, whereby one exchanges the primed coordinates with those that are not primed and substitutes $-v$ for v.

As one would expect, by solving for x' and t' in Eq. (6.30) we return to the usual LT form: Solving the time transformation in Eq. (6.30) for t' we obtain Larmor's format of the time transformation, Eq. (6.25). Solving the second equation in Eq. (6.30) for x' we obtain Larmor's format of the spatial transform, Eq. (6.24).

Example 6.4

Inverting the Lorentz coordinate transformation

Solve the Lorentz coordinate transformation Eq. (6.29) for x, ct, y, z, assuming that the primed coordinates are given. ◀

▶ **Solution** We see that $y = y'$ and $z = z'$. We look at the Lorentz coordinate transformation

1 $x' = \gamma(x - \beta ct), \qquad ct' = \gamma(ct - \beta x)$,

and we form two linear combinations

2
$$\begin{aligned} x' + \beta ct' &= \gamma x(1-\beta^2) - \gamma\beta ct + \gamma\beta ct = \gamma x(1-\beta^2) , \\ ct' + \beta x' &= \gamma\beta x - \gamma\beta x + \gamma ct(1-\beta^2) = \gamma ct(1-\beta^2) . \end{aligned}$$

Since $(1-\beta^2) = 1/\gamma^2$ we multiply by γ to obtain

3 $x = \gamma(x' + \beta ct'), \qquad ct = \gamma(ct' + \beta x')$.

Thus the inverse transformation is achieved by taking $\beta \to -\beta$. This is also the inverse transformation in the sense that transforming $S \to S' \to S'' = S$ is achieved by transforming first with β and next with $-\beta$. ◀

6.5 Lorentz Transformation in Arbitrary Direction

A coordinate transformation can be performed in arbitrary direction $\boldsymbol{v} \equiv \boldsymbol{\beta} c$. The generalization of Eq. (6.21) is as naively expected

$$ct' = \gamma(ct - \boldsymbol{\beta} \cdot \boldsymbol{x}) . \tag{6.31}$$

However, the transformation of spatial coordinates Eq. (6.19) acquires an additional term

$$\boldsymbol{x}' = \gamma(\boldsymbol{x} - \boldsymbol{\beta} ct) - (\gamma - 1)\left(\boldsymbol{x} - \frac{\boldsymbol{\beta} \cdot \boldsymbol{x}\, \boldsymbol{\beta}}{\beta^2}\right) . \tag{6.32}$$

Clearly, the Galilean limit $\gamma = 1$ is not affected by the additional term. This term also vanishes when $\boldsymbol{\beta} \parallel \boldsymbol{x}$ and we obtain back Eq. (6.19). These two limits do not uniquely determine the last term in Eq. (6.32) as we could multiply this term with for example another function of γ. Given Eq. (6.31) the form of Eq. (6.32) is uniquely defined introducing the new insights we gain in the following chapter, see in particular Eqs. (7.4) and (7.6).

Further reading: A thorough analysis of the non-parallel LT is given in the *Special Relativity* by Michael Tsamparlis (Springer 2019). Tsamparlis addresses other SR topics we cannot develop in this introductory book.

Some Consequences of the Lorentz Coordinate Transformation

7

Abstract

The invariance of proper time under Lorentz coordinate transformations is demonstrated. The addition theorem of velocities is presented both for the collinear and for the general case of two arbitrary velocities. The Fresnel drag coefficient of light in moving fluid is shown to be a direct consequence as is the maximum observable body speed $|\boldsymbol{v}_{\max}| < c$. We introduce rapidity and demonstrate its additivity property for collinear motion, akin to nonrelativistic velocity addition.

7.1 Invariance of Proper Time

Consider a familiar three dimensional vector, constrained for simplicity to the xy plane. The important property here is that its length is conserved under a rotation about the z-axis,

$$\boldsymbol{r}^2 = x^2 + y^2 = x'^2 + y'^2 = \boldsymbol{r}'^2 \,. \tag{7.1}$$

As shown in Fig. 7.1, the endpoints of a position vector and its rotated image lie on a circle. We say that the length of a vector is invariant under spatial rotations, which means that all observers agree to the length of this vector.

By analogy, we wish to recognize quantities that under a Lorentz coordinate transformation will transform in a similar fashion to a spatial rotation, and which have a magnitude invariant under LT, retaining the same value for different inertial observers. We call a quantity that does not change under a Lorentz transformation a Lorentz invariant (LI) or just 'invariant'.

We aim to show that the proper time of a body we considered before, see Sect. 4.1 and Eq. (4.1), is just such an invariant that all inertial observers will measure to be the same. To achieve this we compare the square of proper time τ^2 of a body as

J. Rafelski, *Modern Special Relativity*, https://doi.org/10.1007/978-3-030-54352-5_7

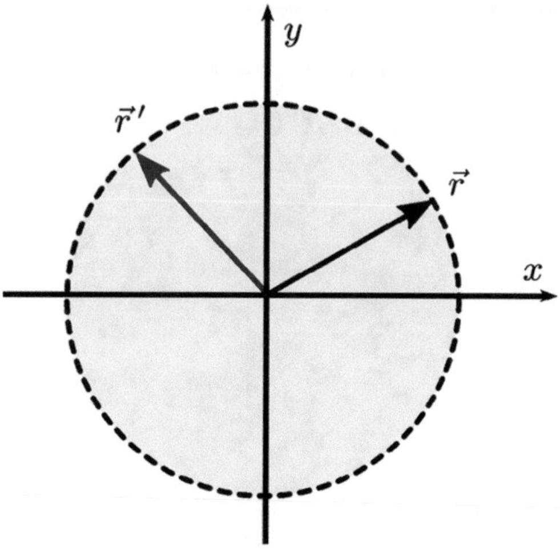

Fig. 7.1 The rotation of a position vector preserves its length: $|\boldsymbol{r}| = |\boldsymbol{r}'|$ is an invariant under rotations

measured by two different observers

$$\begin{aligned} c^2\tau^2 &= s^2 = c^2t^2 - x^2 - y^2 - z^2, \quad \text{with} \\ c^2\tau'^2 &= s'^2 = c^2t'^2 - x'^2 - y'^2 - z'^2 \,, \end{aligned} \tag{7.2}$$

and equivalently for differences

$$\begin{aligned} \Delta s^2 &= c^2(t_2 - t_1)^2 - (x_2 - x_1)^2 - (y_2 - y_1)^2 - (z_2 - z_1)^2, \quad \text{with} \\ \Delta s'^2 &= c^2(t_2' - t_1')^2 - (x_2' - x_1')^2 - (y_2' - y_1')^2 - (z_2' - z_1')^2 \,. \end{aligned} \tag{7.3}$$

It suffices to consider only the LT in the x-direction. If other directions need to be transformed, we can always re-orient our coordinate system so that the x-axis points in the direction of the LT transformation. Therefore in what follows $y^2 + z^2 = y'^2 + z'^2$ in Eq. (7.2), and similarly for the difference between two events, Eq. (7.3).

We employ the LT Eqs. (6.19) and (6.21) to find

$$\begin{aligned} s'^2 &= c^2t'^2 - x'^2 - y'^2 - z'^2 \\ &= \gamma^2(ct - \beta x)^2 - \gamma^2(x - \beta ct)^2 - y^2 - z^2 \\ &= \gamma^2(c^2t^2 + \beta^2x^2 - x^2 - \beta^2c^2t^2) - y^2 - z^2 \,. \end{aligned}$$

Recognizing $\gamma^2 = (1 - \beta^2)^{-1}$ we have

$$s'^2 = \gamma^2(1 - \beta^2)(c^2t^2 - x^2) - y^2 - z^2 = s^2 \,. \tag{7.4}$$

The same transformation property follows for the difference between two events

$$
\begin{aligned}
\Delta s'^2 &= c^2(t'_2 - t'_1)^2 - (x'_2 - x'_1)^2 - \rho'^2 \\
&= \gamma^2[c(t_2 - t_1) - \beta(x_2 - x_1)]^2 - \gamma^2[(x_2 - x_1) - \beta c(t_2 - t_1)]^2 - \rho^2 \\
&= \gamma^2\left[c^2(t_2 - t_1)^2 + (\beta^2 - 1)(x_2 - x_1)^2 - \beta^2 c^2(t_2 - t_1)^2\right] - \rho^2 \\
&= c^2(t_2 - t_1)^2 - (x_2 - x_1)^2 - \rho^2 = \Delta s^2 \,,
\end{aligned}
\tag{7.5}
$$

where we used $\rho^2 = (y_2 - y_1)^2 + (z_2 - z_1)^2 = (y'_2 - y'_1)^2 + (z'_2 - z'_1)^2 = \rho'^2$. Thus for a small separation between events we have

$$
\begin{aligned}
ds'^2 &= c^2 dt'^2 - dx'^2 - dy'^2 - dz'^2 \\
&= c^2 dt^2 - dx^2 - dy^2 - dz^2 = ds^2 \,.
\end{aligned}
\tag{7.6}
$$

One says that s^2 and, equivalently, the increment ds^2, are Lorentz invariant.

Dividing by c we see that $ds/c = d\tau$ and

$$
\begin{aligned}
d\tau'^2 = \frac{ds'^2}{c^2} &= dt'^2 - \frac{dx'^2 + dy'^2 + dz'^2}{c^2} \\
&= dt^2 - \frac{dx^2 + dy^2 + dz^2}{c^2} = \frac{ds^2}{c^2} = d\tau^2 \,.
\end{aligned}
\tag{7.7}
$$

The meaning of τ becomes clear when we consider an observer for whom there is no change of position $dx = 0$. We see that $dt = d\tau$. This observer's clock, which rests in the body frame of reference, measures the proper body time, τ.

We found by deriving Eq. (7.7) that the increment of proper time is an invariant. That means that all observers $S(v)$ having some speed v will agree to how fast a clock ticks inside a body,

$$
\begin{aligned}
d\tau &= \sqrt{dt^2 - dx^2/c^2} = dt\sqrt{1 - dx^2/(cdt)^2} \\
&= dt\sqrt{1 - v^2/c^2} \,,
\end{aligned}
\tag{7.8}
$$

where v is velocity of a body measured by the observer $S(v)$. We see that the proper time increment $d\tau$ is according to Eq. (7.8) shorter compared to the time of any observer $S(v)$.

Important: for all observers the proper time $d\tau$ of each and every body is the same; we say it is a Lorentz invariant; it is the observer's $S(v)$ clock that measures a different and longer time dt depending on how fast the observer measures the speed of the body. More generally we can say that each body in inertial motion has its proper time which is an invariant quantity.

Example 7.1

Proper time of a interstellar probe

The star Alpha Centauri is located 4.4 light years from our solar system. An interstellar probe leaves the solar system, traveling at a constant speed, and reaches Alpha Centauri six years later, according to observers on Earth. How much time passes on a clock traveling with the probe? ◀

▶ **Solution** The use of the invariant τ leads us to an elegant solution of this seemingly complex problem. We evaluate the Lorentz invariant proper time given by:

1 $c^2\Delta\tau^2 = c^2\Delta t^2 - \Delta x^2$.

With $\Delta x = 4.4$ light years and $\Delta t = 6$ years we find

2 $c^2\Delta\tau^2 = (6\,c\mathrm{y})^2 - (4.4c\mathrm{y})^2 = c^2(4.1\,\mathrm{y})^2$.

Thus $\Delta\tau = 4.1$ years. Note that with increasing (average) speed $\Delta x/\Delta t \equiv v \to c$ of the probe, we have $c\Delta t \to \Delta x$ and hence $\Delta\tau \to 0$. A probe moving at ultrarelativistic speed ages very little. ◀

Example 7.2

Positronium annihilation

A metastable configuration of positronium (usually denoted by symbol Ps), the bound state of an electron and its antiparticle, the positron, has a mean proper lifetime of $\tau = 142$ ns before it annihilates into three photons (this is the ortho-positronium[1] where the particle spins align forming spin-1-state). If the mean lifetime observed in the laboratory of the metastable Ps in a mono-energetic beam (that is all Ps at a constant and prescribed speed) in the lab frame is $\Delta t = 300$ ns, what is the mean lab frame distance traveled before annihilation by the Ps 'particle' in the beam? ◀

▶ **Solution** Again, as in Example 7.1 we use the invariant τ

1 $c^2\tau^2 = c^2\Delta t^2 - \Delta x^2$.

If τ is the mean proper lifetime and Δt the mean laboratory lifetime, then the mean distance Ps traveled from the source is

2 $\Delta x = c\sqrt{\Delta t^2 - \tau^2} \simeq 264\,c \cdot \mathrm{ns} = 260\,\mathrm{ft} = 79.2\,\mathrm{m}$.

We used feet as a reminder that the order of magnitude estimate is always 1ns at speed c $\simeq$ 1 ft (see Ref. [2] in Chap. 2).

We note that the reported mean observed lifespan $\Delta t = 300\,\mathrm{ns} > 142\,\mathrm{ns}$ exceeds the proper lifespan due to the effect of time dilation. Without time dilation the maximum average travel distance must be less than 142 ft (corresponding to 142

[1]The other equally bound state of positronium, that annihilates into two photons, the 'para-positronium' with spin-0, has a lifespan of $\tau = 125$ picoseconds, that is more than 1000 times shorter than for the ortho-Ps considered in the example.

ns mean lifespan and a speed which must be below light velocity). This example is thus a redo of the muon traveling from the upper atmosphere to the surface of the Earth, see Example 4.2, and Discussion 6.1 in the different context of a shorter lived positronium formed typically in laboratory experiment.

To be specific we look again at how proper time can be used to evaluate velocity

3 $$\frac{v^2}{c^2} \equiv \frac{\Delta x^2}{(c\Delta t)^2} = \frac{\Delta x^2}{\Delta x^2 + c^2\tau^2} = \frac{1}{1 + (c\tau/\Delta x)^2} ,$$

which was found in Example 4.2.

We now return to Discussion 4.3 with the aim of describing the range Δx an unstable particle can travel in terms of lifespan τ and speed v. We begin with the definition of proper time

4 $$c^2\tau^2 = c^2\Delta t^2(1 - v^2/c^2) , \qquad v = \Delta x/\Delta t .$$

We solve now for Δt and evaluate the range Δx

5 $$\Delta x = \Delta t\, v = \tau\gamma\, v , \qquad \gamma = \frac{1}{\sqrt{1 - v^2/c^2}} .$$

This result shows that Δx is extended by time dilation with the Lorentz factor γ. ◀

7.2 Relativistic Addition of Velocities

Case of Two Parallel Velocities

An observer in frame S' passes with velocity $\boldsymbol{v}$ in the x-direction with respect to another observer located in frame S. An object is seen by the observer in frame S to move at a constant velocity $\boldsymbol{u}$, as depicted in Fig. 7.2. What velocity is measured by the observer in S'? In the nonrelativistic case, the usual answer is obtained from the Galilean coordinate transformation Eq. (1.2)

$$\boldsymbol{u}' = \boldsymbol{u} - \boldsymbol{v} \qquad \text{Galilei} . \tag{7.9}$$

However, this Galilean law of addition of velocities is not consistent with the Lorentz transformation, and it does not respect the requirement that a material body velocity cannot be greater than light velocity.

To obtain the relativistic form of velocity addition we need to remember precisely how velocity relates to position $\boldsymbol{r}$ and time t; in frame S we have

$$\boldsymbol{u} = \frac{\Delta \boldsymbol{r}}{\Delta t} , \tag{7.10}$$

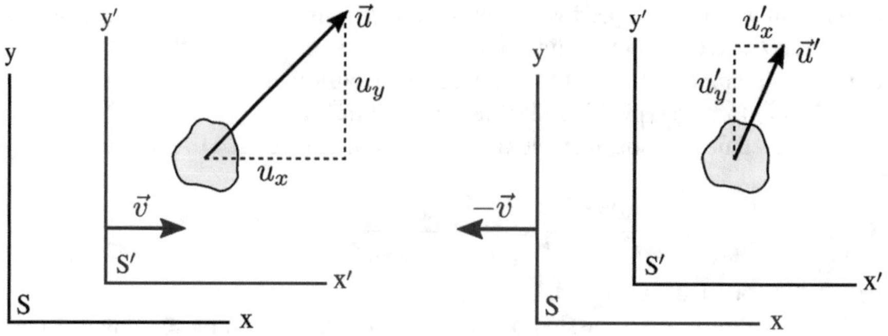

Fig. 7.2 Transformation of velocities: $\boldsymbol{u}$ is the velocity in the frame of reference S; $\boldsymbol{u}'$ is the velocity in the frame of reference S' that is moving with velocity $\boldsymbol{v}$ relative to S. See also Fig. 6.1

where both space and time coordinates are subject to the Lorentz coordinate transformation

$$\boldsymbol{u}' = \frac{\Delta \boldsymbol{r}'}{\Delta t'} , \tag{7.11}$$

rather than $\Delta \boldsymbol{r}' / \Delta t$.

In what follows we orient the coordinate system S' such that the velocity vector $\boldsymbol{v}$, Eq. (7.9) points along the x-axis

$$\boldsymbol{v} \equiv v\hat{i} . \tag{7.12}$$

This choice does not constrain the body velocity vector $\boldsymbol{u}$ as it can point in any arbitrary direction.

We first consider the special case of $\boldsymbol{u}$ being parallel to $\boldsymbol{v}$. To obtain the transformed velocity we employ the Lorentz coordinate transformation in the infinitesimal form, see Example 6.3

$$u'_x = \frac{dx'}{dt'} = \frac{\gamma}{\gamma} \frac{dx - v dt}{dt - \frac{v}{c^2} dx} = \frac{\frac{dx}{dt} - v}{1 - \frac{v}{c^2}\frac{dx}{dt}} . \tag{7.13}$$

We recognize $\frac{dx}{dt}$ as u_x resulting in the relativistic velocity addition relation

$$u'_x = \frac{u_x - v}{1 - u_x v/c^2} , \qquad u'_y = u'_z = 0 , \qquad \boldsymbol{u} \parallel \boldsymbol{v} = v\hat{i} . \tag{7.14}$$

The Galilean relation Eq. (7.9) is corrected by the denominator, which enforces the limit on the maximum allowed speed.

Example 7.3

Checking relativistic velocity addition, case $u \parallel v$

Consider the value of u'_x when $u = 0.9\,c$ and $v = \pm 0.8\,c$. ◄

► **Solution** We will be using the SR result Eq. (7.14) and comparing to the Galilean coordinate transformation Eq. (7.9). For $v = -0.8\,c$, the Galilean coordinate transformation predicts a speed that is 70% higher than the speed of light. However the SR result is

$$u'_x = \frac{0.9\,c + 0.8\,c}{1 + 0.9 \cdot 0.8} = \frac{1.7\,c}{1.72} = 0.9884\,c\,,$$

which comes close to speed of light but clearly respects the limit.

For the case $v = 0.8\,c$ the Galilean coordinate transformation result Eq. (7.9) is that the particle has become nearly nonrelativistic moving with 10% of c. However, the situation is different for the SR case Eq. (7.14); we find

$$u'_x = \frac{0.9\,c - 0.8\,c}{1 - 0.9 \cdot 0.8} = \frac{0.1\,c}{0.28} = 0.36\,c\,.$$

This is more relativistic compared to the result for Galilean coordinate transformation. ◄

Example 7.4

Relativistic relative approach speed: parallel motion

An observer on Earth sees two rockets approach each other with equal and opposite velocities $u_\pm = \pm|u| = \pm 0.6\,c$. For this observer, the distance between the rockets diminishes according to relative speed $u_+ - u_- = 1.2\,c$. Thus the distance between the ships diminishes with a speed that exceeds the speed of light! Does this situation violate relativity? Consider an observer riding along in one of the rockets. What velocity of the other rocket does he report? ◄

► **Solution** Relativity only requires that c be the maximum speed for light and for physical bodies. The relative speed at which coordinate separation between the rockets diminishes as recorded by a third arbitrary observer has nothing to do with the relative speed of two bodies measured by an observer riding one of the rockets. The whole point of the (special) theory of *relativity* is that there is no physical relevance to such an arbitrary third observer.

For an observer S_- riding in one of the rockets, the relative velocity corresponds to the velocity of the other rocket S_+ as observed from this rocket S_-. To find this velocity we must appropriately add individual velocities. For simplicity we consider the observer S_- traveling at u_-. The transformation of the velocity u_+ of the other rocket S_+ into this frame S_- is

1 $$u'_+ = \frac{u_+ - u_-}{1 - u_+ u_-/c^2} = \frac{2|u|}{1 + |u|^2/c^2}\,.$$

With $|u_{\pm}| = 0.6\,c$, we find that the relative velocity is less than the speed of light: $u'_{+} = 15\,c/17 = 0.88\,c$. ◀

Example 7.5

Shuttlecraft rescue

Star Wars scene: The Insurrection base reports to its shuttle traveling with $u_{\mathrm{s}} = 0.3\,c$ that it is being chased by a space torpedo 'T' following it with $u_{\mathrm{T}} = 0.6\,c$, and their rescue spaceship is approaching from the opposite direction traveling with the velocity $u_{\mathrm{S}} = -0.6\,c$; all velocities are parallel. What relative speeds are registered in the shuttlecraft? Are these consistent with Example 7.4? ◀

▶ **Solution** We use subscript 'S' for quantities in the spaceship reference frame and subscript 's' for quantities in the shuttle reference frame; the subscript 'T' refers to the space torpedo. We boost to the shuttle frame 's'. The rescue spaceship approaches the shuttle with relative velocity

$$\mathbf{1}\quad v_{\mathrm{Ss}} = \frac{u_{\mathrm{S}} - u_{\mathrm{s}}}{1 - u_{\mathrm{S}}u_{\mathrm{s}}/c^2} = \frac{-0.9\,c}{1 + 0.18} = -0.763\,c\,.$$

However, the space torpedo is gaining on the shuttle with

$$\mathbf{2}\quad v_{\mathrm{Ts}} = \frac{u_{\mathrm{T}} - u_{\mathrm{s}}}{1 - u_{\mathrm{T}}u_{\mathrm{s}}/c^2} = \frac{0.3\,c}{1 - 0.18} = 0.366\,c\,,$$

so rescue is possible but not assured.

We check if the shuttlecraft crew computations are correct by boosting with $v = v_{\mathrm{Ts}} = -0.366\,c$ from the shuttle frame to the space torpedo frame of reference. This gives the velocity of the spaceship observed in the space torpedo reference frame, independent of what the shuttle is doing, as

$$\mathbf{3}\quad v_{\mathrm{ST}} = \frac{u_{\mathrm{Ss}} - v_{\mathrm{Ts}}}{1 - u_{\mathrm{Ss}}v_{\mathrm{Ts}}/c^2} = \frac{-0.763\,c - 0.366\,c}{1 + 0.763\cdot 0.366} = \frac{-1.129\,c}{1 + 1.28} = -0.88\,c\,.$$

This result for the relative speed v_{TS} can be also obtained as in Example 7.4 without transforming into another reference frame

$$\mathbf{4}\quad v_{\mathrm{ST}} = \frac{u_{\mathrm{S}} - u_{\mathrm{T}}}{1 - u_{\mathrm{T}}u_{\mathrm{S}}/c^2} = \frac{-0.6\,c - 0.6\,c}{1 + 0.36} = -0.88\,c\,.$$

We return to this problem in Example 7.15 after the development of additional tools. ◀

Case of Two Arbitrary Velocities

In the general case where the direction of $\boldsymbol{u}$ is arbitrary it is helpful to decompose $\boldsymbol{u}$ into its component u_x parallel to the relative velocity $\boldsymbol{v}$, Eq. (7.12), and the orthogonal components u_y and u_z:

$$\begin{aligned} \boldsymbol{r} &= x\hat{\imath} + y\hat{\jmath} + z\hat{k} , \\ \boldsymbol{u} &= \frac{dx}{dt}\hat{\imath} + \frac{dy}{dt}\hat{\jmath} + \frac{dz}{dt}\hat{k} = u_x\hat{\imath} + u_y\hat{\jmath} + u_z\hat{k} . \end{aligned} \tag{7.15}$$

Likewise we decompose $\boldsymbol{u}'$ observed in frame S', remembering that the x'- and x-axes are parallel:

$$\boldsymbol{u}' = \frac{dx'}{dt'}\hat{\imath} + \frac{dy'}{dt'}\hat{\jmath} + \frac{dz'}{dt'}\hat{k} = u'_x\hat{\imath} + u'_y\hat{\jmath} + u'_z\hat{k} . \tag{7.16}$$

Even though $dy' = dy$ and $dz' = dz$, we see dt' in the denominator and thus all three velocity components transform; i.e., both the parallel *and* orthogonal components of the velocity are modified when we transform coordinates. Carrying this out we note that the first form of Eq. (7.14) remains valid while for u'_y and u'_z we find

$$\begin{aligned} u'_y &= \frac{dy'}{dt'} = \frac{dy\sqrt{1-(v/c)^2}}{dt - \dfrac{v}{c^2}dx} = \frac{u_y\sqrt{1-(v/c)^2}}{1 - \dfrac{v}{c^2}\dfrac{dx}{dt}} = \frac{u_y\sqrt{1-(v/c)^2}}{1 - \dfrac{vu_x}{c^2}} , \\ u'_z &= \frac{dz'}{dt'} = \frac{u_z\sqrt{1-(v/c)^2}}{1 - \dfrac{vu_x}{c^2}} . \end{aligned} \tag{7.17}$$

In the last step we are recognizing $dx/dt = u_x$.

To summarize, for $\boldsymbol{v}$ along the x-axis, see Eq. (7.12), the relativistic velocity addition equations for parallel and orthogonal addition are, respectively

$$u'_x = \frac{u_x - v}{1 - u_x v/c^2} , \qquad \boldsymbol{v} = v\hat{\imath} , \tag{7.18a}$$

$$u'_y = \frac{u_y}{1 - u_x v/c^2}\sqrt{1 - v^2/c^2} , \quad u'_z = \frac{u_z}{1 - u_x v/c^2}\sqrt{1 - v^2/c^2} . \tag{7.18b}$$

Example 7.6

Relative (rocket) motion: general case

We generalize Example 7.4 to the case of non-parallel motion: An observer on Earth sees two rockets traveling with the velocities $\boldsymbol{u}^{\pm}$. What is the relative speed of the rockets? ◀

► **Solution** We employ the result Eqs. (7.18a) and (7.18b). We align the coordinate x-axis with vector $\boldsymbol{u}^-$ and position the observer on this rocket '−'. Upon the Lorentz coordinate transformation into the reference frame of this rocket; that is, $v = u_x^-$, we find with the help of Eq. (7.18a) the relative velocity vector $\boldsymbol{u}^r = \boldsymbol{u}'^+$

$$u_x^r = \frac{u_x^+ - u_x^-}{1 - u_x^+ u_x^- / c^2} ,$$

1

$$u_y^r = \frac{u_y^+}{1 - u_x^+ u_x^- / c^2} \sqrt{1 - (u_x^- / c)^2} ,$$

$$u_z^r = \frac{u_z^+}{1 - u_x^+ u_x^- / c^2} \sqrt{1 - (u_x^- / c)^2} .$$

This is the relative velocity vector expressed in a coordinate system of which the orientation of the x-axis is parallel to the laboratory rocket '−' velocity vector. Thus u_x^r is what the observer in rocket '−' sees as the velocity of rocket '+' in approach, while u_y^r and u_z^r are components that are normal to the x-axis: $\hat{i} \parallel \boldsymbol{u}^-$. ◄

Example 7.7

Relativistic forward projection of direction of motion

We consider two sources of photons S and S'. S is at rest with respect to the laboratory observer while S' moves with speed $v' = 0.6\,c$ relative to S along the x-axis. In the rest-frame of both systems one measures an angle of emission $\vartheta = 60°$ between the direction of movement of the photon and the direction of movement of S' as depicted in Fig. 7.3. What is the angle of emission that the laboratory observer reports for the photon originating in the moving source S'? ◄

► **Solution** Geometry: The source S' moves along the x-axis, which is chosen to be parallel to the x'-axis; the movement of the photon is confined to the xy-plane, Fig. 7.3. The y and y' axes are also parallel.

The velocity of the photon in system S is therefore

1

$$\boldsymbol{v} = c \cos 60° \hat{i} + c \sin 60° \hat{j} = u_x \hat{i} + u_y \hat{j} ;$$

that is, $u_x = 0.5c,\ u_y = 0.866\,c$.

We calculate the component u'_x for the photon emitted by moving source using the addition of velocity theorem

2

$$u'_x = \frac{u_x + v'}{1 + u_x v' / c^2} .$$

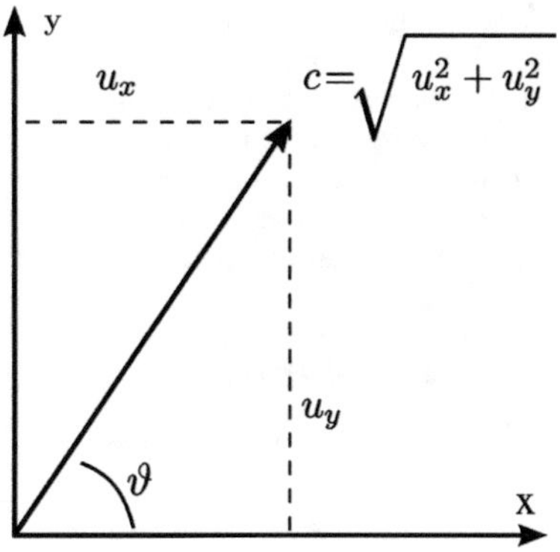

Fig. 7.3 Photon moving in S at angle ϑ to the x-axis. See Example 7.7

Inserting $u_x = 0.5\,c$ and $v' = 0.6\,c$, one obtains $u'_x = 0.846\,c$.

To obtain u'_y we consider the requirement

3 $c = c' = \sqrt{u'^2_x + u'^2_y}\ .$

Therefore

4 $u'_y = \sqrt{c^2 - u'^2_x} = \sqrt{1 - 0.846^2}\,c = 0.533\,c\ .$

We recall that this transverse to motion component of velocity of light is changed solely due to the transformation of laboratory time.

We measure in the laboratory frame S the angle ϑ' between the direction of photon motion and the direction of motion of the system S'

5 $\tan\vartheta' = \frac{u'_y}{u'_x} = \frac{0.533\,c}{0.846\,c} = 0.630\ ,$

and therefore: $\vartheta' = 32.2°$.

We performed this computation for photons, but the principles apply equally to the emission of all relativistic particles. Therefore this example shows an important phenomenon often observed in particle production experiments. Newly created particles are relativistically focused, pointing forward along the direction of motion of the source.

Such a situation arises for example when a fast cosmic particle hits the upper atmosphere. Particles are produced in a rest-frame intrinsic to the reaction, often in a spherical distribution. However, the source moves at a high speed in the same direction as the primary cosmic ray. Our computation explains why the secondary particles observed in the Earth frame are focused along the direction of the primary cosmic particle.

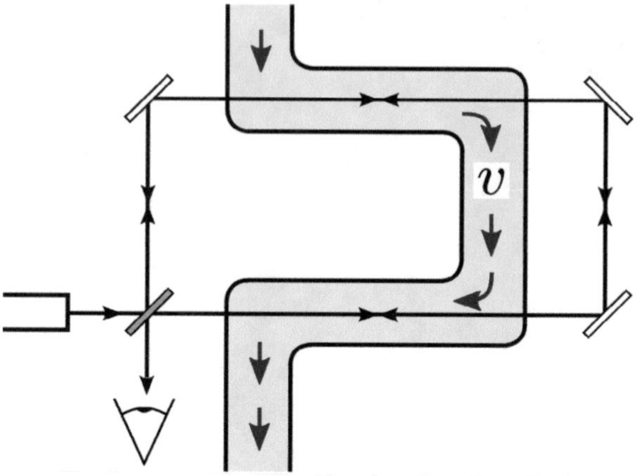

Fig. 7.4 Illustration of an interference experiment to measure the Fresnel's drag coefficient: light carried along with the flow interferes with the light that is moving against the flow direction; one observes a change in the interference with changing flow velocity v of a fluid, see Example 7.8

The study of light aberration leads to a related and more general result, see Example 13.2. A complete discussion of the relativistic focusing effect is presented in Example 15.11. ◀

Example 7.8

Fresnel drag coefficient

The speed of light in a medium with a refraction index $n > 1$ is known to be $\tilde{c} < c$. Furthermore, $\tilde{c}$ is dependent on the velocity v of the medium relative to a (laboratory) observer. Obtain the lowest correction in v: $\tilde{c}(v) = \tilde{c}_0 + vf(n)$ as measured by the laboratory observer, where f is the Fresnel drag coefficient (Fresnel,[2] 1818). ◀

▶ **Solution** The speed of light is reduced by the index of refraction n in a stationary medium. It is known that the Maxwell equations allow one to deduce the relation

1 $$\tilde{c}_0 = \frac{c}{n}\,.$$

We wish to understand what happens when the fluid medium as shown in Fig. 7.4 is set in motion where v is medium velocity relative to the lab frame, and the beam of light is propagating parallel or antiparallel to v. There is a fluid comoving observer $S(v)$, moving with $\pm v$ with respect to the laboratory observer S, for whom the light can be traveling through a stationary medium in each of the arms of interferometer

[2]Augustin-Jean Fresnel (1788–1827), luminary French physicist and founder of the theory of wave optics.

seen in Fig. 7.4. To find the speed of light $\tilde{c}(v)$ as measured by the observer in the lab frame, we can thus use the addition of velocities theorem, Eq. (7.18a):

2
$$\tilde{c}(v) = \frac{\tilde{c}_0 + v}{1 + \tilde{c}_0 v/c^2} = \frac{c/n + v}{1 + v/(nc)} = \left(\frac{c}{n}\right)\left(\frac{1 + nv/c}{1 + v/(nc)}\right) .$$

Any velocity that can be achieved in a laboratory satisfies $v \ll c$, thus we can expand

3
$$\begin{aligned} \tilde{c}(v) &\simeq \frac{c}{n}\left(1 + \frac{nv}{c}\right)\left(1 - \frac{v}{nc}\ldots\right) \simeq \frac{c}{n}\left(1 + \frac{nv}{c}\left(1 - \frac{1}{n^2}\right)\right) \\ &\simeq \frac{c}{n} + v\left(1 - \frac{1}{n^2}\right) . \end{aligned}$$

This result applies to both v positive and negative.

The Fresnel drag coefficient is therefore

4
$$f = 1 - \frac{1}{n^2} .$$

For the special case of a vacuum ($n = 1$), we find that $f = 0$, confirming the drag limit in the vacuum where the speed of light is independent of the motion of the observer.

Historical remarks: The first experimental demonstration of Fresnel drag was carried out by Fizeau in 1851, and much experimental and theoretical work followed in the second half of the nineteenth century. Fresnel drag was a cornerstone precursor to the development of SR; the concept of æther motion does not enter the present discussion, a topic that had been much a part of Fresnel drag experiments prior to Einstein's development of SR. The solution of Fresnel drag using SR principles was presented by Max von Laue.[3] ◀

[3]Max von Laue, "Die Mitführung des Lichtes durch bewegte Körper nach dem Relativitätsprinzip, (The Entrainment of Light by Moving Bodies According to Principle of Relativity)" *Annalen der Physik* **328** 989–990 (1907). Max von Laue (1879–1960), won the Nobel prize in 1914 "for his discovery of the diffraction of X-rays by crystals". A man with passion for truth in life as in physics.

Example 7.9

Light speed as upper speed limit

A body moves at a velocity of magnitude $\boldsymbol{u}$ ($|\boldsymbol{u}\,| < c$) in an arbitrary direction as observed in frame S. Show that the velocity $\boldsymbol{u}'$ observed in frame S' moving relative to S at velocity $\boldsymbol{v}$ along the x-axis is less than the speed of light, $|\boldsymbol{u}'\,| < c$. ◀

▶ **Solution** We want to prove the relationship

$$\mathbf{1}\quad \boldsymbol{u}'^{\,2} < c^2\,.$$

We can rewrite $\boldsymbol{u}'^{\,2}$ in components

$$\mathbf{2}\quad \boldsymbol{u}'^{\,2} = u_x'^{\,2} + u_y'^{\,2} + u_z'^{\,2}\,,$$

where u_x' is the component of the velocity parallel to the x- and x'- axes, and the u_y' and u_z' are the two transverse components. Using Eqs. (7.18a) and (7.18b) to express these components in terms of u and v we obtain, after some simple algebraic manipulations

$$\mathbf{3}\quad \begin{aligned} \boldsymbol{u}'^{\,2} &= \frac{(u_x - v)^2 + (u_y^2 + u_z^2)(1 - v^2/c^2)}{(1 - vu_x/c^2)^2} = \frac{u^2 + v^2 - 2u_x v - (u_y^2 + u_z^2)v^2/c^2}{(1 - vu_x/c^2)^2} \\ &= c^2\left(1 - \frac{(1 - vu_x/c^2)^2 - u^2/c^2 - v^2/c^2 + 2u_x v/c^2 + (u_y^2 + u_z^2)v^2/c^4}{(1 - vu_x/c^2)^2}\right), \end{aligned}$$

where $u^2 = u_x^2 + u_y^2 + u_z^2$. Regrouping all terms we obtain

$$\mathbf{4}\quad \boldsymbol{u}'^{\,2} = c^2\left(1 - \frac{(1 - \boldsymbol{u}^{\,2}/c^2)(1 - \boldsymbol{v}^{\,2}/c^2)}{(1 - \boldsymbol{v}\cdot\boldsymbol{u}/c^2)^2}\right),$$

where for $\boldsymbol{v} = v\hat{i}$ we used $v^2 = \boldsymbol{v}^{\,2}$ and $vu_x = \boldsymbol{v}\cdot\boldsymbol{u}$. This is the relativistic addition theorem of velocities when the direction of motion is not aligned with the direction of LT. Note that the simplicity of the result is due to our interest in speed only. We did not determine the direction of motion.

Inspecting the large brackets above, we see that a manifestly positive term is always subtracted from unity. This term becomes arbitrarily small when either of the speeds, u, v approaches speed of light, yet always

$$\mathbf{5}\quad \boldsymbol{u}'^{\,2} < c^2\,.$$

Thus we confirm that, irrespective of the direction of motion, the addition theorem of velocities derived from the Lorentz coordinate transformation is consistent with c being the universal and highest speed which cannot be exceeded.

It is of interest to consider the special case of a light ray with $\boldsymbol{u}^2 = c^2$ observed in the moving frame S'. We find

6 $\boldsymbol{u'}^2 = c^2, \qquad \text{for} \quad \boldsymbol{u}^2 = c^2 .$

This result is of use in Sect. 13.3.

Another cross-check of interest is verification of the result Eq. (7.14), corresponding to the case $u_y = u_z = 0$; that is, $u^2 = u_x^2$. We form the common denominator in Eq. 4 to obtain

7
$$\boldsymbol{u'}^2 = c^2 \frac{1 - 2vu_x/c^2 + v^2u_x^2/c^4 - 1 - v^2u_x^2/c^4 + u_x^2/c^2 + v^2/c^2}{(1 - vu_x/c^2)^2}$$
$$= \frac{(u_x - v)^2}{(1 - vu_x/c^2)^2} ,$$

which is the square of Eq. (7.14). This cross-check confirms the principal result of this example. ◀

7.3 Two Lorentz Coordinate Transformations in Sequence

We consider two subsequent Lorentz coordinate transformations:

(a) the event (t, x, y, z) in S transformed into (t', x', y', z') in S' using v_1; and
(b) the event (t', x', y', z') in S' transformed into (t'', x'', y'', z'') in S'' moving with v_2 in reference to S'.

We consider the case where both transformations are in the x-direction. We now show that the outcome is as if there was one transformation from S to S'' by v_x, which arises from the relativistic addition of the two velocities.

We begin with

$$x'' = \gamma_2(x' - \beta_2 ct'), \quad ct'' = \gamma_2(ct' - \beta_2 x') , \tag{7.19}$$

where we insert

$$x' = \gamma_1(x - \beta_1 ct), \quad ct' = \gamma_1(ct - \beta_1 x) , \tag{7.20}$$

which results in

$$x'' = \gamma_2\gamma_1(x - \beta_1 ct - \beta_2(ct - \beta_1 x)) ,$$
$$ct'' = \gamma_2\gamma_1(ct - \beta_1 x - \beta_2(x - \beta_1 ct)) . \tag{7.21}$$

A reorganization of terms produces

$$x'' = \gamma_2\gamma_1(x(1+\beta_2\beta_1) - (\beta_2+\beta_1)ct)\,,$$
$$ct'' = \gamma_2\gamma_1(ct(1+\beta_2\beta_1) - (\beta_2+\beta_1)x)\,. \tag{7.22}$$

This is another Lorentz coordinate transformation from $S \to S''$ if

$$\gamma = \gamma_2\gamma_1(1+\beta_1\beta_2), \quad \gamma\beta = \gamma_2\gamma_1(\beta_1+\beta_2)\,. \tag{7.23}$$

Dividing these two equations by each other we obtain again the velocity addition theorem

$$\beta = \frac{\beta_1+\beta_2}{1+\beta_1\beta_2}\,; \quad \text{that is,} \quad v_x = \frac{v_1+v_2}{1+v_1v_2/c^2}\,, \tag{7.24}$$

where $v_x = c\beta$. We see that two Lorentz coordinate transformations in sequence in the x-direction are described by a transformation with the velocity obtained according to the relativistic addition theorem for the two velocities of these transformations.

Comparing to our earlier results Eqs. (7.14) and (7.18a), we note a change in sign: In the prior result the velocity v was that of a moving body. In the present consideration β_2 describes a change of coordinate system. Thus we have shown the equivalence of both passive and active relativistic velocity addition theorems.

We next check if $\beta^2 < 1$. This is best done evaluating

$$1/\gamma^2 \equiv 1-\beta^2 = \frac{(1-\beta_1^2)(1-\beta_2^2)}{(1+\beta_1\beta_2)^2}\,, \tag{7.25}$$

which is positive and smaller than unity for any $\beta_1^2, \beta_2^2 < 1$.

For $\beta_1 = -\beta_2$ we find from Eq. (7.24) as well as from Eq. (7.25) that with the second transformation we transformed back to the original system $S'' = S$, since $\beta = 0, \gamma = 1$.

Example 7.10

Two sequential LTs in orthogonal directions

Find the transformation describing the consequence of two Lorentz coordinate transformations carried out in sequence, the first one in the x-direction with v_{1x} and the second one in the y-direction with v_{2y}. What happens when the sequence of transformations is exchanged? ◀

▶ **Solution** Since $z = z' = z''$, we will ignore this coordinate. The first transformation is from system S to S' along the x-axis

$$x' = \gamma_{1x}(x - \beta_{1x}ct)\,, \qquad y' = y\,, \qquad ct' = \gamma_{1x}(ct - \beta_{1x}x)\,, \tag{1}$$

and the second from S' to S'' along the y-axis

$$x'' = x'\,, \qquad y'' = \gamma_{2y}(y' - \beta_{2y}ct')\,, \qquad ct'' = \gamma_{2y}(ct' - \beta_{2y}y')\,. \tag{2}$$

By substitution we obtain the double primed coordinates expressed in terms of the not primed coordinates

$$\begin{aligned} x'' &= \gamma_{1x}(x - \beta_{1x}ct)\,, \\ y'' &= \gamma_{2y}(y - \beta_{2y}\gamma_{1x}(ct - \beta_{1x}x)) = \gamma_{2y}(y + \gamma_{1x}\beta_{2y}\beta_{1x}x) - \gamma_{1x}\gamma_{2y}\beta_{2y}ct\,, \\ ct'' &= \gamma_{2y}(\gamma_{1x}(ct - \beta_{1x}x) - \beta_{2y}y) = \gamma_{2y}\gamma_{1x}ct - \gamma_{2y}\gamma_{1x}\beta_{1x}x - \gamma_{2y}\beta_{2y}y\,. \end{aligned} \tag{3}$$

We see that this format is very different from the usual Lorentz coordinate transformation.

We now check if the sequence of two non-parallel Lorentz coordinate transformations matters. This means that we consider first the transformation in the y-direction with v_{2y}, followed by a transformation in x-direction with v_{1x}. This result is obtained exchanging both $x \Leftrightarrow y$ and the indices $1x \Leftrightarrow 2y$

$$\begin{aligned} x'' &= \gamma_{1x}(x - \beta_{1x}\gamma_{2y}(ct - \beta_{2y}y)) = \gamma_{1x}(x + \gamma_{2y}\beta_{1x}\beta_{2y}y) - \gamma_{2y}\gamma_{1x}\beta_{1x}ct\,, \\ y'' &= \gamma_{2y}(y - \beta_{2y}ct)\,, \\ ct'' &= \gamma_{1x}(\gamma_{2y}(ct - \beta_{2y}y) - \beta_{1x}x) = \gamma_{1x}\gamma_{2y}ct - \gamma_{1x}\gamma_{2y}\beta_{2y}y - \gamma_{1x}\beta_{1x}x\,. \end{aligned} \tag{4}$$

We see that the two results we obtained differ. The sequence in which one carries out transformations is relevant. One says, the non-parallel Lorentz transformations do not commute. ◀

Example 7.11

Relativistic speed – addition theorem for magnitude of arbitrary velocity vectors

State the magnitude of the speed obtained in Example 7.9 in vector notation and explore the case of two orthogonal input velocities explicitly. ◀

▶ **Solution** We recall that the velocity v appearing in Example 7.9 has one single component, thus $\boldsymbol{\beta}_2 = -(v/c, 0, 0)$. We call now $\boldsymbol{\beta}_1 = (u_x/c, u_y/c, u_z/c)$. In this notation the result obtained in Example 7.9 reads

$$\beta^2 \equiv \boldsymbol{u}'^2/c^2 = \left(1 - \frac{(1 - \boldsymbol{\beta}_1^2)(1 - \boldsymbol{\beta}_2^2)}{(1 + \boldsymbol{\beta}_1 \cdot \boldsymbol{\beta}_2)^2}\right)\,. \tag{1}$$

Simplifying with a common denominator we obtain

$$2\quad \beta^2 = \frac{1+2\boldsymbol{\beta}_1\cdot\boldsymbol{\beta}_2+(\boldsymbol{\beta}_1\cdot\boldsymbol{\beta}_2)^2-1+\boldsymbol{\beta}_1^2+\boldsymbol{\beta}_2^2-\boldsymbol{\beta}_1^2\boldsymbol{\beta}_2^2}{(1+\boldsymbol{\beta}_1\cdot\boldsymbol{\beta}_2)^2}\,.$$

Noting $(\boldsymbol{\beta}_1\times\boldsymbol{\beta}_2)^2 = \boldsymbol{\beta}_1^2\boldsymbol{\beta}_2^2-(\boldsymbol{\beta}_1\cdot\boldsymbol{\beta}_2)^2$ in the numerator we find

$$3\quad \beta^2 = \frac{(\boldsymbol{\beta}_1+\boldsymbol{\beta}_2)^2-(\boldsymbol{\beta}_1\times\boldsymbol{\beta}_2)^2}{(1+\boldsymbol{\beta}_1\cdot\boldsymbol{\beta}_2)^2}\,,\quad \text{or}\quad \boldsymbol{v}^2 = \frac{(\boldsymbol{v}_1+\boldsymbol{v}_2)^2-(\boldsymbol{v}_1\times\boldsymbol{v}_2)^2/c^2}{(1+\boldsymbol{v}_1\cdot\boldsymbol{v}_2/c^2)^2}\,.$$

Aside of the expected $\boldsymbol{v}_1 \parallel \boldsymbol{v}_2$ term we see above a further contribution vanishing in the limit $\boldsymbol{v}_1 \parallel \boldsymbol{v}_2$. In this case we retain in the nominator only the first nonrelativistic term.

We note that an interesting special case arises for orthogonal $\boldsymbol{v}_1\cdot\boldsymbol{v}_2 = 0$ velocities

$$4\quad \boldsymbol{v}^2 = \boldsymbol{v}_1^2+\boldsymbol{v}_2^2-\boldsymbol{v}_1^2\boldsymbol{v}_2^2/c^2\,,\qquad \boldsymbol{v}_1\perp\boldsymbol{v}_2\,.$$

One can easily see that as long as $\boldsymbol{v}_1^2 < c^2$ and $\boldsymbol{v}_2^2 < c^2$, then $\boldsymbol{v}^2 < c^2$. Comparing this little known result with our study of the orthogonal LT in Example 7.10 we can better appreciate this significant simplification. ◀

7.4 Rapidity

We now introduce a new way to characterize speed within the Lorentz coordinate transformation through a function[4] $y_\mathrm{r}(\beta)$. We are motivated by the desire to find a new variable which, unlike v, is not bounded by c. Therefore, it can more accurately characterize motion at ultrarelativistic speeds. We also want that in the nonrelativistic limit

$$y_\mathrm{r} = \beta \qquad \text{for } \beta \ll 1\,. \tag{7.26}$$

This relation suggests that we call y_r something that relates to speed, and the word rapidity is in common use.

We further recall that in the nonrelativistic limit two velocities add vectorially. If these velocity vectors are parallel, they add as numbers in nonrelativistic limit; that is, the two speeds add. Among many functions $y_\mathrm{r}(\beta)$ we choose a unique one

[4] It is inconvenient at this point to introduce rapidity using the conventional symbol y which can be confounded in this book with a coordinate. We thus introduce a subscript, here y_r, with 'r' for rapidity.

such that for parallel relativistic motion the rapidities sum just like speeds do for nonrelativistic motion. How this could work is easily recognized by recalling the addition theorem for tanh

$$\tanh(y_{r\,1} + y_{r\,2}) = \frac{\tanh y_{r\,1} + \tanh y_{r\,2}}{1 + \tanh y_{r\,1} \tanh y_{r\,2}}, \tag{7.27}$$

which we compare with the addition theorem of relativistic parallel speeds

$$\beta_{12} = \frac{\beta_1 + \beta_2}{1 + \beta_1 \beta_2}. \tag{7.28}$$

Thus we explore a new variable y_r that satisfies

$$\beta = \tanh y_r = \frac{e^{y_r} - e^{-y_r}}{e^{y_r} + e^{-y_r}} < 1. \tag{7.29}$$

We find the useful relation

$$e^{y_r} = \sqrt{\frac{1+\beta}{1-\beta}} = \gamma(1+\beta), \qquad e^{-y_r} = \sqrt{\frac{1-\beta}{1+\beta}} = \gamma(1-\beta), \tag{7.30}$$

and hence

$$y_r = \ln\sqrt{\frac{1+\beta}{1-\beta}} = \frac{1}{2}\ln\left(\frac{1+\beta}{1-\beta}\right) \equiv \operatorname{artanh}(\beta). \tag{7.31}$$

Considering that hyperbolic exponential functions can be written in terms of the exponential function

$$\cosh y_r = \frac{e^{y_r} + e^{-y_r}}{2}, \qquad \sinh y_r = \frac{e^{y_r} - e^{-y_r}}{2}, \tag{7.32}$$

a short computation shows

$$\cosh y_r = \gamma, \qquad \sinh y_r = \gamma\beta. \tag{7.33}$$

We cross-check our computation

$$\cosh^2 y_\mathrm{r} - \sinh^2 y_\mathrm{r} = \gamma^2(1-\beta^2) = 1\,, \tag{7.34}$$

and furthermore

$$\tanh y_\mathrm{r} = \frac{\sinh y_\mathrm{r}}{\cosh y_\mathrm{r}} = \frac{\gamma\beta}{\gamma} = \beta\,, \tag{7.35}$$

as expected, see Eq. (7.29).

Expanding Eq. (7.35) in a power series we see that

$$\beta = \tanh y_\mathrm{r} = y_\mathrm{r} - \frac{1}{3}y_\mathrm{r}^3 + \frac{2}{15}y_\mathrm{r}^5 - \dots\,, \tag{7.36}$$

and from Eq. (7.31)

$$\begin{aligned} y_\mathrm{r} &= \tfrac{1}{2}[\ln(1+\beta) - \ln(1-\beta)] \\ &= \frac{1}{2}\left[\left(\beta - \frac{\beta^2}{2} + \frac{\beta^3}{3} - \dots\right) - \left(-\beta - \frac{\beta^2}{2} - \frac{\beta^3}{3} - \dots\right)\right] \\ &= \beta + \tfrac{1}{3}\beta^3 + \tfrac{1}{5}\beta^5 + \dots\,. \end{aligned} \tag{7.37}$$

Furthermore, for $\beta \to 1$ we use Eq. (7.29) followed by a Taylor series

$$1-\beta = 1 - \frac{1-e^{-2y_\mathrm{r}}}{1+e^{-2y_\mathrm{r}}} = \frac{2}{1+e^{2y_\mathrm{r}}} = 2\left(e^{-2y_\mathrm{r}} - e^{-4y_\mathrm{r}} + e^{-6y_\mathrm{r}} - \dots\right). \tag{7.38}$$

It is interesting to note how rapidity describes the approach to the speed of light: $y_\mathrm{r} = 10$ corresponds to $1-\beta = 4\cdot 10^{-9}$, thus a deviation from the speed of light by 1.2 m/s.

We now write the Lorentz coordinate transformation using rapidity in the form

$$\begin{aligned} x' &= \frac{x - vt}{\sqrt{1-(v/c)^2}} = \gamma(x - \beta ct) = x\cosh y_\mathrm{r} - ct\,\sinh y_\mathrm{r}\,, \\ ct' &= \frac{ct - (v/c)x}{\sqrt{1-(v/c)^2}} = \gamma(ct - \beta x) = ct\cosh y_\mathrm{r} - x\,\sinh y_\mathrm{r}\,. \end{aligned} \tag{7.39}$$

The form of Eq. (7.39) is similar to a rotation around the z-axis:

$$\begin{aligned} x' &= x\cos\phi + y\sin\phi\,, \\ y' &= y\cos\phi - x\sin\phi\,, \end{aligned} \tag{7.40}$$

but with one sign being different and a hyperbolic angle of rotation. This takes into account the different properties of space-time: While the usual rotation e.g. around the z-axis leaves $\rho^2 = x^2 + y^2$ invariant, the Lorentz coordinate transformation leaves invariant $s^2 = (ct)^2 - \boldsymbol{x}^2$, compare Eq. (7.6).

With the help of rapidity we thus recognize the Lorentz coordinate transformation to be a new form of rotation in what one calls Minkowski space-time. LT is characterized by an angle which is hyperbolic and not trigonometric. This difference arises from the unbounded character of y_r compared to $0 \leq \phi \leq 2\pi$, where ϕ is a regular rotation angle.

We have constructed the rapidity variable to be additive, which means that when we consider two LT in sequence that the rapidities add just like two angles add in case of regular rotations

$$y_\mathrm{r} = y_{\mathrm{r}\,1} + y_{\mathrm{r}\,2} \,. \tag{7.41}$$

This will be the content of the Example 7.14 below. It is important to remember that Eq. (7.41) is only true for two Lorentz coordinate transformations in the same spatial direction. Even so, Eq. (7.41) is a pivotal property of rapidity, which behaves just like the nonrelativistic addition of velocity does. This makes rapidity an extraordinarily important tool in the study of many problems in physics.

For example, the rapidity formulation of LT Eq. (7.33) allows us to show the invariance of proper time just as we show that length of a vector remains the same under rotations:

$$\begin{aligned} s'^2 &= c^2 t'^2 - x'^2 \\ &= (ct \cosh y_\mathrm{r} - x \sinh y_\mathrm{r})^2 - (x \cosh y_\mathrm{r} - ct \sinh y_\mathrm{r})^2 \\ &= (\cosh^2 y_\mathrm{r} - \sinh^2 y_\mathrm{r}) c^2 t^2 + (\sinh^2 y_\mathrm{r} - \cosh^2 y_\mathrm{r}) x^2 \\ &= c^2 t^2 - x^2 = s^2 \,, \end{aligned} \tag{7.42}$$

which is the same result as obtained in Sect. 7.1, see Eq. (7.4).

Example 7.12

Rotations in Minkowski space-time

Obtain the transformation matrix representation in the four-dimensional Minkowski space-time for the rotations Eq. (7.40). Use this new representation of rotations to demonstrate invariance of proper time under rotations. ◀

► **Solution** According to Eq. (7.40) we have for the rotation around the x-axis

1
$$\begin{aligned} ct' &= ct\ , \\ x' &= x\ , \\ y' &= \ \ y\ \cos\phi + z\ \sin\phi\ , \\ z' &= -y\ \sin\phi + z\ \cos\phi\ . \end{aligned}$$

We seek to write this in form of a matrix multiplication. This can be done using a rotation matrix R_x

2 $$X' = R_x X \ \Rightarrow \qquad \begin{pmatrix} ct' \\ x' \\ y' \\ z' \end{pmatrix} = \begin{pmatrix} 1 & 0 & 0 & 0 \\ 0 & 1 & 0 & 0 \\ 0 & 0 & \cos\phi & \sin\phi \\ 0 & 0 & -\sin\phi & \cos\phi \end{pmatrix} \begin{pmatrix} ct \\ x \\ y \\ z \end{pmatrix} .$$

This differs by the introduction of ct, the fourth time coordinate from the well-known form.

We now explore the behavior of the proper time subject to rotation using the rotation matrix R_x. Since we need for the spatial components a minus-sign, we need to introduce an additional matrix G

3 $$(c\tau)^2 = (ct)^2 - x^2 - y^2 - z^2 \equiv X^{\mathrm{T}}\ G\ X \ \Rightarrow \qquad G = \begin{pmatrix} 1 & 0 & 0 & 0 \\ 0 & -1 & 0 & 0 \\ 0 & 0 & -1 & 0 \\ 0 & 0 & 0 & -1 \end{pmatrix} .$$

Further, let 'T' denote the transposed matrix, that is

4 $$X^{\mathrm{T}} = (ct,\ x,\ y,\ z)\ .$$

We are now ready to evaluate the transformation property of the proper time under rotations. Using the matrix notation the transformed proper times becomes

5 $$(c\tau')^2 = X'^{\mathrm{T}}\ G\ X' = X^{\mathrm{T}} R_x^{\mathrm{T}}\ G\ R_x X\ .$$

Comparing with the prior form of $(c\tau)^2$ we recognize that proper time is an invariant under the condition

6 $$R_x^{\mathrm{T}}\ G\ R_x = G\ .$$

We evaluate the product of these three matrices

$$R_x^{\mathrm{T}}\, G\, R_x =$$

$$= \begin{pmatrix} 1 & 0 & 0 & 0 \\ 0 & 1 & 0 & 0 \\ 0 & 0 & \cos\phi & -\sin\phi \\ 0 & 0 & \sin\phi & \cos\phi \end{pmatrix} \begin{pmatrix} 1 & 0 & 0 & 0 \\ 0 & -1 & 0 & 0 \\ 0 & 0 & -1 & 0 \\ 0 & 0 & 0 & -1 \end{pmatrix} \begin{pmatrix} 1 & 0 & 0 & 0 \\ 0 & 1 & 0 & 0 \\ 0 & 0 & \cos\phi & \sin\phi \\ 0 & 0 & -\sin\phi & \cos\phi \end{pmatrix}$$

7

$$= \begin{pmatrix} 1 & 0 & 0 & 0 \\ 0 & 1 & 0 & 0 \\ 0 & 0 & \cos\phi & -\sin\phi \\ 0 & 0 & \sin\phi & \cos\phi \end{pmatrix} \begin{pmatrix} 1 & 0 & 0 & 0 \\ 0 & -1 & 0 & 0 \\ 0 & 0 & -\cos\phi & -\sin\phi \\ 0 & 0 & \sin\phi & -\cos\phi \end{pmatrix}$$

$$= \begin{pmatrix} 1 & 0 & 0 & 0 \\ 0 & -1 & 0 & 0 \\ 0 & 0 & -\cos^2\phi - \sin^2\phi & \cos\phi\sin\phi - \cos\phi\sin\phi \\ 0 & 0 & \cos\phi\sin\phi - \cos\phi\sin\phi & -\cos^2\phi - \sin^2\phi \end{pmatrix} = G\,.$$

This completes the demonstration of proper time invariance under rotations.

In this example we have rederived in a more elaborate fashion a result well-known to us. This allowed us to learn how we need to treat in the matrix representation of LT the space-time variables, assigning the correct signs to space $(-)$ and time $(+)$ in the convention employed in this book. Specifically, we saw the need to introduce a matrix G which characterizes the metric of the Minkowski space-time. ◀

Example 7.13

Boosts in Minkowski space

Considering the similarity of Eq. (7.39) with Eq. (7.40), present LT using rotation-like matrices as a new type of rotation in Minkowski space-time. Using this result demonstrate the invariance of proper time under this transformation. ◀

► **Solution** We consider LT in x-direction according to Eq. (7.39)

1

$$ct' = \frac{ct - (v/c)x}{\sqrt{1-(v/c)^2}} = \gamma(ct - \beta x) = ct\,\cosh y_{\mathrm{r}} - x\,\sinh y_{\mathrm{r}}\,,$$

$$x' = \frac{x - vt}{\sqrt{1-(v/c)^2}} = \gamma(x - \beta ct) = x\,\cosh y_{\mathrm{r}} - ct\,\sinh y_{\mathrm{r}}\,,$$

$$y' = y\,, \quad z' = z\,.$$

We seek to write this in the format of a matrix multiplication. We need for this purpose a boost-matrix Λ_x

2 $X' = \Lambda_x X \Rightarrow \qquad \begin{pmatrix} ct' \\ x' \\ y' \\ z' \end{pmatrix} = \begin{pmatrix} \cosh y_r & -\sinh y_r & 0 & 0 \\ -\sinh y_r & \cosh y_r & 0 & 0 \\ 0 & 0 & 1 & 0 \\ 0 & 0 & 0 & 1 \end{pmatrix} \begin{pmatrix} ct \\ x \\ y \\ z \end{pmatrix} .$

We note that some (relative) signs of elements within the boost matrix Λ_x differ from those seen in the rotation matrix R_x in Example 7.12.

We turn to consider the transformation of proper time under a boost. Since we need to introduce a minus sign in spatial components we will need to include the metric matrix G, Example 7.12

3 $(c\tau)^2 = X^{\mathrm{T}}\, G\, X .$

The transformed proper time is

4 $(c\tau')^2 = X'^{\mathrm{T}}\, G\, X' = X^{\mathrm{T}} \Lambda_x^{\mathrm{T}}\, G\, \Lambda_x X .$

Comparing these two results for $(c\tau')^2$ it is evident that in analogy to Example 7.12 the invariance of proper time requires

5 $\Lambda_x^{\mathrm{T}}\, G\, \Lambda_x = G .$

We evaluate the product of the three matrices

$$\Lambda_x^{\mathrm{T}}\, G\, \Lambda_x =$$

$$= \begin{pmatrix} \mathrm{ch}\, y_r & -\mathrm{sh}\, y_r & 0 & 0 \\ -\mathrm{sh}\, y_r & \mathrm{ch}\, y_r & 0 & 0 \\ 0 & 0 & 1 & 0 \\ 0 & 0 & 0 & 1 \end{pmatrix} \begin{pmatrix} 1 & 0 & 0 & 0 \\ 0 & -1 & 0 & 0 \\ 0 & 0 & -1 & 0 \\ 0 & 0 & 0 & -1 \end{pmatrix} \begin{pmatrix} \mathrm{ch}\, y_r & -\mathrm{sh}\, y_r & 0 & 0 \\ -\mathrm{sh}\, y_r & \mathrm{ch}\, y_r & 0 & 0 \\ 0 & 0 & 1 & 0 \\ 0 & 0 & 0 & 1 \end{pmatrix}$$

6

$$= \begin{pmatrix} \mathrm{ch}\, y_r & -\mathrm{sh}\, y_r & 0 & 0 \\ -\mathrm{sh}\, y_r & \mathrm{ch}\, y_r & 0 & 0 \\ 0 & 0 & 1 & 0 \\ 0 & 0 & 0 & 1 \end{pmatrix} \begin{pmatrix} \mathrm{ch}\, y_r & -\mathrm{sh}\, y_r & 0 & 0 \\ \mathrm{sh}\, y_r & -\mathrm{ch}\, y_r & 0 & 0 \\ 0 & 0 & -1 & 0 \\ 0 & 0 & 0 & -1 \end{pmatrix}$$

$$= \begin{pmatrix} \mathrm{ch}^2\, y_r - \mathrm{sh}^2\, y_r & \mathrm{sh}\, y_r\, \mathrm{ch}\, y_r - \mathrm{sh}\, y_r\, \mathrm{ch}\, y_r & 0 & 0 \\ \mathrm{sh}\, y_r\, \mathrm{ch}\, y_r - \mathrm{sh}\, y_r\, \mathrm{ch}\, y_r & -\mathrm{ch}^2\, y_r + \mathrm{sh}^2\, y_r & 0 & 0 \\ 0 & 0 & -1 & 0 \\ 0 & 0 & 0 & -1 \end{pmatrix} = G , \ .$$

where we used the abbreviated notation $\cosh \equiv \mathrm{ch}$, $\sinh \equiv \mathrm{sh}$.

This demonstration of the invariance of the proper time shows that it is meaningful to interpret boosts as a new type of rotation in Minkowski space-time, and that representation in terms of transformation matrices is meaningful: all steps of our demonstration equal those we had developed and studied in Example 7.12 in the context of the spatial rotations. This will become even more evident in the following example. ◀

Example 7.14

Addition of rapidity

Demonstrate that rapidities for two Lorentz coordinate transformations carried out in sequence add akin to the situation with rotation angles. Repeat this computation using the matrix representation of boosts, see Example 7.13. ◀

▶ **Solution** We now carry out two Lorentz coordinate transformations in sequence, see Sect. 7.3, employing the rapidity format. We have

1
$$\begin{aligned} x' &= x \ \cosh y_{\mathrm{r}\,1} - ct \ \sinh y_{\mathrm{r}\,1}\,, \quad & ct' &= ct \ \cosh y_{\mathrm{r}\,1} - x \ \sinh y_{\mathrm{r}\,1}\,, \\ x'' &= x' \ \cosh y_{\mathrm{r}\,2} - ct' \ \sinh y_{\mathrm{r}\,2}\,, \quad & ct'' &= ct' \ \cosh y_{\mathrm{r}\,2} - x' \ \sinh y_{\mathrm{r}\,2}\,. \end{aligned}$$

We insert the first transformation into the second

2
$$\begin{aligned} x'' &= (x \ \cosh y_{\mathrm{r}\,1} - ct \ \sinh y_{\mathrm{r}\,1}) \cosh y_{\mathrm{r}\,2} - (ct \ \cosh y_{\mathrm{r}\,1} - x \ \sinh y_{\mathrm{r}\,1}) \sinh y_{\mathrm{r}\,2}\,, \\ ct'' &= (ct \ \cosh y_{\mathrm{r}\,1} - x \ \sinh y_{\mathrm{r}\,1}) \cosh y_{\mathrm{r}\,2} - (x \ \cosh y_{\mathrm{r}\,1} - ct \ \sinh y_{\mathrm{r}\,1}) \sinh y_{\mathrm{r}\,2}\,. \end{aligned}$$

Reordering terms we find

3
$$\begin{aligned} x'' &= x\,(\mathrm{ch}\,y_{\mathrm{r}\,1}\,\mathrm{ch}\,y_{\mathrm{r}\,2} + \mathrm{sh}\,y_{\mathrm{r}\,1}\,\mathrm{sh}\,y_{\mathrm{r}\,2}) - ct\,(\mathrm{ch}\,y_{\mathrm{r}\,1}\,\mathrm{sh}\,y_{\mathrm{r}\,2} + \mathrm{sh}\,y_{\mathrm{r}\,1}\,\mathrm{ch}\,y_{\mathrm{r}\,2})\,, \\ ct'' &= ct\,(\mathrm{ch}\,y_{\mathrm{r}\,1}\,\mathrm{ch}\,y_{\mathrm{r}\,2} + \mathrm{sh}\,y_{\mathrm{r}\,1}\,\mathrm{sh}\,y_{\mathrm{r}\,2}) - x\,(\mathrm{ch}\,y_{\mathrm{r}\,1}\,\mathrm{sh}\,y_{\mathrm{r}\,2} + \mathrm{sh}\,y_{\mathrm{r}\,1}\,\mathrm{ch}\,y_{\mathrm{r}\,2})\,, \end{aligned}$$

where we again used the abbreviated notation $\cosh \equiv \mathrm{ch}$, $\sinh \equiv \mathrm{sh}$. It is well known that

4
$$\begin{aligned} \cosh(y_{\mathrm{r}\,1} + y_{\mathrm{r}\,2}) &= \cosh y_{\mathrm{r}\,1} \cosh y_{\mathrm{r}\,2} + \sinh y_{\mathrm{r}\,1} \sinh y_{\mathrm{r}\,2}\,, \\ \sinh(y_{\mathrm{r}\,1} + y_{\mathrm{r}\,2}) &= \cosh y_{\mathrm{r}\,1} \sinh y_{\mathrm{r}\,2} + \sinh y_{\mathrm{r}\,1} \cosh y_{\mathrm{r}\,2}\,, \end{aligned}$$

which can be also checked using

5
$$\cosh y_{\mathrm{r}} = \frac{e^{y_{\mathrm{r}}} + e^{-y_{\mathrm{r}}}}{2}\,, \qquad \sinh y_{\mathrm{r}} = \frac{e^{y_{\mathrm{r}}} - e^{-y_{\mathrm{r}}}}{2}\,.$$

The combined LT thus have the form

6
$$x'' = x \ \cosh y_{\mathrm{r}} - ct \ \sinh y_{\mathrm{r}}\,, \quad ct'' = ct \ \cosh y_{\mathrm{r}} - x \ \sinh y_{\mathrm{r}}\,,$$

where

$$y_{\mathrm{r}} = y_{\mathrm{r}\,1} + y_{\mathrm{r}\,2}\,. \tag{7}$$

The rapidities add under Lorentz coordinate transformations carried out in the same direction, just like the case with speeds of two Galilean transformations carried out in sequence in the direction of two parallel velocities. This result is often called *addition theorem of rapidities*.

We noted above that the rapidity is a (hyperbolic) rotation between one of the spatial axis and the time axis. Now we have shown that the rapidity is additive just like consecutive rotation angles for rotations around a fixed spatial axis. This will be more visible when we rewrite our results using boost matrices (and using again the abbreviated notation $\cosh \equiv \mathrm{ch},\ \sinh \equiv \mathrm{sh}$)

$$\begin{aligned}
&\Lambda_x(y_{\mathrm{r}\,1})\,\Lambda_x(y_{\mathrm{r}\,2}) = \\
&= \begin{pmatrix} \mathrm{ch}\,y_{\mathrm{r}\,1} & -\mathrm{sh}\,y_{\mathrm{r}\,1} & 0 & 0 \\ -\mathrm{sh}\,y_{\mathrm{r}\,1} & \mathrm{ch}\,y_{\mathrm{r}\,1} & 0 & 0 \\ 0 & 0 & 1 & 0 \\ 0 & 0 & 0 & 1 \end{pmatrix} \begin{pmatrix} \mathrm{ch}\,y_{\mathrm{r}\,2} & -\mathrm{sh}\,y_{\mathrm{r}\,2} & 0 & 0 \\ -\mathrm{sh}\,y_{\mathrm{r}\,2} & \mathrm{ch}\,y_{\mathrm{r}\,2} & 0 & 0 \\ 0 & 0 & 1 & 0 \\ 0 & 0 & 0 & 1 \end{pmatrix} \\
&= \begin{pmatrix} \mathrm{ch}\,y_{\mathrm{r}\,1}\,\mathrm{ch}\,y_{\mathrm{r}\,2} + \mathrm{sh}\,y_{\mathrm{r}\,1}\,\mathrm{sh}\,y_{\mathrm{r}\,2} & -\mathrm{ch}\,y_{\mathrm{r}\,1}\,\mathrm{sh}\,y_{\mathrm{r}\,2} - \mathrm{sh}\,y_{\mathrm{r}\,1}\,\mathrm{ch}\,y_{\mathrm{r}\,2} & 0 & 0 \\ -\mathrm{ch}\,y_{\mathrm{r}\,1}\,\mathrm{sh}\,y_{\mathrm{r}\,2} - \mathrm{sh}\,y_{\mathrm{r}\,1}\,\mathrm{ch}\,y_{\mathrm{r}\,2} & \mathrm{ch}\,y_{\mathrm{r}\,1}\,\mathrm{ch}\,y_{\mathrm{r}\,2} + \mathrm{sh}\,y_{\mathrm{r}\,1}\,\mathrm{sh}\,y_{\mathrm{r}\,2} & 0 & 0 \\ 0 & 0 & 1 & 0 \\ 0 & 0 & 0 & 1 \end{pmatrix} \\
&= \begin{pmatrix} \mathrm{ch}(y_{\mathrm{r}\,1} + y_{\mathrm{r}\,2}) & -\mathrm{sh}(y_{\mathrm{r}\,1} + y_{\mathrm{r}\,2}) & 0 & 0 \\ -\mathrm{sh}(y_{\mathrm{r}\,1} + y_{\mathrm{r}\,2}) & \mathrm{ch}(y_{\mathrm{r}\,1} + y_{\mathrm{r}\,2}) & 0 & 0 \\ 0 & 0 & 1 & 0 \\ 0 & 0 & 0 & 1 \end{pmatrix} = \Lambda_x(y_{\mathrm{r}\,1} + y_{\mathrm{r}\,2})\,.
\end{aligned} \tag{8}$$

We have obtained the addition theorem of rapidities using the method of matrix multiplication as we found previously. Arguably, the matrix multiplication method is a more effective path, and it demonstrates the relationship between spatial rotations and boosts. ◀

Example 7.15

Shuttlecraft rescue: rapidity method

We return here to the Example 7.5 employing instead of speeds/velocities the associated rapidities. We recall that in the *Star Wars* scene the insurrection base reports to its shuttle traveling with $u_{\rm s} = 0.3\,c$ that it is being chased by a space torpedo T following it with $u_{\rm T} = 0.6\,c$, and their rescue spaceship is approaching from the opposite direction traveling with the velocity $u_{\rm S} = -0.6\,c$; all velocities are parallel. What relative speeds are registered in the shuttlecraft? Are these consistent with Example 7.5? ◀

▶ **Solution** We first establish the appropriate rapidities using Eq. (7.31). We find:

1
$$y_{\rm r\,T} = 0.5 \ln \frac{1.6}{0.4} = 0.693\,, \qquad y_{\rm r\,s} = 0.5 \ln \frac{1.3}{0.7} = 0.310\,,$$
$$y_{\rm r\,S} = 0.5 \ln \frac{0.4}{1.6} = -0.693\,.$$

The values presented are with respect to the base, the rapidities of all four bodies are depicted in Fig. 7.5. A difference between any two values is the relative rapidity which is of relevance physically.

We use this now: We boost to the shuttle frame. The spaceship S approaches the shuttle with rapidity

2 $$y_{\rm r\,Ss} = y_{\rm r\,S} - y_{\rm r\,s} = -1.003\,, \qquad \to v_{\rm Ss} = 0.763\,c\,,$$

where we use Eq. (7.29) to compute velocity v. However, the torpedo T is gaining on the shuttle with rapidity

3 $$y_{\rm r\,Ts} = y_{\rm r\,T} - y_{\rm r\,s} = 0.383\,, \qquad \to v_{\rm Ts} = 0.366\,c\,,$$

so rescue is possible but not assured. We check if shuttlecraft crew computations are correct by boosting with $v = v_{rs} = -0.366\,c$ from the shuttle frame to the torpedo T frame of reference. We obtain the rapidity of the spaceship S seen from the torpedo, independent of what the shuttle is doing

4 $$y_{\rm r\,ST} = y_{\rm r\,Ss} - y_{\rm r\,Ts} = -1.003 - 0.383 = -1.386\,, \qquad \to v_{\rm ST} = 0.88\,c\,.$$

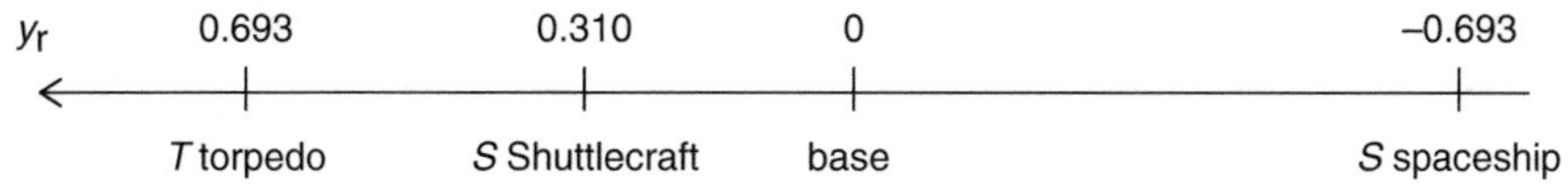

Fig. 7.5 The rapidities of the three speeding space bodies with respect to the Base at zero rapidity: T = (space)torpedo, s = shuttlecraft and S = spaceship. See Example 7.15

This is the expected result

5 $y_{\mathrm{r\,ST}} = y_{\mathrm{r\,S}} - y_{\mathrm{r\,T}} = -0.693 - 0.693 = -1.386\,,$

since trivially

6 $y_{\mathrm{r\,ST}} = y_{\mathrm{r\,Ss}} - y_{\mathrm{r\,Ts}} = y_{\mathrm{r\,S}} - y_{\mathrm{r\,s}} - (y_{\mathrm{r\,T}} - y_{\mathrm{r\,s}}) = y_{\mathrm{r\,S}} - y_{\mathrm{r\,T}}.$

These results demonstrate how the additivity of rapidity can help us understand the relative relativistic motion and simplify the solution path. The advantage of the use of rapidity in the *Star Wars* context, and for that matter all science fiction conforming to Einstein's relativity, is now evident. However, this 'rapidity' variable has yet to be discovered by movie makers. ◀

Example 7.16

Lorentz coordinate transformations in a sequence: factor γ_{12}
Obtain using rapidity the Lorentz factor γ_{12} for two Lorentz coordinate transformations carried out in sequence along the same axis. ◀

▶ **Solution** We take advantage of the fact that rapidities of these two transformations add

1 $\gamma_{12} = \cosh y_{12} = \cosh(y_{\mathrm{r}\,1} + y_{\mathrm{r}\,2})\,.$

Recalling $\cosh(a + b) = \cosh a \cosh b + \sinh a \sinh b$, we obtain

2 $\gamma_{12} = \cosh y_{\mathrm{r}\,1} \cosh y_{\mathrm{r}\,2}(1 + \tanh y_{\mathrm{r}\,1} \tanh y_{\mathrm{r}\,2})\,.$

We use Eqs. (7.29) and (7.33) to obtain

3 $\gamma_{12} = \gamma_{\mathrm{r}\,1}\gamma_{\mathrm{r}\,2}(1 + \beta_{\mathrm{r}\,1}\beta_{\mathrm{r}\,2})\,.$

Naturally, this result follows in a more cumbersome computation using velocity addition and evaluating γ_{12} explicitly, as we have shown deriving Eq. (7.23). However, the use of rapidity simplified the evaluation considerably. ◀

Part IV

Measurement of Body Properties

In Part IV: We study the relation of relativistic Lorentz coordinate transformation (LT) to the process of the measurement of body properties. The consistency of body properties, i.e. time dilation and body contraction with LT, is shown. Different ways to perform time and space measurements are studied and shown to produce consistent results. When a body transits from one inertial reference frame to another by application of an acceleration, the history of body contraction can be recorded.

Introductory Remarks to Part IV

We explore the consistency of the Lorentz coordinate transformation with the relativistic body properties, time dilation and the Lorentz-FitzGerald body contraction. The objective is to introduce the definition of spatial separation measurement which, when combined with the coordinate transformation, produces time dilation and the Lorentz-FitzGerald body contraction, assuring consistency of body properties with coordinate transformations.

We introduce a graphic method to characterize the relation of Lorentz coordinate transformation to the process of the measurement of body properties such as time dilation and the Lorentz-FitzGerald body contraction. Regarding the measurement of time we describe the consistency of time dilation with the Lorentz coordinate transformation. We study how the time difference between events depends on the observer's frame of reference and show how a traveling clock results in the phenomenon of time dilation.

A clock attached to a body scores the time always at the one and the same body location. This clock measures accumulated body proper time that is dilated by any motion imparted on the body, which makes the time dilation effect 'real', and has been experimentally observed.

Temporal simultaneity of two events can only occur in one particular observer reference frame, which creates a unique measuring process. We consider the ends of a moving body, which simultaneous measurement produces within the context of the Lorentz coordinate transformation a consistent interpretation of the Lorentz-FitzGerald body contraction.

Since a body does not on its own retain the record of the Lorentz-FitzGerald body contraction, and nobody has built a body-contraction 'clock', some think body contraction is not real. However, we show that a contraction clock can be in principle constructed. This demonstrates that the body contraction is as real as is the time dilation.

John S. Bell adopted an example of two independently propelled identical rockets that maintain their spatial separation distance as measured at their start. Tracking separation of two such beacon rockets floating within a vacuum chamber within the body, it is possible to preserve in this separation the original unit of length which is non-material and thus not subject to the Lorentz-FitzGerald body contraction. We can imagine that these rockets contained in a vacuum chamber float within the larger ship, duplicating exactly the acceleration processes. This observation allows the construction of a device that records the Lorentz-FitzGerald body contraction as it occurs in the moving body.

This device is a Lorentz-FitzGerald body contraction 'clock' which can record the history of body contraction along with the associated process of acceleration. In a similar way we can keep record of the time dilation of rocket proper time compared to laboratory time. To achieve this we imagine that these two beacon rockets could also be bouncing a light signal between them, thus allowing the comparison of the proper time in the rocket to a preserved standard unit of time, allowing the determination of the time dilation as it occurs.

We learn: The measurement of the Lorentz-FitzGerald body contraction and/or time dilation *as it occurs* means that body property changes caused by a (gentle) acceleration are being observed at the time of their creation. These body changes arise while the body is accelerated. The principle of relativity assures that we can also obtain this result just considering the relative speed of the bodies as was shown by Albert Einstein.

8 Time Measurement and Lorentz Coordinate Transformations

Abstract

We introduce graphic representation of the Lorentz coordinate transformation. The consistency of time dilation with LT is illustrated, and we show: (1) The time difference between events depends on the observer's frame of reference; (2) A traveling clock always ticks slower in its rest-frame compared to any laboratory clock; this is called time dilation; (3) For two events temporal simultaneity can only occur in one particular observer reference frame.

8.1 Graphic Representation of Lorentz Transformation

We consider an observer at rest in the frame of reference S, in which a clock moves in the direction of the x-axis with $\boldsymbol{v} = v\hat{i}$. This movement is depicted by the line ℓ in the space-time diagram Fig. 8.1. ℓ is the world line of the clock.

The world line ℓ creates an angle ϑ with the x-axis, and we have

$$\cot\vartheta = \frac{x - x_1}{ct} = \frac{v}{c}\,,$$

and therefore

$$\vartheta = \operatorname{arcot}\frac{v}{c}\,. \tag{8.1}$$

Since $ct > vt$, the world line of any object must have a slope steeper than 45°. With 90° representing a clock at rest, the range for allowed motion is from left to right $135° > \vartheta > 45°$.

Let S_0 be the proper coordinate system in which a clock is at rest. S_0 has the velocity $+\boldsymbol{v}$ with respect to S. We now find the new coordinates ct_0 and x_0 and draw

J. Rafelski, *Modern Special Relativity*, https://doi.org/10.1007/978-3-030-54352-5_8

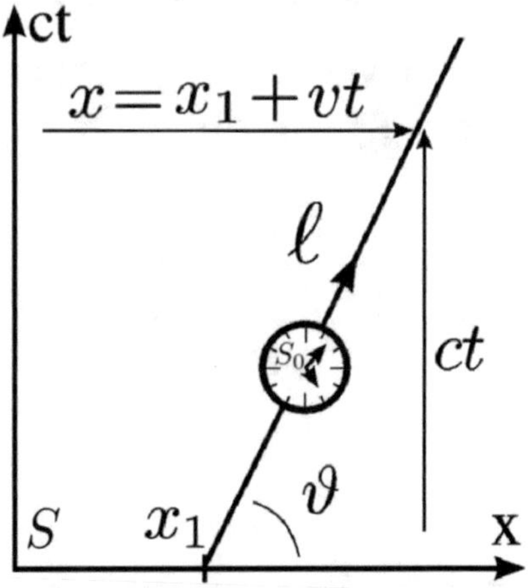

Fig. 8.1 The moving clock as observed in the system S

these along with the coordinate axes ct and x. The proper coordinates of the clock in S are $ct_0,\ x_0$.

In order to incorporate this new coordinate system in our original diagram Fig. 8.1, we draw the t_0-axis parallel to ℓ, as the clock measures the time t_0, see Fig. 8.2. The x_0-axis will be where the coordinate $t_0 = 0$. The LT relation, which describes the coordinate time t_0 in terms of ct and x, is

$$ct_0 = \frac{ct - (v/c)x}{\sqrt{1-(v/c)^2}} = \gamma(ct - \beta x)\ . \tag{8.2}$$

Setting $t_0 = 0$ in Eq. (8.2) yields the line

$$ct = \frac{v}{c}x\ , \tag{8.3}$$

which describes the spatial x_0-axis in the $(ct,\ x)$ diagram of system S depicted in Fig. 8.2. The x_0-axis and the x-axis make an angle φ, which is given by

$$\varphi = \arctan\frac{v}{c} = 90^\circ - \vartheta\ . \tag{8.4}$$

The last relation follows from Eq. (8.1) and can be recognized inspecting Fig. 8.2. We further see that the axes ct_0 and x_0 of the system S_0 lie within the axes ct, x of the system S. The ct_0 axis is always above the 45° boundary of the 'light cone' (dashed), and the x_0 below, see Fig. 8.2.

We now consider an event E_1 located on the world line ℓ of the clock: for example, the striking of the hour. We obtain the coordinates of this event in both systems by projecting parallel to the axes of the frame of reference as shown in Fig. 8.3a. In system S the event has coordinates x_{E_1} and t_{E_1}; in S_0 the event has the coordinates x_{0E_1} and t_{0E_1}. To clarify, the projection of the position onto the x_0 axis

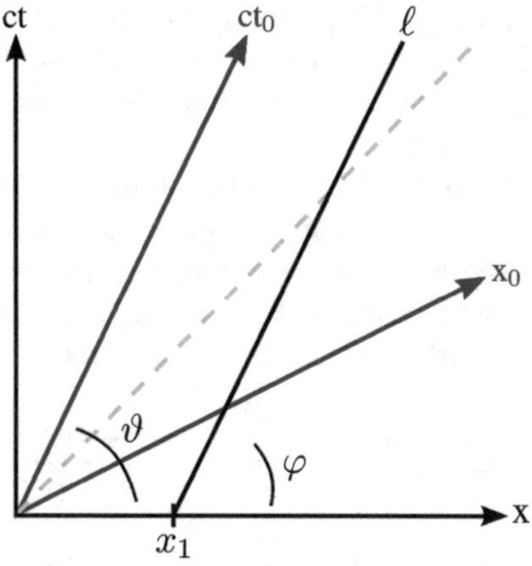

Fig. 8.2 The t_0-axis of the rest-frame of the clock S_0 is parallel to ℓ, the x_0-axis inclines with reference to x-axis in an opposite manner compared to the t_0-axis, see text

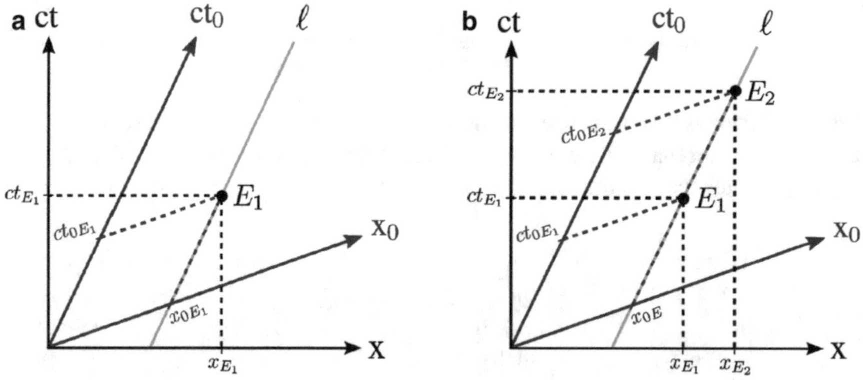

Fig. 8.3 (**a**): The coordinates of event E_1 in systems S and S_0. (**b**): Two events E_1 and E_2 occurring at rest in S_0, i.e., with $\Delta x_0 = 0$ are shown. Coordinate projections in systems S_0 and S are shown by dashed lines

occurs in the direction of the ct_0 axis, and the spatial position of the clock must be constant in S_0 where the clock is at rest.

8.2 Time Dilation and Simultaneity

Time Dilation

We consider the same clock and let it strike twice, creating the events E_1 and E_2 shown in Fig. 8.3b. In S the clock has moved between the two events and the new clock coordinates are x_{E_2} and t_{E_2}. In contrast, in system S_0 only the time between events has moved forward, $t_{0E_2} > t_{0E_1}$; the event location has not changed, since S_0 is the rest-frame of the clock. The events occur at the coordinates $x_{0E_2} = x_{0E_1} \equiv x_{0E}$; that is, $\Delta x_0 \equiv x_{0E_2} - x_{0E_1} = 0$. Two events occurring at the same location at different times are called *timelike*.

For an observer at rest in system S, the time difference $t_{E_1} - t_{E_2}$ is, in general, different from the time difference $t_{0E_1} - t_{0E_2}$, as is clearly seen in Fig. 8.3b. To determine the relationship between these two time intervals we use the inverse Lorentz coordinate transformation Eq. (6.30), where S moves with velocity $-v$ relative to S_0. The time coordinates of the two events are

$$t_{E_1} = \frac{t_{0E_1} + (v/c^2)x_{0E}}{\sqrt{1-(v/c)^2}}\,, \qquad t_{E_2} = \frac{t_{0E_2} + (v/c^2)x_{0E}}{\sqrt{1-(v/c)^2}}\,. \tag{8.5}$$

Because the two events occur at the same location x_0 in S_0, when we form the difference, the last terms in the two forms Eq. (8.5) cancel. The relation between the two time differences as measured in S and S_0 is

$$t_{E_2} - t_{E_1} = \frac{t_{0E_2} - t_{0E_1}}{\sqrt{1-(v/c)^2}} = \gamma(t_{0E_2} - t_{0E_1})\,. \tag{8.6}$$

This result shows that the proper (measured at rest in a body) time interval $t_{0E_2} - t_{0E_1}$ is always shorter than the time interval measured in any other frame S. The Lorentz factor γ relates the two time interval measurements.

Lesson 1: The time difference between timelike events is recognized to be dependent on the observer's frame of reference. Since γ is positive the sequence of events remains unchanged.

We have found the time dilation as discussed earlier, see in particular Chap. 4. Here we recognize how time dilation originates in:

(i) a measurement prescription defining the proper time of a body, combined with
(ii) the relativistic coordinate transformation.

We have thus established the consistency of time dilation with the Lorentz coordinate transformation.

Who among two observers will be younger when they meet again requires evaluation of the full travel history following synchronization of their watches, which circumstance we already have mentioned before, see Sects. 4.1 and 6.1. We return to this matter in greater detail in Sect. 12.2. Here we remember that according to Langevin a measurement of time dilation cannot be reciprocated; one of the observers will always be younger when they reunite, except if they always traveled together. This leads on to the insight:

Lesson 2: A traveling clock ticks in its rest-frame slower compared to any laboratory clock; the effect is called time dilation.

P. Langevin writes:[1]

> The time interval between two events which succeed one another and are coincident in space for a certain system of reference, is less for this one than for any other inertial system in any relative motion to it.

Simultaneity

Let us now consider two other events E'_1 and E'_2 as shown in Fig. 8.4. These events are now set to be simultaneous in the frame S_0; that is, $\Delta t_{0\,E'} \equiv t_{0\,E'_1} - t_{0\,E'_2} = 0$. These events could for example correspond to the striking of two clocks separated by a fixed distance and represented by world lines ℓ_1 and ℓ_2 in Fig. 8.4. Two simultaneous events occurring at different locations are called *spacelike*. In the following Chap. 9 we consider such a simultaneous measurement of the ends of a moving body and show that the Lorentz-FitzGerald body contraction is consistent under this measurement with the Lorentz coordinate transformation.

The projection onto the ct and ct_0 axes in Fig. 8.4 shows that although these events are simultaneous in frame S_0, they cannot be simultaneous in S. We now ask how two simultaneous events in S_0 will be observed in S, and specifically, what is the time difference recorded in S seen in Fig. 8.4. To answer this question we use as in Eq. (8.5) the inverse Lorentz coordinate transformation Eq. (6.30). The two events occur at the following times

$$t_{E'_1} = \frac{t_{0E'} + (v/c^2)x_{0E'_1}}{\sqrt{1-(v/c)^2}}\,, \qquad t_{E'_2} = \frac{t_{0E'} + (v/c^2)x_{0E'_2}}{\sqrt{1-(v/c)^2}}\,. \tag{8.7}$$

Because the two events occur at the same time $t_{0E'}$ in S_0, when we form the time difference measured in S, the first terms in two forms Eq. (8.7) cancel.

$$t_{E'_2} - t_{E'_1} = (v/c^2)\frac{x_{0E'_2} - x_{0E'_1}}{\sqrt{1-(v/c)^2}} = (v/c^2)\gamma(x_{0E'_2} - x_{0E'_1})\,. \tag{8.8}$$

[1]In original (see Ref. [18]): "L'intervalle de temps entre deux événements qui coincident dans l'espace, qui se succèdent en un même point pour un certain système de référence, est moindre pour celui-ci que pour tout autre en translation uniforme quelconque par rapport au premier."

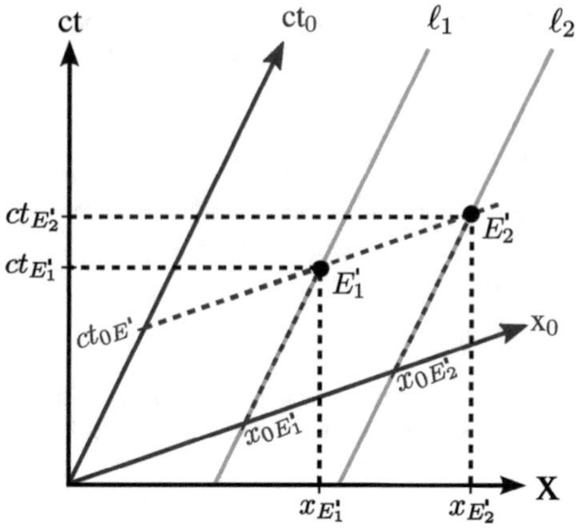

Fig. 8.4 Two events E_1' and E_2', simultaneous in S_0, i.e. with $\Delta t_{0\,E'} = 0$, are shown. Coordinate projections in systems S_0 and S are shown by dashed lines

We note, comparing Fig. 8.3b with Fig. 8.4, that there are two different measurements of time difference corresponding, respectively, to timelike, Eq. (8.6), and spacelike, Eq. (8.8), event separation.

Even if two spacelike events occur simultaneously in one reference frame, they will always be observed as non-simultaneous in any other reference frame. We further note that now the sign of the observed time difference in Eq. (8.8) is not fixed. We will return to this topic in Chap. 12 and in Example 12.1.

Lesson 3: Two spacelike events are measured to be simultaneous in one special reference frame. This generates a unique measurement prescription for such events.

Example 8.1

Coordinate time units in S_0 and S

We see in Fig. 8.3b that the time increment between two events measured in S_0, $\Delta t_0 = t_{0E_2} - t_{0E_1}$, are represented by a longer geometric distance compared to $\Delta t = t_{E_2} - t_{E_1}$ in S. However, we have shown in Eq. (8.5), that Δt_0 is shorter, $\Delta t_0 \leq \Delta t$. This implies that the time units, for example a second, is represented by different geometric line increments, with the increment e_0 on the ct_0-axis being larger compared to the unit of length e on ct-axis. What is the relationship between e_0 and e? ◀

▶ **Solution** Upon the co-location of t_{E_1} with t_{0E_1} shown in Fig. 8.3b, arises the triangle in Fig. 8.5. We recall Eq. (8.6)

$$\mathbf{1}\quad \Delta t = \frac{\Delta t_0}{\sqrt{1 - v^2/c^2}} \,.$$

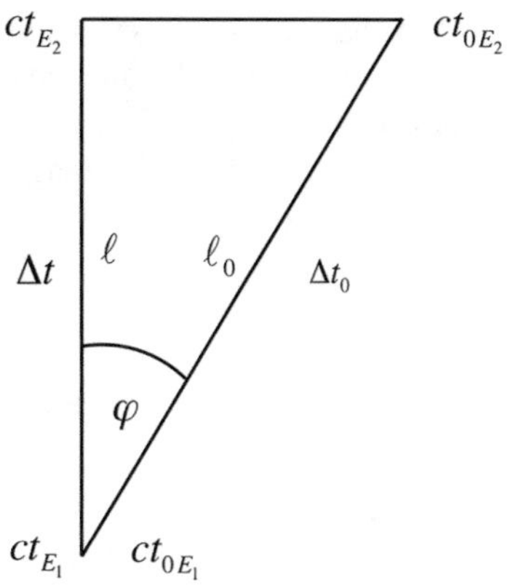

Fig. 8.5 Time units in two inertial systems, see Example 8.1

The lengths ℓ and ℓ_0 seen in Fig. 8.5 can be stated using the units e_0 on the ct_0 and e on the ct-axis respectively.

2 $\ell_0 = e_0 c \Delta t_0 , \qquad \ell = e c \Delta t ,$

and thus

3 $$\ell = e \frac{c\Delta t_0}{\sqrt{1 - v^2/c^2}} = \frac{e}{e_0} \frac{e_0 c \Delta t_0}{\sqrt{1 - v^2/c^2}} = \frac{e}{e_0} \frac{\ell_0}{\sqrt{1 - v^2/c^2}} .$$

On the other hand we know the geometric length relation between ℓ and ℓ_0

4 $$\ell_0 = \frac{\ell}{\cos \varphi} = \ell \sqrt{1 + \tan^2 \varphi} .$$

Considering Fig. 8.1 again we further recognize

5 $$\tan \varphi = \frac{v}{c} , \quad \text{or} \quad \ell_0 = \ell \sqrt{1 + v^2/c^2} .$$

This result is evident considering the triangle in Fig. 8.5.

We combine these results to obtain for the ratio of time units in S_0 and S

6 $$\frac{e_0}{e} = \frac{\sqrt{1 + v^2/c^2}}{\sqrt{1 - v^2/c^2}} .$$

This ratio $e_0/e \geq 1$ incorporates both the effect of the longer optical path (Lorentz factor $1/\sqrt{1 - v^2/c^2}$), which we recognized and described in Sect. 4.2 below Eq. (4.5), as well as the effect of the geometric projection (factor $\sqrt{1 + v^2/c^2}$). An analogous study of the units on the x and x_0-axes, respectively, shows that the last result applies to this case. This insight is also obtained remembering that the world line of a light pulse emitted at time axes origin $t = t_0 = 0$ forms in both coordinate systems S_0 and S the same angle with time and space coordinates.

The changes of scales we described are known from geography: For example, the Mercator-projection retains the correct angular relations at the cost of introducing scale factors except at the equator. This effect is particularly visible in the Mercator representation of the pole areas. ◀

Different Methods of Measuring Spatial Separation

9

Abstract

We observe signals emitted from the ends of a material body. This allows us to perform the evaluation of spatial separation between these ends in the observer's reference frame, as well as in the rest-frame of the body. We show that this measurement is consistent with the Lorentz-FitzGerald body contraction only if we measure the spatial separation at equal time in an observer's reference frame. In the third example the spatial separation of body ends is measured by applying an illumination signal from the rest-frame of the observer.

9.1 Introductory Remarks

The process of measuring the spatial separation interval Δx between two events is considerably more involved than time measurement and we need to establish a clear operational approach at the outset. There are two causes for this circumstance:

1. When we discuss spatial separation of events, time separation must be also considered. The temporal simultaneity of a measurement in the reference frame of the observer will be identified as a defining constraint for the measurement of spatial separation of events.
2. As we begin our discussion we cannot propose a device one could call a body contraction of body length 'clock'. However, based on our findings in this and in the following chapter, we will be able to describe in Sect. 10.3 how it is in principle possible to build such a length 'clock'.

We begin with exploration of the Lorentz-FitzGerald body contraction using the Lorentz coordinate transformation. We now assume that the observer is at rest in the system S. The observed body is at rest in system S_0. Our objective is to determine the magnitude of the spatial separation between events characterizing the two ends

J. Rafelski, *Modern Special Relativity*, https://doi.org/10.1007/978-3-030-54352-5_9

of the observed body. We discuss the following three different possible methods of measurement:

(a) The body's ends are lit permanently and we photograph them with a camera located in system S. This means that the observation is simultaneous in S, but not in S_0.
(b) The ends of the body light up for a brief instant in the body rest-frame, and their pictures are taken by a camera that has a permanently open lens. The measurement of any two points on the body is therefore simultaneous in S_0 but not simultaneous in S.
(c) We place mirrors at the two body ends and illuminate the body for a short instant by a flash originating in a location symmetric with respect to the ends of the to be observed body. The light is scattered and returns from the two body ends to a camera located in S with a permanently open objective.

In the examples presented in Sect. 9.2 we record events where they occur – every reference system is equipped everywhere with a proper clock. Therefore we do not need to consider the propagation time of signals. However, the return travel time of light emitted by the observer in case (c) is implicit in this measurement.

9.2 Determination of Spatial Separation

Signal Synchronized in S, the Rest-Frame of the Observer

Case (a), Sect. 9.1, is depicted in Fig. 9.1. The camera's aperture opens and closes instantaneously at time $t_E \equiv t_{E_1} = t_{E_2}$ in system S. The incoming observed light is therefore sent out from system S_0 at different times t_{0E_1} and t_{0E_2}. We define the spatial separation in the two frames of reference

$$l_0 = x_{0E_2} - x_{0E_1} , \qquad l^{\mathrm{sim}} = x_{E_2} - x_{E_1} ,$$

where the superscript 'sim' reminds us in which frame of reference the measurement is simultaneous.

The Lorentz coordinate transformation Eq. (6.29) yields

$$x_{0E_1} = \frac{x_{E_1} - vt_E}{\sqrt{1-(v/c)^2}} , \qquad x_{0E_2} = \frac{x_{E_2} - vt_E}{\sqrt{1-(v/c)^2}} .$$

Taking the difference of the two equations we obtain

$$l_0 = \frac{l^{\mathrm{sim}}}{\sqrt{1-(v/c)^2}} , \quad \text{or} \quad l^{\mathrm{sim}} = l_0\sqrt{1-\left(\frac{v}{c}\right)^2} . \tag{9.1}$$

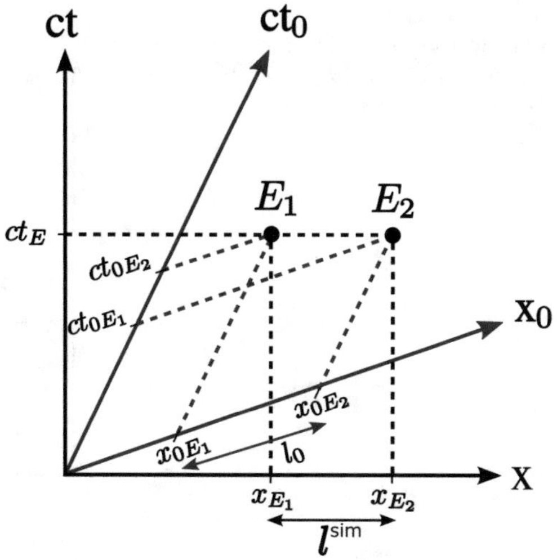

Fig. 9.1 Case (a): Simultaneous measurement of events E_1 and E_2 in S, the rest-frame of the observer

The result: The distance l^{sim} measured according to method (a) is shorter than the distance l_0.

Lesson 4a: The spatial separation l^{sim} of two events evaluated in the frame $S = S^{\text{sim}}$ in which the two events are simultaneous is shorter in the direction of motion by the factor $\sqrt{1-(v/c)^2} = \gamma^{-1}$ when compared to the same spatial separation measurement carried out in a non-simultaneous manner in a body proper frame S_0 moving with the speed v.

A body is introduced in Lesson 4a in order to demonstrate that a measurement prescription is compatible with the body contraction. Even though event separation measurement could be seen as being independent of a material body, it would be necessary to move position lights or mirrors to a specific event location in order to imitate the behavior of a material body.

Signal Synchronized in S_0, the Rest-Frame of a Body

We consider the case (b), Sect. 9.1, illustrated in Fig. 9.2. The events E_1 and E_2 occur at the two ends of a body K, emitting a light flash simultaneously in the body rest-frame S_0 at the same time $t_{0\,E} \equiv t_{0\,E_1} = t_{0\,E_2}$. The two events are not simultaneous in S.

The distance between the two events in Fig. 9.2 is

$$l = x_{E_2} - x_{E_1}\,, \qquad l_0^{\text{sim}} = x_{0E_2} - x_{0E_1}\,.$$

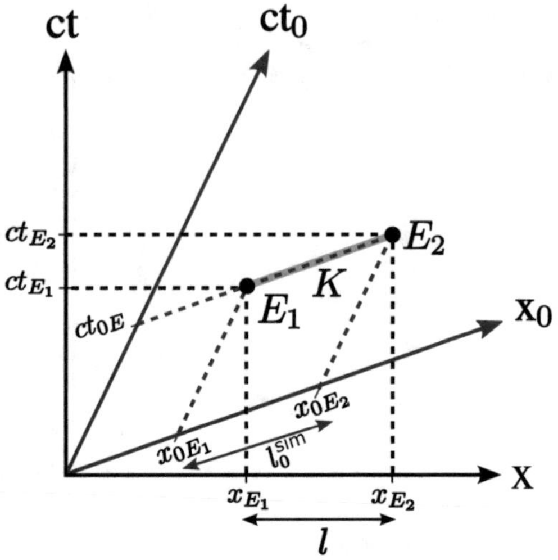

Fig. 9.2 Case (b): Simultaneous measurement of two events E_1 and E_2 inS_0, the moving reference frame of the body

We determine x_{E_2} and x_{E_1} using the inverse Lorentz coordinate transformation Eq. (6.30)

$$x_{E_1} = \frac{x_{0E_1} + vt_{0E}}{\sqrt{1-(v/c)^2}}\,, \qquad x_{E_2} = \frac{x_{0E_2} + vt_{0E}}{\sqrt{1-(v/c)^2}}\,.$$

By taking the difference we obtain a result with same physical content as Eq. (9.1)

$$l = \frac{l_0^{\text{sim}}}{\sqrt{1-(v/c)^2}}\,, \quad \text{or} \quad l_0^{\text{sim}} = l\sqrt{1-\left(\frac{v}{c}\right)^2}\,. \tag{9.2}$$

Lesson 4b: The spatial separation l_0^{sim} of two events evaluated in the frame S^{sim} in which the two events are simultaneous, even if this is the rest-frame S_0 of the moving body, is shorter by the factor $\sqrt{1-(v/c)^2} = \gamma^{-1}$ when compared to such non-simultaneous separation measurement carried out by an observer moving with the speed v with reference to S^{sim}.

A body is introduced in Lesson 4b as in Lesson 4a in order to demonstrate that a measurement prescription is compatible with the body contraction. Even though event separation measurement could be seen as being independent of a material body, it would be necessary to move position lights or mirrors to a specific event location in order to imitate the behavior of a material body.

Body Length in Einstein's Didactics

We interpret the separation between two events as body length. What we call l_0 or l is, as we have seen, not relevant. Thus the textbook statement $l = l_0\sqrt{1-(v/c)^2}$ must be replaced by the more accurate

$$l^{\text{sim}} = l\sqrt{1-\left(\frac{v}{c}\right)^2} . \tag{9.3}$$

The superscript 'sim(ultaneous)' reminds us how the measurement of the contracted body length is carried out.

This clarifies that (i) absolute speed is not measurable, but only the speed with respect to a unique observer 'sim', and, (ii) any two observers claiming that the other is Lorentz-FitzGerald body contracted are performing two very different measurements in regard to the treatment of time. This resolves the frequently encountered interpretational inconsistency in regard to the Lorentz-FitzGerald body contraction. The conclusion is:

Lesson 5: Relevant for the understanding of the Lorentz-FitzGerald body contraction is the recognition of the reference frame in which the events that define body ends are measured in simultaneous manner. This reference system is characterized by superscript 'sim'. The spatial separation between two events l^{sim} measured in the simultaneous manner is always shorter compared to any other event spatial separation measured at unequal times.

Einstein interpreted the Lorentz-FitzGerald body contraction in this manner: The body contraction in Einstein didactics does not require the existence of a body that is contracted. It relies on coordinate transformations. John S. Bell in a letter to the author (see Preface to this book) calls this approach *perfectly sound, and very elegant and powerful, but pedagogically dangerous.*

Example 9.1

Special character of the equal time observer

In Sect. 7.1 we introduced the invariance of $(\Delta s)^2$. Use this to demonstrate that the S^{sim} observer always will see a contracted event separation. ◀

▶ **Solution** The invariance under the Lorentz coordinate transformation, Eq. (7.5), can be written as

1 $(\Delta \boldsymbol{x})^2 - (\Delta ct)^2 = (\Delta \boldsymbol{x}')^2 - (\Delta ct')^2$.

The two event separations $\Delta \boldsymbol{x}$, Δct and $\Delta \boldsymbol{x}'$, $\Delta ct'$ are measured in the reference frames S and S' respectively. We orient the coordinate x- and x'-axes in the direction of relative motion between S and S'. Transverse to the direction of motion we have $y = y'$ and $z = z'$, which simplifies the RHS above.

Let $S \to S^{\rm sim}$ be the equal time observer; that is, $\Delta ct = 0$. We then have the identity

2 $(\Delta \boldsymbol{x}^{\rm sim})^2 = (\Delta \boldsymbol{x}')^2 - (\Delta ct')^2$.

We have now demonstrated unequivocally the special character of the equal time observer $S^{\rm sim}$: $\Delta \boldsymbol{x}^{\rm sim} < \Delta \boldsymbol{x}'$ where S' is any other observer for whom $\Delta ct' \neq 0$.

Caution: If you are interested in rederiving the Lorentz-FitzGerald body contraction Eq. (9.3), you need to compute $\Delta ct'$ with the help of Lorentz transformation, that is

3 $(\Delta ct')^2 = \gamma^2 \beta^2 (\Delta \boldsymbol{x}')^2$,

since $\Delta ct = 0$. Solving for $\Delta \boldsymbol{x}^{\rm sim}$

4 $\Delta \boldsymbol{x}^{\rm sim} = \Delta \boldsymbol{x}' \sqrt{1 - v^2/c^2}$.

Only this approach leads to the appropriate relationship between $\Delta \boldsymbol{x}^{\rm sim}$, $\Delta \boldsymbol{x}'$ and the relative velocity of the two reference systems.

If and when writing

5 $(\Delta \boldsymbol{x}^{\rm sim})^2 = (\Delta \boldsymbol{x}')^2 \left(1 - \frac{(\Delta ct')^2}{(\Delta \boldsymbol{x}')^2}\right)$,

you should not consider the ratio $\Delta x'/\Delta t'$ as a velocity, since the space-time intervals $\Delta x'$, $\Delta ct'$ are not related to relative motion of two reference frames. ◀

9.3 Spatial Separation Measurement by Illumination from the Rest-Frame of the Observer

Finally we consider the more complex case (c), as illustrated in Fig. 9.3. Note that the body is at rest in S_0 as the body ends project onto values $x_{0\,E_1}$ and $x_{0\,E_2}$ for all time ct_0. The flash of light occurs at $t = 0$ and $x = 0$ in S and propagates along the path $x = ct$, illuminating the moving body at different times and positions. In both frames of reference the observation of the body is made at unequal times; that is, $\Delta t = t_{E_2} - t_{E_1} \neq 0$ and $\Delta t_0 = t_{0E_2} - t_{0E_1} \neq 0$, see Fig. 9.3.

The light reaches the ends of the body at events E_1 and E_2 clearly at different times in the two frames S_0 and S. However, $x_{E_1} = ct_{E_1}$ and $x_{E_2} = ct_{E_2}$. We insert these constraints into the Lorentz coordinate transformation Eq. (6.29) and obtain for the positions in S_0

$$x_{0E_1} = \frac{x_{E_1}(1 - v/c)}{\sqrt{1 - (v/c)^2}} , \qquad x_{0E_2} = \frac{x_{E_2}(1 - v/c)}{\sqrt{1 - (v/c)^2}} .$$

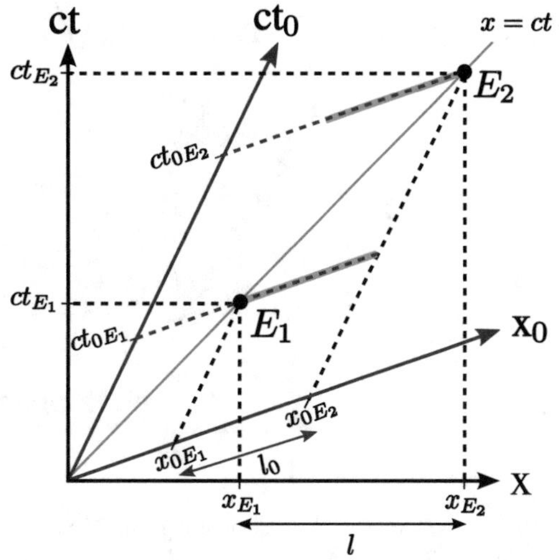

Fig. 9.3 Case (c): Active observation, the moving body is lit with a flash emitted by observer S

By evaluating the difference we find a relation between the event separation l_0 and the laboratory observed event separation l

$$l_0 = l\frac{(1 - v/c)}{\sqrt{1 - (v/c)^2}} = l\sqrt{\frac{1 - v/c}{1 + v/c}}\,. \tag{9.4}$$

This third measurement case is reminiscent of the relativistic Doppler wavelength shift for collinear motion, but conceptually it is not related to it.[1] We return to discuss the relativistic Doppler effect in Chap. 13.

Example 9.2

Quasi-rotation of a moving body for a special observer

A rectangular prism, with edge lengths a_0, b_0, and c_0, moves with a velocity v parallel to the edge a_0 past an observer S at rest. A light shines from each corner of the prism, and the observer S takes a snapshot of it as it flies past. What picture does the observer obtain and how can she interpret it? ◄

► **Solution** The light from each corner must arrive simultaneously in S at the camera, and thus must be sent from the different corners of the prism at different times in order to be recorded simultaneously. Inspection of Fig. 9.4 shows that the light emanating from corner 1 must travel the extra stretch b_0 compared to the light from corner 4.

[1] J. Terrell "Invisibility of the Lorentz Contraction" *Phys. Rev.* **116**, 1041 (1959) writes in the abstract: "Observers photographing the meter stick simultaneously from the same position will obtain precisely the same picture, except for a change in scale given by the Doppler shift ratio."

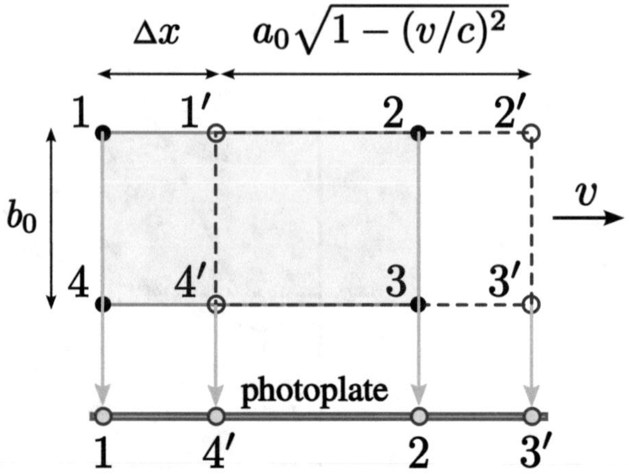

Fig. 9.4 A side view of a rectangular prism. The paths of light originating in corners of the body, the thick dots at 1, 2, 3, 4, and hitting the photo plate are shown. See Example 9.2

Compared to corner 4, the light from corner 1 must be sent out earlier by

1 $\Delta t = \frac{b_0}{c}$.

During this time the rectangular prism moves, as shown in Fig. 9.4, an additional distance

2 $\Delta x = v\Delta t = \frac{v}{c} b_0$.

Thus in order to arrive at the same time on the photo plate, emission occurs from points 1 and 4' and similarly 2 and 3'. We note also that a Lorentz-FitzGerald body contraction applies only in the direction of motion and thus only to the side a_0.

A possible interpretation of the photo plate picture is shown in Fig. 9.5. The corners of the moving rectangular prism appear at locations which are the same as would be seen considering a projection of a prism rotated out of the photo plate plane around an axis in the direction normal to the velocity $\boldsymbol{v}$ and normal to the normal of the photo surface by the angle ϑ, where

3 $\frac{v}{c} = \sin\vartheta$, $\sqrt{1 - \left(\frac{v}{c}\right)^2} = \cos\vartheta$.

This also means

4 $\Delta x = b_0 \sin\vartheta$, $a = a_0 \cos\vartheta$.

The observed quasi-rotation originates in two effects: (i) the time difference required to see signals originating from the back and the front of the moving body simultaneously in observer frame S; (ii) the Lorentz-FitzGerald body contraction.

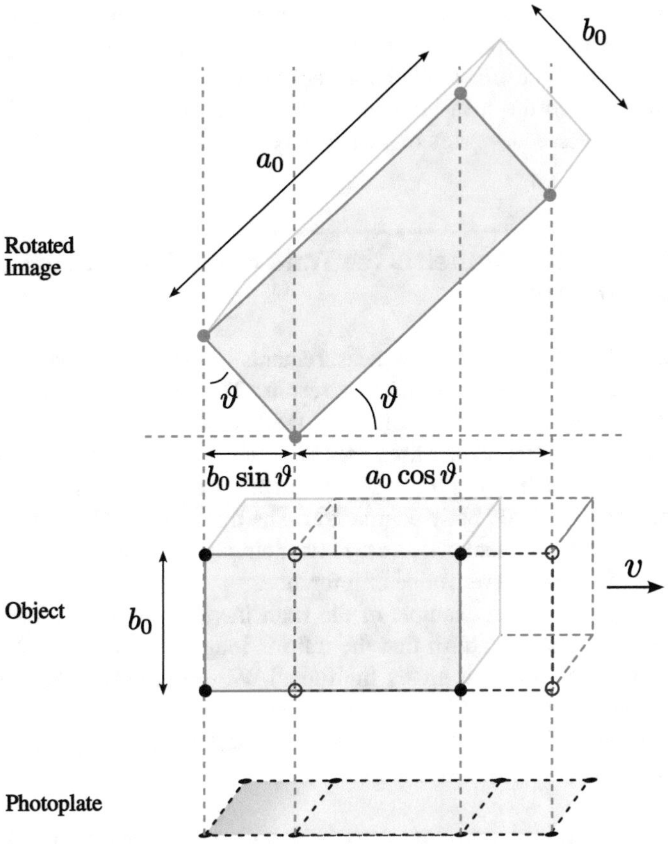

Fig. 9.5 The snapshot of the rectangular prism seen on the photo plate can be construed as a rotation of the prism by ϑ. See Example 9.2

What we observe is clearly dependent on the manner of measurement. The quasi-rotation interpretation with angle given in the above result is specific to the body motion normal to the line of observation. This result thus is found in a specific method of inspection of the body during one particular observational instant depicted in Fig. 9.5. At all other times light travel time is different and the resulting image is different. Consideration of this situation and generalization of our results will not be further studied here.

Further reading: The rotation effect introduced in the article by J. Terrell, see Ref. [1], stimulated considerable discussion and several further presentations, for most readable see contributions by Victor F. Weisskopf[2] and by Mary L. Boas.[3] ◂

[2]V.F. Weisskopf, "The visual appearance of rapidly moving objects," *Physics Today*, pp 24–27 (September 1960).

[3]M.L. Boas, "Apparent shapes of large objects at relativistic speeds," *American J. Physics* **29**, 283 (1961).

We have seen that the Lorentz-FitzGerald body contraction is an input into any process of body measurement, irrespective of the inspection method, which further needs to account for the effect of the finite speed of light. What we observe depends on how we perform the measurement. The change in measuring method which impacts what we see does not impact in any way the Lorentz-FitzGerald body contraction.

9.4 Train in the Tunnel: Is the Train or the Tunnel Contracted or Expanded?

We have considered three different measurements of the distance between coordinate events and obtained three different results. The first two cases describe the Lorentz-FitzGerald body contraction using the Lorentz coordinate transformation and show that the observer for whom the time interval between the two coordinate events vanishes observes the shortest separation of the events, the same as we expect for the Lorentz-FitzGerald body contraction. The third case (c) leads to a different result since the measurement events are neither simultaneous in the source frame of reference, nor in the observer frame of reference.

We now reconsider the example of the train traveling through a gated tunnel. Before the journey we establish that the train is longer than the tunnel. After the train reaches a high speed it enters the tunnel. We observe that the gates at the tunnel ends close at the same time. Will the train fit into the tunnel? This situation can be looked at in both the context of the Lorentz-FitzGerald body contraction and of the Lorentz coordinate transformation.

- Lorentz-FitzGerald body contraction advocates note that the length of the moving train is contracted and the train will fit easily in the tunnel. However, they start to worry when Simplicius points out a seemingly equally valid view of the situation: the principle of relativity should mean that considering a frame of reference where the train is at rest and the tunnel is moving; the tunnel would be Lorentz-FitzGerald body contracted and the train would not fit. Fortunately, we have just seen that there is no trivial reciprocity; it is important to define the measurement process and remember that time in the context of measurement is relevant.
- Coordinate transformation experts note, much to the relief of the concerned Lorentz-FitzGerald body contraction advocates, that in the rest-frame of the train (rather than the tunnel) the front and rear tunnel gates no longer drop down at the same time, and thus when the front of the train reaches the tunnel exit, at the back there is still time to escape the slam of that rear gate.

As we have demonstrated, determining the outcome of measurement is the type of measurement process. In either frame of reference – the train or the mountain – coordinate transformation experts determine exactly the same Lorentz-FitzGerald body contracted length for the train since the measurement process determines which body is contracted. The Lorentz coordinate transformation is therefore consistent with the Lorentz-FitzGerald body contraction.To recognize the train contraction clearly it is best to look at the (slowly) accelerated train from the reference frame of the train station where the tunnel is at rest. Observing the train ends at equal time we see it become shorter and shorter, since it moves faster and faster. This clarifies which measurement process is to be considered.

Simplicius, who claimed that in the rest-frame of the train the tunnel is observed to be contracted, performed a different experiment which has nothing to do with the train traveling through a tunnel. An expanded tunnel so that the train would fit could be measured from the moving train performing a special measurement such that the two ends of the tunnel are observed at equal time in the tunnel rest-frame.

Recall here again the Michelson-Morley experiment, in which we compare two lengths, one contracted (parallel to motion), and the other not (perpendicular to motion). The contraction in the direction of motion is needed to assure that the optical path is equal in all directions irrespective of the motion that the experimental table experiences. This signifies that the relative velocity vector with respect to the æther, the carrier of light waves, is unobservable. No matter what *relative* velocity there is between the experimental table with the MM interferometer and the æther, the outcome of the experiment is null; the magnitude of the relative velocity is unmeasurable. However, it is not the æther which is rendered nonexistent; it is the velocity of the æther that has been shown to be inaccessible to observation, as Einstein explained in 1920, see Sect. 2.3 and Ref. 17 in Chap. 2.

The unobservability of the relative motion with the æther does not contradict the assertion that we can observe the Lorentz-FitzGerald contracted body length; we can directly measure the Lorentz-FitzGerald body contraction as it occurs, e.g., for the train that starts from a station, accelerates, and once at a required speed, fits into the mountain tunnel. In the next chapter we will demonstrate that the classic Lorentz-FitzGerald body contraction of a body that starts to move is indeed a real effect, in that an experimental arrangement can be built to measure the effect of contraction and keep it in memory even after the motion has stopped and the body is no longer contracted.

Discussion 9.1 Body contraction and coordinate transformation

About the topic: We take a second look at how one connects Lorentz coordinate transformation with the Lorentz-FitzGerald body contraction.

- *Simplicius:* Without doubt there must be a way to understand the Lorentz-FitzGerald body contraction using the Lorentz coordinate transformation.
- *Professor:* Of course this is the case. However, it is unfortunate that 'Lorentz' appears twice in this statement as it often creates confusion. Lorentz was seeking a coordinate transformation that would be consistent with his explanation of the MM experiment in terms of the Lorentz-FitzGerald body contraction, yet remain consistent with Maxwell

(continued)

electromagnetism. Ultimately this problem was solved by others. Larmor and Poincaré remembered Lorentz's effort and named the coordinate transformation after him. It would have been far better to avoid the 'Lorentz' confusion given that 'contraction' and 'transformation' are two different concepts. In order to amplify this difference we speak always in full expression when referring to the Lorentz-FitzGerald body contraction.

- *Student:* Lorentz recognized that both effects named today after him must be consistent with each other.
- *Simplicius:* So how does that consistency work for the case of a train in the tunnel?
- *Student:* We measure at the station the train length $L_0 = x_2 - x_1$ with a measuring stick, which means we made simultaneous observations in the station frame of reference of both train ends, measuring at $t_2^{\mathrm{st}} - t_1^{\mathrm{st}} = 0$. For the observer in the train once the train departs the station the train length remains the same since the measurement process is the same. This must be so since an absolute speed is unmeasurable. Another way to explain this is to say that for an observer at station, and more generally observing from outside, both the measuring stick and the train change in the same way.
- *Professor:* The important point here is that a measurement which does not show an effect of contraction is carried out in the train reference frame at $t_2^{\mathrm{train}} - t_1^{\mathrm{train}} = 0$. To notice something happened to the moving train length we must measure insisting on equal time measurement in the station frame of reference, $t_2^{\mathrm{st}} - t_1^{\mathrm{st}} = 0$. Thus we measure in the same way as compared to before the train is set in motion. The result we find is that a train is observed to be contracted by an observer who sees it moving, e.g. observing from the station frame of reference. This measurement is at equal time in the reference frame in which the length of the tunnel remains the same as before.
- *Simplicius:* If the contraction depends on the process of measurement, it seems to me that it is not 'real'.
- *Professor:* No, the contraction is real to the observer at the station. We have presented this discussion to illustrate the consistency with the Lorentz coordinate transformation. We proceeded with extra caution and considered the time difference involved in length measurement. We also need to remember that two different observers will in general obtain different Lorentz-FitzGerald body contractions since the magnitude of the effect depends on relative speed between body and observer.
- *Student:* We have seen before that this is just like with kinetic energy: Two different Galilean observers perceive two different values of the body kinetic energy. The comoving observer does not observe any kinetic energy of a body.

(continued)

- *Professor:* Agreed, and we recall further Einstein derived the Lorentz coordinate transformation from basic principles. The connection and interpretation of a measurement process leading to an understanding of the Lorentz-FitzGerald body contraction in terms of the Lorentz coordinate transformation is another matter. This has created some misunderstandings in Einstein's times, forcing Einstein to clarify how the Lorentz-FitzGerald body contraction is to be understood. In Sect. 3.3 we discussed this topic; let us hear Einstein's words again: "...whether the Lorentz-FitzGerald contraction is a physical phenomenon or not can lead to a misunderstanding. For a comoving observer it is not present and as such it is not observable; however, it is real and in principle observable by physical means by any non-comoving observer."
- *Student:* This quote argues just the way our example with kinetic energy of a body works.
- *Simplicius:* I find it very odd that a young clerk in a patent office came across the solution of such a fundamental riddle that the greatest minds of the epoch Larmor, Lorentz, Poincaré struggled with for many years.
- *Professor:* Lorentz became a great admirer and friend of Einstein's.
- *Simplicius:* But how and why could the young Einstein create SR and in the process find the solution to the Lorentz coordinate transformation riddle?
- *Professor:* My personal view in regard to this often posed question is that Einstein was in a unique, and even, privileged circumstance. He had to evaluate numerous patents written in the wake of Maxwell, Hertz, Edison, Tesla, and others. These patent applicants probably did not always argue their cases correctly. Confronted with many misunderstandings of electromagnetism, facing the need to reconcile the different contrasting views, and acting also out of necessity to do his patent clerk job effectively, Einstein formulated a new way to think about coordinate transformation in Maxwell-Hertz electromagnetism. In his first 1905 relativity paper Einstein acknowledges his lifelong friend Michelangelo Besso, an engineer, whom he had known since 1896, and with whom he walked nearly everyday to work in Bern. I believe that this acknowledgment of Besso is not only of the person, but also of the technical-engineering environment that was Einstein's intellectual home when he created SR. I wonder what would have happened with SR had Einstein been in a different environment?

Example 9.3

Compounding of two Lorentz-FitzGerald body contractions

We seek the relation between two measurements by two different observers of the Lorentz-Fitzgerald body contraction of a body oriented in the direction of motion. The reference frames S_1 and S_2 move with the speeds v_1 and v_2 measured from an observer reference frame S_S. Let $v_2 < v_1$, and the relative speed between

S_1 and S_2 be v_{12}. The observer S_S measures the size $L_1(v_1)$ of a contracted body located in the reference frame S_1. We want to write this in terms of $L_{12}(v_{12})$, the Lorentz contraction reported by the moving observer S_2. The measurements of body length are performed at equal time by each observer. ◀

▶ **Solution** We will rely in our evaluation on the earlier discussion of a train leaving the station and heading for a tunnel. Let the observer S_S be at rest in the train station from which each of the two trains S_i starts accelerating towards a tunnel, achieving velocity v_i, where $i = 1, 2$. Let the second slower train follow the first. The relative speed between S_2, and S_1 is, according to Eq. (7.24)

1 $$\beta_{12} = \frac{\beta_1 - \beta_2}{1 - \beta_1\beta_2}\,, \qquad y_{r\,12} = y_{r\,1} - y_{r\,2}\,.$$

We used above β_i, the velocities measured in units of c, and the additivity of rapidities $y_{r\,i}$ for motion in the same direction. We need below

2 $$\cosh(y_{r\,1} - y_{r\,2}) = \gamma_{12}\,, \quad \tanh(y_{r\,1} - y_{r\,2}) = \beta_{12}\,, \quad \sinh y_{r\,2} = \beta_2\gamma_2\,.$$

The subscripts $r\,1$ and $r\,2$ refer to the rapidities in reference to station S_S.

The proper length L_0 of a body in reference frame S_1 can be computed independently by two equal time observers (here S_S and S_2) who see it in motion at two different rapidities $y_{r\,1}$ and $y_{r\,12}$

3 $$L_0 = \cosh y_{r\,1}\, L_1\,, \quad L_0 = \cosh y_{r\,12}\, L_{12}\,, \Rightarrow \cosh y_{r\,12}\, L_{12} = \cosh y_{r\,1}\, L_1\,.$$

The length of train '1' measured by the observer S_2 whose speed (rapidity) with reference to the train station is known can now be written in the form

4 $$L_{12} = \frac{\cosh(y_{r\,1} - y_{r\,2} + y_{r\,2})}{\cosh(y_{r\,1} - y_{r\,2})}\, L_1\,,$$

where we extended trivially the argument of cosh in the nominator. Proceeding as in Example 7.16 we obtain

5 $$L_{12} = [\cosh y_{r\,2} + \tanh(y_{r\,1} - y_{r\,2}) \sinh y_{r\,2}] L_1\,.$$

Reinserting now the Lorentz factor γ_{12} and β_2 we obtain

6 $$L_{12} = \gamma_2(1 + \beta_{12}\beta_2) L_1\,,$$

and thus

7 $$\sqrt{1 - \beta_2^2}\, L_{12} = (1 + \beta_{12}\beta_2) L_1\,.$$

To interpret our result we replace these two lengths L_{12} and L_1 in terms of their relation with L_0 and cancel the common factor L_0 to obtain

8 $$\frac{\sqrt{1-\beta_{12}^2}\sqrt{1-\beta_2^2}}{1+\beta_{12}\beta_2} = \sqrt{1-\beta_1^2}\,.$$

Seeing the denominator we recognize that a compounded Lorentz-FitzGerald body contraction factor is not a product of two individual Lorentz-FitzGerald body contraction factors.

This result is, however, not a surprise. This is what we saw before in the study of velocity addition looking at the compounded speed, see Example 7.11. We can use the format of β seen Example 7.11 to derive a generalization allowing non-parallel $\boldsymbol{\beta}_i, i = 1, 2$

9 $$\frac{\sqrt{1-\boldsymbol{\beta}_{12}^2}\sqrt{1-\boldsymbol{\beta}_2^2}}{1+\boldsymbol{\beta}_{12}\cdot\boldsymbol{\beta}_2} = \sqrt{1-\boldsymbol{\beta}_1^2}\,.$$

Now we also see that when $\boldsymbol{\beta}_{12}$ and $\boldsymbol{\beta}_2$ are orthogonal, the product of contractions is the compound Lorentz-FitzGerald body contraction factor.

We obtained these results using the relativistic addition of velocities, and assuming that both observers measure the same proper (rest-frame) length L_0 of the body ('length of a train standing at the station'). ◀

10 The Bell Rockets

Abstract

We study in the context of SR the difference between space and material bodies. We consider the example of two rockets set in motion in a synchronous way, connected by a fragile thread. These rockets are allowed to move faster and faster, while the spatial distance between the rockets remains unchanged, breaking the Lorentz-FitzGerald contracted thread. Just as a clock can be used to record the history of time dilation effects, we can design a Lorentz-FitzGerald body-contraction measuring device to record the history of body-contraction effects.

10.1 Rockets Connected by a Thread

Imagine a physicist in a relativistic rocket who wants to measure the Lorentz-FitzGerald body contraction. She would find that all body lengths remain the same, as all measuring instruments would also be subject to the Lorentz-FitzGerald body contraction. Because of this difficulty, it would seem that we cannot prepare and maintain the standard of length between two different reference frames. Faced with one simple argument, many conclude prematurely that measurement of the Lorentz-FitzGerald body contraction is, in principle, impossible.

To remedy this pessimistic view, we recall that the Lorentz-FitzGerald body contraction was discovered in order to explain the Michelson-Morley experiment; it shows no fringe-shift because, according to FitzGerald, and independently, to Lorentz, the material body of the experiment is contracted in the direction of motion. The MM is a laboratory demonstration of the Lorentz-FitzGerald body contraction of an unknown magnitude. The magnitude is unknown since an absolute velocity is in principle non-measurable within the context of Einstein's theory of

J. Rafelski, *Modern Special Relativity*, https://doi.org/10.1007/978-3-030-54352-5_10

relativity. For this reason Einstein saw the MM-experiment in his 1905 work not as a demonstration of body contraction but as a demonstration that an absolute velocity cannot be determined.

John Stuart Bell ***British-Irish physicist, 1928–1990***

From
©CERN-Photo-73-4-271
From: CERN Annual Report 1972, p. 75

John S. Bell, FRS (1972), was born in Belfast, where he received a B.Sc. in Experimental Physics in 1948, and one in Mathematical Physics a year later from Queens University. He joined the Atomic Energy Research Establishment at Harwell; on leave from Harwell he worked on his PhD (1956) with Paul Mathews and Rudolf Peierls at Birmingham. Bell joined in 1960 the CERN Theory Division in Geneva where he worked for the following 30 years. Bell is world renowned for the **formulation of the Bell inequality in 1964**, making it possible to test experimentally for hidden variables in quantum mechanics. He made pioneering contributions in high energy physics, teaching of SR (discussed in this book), and relativistic particle accelerator theory.[1]

However, the Lorentz-FitzGerald interpretation of the MM-experiment creates the need to explore possible modification of the standards of length between two reference systems with a known relative velocity. For this to work we need to develop a method to transfer a unit of body length from one inertial system to another, allowing for the comparison to be made of two material body lengths.

We proceed to present the method advanced by John S. Bell, which involves an apparatus that will keep a 'memory' of the history of the contraction. J.S. Bell's

[1]For further appreciation see: R. Jackiw and A. Shimony, "The depth and breadth of John Bell's physics," *Phys. Perspect.* **4** 78 (2004); arXiv:physics/0105046 [physics.hist-ph]; and the special issue of *Europhysics News,* **22** No. 4 pp. 65–80 (April 1991); Andrew Whitaker *John Stewart Bell and Twentieth-Century Physics: Vision and Integrity,* (Oxford University Press 2016).

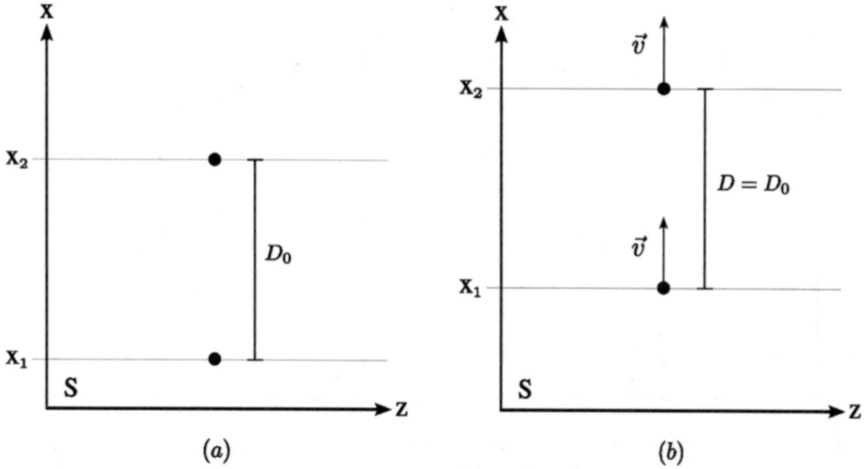

Fig. 10.1 Two point particles: (**a**) initially at rest separated by distance $x_2 - x_1 = D_0$; (**b**) moving both with the same velocity $\boldsymbol{v}$ acquired at a later time. Their spatial separation does not depend on time, $D = D_0$

article, "How to teach special relativity"[2] posed a question based on these concepts,[3] which has come to be known as the "Bell Rocket Paradox".

To explain the method we consider the example depicted in Fig. 10.1 of two point particles initially separated in space by a distance $D = x_2 - x_1 = D_0$ and starting to move from rest in the laboratory frame S. The particles are accelerated uniformly, very weakly, and identically for all time $t > 0$. Their world lines are obtained solving equations of motion; for their difference we obtain

$$m\frac{d^2}{dt^2}(\boldsymbol{x}_1 - \boldsymbol{x}_2) = 0\,, \rightarrow\ \frac{d}{dt}(\boldsymbol{x}_1 - \boldsymbol{x}_2) = \Delta\boldsymbol{v}\,, \rightarrow\ (\boldsymbol{x}_1 - \boldsymbol{x}_2) = \Delta\boldsymbol{v}\,t + \Delta\boldsymbol{x}_0\,. \tag{10.1a}$$

[2] J.S. Bell, "How to teach special relativity," in A. Zichichi, ed. *Progress in Scientific Culture* **1** No. 2 pp. 135–148 (1976); "I did not invent this situation ..." from a note by J.S. Bell to J. Rafelski, reprinted with references, see Ref. 3 in: J.S. Bell, *Speakable and unspeakable in quantum mechanics* Chapter 9, pp. 67–80, Cambridge University Press (1987); reprinted in M. Bell, K. Gottfried, and M. Veltman, edts. *Quantum Mechanics, High Energy Physics and Accelerators,* World Scientific, Singapore (1995).

[3] Bell's article, Ref. 2 elaborated a discussion topic that begun with a short note by E. Dewan and M. Bernan, "Note on stress effects due to relativistic contraction," *Am. J. Phys.* **27**, 517 (1959); Bell gives also credit to: A.A. Evett and R.K. Wangsness, "Note on the separation of relativistically moving rockets," *Am. J. Phys.* **28**, 566 (1960); E.M. Dewan, "Stress effects due to Lorentz contraction," *Am. J. Phys.* **31**, 383 (1963); A.A. Evett, "A relativistic rocket discussion problem," *Am. J. Phys.* **40**, 1170 (1972).

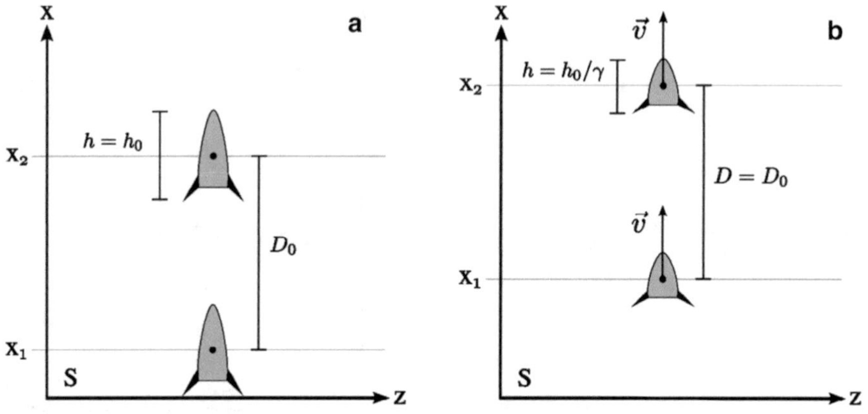

Fig. 10.2 Two rockets of length h separated by distance $D = x_2 - x_1 = D_0$. (**a**) at rest, and in case (**b**) moving at velocity $\boldsymbol{v}$ acquired at a later time

We see that if we start accelerating our particles simultaneously from rest, i.e., with $\Delta\boldsymbol{v} = 0$, their spatial separation

$$D = D_0 \equiv |\Delta\boldsymbol{x}| = |\boldsymbol{x}_1 - \boldsymbol{x}_2| = \Delta x_0 , \tag{10.1b}$$

remains constant.

This is so since the spatial separation between the two accelerated bodies is not subject to a Lorentz-FitzGerald body contraction. Therefore we recognize that the independent but identical and weakly accelerated motion of two point bodies transfers parallel to motion the standard of spatial distance between two different inertial frames of reference, here between the initial S and any other reference frame that attains relative velocity by means of an acceleration.

If we consider along with J.S. Bell two rockets instead of two point particles, as in Fig. 10.2, we find that the centers of mass of the rockets also maintain a constant separation $D = D_0$. However, because the rockets themselves are rigid bodies, the measured length of either rocket will surely contract about its center of mass by a factor of γ. Recall, the Lorentz-FitzGerald body contraction has an effect only on the physical length of each rocket, but not on the spatial separation between the two centers of mass of the rigid bodies.

10.2 The Thread Breaks

Next we add an arbitrarily weak, thin thread of length $l = D_0$, just long enough when the rockets are at rest to span the gap connecting the two rockets by their centers of mass, as is shown in Fig. 10.3. Bell's question was, what happens to this thread as the rockets gently accelerate? We have already considered the separation in space of the two independent rockets and have determined that it must remain

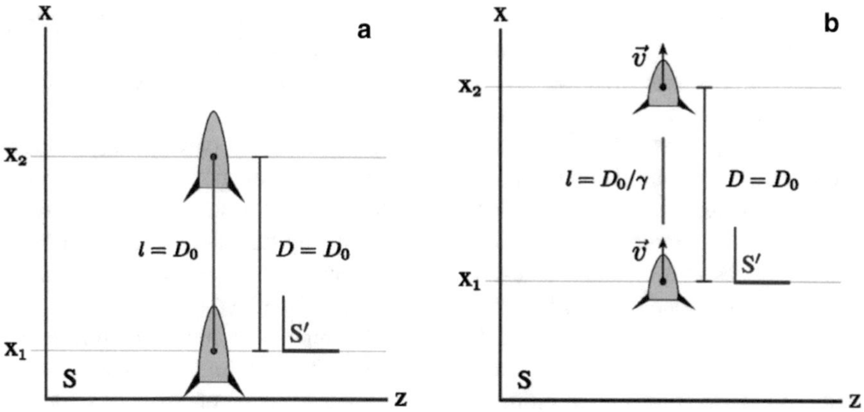

Fig. 10.3 Two rockets separated by distance $D = x_2 - x_1 = D_0$ and connected by a thin thread: (**a**) at rest, and in case (**b**) moving at velocity $\boldsymbol{v}$ acquired at a later time

constant, since the rockets accelerate independently. We have already discussed several times that any object, such as a train in a tunnel or here a thread shown in Fig. 10.3 in the conditions here described will be subject to a Lorentz-FitzGerald body contraction

$$l = \frac{D_0}{\gamma} \to l < D_0 \,, \qquad D(t) = D_0 \,. \tag{10.2}$$

The obvious and correct answer, that *the Lorentz-FitzGerald body contraction causes the string to break*, is often met with skepticism. Indeed one can see that if the thread was sufficiently strong it could pull the rockets together despite the presence of two independent rocket engines, and the entire system should contract as one rigid body. However, one can always choose a thread that is as weak as needed to allow the rockets to move independently. Such a string must break.

The deep physical relevance of the above statement is rooted in the irreversible change in the state of a material body generated by the Lorentz-FitzGerald body contraction. This is irrefutable evidence that the Lorentz-FitzGerald body contraction is a 'real' effect.

10.3 The Thread Spools

We now refine Bell's 'two rockets connected by a thread experiment' by providing two ratcheted spools, one with thread allowed to unroll on one rocket, and another roll-up thread spool on the other rocket. The spool with thread unwinds, adding more thread when we accelerate the rockets, allowing the Lorentz-FitzGerald body contracted thread not to break but always connecting the two rockets. In this way we can read off and gauge the body contraction at any speed. When the rockets slow down, the roll-up spool picks up the slack in the connecting thread. The roll-up spool assures permanent memory of changing body contraction effects. Bell's 'two rockets with two ratcheted spools' allow measurement of both transient contraction effects and cumulative body contraction effects.

The roll-up thread spool assures that a lasting memory of ongoing Lorentz-FitzGerald body contraction is recorded. Bell's rockets allow both the measurement of instantaneous Lorentz-FitzGerald body contraction as it occurs, and the measurement of the accumulated contraction. Thus the roll-up thread spool has the same characteristics as a common clock, which will demonstrate the time dilation after the trip is completed because it records the time dilation effect as it occurs.

Even though this Bell's spooled thread method appears extremely cumbersome, it establishes the in-principle observability of the effect of the Lorentz-FitzGerald body contraction based on a comparison with a spatial separation distance standard. Therefore it clarifies the physics contents of one of the most contentious phenomena of special relativity.

We now return to our original question: *Can we transport between different reference frames the standard unit of length without contraction, and if so how?* We see now that the Bell rocket example offers a method by comparing (contracted) material bodies with a (constant in time) spatial separation. One can envision a box carried on the relativistic train. The box contains two miniature rockets, each accelerating independently at the same rate, and exactly the same rate as the relativistic train. The constant separation of the floating rockets then preserves the unit distance providing all along the entire travel path the reference unit of length against which one determines instantaneous contraction of the relativistic train.

This instrument does not measure the absolute velocity, but the relative velocity achieved since the travel started, and does this by means of the associated Lorentz-FitzGerald body contraction. Like relativistic trains and rockets, Bell's device remains in the realm of distant technology.

Example 10.1

Bell's example considered from one of the rockets

We consider how an observer S', located on the rear rocket as shown in Fig. 10.3, evaluates the situation with Bell's rockets. ◀

▶ **Solution** We call the measurements of the positions of the spaceships in S', x'_1 and x'_2, and the positions of either end of the string in S', l'_1 and l'_2. The separation of the ships, D', and the length of the string, l', are

$$\textbf{1}\quad D' = |x'_2 - x'_1|\,, \qquad l' = |l'_2 - l'_1|\,.$$

Measurements in S' take place at $t'_2 - t'_1 = 0$. Assuming the string is accelerated by the rear rocket, independently of the front rocket, we know that its length in S' will be constant: $l' = D_0$.

To determine the separation between the rockets, we first look at the temporal back transformation into S

2 $t_2 - t_1 = \gamma((t'_2 - t'_1) + \frac{v}{c^2}(x'_2 - x'_1))\,, \qquad t_2 - t_1 = \gamma \frac{v}{c^2} D'\,.$

The right-hand side of the last equation is positive; thus it is clear that a measurement of the rockets simultaneously in S' corresponds to a measurement in S of the rear rocket's position prior to the measurement of the front rocket's position. The front rocket will in the available time move ahead and thus the string connecting the rockets, which have moved further apart, must break, unless it was strong enough to pull the two rockets together, defeating the independent propulsion systems of the two rockets. ◀

Example 10.2

Observing the Bell rockets: laboratory system distance measurement

Another way to reconsider the Bell rocket problem is to perform a measurement simultaneous in the laboratory frame S; that is, to study the two rockets at equal S-time: $t_2 - t_1 = 0$. Given this context, calculate the separation of the rockets in S', $D' = x'_2 - x'_1$. ◀

▶ **Solution** A measurement simultaneous in S corresponds to a measurement with time difference in S'

1 $t'_2 - t'_1 = \gamma((t_2 - t_1) - \frac{v}{c^2}(x_2 - x_1)) = -\gamma \frac{v}{c^2}(x_2 - x_1) = -\gamma \frac{v}{c^2} D_0\,.$

We now put this time difference into the spatial inverse transformation from S' to S

2 $x_2 - x_1 = \gamma((x'_2 - x'_1) + v(t'_2 - t'_1)) = \gamma((x'_2 - x'_1) - \gamma \frac{v^2}{c^2}(x_2 - x_1))\,.$

Solving for $x'_2 - x'_1$ yields

3 $D' = (x'_2 - x'_1)|_{t_2=t_1} = \dfrac{D_0}{\sqrt{1 - v^2/c^2}} > D_0\,.$

We also obtain this result applying the Lorentz transformation

4 $D' = (x'_2 - x'_1)|_{t_2=t_1} = \gamma\,(x_2 - x_1 - v(t_2 - t_1))\,|_{t_2=t_1} = \gamma\, D_0\,,$

which confirms our considerations; it takes more than a cursory inspection to recognize in above result that the origin of the effect is the time difference in S', our point of departure.

We conclude that the two rocket separation D' observed from the reference frame of the rear rocket S' is longer compared to the common separation D measured at equal time in the laboratory system S. We are also not surprised that the ratio of both measurements

$$\frac{D'}{D_0} = \gamma \,,$$

contains the Lorentz-factor γ. On the other hand the observer S' cannot observe a change of the thread length, which he thus thinks is always $l' = D_0$. Thus we have demonstrated that the thread comoving with S' should rip.

We summarize the insights gained in this and the previous example as follows:

- When observed from laboratory S at equal time $t_1 = t_2$ the separation distance between rockets does not change; the thread connecting the rockets is observed to be shorter by the (instantaneous) Lorentz-factor γ; this corresponds to the expected Lorentz-FitzGerald body contraction.
- When we explore this situation in the reference frame of one of the rockets, we see that the comoving thread is unchanged in its length but the rocket separation has increased by the same γ-factor.

The outcome for the fate of the thread is of course the same, independent of the method how we carry out the measurement. ◀

Discussion 10.1 Rigid bodies, relativity and 'length clock'

About the topic: The thread breaks only if two rockets connected by a thread do not constitute a single rigid body. But when is a body not rigid? Is the theory of relativity able to describe the behavior of all extended material bodies? Many of these and similar questions can be debated. Our conversation today includes a colleague 'Iwo' who will intervene only to keep the debate 'honest'. This discussion picks up the argument we embarked on in Discussion 5.2.

- *Iwo:* The detailed analysis of the 'two rockets riddle' does not contain an essential ingredient: the discussion of the concept of a rigid body in the framework of relativity theory.
- *Simplicius:* Right, I share your concern. Our discussion relies on the physical properties of the connecting thread. We expect that the thread length changes, and if the thread is strained it must be able to fail. This means we have applied a model of the material body. Are these rooted in principles of relativity?
- *Student:* Agreed, the situation we discussed depends on the physical properties of the connecting thread. We expect a change in the thread length. When the thread is stressed strongly it is able to rip. All this points to study of material properties rather than to study of SR.
- *Professor:* I think this topic is moot, for two reasons: (i) We introduce a spool of thread which means we can and indeed should use a thread that will not break or stretch but which unwinds. (ii) Any study of the

(continued)

Lorentz-FitzGerald body contraction must consider extended objects; there is nothing new in looking at two rockets connected by thread.

- *Simplicius:* Professor, could you please explain how the Lorentz-FitzGerald body contraction, here of the thread, arises from the physical structure of the thread?
- *Professor:* I believe that material bodies of finite extent exist solely due to quantum physics.
- *Simplicius:* So how exactly does quantum physics help the formation of extended and rigid bodies?
- *Student:* Crystals are a good example.
- *Simplicius:* How does a crystal rod connecting two crystal rockets contract?
- *Professor:* We need to determine how the electromagnetic forces that shape the quantum crystal structure, in terms of quantum physics and electromagnetic interactions, change for different observers.
- *Simplicius:* But Maxwell's electromagnetism is a classical theory and inherently relativistic!
- *Professor:* Maxwell's EM theory is today part of quantum physics. Decisive is how the quantum electrons in the crystal adjust to the fact that the charged atomic nuclei (we can think of as being classical) move.
- *Simplicius:* Moving charges make currents.
- *Student:* And in turn, electrical currents induce magnetic fields. Thus, the moving atomic nucleus is a source of both electric and magnetic fields.
- *Professor:* We use the fields of moving nuclei to find by solving the quantum dynamical equations the needed deformation of the electron orbitals. This done, we find that the crystal will be Lorentz-Fitzgerald body contracted.
- *Simplicius:* Can you get a large body contraction in this way?
- *Professor:* The computation is difficult if we want to make a large contraction in one step. Instead, we make very many small contraction steps, and use addition theorems. Thus all we need to show is that for a very small speed we obtain the expected tiny contraction of the crystal.
- *Simplicius:* Again, how do you get a large contraction?
- *Professor:* After a first small contraction I can add a second small one, corresponding to small relative velocity and so on. We have seen that the effects add according to the principles of two Lorentz coordinate transformations with respective speeds, see Example 9.3. Similarly we can compute the expected summed up contraction comprising many small contractions. After accounting for the effect of all the small steps, we must find the expected result, a contraction of any magnitude, depending on the total velocity.

(continued)

- *Simplicius:* Even if crystals contract this does not mean that a piece of wood, liquids, or a heap of sand will do so.
- *Professor:* Sand is a heap of individual crystalline grains of silicon, and we actually only expect contraction of each crystalline particle. So in the absence of other forces in microgravity a heap of sand will not contract as this would mean contraction of space. This is like having many rockets chase each other, not only two as considered in Bell's example. Each sand-rocket is contracted but the space between them is not. However, if you connect rockets strongly while the rocket engines are very weak, the entire system could contract.
- *Simplicius:* What about a wooden stick?
- *Professor:* Wood is made of complex macro molecules and therefore a numerical model is entirely intractable on my computer. However, as electron orbitals will contract and compress the long molecules just like the case of a crystal, I am convinced a piece of wood contracts just like a crystalline rigid body.
- *Simplicius:* There are easily deformable substances such as water. I doubt that the Lorentz-FitzGerald body contraction is observable in this intermediate case.
- *Professor:* The question is if 'soft bodies' experience a change in 'packing' in the context of body contraction. After some thought I am prepared to defend the following point of view: *Any bound body*; that is, a body where removal of atoms or molecules costs noticeable energy, and that includes water and all liquids but few if any gases, will be subject to one and the same Lorentz-FitzGerald body contraction. I think a loose heap of matter will not contract. In between there can be weakly bound objects which are difficult to understand.
- *Student:* I still have one issue to resolve: In many relativity texts I see that there is a causality problem considering rigid bodies. Does this create another problem for Bell's rockets?
- *Simplicius:* I also read about quantum physics and conflicts with causality.
- *Professor:* The problem that these books note is that an infinitely rigid body could transmit a signal at superluminal speed: You push at one rod end, and the other end of a long stick moves at the same instant even if it is far away. Some observers see the other end move before the front end is pushed.
- *Student:* I was taught that if I push at one end of the quantum crystal, the action propagates in the crystal such that the other end moves away later. The action propagates no faster than the crystal's light velocity which is known to be lower than light velocity in free space.
- *Professor:* This is so since crystals actually are not infinitely rigid; they can fail when forced, and they support wave propagation, compression,

(continued)

density oscillations and all that is needed to assure signal delay required by causality and Lorentz-FitzGerald body contraction.

- *Simplicius:* Yet some claim quantum physics can be acausal; is this possible?
- *Professor:* This is loose talk. Quantum electrodynamics, the relativistic quantum field theory of charged particles achieves causality, for further discussion see Chap. 12.
- *Iwo:* All this is nice and good but again, accelerated extended rigid bodies have no place in special relativity!
- *Professor:* Agreed as it concerns considerations developed in the first years of special relativity before quantum physics was invented. Since then, we have learned how to deal with such circumstances. For example we know that we do not need to worry about the effect of a weak acceleration.
- *Iwo:* Do you claim that forces acting in the LHC[4] are insignificant?
- *Professor:* The electromagnetic forces keeping particles in a LHC orbit, which while large compared to daily experience, are still extremely weak on a natural scale. Were it not so, the charged particles kept by the magnetic field in the accelerator would radiate and lose energy upon traveling quantum scale distance! This is the meaning of 'strong' acceleration which breaches the physics discussed in this book. We will return to this topic in Chap. 20.
- *Simplicius:* So what about "extended rigid bodies have no place in relativity"?
- *Student:* The matter was settled with the example of a crystal. Naturally there are complicated semi-rigid bodies which would need to be studied in depth when and if we have time to do that.

[4]The Large Hadron Collider (LHC) at the CERN laboratory is the world's largest and highest-energy particle accelerator 27 kilometers in circumference; it lies in a tunnel as deep as 175 meters beneath the Franco-Swiss border near the city of Geneva.

Part V

Space, Time, and the Doppler Shift

Part V of the book introduces new concepts about the 3+1 dimensional space-time of SR. The light cone allows the subdivision into timelike and spacelike event separation. The concepts, future, past, and causality, are characterized and studied. We question: How the Universe can be homogeneous? Do faster than light particles (tachyons) make sense? Can causality be questioned considering the quantum non-locality? We show the origin and workings of light aberration and explain the origin and workings of the relativistic Doppler effect.

Introductory Remarks to Part V

The emphasis of this part of the book is on time, and the consistency of SR when space-time properties are studied. Accordingly we describe many new concepts and associated vocabulary. We begin with an introduction of the light cone and the associated division into timelike and spacelike event separation. How the precise meaning of the future and the past arises from this is reviewed. We look at near speed of light motion and introduce the light cone coordinates.

We follow up with discussion of temporal sequence of events, which amplifies the importance of the difference between spacelike and timelike event separation. These topics connect to and illuminate pivotal questions such as: Why is superluminal speed of travel not possible? What is the meaning of causality? Is event causality always present? Is travel in time possible? Of course everything that follows here is based on principles of SR. Still, there is much to discuss, and these considerations lay the foundations for several topics we discuss later in this book.

We next explain how and why the acquired time dilation effect is not reversible; that is, specifically the 'traveling twin' stays younger compared to the other remaining in the inertial laboratory. By sending one twin on a space trip we assure by process of rocket acceleration that the path of the twin will reduce its proper time below that recorded by the inertial observer; i.e. the 'twin at rest at base'. We extend the argument to include the case of triplets, and argue which of several travelers will be youngest compared to the base reference observer, and explain why.

Causality and quantum physics are the subjects of an essay extending and connecting the ideas of special relativity into the vibrantly growing field of quantum non-locality. We describe the current status of the field and argue that there is no conflict with special relativity. What happens is prescribed by a causal event sequence. We cannot, however, resolve the quantum non-locality questions in the context of the local realism of special relativity.

We then turn to discuss the Doppler light (wavelength or frequency) shift. Compared to the Doppler effect of sound we must argue very differently: We cannot tell 'who' is moving, the source or the observer; as in SR only relative velocity matters. Demanding the Lorentz invariance of the light wave phase we derive the relativistic Doppler effect.

The aberration effect, see Sect. 13.3, is an essential element required for understanding the relativistic Doppler shift of light. Aberration alters the direction from which we see the source of light. We derive the relativistic aberration of line of sight with the help of the Lorentz transformation, including the vectorial character of the effect. We show that nonrelativistic aberration exists only transverse to the true geometric line of sight.

Our derivation of the Doppler shift of light demonstrates that the effect is reciprocal. This means that each of the source-observers measures the same Doppler shift when observing the other. This implies that each of two observers will find the same relative velocity vector. This implies that, unlike the common argument, time dilation does not enter into the discussion or characterization of the relativistic Doppler effect.

The Light Cone

11

Abstract

Light lives on the surface of the light cone and divides the space-time into two domains of different physical character: Within the light cone we have the 'future' that we can influence by our actions; and the 'past' which can influence our present state. Outside the light cone is the acausal space-time domain we cannot communicate with at below the speed of light. We will explore why and when signals propagating with superluminal speed contradict causality.

11.1 The Future

We know that between any two events in a vacuum a signal can propagate at a maximum speed of c. We place ourselves at the event point $x = y = z = 0$, at $t = 0$ and emit a light signal. Let us first consider a highly focused laser pulse propagating only along the x-axis. The signal reaches all event points that lie along the lines $x = ct$, and also $-x = ct$, should we point a second laser in the opposite direction. These two lines are shown in Fig. 11.1. Similarly, we can accelerate material particles, shooting these with speeds $v < c$ along direction $\pm x$. Their paths will fill the region above these two lines in Fig. 11.1. The unshaded domain for $ct > 0$ cannot be reached by either light originating at $x = y = z = 0$ or particles moving with $v < c$.

A signal transmitted from the origin, $x = y = z = 0$, could be emitted in any direction in the three spatial dimensions. If the signal were a pulse of light, it could reach the points described by

$$\boldsymbol{x} = (ct)\hat{n} \ , \tag{11.1}$$

J. Rafelski, *Modern Special Relativity*, https://doi.org/10.1007/978-3-030-54352-5_11

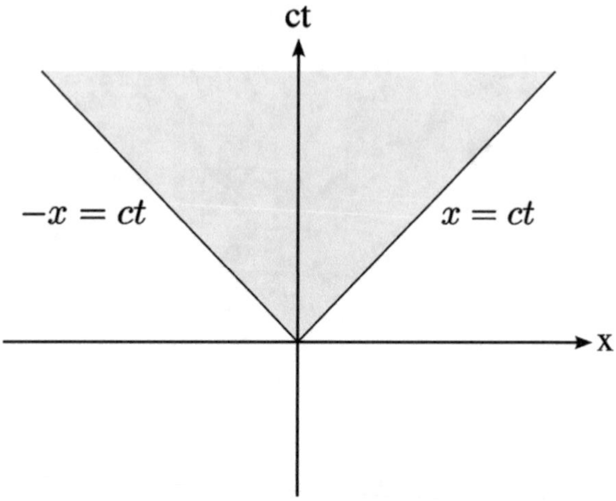

Fig. 11.1 The light cone in one spatial dimension: the region of the future for $t > 0$, see text

where $\hat{n}$ is a unit vector pointing in any direction. We can square this equation

$$\boldsymbol{x}^2 = (ct)^2 . \tag{11.2}$$

We write out Eq. (11.2) showing all coordinates explicitly

$$(ct)^2 = x^2 + y^2 + z^2 . \tag{11.3}$$

We recognize that for this case of a flash in all directions, light traces not only the two boundary lines as shown in Fig. 11.1, but a much larger space domain. We explore the situation setting $z = 0$; that is, allowing vector $\hat{n}$ to point in the x-y-plane with $x, y \neq 0$. We obtain a cone in the space-time-manifold

$$(ct)^2 = x^2 + y^2 , \tag{11.4}$$

shown in Fig. 11.2.

This is the boundary of the future domain of space-time. In general the light cone is the three dimensional domain of space-time that light reaches in the general case $z \neq 0$, see Eq. (11.3). We see this for one spatial dimension in Fig. 11.1, and for two spatial dimensions in Fig. 11.2. The general case of three spatial dimensions cannot be easily depicted graphically, as we cannot represent a four dimensional manifold projected onto two dimensional paper plane.

To understand better where the spherical light flash emitted at $x = y = z = 0$, $t = 0$ can go we consider it again allowing now $z \neq 0$. We write

$$(ct)^2 - (x^2 + y^2) = z^2 . \tag{11.5}$$

The left side of Eq. (11.5) for $z \neq 0$ is always positive, and thus it belongs to some spatial point within the two dimensional light cone, as shown by shaded area

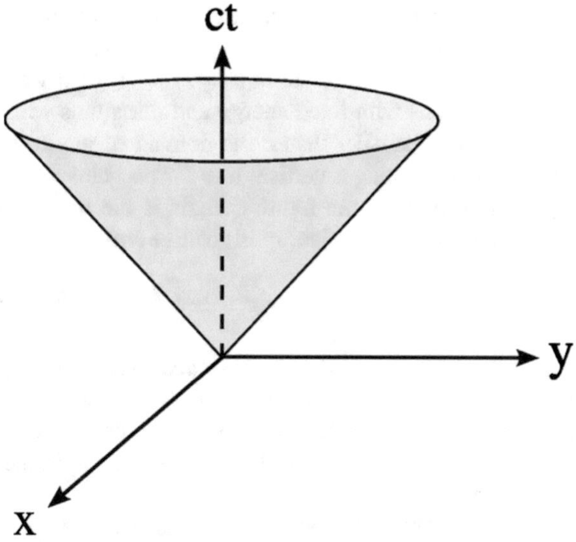

Fig. 11.2 The future light cone $t > 0$ showing time and two spatial dimensions, see text

in Fig. 11.2. This means that projecting the three dimensional light cone into the two dimensional representation Fig. 11.2 we fill the entire domain bounded by the two dimensional light cone shown in Fig. 11.2. An analogous situation arises when projecting the cone surface in Fig. 11.2 onto the plane in Fig. 11.1.

This consideration shows that for $z \neq 0$ the light flash reaches the entire space-time contents of the light cone in Fig. 11.2. This is so since from Eq. (11.3) it follows that the light flash can be seen from anywhere but at different times: If time permits, it can reach all spatial points in the future domain within the light cone.

Another way to explain this situation is to realize that after the time t a spherical light pulse reaches the surface of a sphere of radius $|\boldsymbol{x}| = ct$. Thus with passage of time the spherical light pulse will illuminate all and every point in the future light cone. That is why, when we look at the sky, we see the light emitted by all stars. However, the starlight is emitted at different times in the past, depending on the distance to the source.

The spherical light cone as introduced in Eq. (11.2) illuminates the space-time domain in Minkowski space with

$$s^2 = (ct)^2 - \boldsymbol{x}^2 = 0, \qquad t > 0\,. \tag{11.6}$$

s^2 does not change under a Lorentz coordinate transformation, see Sect. 7.1. s^2 has the same value for any inertial observer. This means that the domain within the future light cone is not affected by a proper Lorentz coordinate transformation change of coordinates. The meaning of 'future' is thus a Lorentz invariant concept given the Lorentz transformations of physical interest to us: We consider the 'proper' Lorentz coordinate transformation which e.g. preserves the sequence of events. As we noted briefly in the discussion following Eq. (6.16), the inclusion of 'improper' LT is necessary when considering the full set of space-time transformations called the Poincaré group.

Imagine that an observer at $x = y = z = 0$ were to shoot material particles in all directions with fixed energy and mass thus with a fixed speed $v < c$. For a signal velocity smaller than c the domain of world lines forms a cone within the light cone, becoming a vertical line for particles with $v \to 0$. Thus for a material signal transmission, the future domain is the region inside the light cone defined where the Lorentz invariant s^2 is positive and

$$s^2 = c^2 t^2 - \boldsymbol{x}^2 > 0, \qquad t > 0\,. \tag{11.7}$$

The light limit $s^2 = 0$ is approached asymptotically as the energy of material particles is increased. Since speed $v > c$ is not allowed, the space-time diagram such as Fig. 11.2 excludes for $t > 0$ much of the domain outside the future light cone due to the finite value of c. We are unable to influence observers outside of our light cone.

Without going into technical details we note that in the context of quantum physics material particles explore a very, very small domain outside the light cone; the range is exponentially small with characteristic quantum distance, the Compton wavelength $\lambda_C = h/mc$, h being the Planck constant.

The quantity s^2 is closely related to the proper time of a body, see Sect. 7.1. By considering an observer always at rest at the origin of a reference frame we have

$$\sqrt{\frac{s^2}{c^2}} = \sqrt{t^2 - \frac{(\boldsymbol{x} = 0)^2}{c^2}} = \tau\,, \tag{11.8}$$

where the use of τ denotes the fact that time measured by this observer at rest is the proper time in this reference frame. For another observer the values of t and $\boldsymbol{x}$ change according to the Lorentz coordinate transformation; however the value of τ does not change, see Eq. (7.7).

Example 11.1

Light cone coordinates

Normal time and space coordinates are not the optimal coordinates for describing physical phenomena involving motion at a speed close to that of light. Such motion can be more accurately characterized with the help of the light cone coordinates $x_- = ct - x$, $x_+ = ct + x$. How do these coordinates transform under Lorentz transformations? Consider further the rest-frame of a particle to recognize physical relevance of your results. ◀

▶ **Solution** Ultrarelativistic motion occurs near to the light cone and can be described by light cone coordinates

1 $x_- = ct - x, \qquad x_+ = ct + x\,, \qquad y = y\,, \quad z = z\,,$

and we further have for the invariant s^2

2 $s^2 = (ct)^2 - x^2 - y^2 - z^2 = (ct - x)(ct + x) - y^2 - z^2 = x_-x_+ - y^2 - z^2$,

where we recognize the product x_-x_+. This implies that under a Lorentz coordinate transformation while coordinates y, z remain unchanged, the two new coordinates must transform inversely to each other. Moreover, depending on in which direction a particle moves, we have either $x \simeq ct$ or $x \simeq -ct$. Therefore, one of the coordinates $x_\pm$ is very small, and the other is very large.

Using the explicit form of the Lorentz coordinate transformation we obtain

3 $x'_\pm = \gamma(ct - \beta x \pm x \mp \beta ct) = (ct \pm x)\gamma(1 \mp \beta)$.

A straightforward computation, see also Eq. (7.30), yields

4 $\gamma(1 \mp \beta) = \dfrac{1 \mp \beta}{\sqrt{(1 \pm \beta)(1 \mp \beta)}} = \sqrt{\dfrac{1 \mp \beta}{1 \pm \beta}} = \exp\left\{\ln\sqrt{\dfrac{1 \mp \beta}{1 \pm \beta}}\right\} = \mathrm{e}^{\mp y_r}$.

Thus in rapidity representation the light cone coordinates transform as

5 $x'_+ = \mathrm{e}^{-y_r}x_+\ , \qquad x'_- = \mathrm{e}^{y_r}x_-\ , \qquad x'_+x'_- = x_+x_-$.

We consider a fast particle moving to the right such that $x \simeq ct$. The coordinates measured by an observer moving along with this particle are

6 $x' = 0 \quad \to \quad x'_\pm = ct' = c\tau$.

Here is τ the particle proper time. We can cast our final result into the simpler form

7 $x_\pm = \mathrm{e}^{\pm y_p}c\tau$.

In this equation y_p is the rapidity of the particle with respect to the laboratory rest-frame in which we evaluated the particle proper time τ. Since τ is invariant our result relates the light cone coordinate to particle rapidity with reference to the observer, and particle proper time.

Our present discussion addresses inertial motion and related light cone coordinates. Considering the context of accelerated particle motion we return in Sect. 23.4 to light cone coordinates and obtain for the increments $dx_{\pm}$ and $d\tau$ an equivalent result. ◀

11.2 The Past

The same line of argument developed above for the future light cone, case $t > 0$, also applies to the past domain of the light cone, case $t < 0$. The region

$$s^2 > 0, \quad t < 0\,, \tag{11.9}$$

is shown in one dimensional representation in Fig. 11.3. We can receive messages from all event points in the region bounded by the past light cone; that is to say, the present can be influenced by any event in the shaded region in Fig. 11.3.

When we look at the stars we see light from within the past light cone, since each star shines in all directions. However the starlight we see can be of very different ages, determined by the distance light needs to travel to be recorded today. The domain of the Universe we can observe is larger and larger as we look at older and older light. We will address this situation in Sect. 12.1.

This region Eq. (11.9) contains all space-time events in our past. Signals from our past are *causal*: for all observers with speed $v < c$ the signal is created before it is observed. However, situation changes if we were to have superluminal signals, $v > c$, see Example 11.2 below. The existence of faster than light particles would upend many concepts of SR.

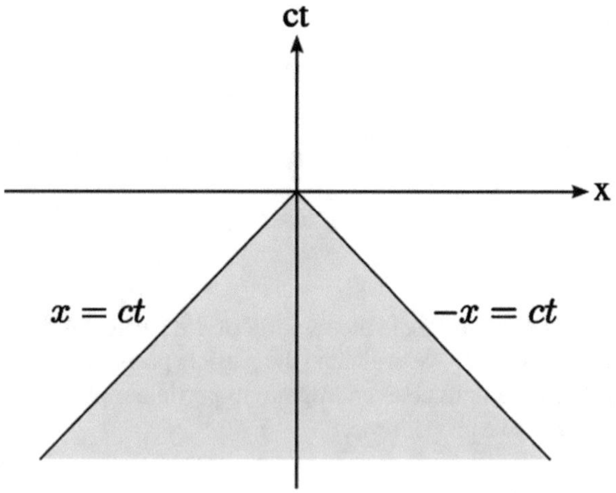

Fig. 11.3 The light cone in one spatial dimension: the region of the past for $t < 0$, see text

Discussion 11.1 Tachyons?

About the topic: Can particles that travel with superluminal velocities exist? We return to the topic we addressed briefly in Discussion 5.3.

- *Simplicius: Star Trek* frequently mentions particles that are faster than light, called tachyons. There is even a book, *The Physics of Star Trek.*[1] Can tachyons exist?
- *Student:* I checked the book. No tachyons there. The author, a well regarded theorist, points out large effects of time dilation would create situations that would make the show impossible, since the actors remaining at the base would need to age with a faster rate compared to traveling actors. The warp drive was introduced to allow travel over galactic distances without the need to approach light velocity and to battle time dilation. If and how this could be compatible with SR was not addressed; most important was the absence of age dilation effects which would make the filming of the show impossible.
- *Simplicius:* But traveling faster than light over a large distance will for sure create serious problems with causality, right?
- *Professor:* Indeed: This being the case, for certain observers the effect precedes in time the cause.
- *Student:* Any signal that travels faster than c including the *Star Trek* spaceship will spell problems for relativity, and the easiest to recognize are the challenges to causality as we just considered. On the other hand, we can speculate how we could sidestep the problems that warp drive introduces.
- *Professor:* I think we will be in trouble unless there is another very large synchronization speed, just like light, which helps establish the temporal sequence in our world. I do not know how this could work.
- *Simplicius:* Science fiction literature proposes an extension of space-time to a larger dimensional context, say 'sub-space'.
- *Student:* This speed would need to be billions of times greater than that of light.
- *Simplicius:* A plane can fly with a speed in excess of the speed of sound; maybe there is some way to pick us up and transport with a speed in excess of the speed of light, just as the warp drive we see watching *Star Trek*.
- *Student:* In other science fiction TV shows there is the 'FTL-drive' (**F**aster **T**han **L**ight) where the spaceship is not 'picked up' but finds its own jump-path in subspace.
- *Professor:* As you explained, FTL is science fiction. Let me reiterate: Everything we learn in this book about special relativity is applicable only

(continued)

[1] L.M. Krauss, *The Physics of Star Trek*, Basic Books (1995).

to the realm of our space-time and matter we are made of, matter made of 'light', in the sense that its rest energy is $E = mc^2$ (note the c). In the absence of a shred of experimental evidence we can only speculate about the existence of another realm where everything can move much faster.

- *Simplicius:* This is close to my still open question: Do you believe tachyons could exist?
- *Student:* We already agreed tachyons are not around, see Discussion 5.3.
- *Professor:* In context of this discussion I would like to explain how real science differs from pseudo-science contents of science fiction. Science is based on the observation of phenomena present around us, or on proposals which, however difficult, can be decided by experiment. The question about tachyons reaches beyond the science paradigm[2] of the present day. A scientific tachyon revolution can occur in the future, as a paradigm shift. In that case based on a new experimental observation and/or compelling theoretical framework, we will face the tachyon question.
- *Simplicius:* So maybe there are tachyons?
- *Student:* No, our Professor clearly said, neither yes, nor no, and not even maybe. The only answer today is 'nobody knows'.
- *Professor:* This is a good moment to tell an anecdote. A few years ago there was some commotion about the possibility that neutrinos could be just a little faster than the speed of light. The subject had deeper roots: it began with scientific publications proposing the neutrino as a tachyon,[3] and when the speed of neutrinos was finally for the first time precisely measured, it seemed in excess[4] of c. The appearance of such an experimental possibility would have been just such a paradigm-shifting result. Along with many colleagues I dropped ongoing research projects as we turned to think about this topic. But it soon became clear someone did not plug in a wire!
- *Simplicius:* Too bad this experiment was wrong.

(continued)

[2]Thomas Kuhn, *The Structure of Scientific Revolutions* (University of Chicago Press, 1962) characterized a paradigm shift (or revolutionary science) as a change in one of key basic assumptions ruling science. A paradigm is a concrete scientific position accepted at a given time by the pertinent scientific community. Nearly all scientific knowledge is based on contextual paradigms. Einstein's paradigm shift can be characterized succinctly as follows: old $t = t', c \neq c'$; new $t \neq t', c = c'$.

[3]A. Chodos, A.I. Hausera, V.A. Kostelecky, "The Neutrino as a Tachyon" *Physics Letters B* **150** 431 (1985).

[4]OPERA Collaboration (T. Adam et al.), "Measurement of the neutrino velocity with the OPERA detector in the CNGS beam", September 2011. 24 pp. e-Print: arXiv:1109.4897, see v1, result retracted a few months later.

- *Professor:* Actually, it felt good to try and fail: We have looked for a loophole in SR seeking a new formulation, and we failed. We learned from this how we should keep testing SR, and we will meet in this book this question in Chap. 19. However, always remember that the end of Newtonian mechanics was only in part brought about by testing Newton's laws; just as important if not more decisive was the study of electromagnetism, a new physics domain, which seemed at first in conflict with classical physics.
- *Student:* Is there anything new like electromagnetism we can hope to discover today?
- *Professor:* The Universe is full of stuff we have not seen, literally. We call the material part 'dark matter' and there is four times more of that around than of the matter we observe. Then there is the yet more mysterious 'dark energy': three times the amount of matter, acting akin to Einstein's 'cosmological constant'.
- *Simplicius:* How can one recognize the existence of something invisibly dark?
- *Professor:* We detect dark matter by checking the action of gravity in the Universe at large distances and comparing it to GR. At first it was not clear if instead of dark matter a modification of GR was needed. However, ongoing astronomical discoveries rule this out.
- *Student:* Even so, I heard GR needs improvement.
- *Professor:* There are well-known problems in reconciling gravity and quantum physics, and even at the level of classical electromagnetism we do not understand the physics of large acceleration, see Chap. 20. The list of mysteries around us is long.
- *Simplicius:* Please do not forget to add tachyons to your list.
- *Student:* No. Our Professor mentioned a few established scientific riddles originating in either repeatable experimental results and/or theoretical inconsistencies in established theories for which either no physical explanation is yet known or where we keep arguing about which explanation is correct. Tachyons have no place on this list as nothing indicates their presence in nature.

Discussion 11.2 Causality

About the topic: What exactly is causality and why is this a topic we discuss in SR?

- *Simplicius:* What exactly is causality?
- *Student:* Causality refers to a fixed sequence of cause and effect.

(continued)

- *Professor:* Considering that in SR time is a coordinate which changes from one observer to another, this concept is not as self-evident as one may think. We expect but need to verify that all observers will observe the same sequence of events.
- *Simplicius:* Is it really possible, with a Lorentz coordinate transformation, to change the sequence of cause and effect?
- *Professor:* If you could travel faster than light, you could overtake messages sent at the speed of light. This means that certain observers would see the causal sequence reversed.
- *Simplicius:* How would traveling faster than light change causality? No matter how fast I move I can't arrive somewhere before I have left.
- *Professor:* True, but if you were able to accelerate to reach trans-light-speed, two events along your path, for example your birth and your death, could be seen as occurring in varying sequence in certain other reference frames – this is illustrated in Fig. 11.4.
- *Student:* Breaking causality would contradict much of what we now understand about physics.
- *Professor:* One can use your argument to exclude faster than light motion (tachyons), see Discussion 11.1, as unphysical on the grounds that causality could be violated.
- *Simplicius:* So, if anything exceeds the speed of light, the causal element in the laws of physics breaks?
- *Student:* Indeed.
- *Professor:* It is for this reason that the speed of light is the synchronization speed of the Universe. It is the maximum speed we can reach before breaking causality.
- *Simplicius:* What do you mean by synchronization speed?
- *Professor:* Different reference frames will have slightly different perspectives on the length of time between two events, although the sequence of events remains. A synchronization signal sent by some observer in the past allows us to establish the correct sequence of events anywhere the signal reaches, see Chap. 12.
- *Student:* Because of synchronization, any inertial observer will view a series of events in the correct sequence. A synchronized clock keeps the time sequence of the original synchronization signal clock. This assures us that for all observers time flow is unidirectional and all events have the same sequence.
- *Simplicius:* What about quantum entanglement? I heard that when two quantum particles travel apart and one is observed one knows what the other particle will do. Sometimes with certainty. Does this not mean that there was faster than light motion? Then all this synchronization is gone.

(continued)

- *Student:* Nothing is traveling in quantum entanglement measurement, so nothing is traveling with a speed faster than light.
- *Professor:* Information is not being exchanged between these particles at the instant of the measurement of one, and thus also not at a speed faster than light. In fact when these particles were together and became entangled, a quantum table of probabilities regarding their entanglement was created. When we observe a part of any entangled quantum system, the unobserved part is also observed. Quantum entanglement illustrates quantum-nonlocality, an idea violating our instincts, but this does not impact the principles we implement developing special relativity. We return to this topic in the Essay: *Quantum entanglement and causality* in Sect. 12.3.

Example 11.2

Faster than light signals

An inertial observer S originates a 'tachyonic' signal that allegedly propagates faster than the speed of light (i.e., superluminal); that is, $\dfrac{\Delta x}{\Delta t} \equiv v > c$. Considering another inertial observer, S', moving relative to S with the velocity v', determine the range of v' for which time sequence of events changes; that is, the range in which causality is violated. ◀

▶ **Solution** The world line in S of the emitted signal has allegedly $v > c$. The observer S' moves relative to S with velocity $v' < c$. This observer measures the time interval $\Delta t'$, which can be determined with the help of Lorentz coordinate transformation from S to S'

1 $$\Delta t' = \frac{\Delta t - v'\Delta x/c^2}{\sqrt{1 - v'^2/c^2}} = \Delta t \frac{1 - (\Delta x/\Delta t)(v'/c^2)}{\sqrt{1 - v'^2/c^2}} \,.$$

Here we take $\Delta x/\Delta t$ to be the world line slope of the signal emitted by S that is spreading faster than light; this is $v > c$. In Fig. 11.4 this world line is denoted as W.

We are looking for the condition such that $\Delta t'$ above reverses sign compared to Δt. Such reversal occurs for

2 $$vv'/c^2 > 1 \,, \quad v > c \,, \quad v' < c \,.$$

This is normally impossible, but we are considering now consequences of the assumption $v > c$, which is also impossible, so we proceed anyway. Dividing above by v/c^2 we obtain

3 $$c > v' > c^2/v \,.$$

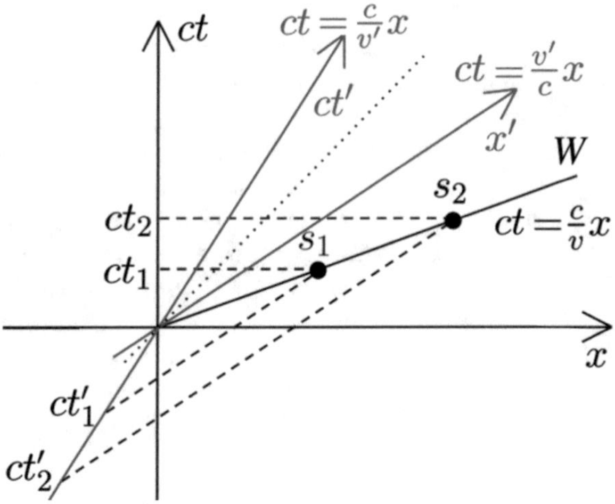

Fig. 11.4 Two events s_1 and s_2 are seen on the superluminal world line W with $ct = \frac{c}{v}x,\ v > c$. The space-time axes ct, x are of the laboratory observer S, the axes ct', x' define an observer S' moving with relative velocity v'. The condition $\frac{v'}{c} > \frac{c}{v}$ is satisfied. Therefore the time sequence of events reported by S' is opposite to that set up by S

Thus when a signal travels at $v > c$, for all observers S' traveling with respect to the signal source at speed

4 $v' > c\left(\frac{c}{v}\right)$,

the sequence of events is violated. This is demonstrated graphically in Fig. 11.4. Here we note that the coordinate axis of S' seen in the x-ct-diagram are represented by the following straight lines: The x'-axis is determined setting $ct' = 0$

5 $ct' = \gamma\left(ct - \frac{v'}{c}x\right) = 0 \quad \to ct = \frac{v'}{c}x$,

and the ct'-axis is determined setting $x' = 0$

6 $x' = \gamma\left(x - \frac{v'}{c}ct\right) = 0 \quad \to ct = \frac{c}{v'}x$.

In Fig. 11.4 two events s_1 and s_2 are located on the superluminal world line W with $ct = \frac{c}{v}x,\ v > c$. Upon projection of these events onto respective time axes we can read off their time coordinates t_i and t'_i. We recognize the reversal of time sequence between the observers S and S'. A superluminal signal could reach observer S' before this observer could possibly observe (in his reference system) the cause of the signal. This means that the causality is violated, see Chap. 12. ◀

Space-Time

12

Abstract

Space-time consists of domains with timelike and spacelike separation from the coordinate origin. The future and the past are reconsidered in this context. Time dilation is revisited and studied in depth. The relative age of several travelers is explored. In an essay we address macroscopic quantum entanglement showing that causality is preserved.

12.1 Timelike and Spacelike Event Separation

We divide the entire space-time as seen from the origin of the coordinate system into two event regions.

1. All events in the region

$$s^2 > 0\,, \quad \text{i.e.} \quad c^2t^2 - \boldsymbol{x}^{\,2} > 0\,, \tag{12.1}$$

   are called *timelike*. This includes the regions within the past and future light cones.
2. In contrast, the remainder of space-time, a region where

$$s^2 < 0\,, \quad \text{i.e.} \quad c^2t^2 - \boldsymbol{x}^{\,2} < 0\,, \tag{12.2}$$

   is called the *spacelike* region. An observer at the origin of the coordinate system cannot influence events in the spacelike region. Conversely, an action by a spacelike region observer will be noted in the timelike domain with delay. For the observer at rest at the origin of the coordinate system this delay is at least as long as the light takes to travel the spacelike separation distance.

J. Rafelski, *Modern Special Relativity*, https://doi.org/10.1007/978-3-030-54352-5_12

3. The surface separating the two regions

$$s^2 = 0\,, \quad \text{i.e.} \quad c^2t^2 - \boldsymbol{x}^{\,2} = 0\,, \tag{12.3}$$

is *lightlike*, as only light can be found there. Events can occur on this light cone surface: e.g. a mirror located at $|\boldsymbol{x}| = ct$ deflects a light ray into a different propagation direction.

The naming of the two regions follows the larger of coordinate values; that is, for $ct > |\boldsymbol{x}|$ the region is timelike and where $|\boldsymbol{x}| > ct$ it is spacelike.

In order to further generalize this discussion of the invariant quantity s^2, we now consider an observer away from the origin of the coordinate system. The more general definition of the space-time interval between two events E_1 and E_2 is

$$s^2 = c^2(t_1 - t_2)^2 - (\boldsymbol{x}_1 - \boldsymbol{x}_2)^2\,. \tag{12.4}$$

Once we shift our origin to one of the event points, Eq. (12.4) is equivalent to our previous definition of s. Equation (12.4) thus defines the invariant 'distance' between two events; it is invariant in the sense that it does not change when events are subject to LT. For $s^2 \geq 0$, the distance is called timelike and the event with the smaller t can influence the event with the larger t. On the other hand, for $s^2 < 0$, the two events are separated by a spacelike distance and they cannot communicate in a causal way since $v \leq c$.

At first glance it would seem that there are large domains of the Universe which not only are 'out of sight' but also will forever remain so, and thus could be subject to conditions that we cannot influence or ever observe. However, this, in general, need not be the case, since we can recognize two cases where the influence of the past could exceed the limits imposed by the light cone definition:

(1) It was possible in the distant past, $t \to -\infty$, to furnish a large future region with measuring sticks and clocks synchronized by a light signal, starting from $x = 0$ as indicated in the larger light cone in Fig. 12.1. This for example allows

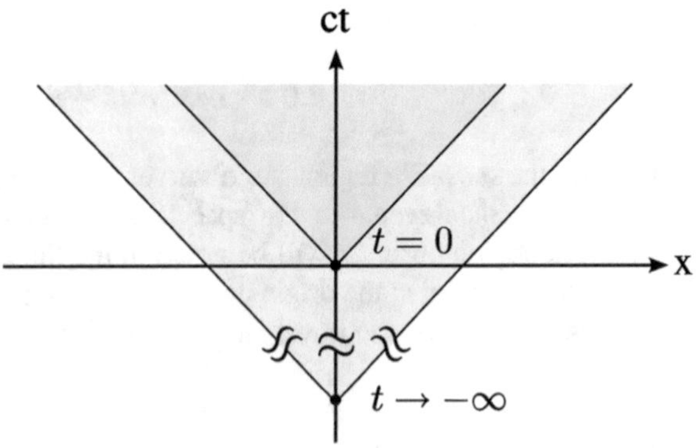

Fig. 12.1 Synchronization of space from $t \to -\infty$: a signal originating at $t \to -\infty$ reaches the entire space-time domain of the Universe

introduction of common units across this domain. The future light cone of the initial synchronization signal at $t \to -\infty$ is far larger and includes much more space-time than is found within our future cone today – this means that an observer in the distant past could make sure that the Universe is the same on a scale vastly larger than is accessible to us today;

(2) The future domain of a light cone of observers far outside of our future light cone and at rest or in motion with speed $v < c$ with respect to us today moves into our light cone domain at some later time – thus, sometime in the future we can be subject to actions that this observer had set up. Events originating today outside of our light cone will ultimately intersect our cone, and thus can influence us later.

Discussion 12.1 The Structure-Homogeneous Universe

About the topic: We now discuss the beginning of time, the Big-Bang, and how this can be consistent with a homogeneous Universe.

- *Simplicius:* I have a problem with the ideas presented here: I heard that there was a beginning in time of the Universe, a Big-Bang. So nobody can go back further, and thus not all of the Universe is in the future light cone, ever. There must be spatial areas of the current Universe that do not know about the Big-Bang.
- *Professor:* Indeed, the Universe is about 13.8 billion years old. All events follow this beginning. However, . . .
- *Student:* . . . I also have a problem with that. Today I can look to the right and to the left and see stars from far and away. If they are older than $13.8/2 = 6.9$ billion years and are seen in opposite directions, I believe these stars are so far apart, more than 13.8 billion years, that they could not have been causally connected by a signal emitted just after the Big-Bang at $t = 0$. This means that at least one of these stars is outside of the future domain of the Big-Bang observer.
- *Simplicius:* That is indeed my problem! For me it is very unsettling to recognize that no matter where we look, we see the same Universe. How can this be unless some of these domains are communicating with each other?
- *Student:* Indeed, if there is a beginning of time, one cannot go back far enough in time in order to synchronize the behavior of today's visible Universe.
- *Simplicius:* However, somehow there must have been a coordination of these events, i.e. a synchronization, which has little to do with our context.
- *Professor:* Thank you for setting up this important conceptual Big-Bang challenge. The situation you have recognized, that the Universe to the 'right' and to the 'left' is the same, calls for new ideas and principles. This in particular calls for a closer look at what happens to space after the

(continued)

Big-Bang. A Big-Bang implies that there was a spatial expansion of the Universe. Such an expansion means that the light cone is opening wider; it is not the straight line you have implicitly assumed. This means that a small spatial domain from the past could create the vast present day Universe. Specifically, if the Universe spatial expansion were at a constant rate, we could argue that the stars we see back 6.9 billion years on the left and on the right had been before a factor two closer to each other compared to our current expectation. Continuing in this line of thought we see that the Universe could be synchronized in its properties at the time of the Big-Bang as it would originate in a singular spatial point. The models that are considered today are more complex but the principle is the same – all of the existent Universe has one and the same spatial origin; that is the meaning of the Big-Bang.

- *Simplicius:* You describe how within the Big-Bang model of the Universe large space domains are synchronized, domains which within the SR framework should not know about each other. Does this imply that at some stage in Universe evolution there has been superluminal communication?
- *Professor:* If at all, the SR principles are violated at the birth of the Universe. In an inflationary model the space grew exponentially. This spatial inflation process assures that all of the space in the Universe originates at the Big-Bang instant, allowing that the nature of the Universe is everywhere the same.
- *Simplicius:* Does exponential spatial growth actually mean superluminal growth?
- *Student:* This does not worry me. I do not believe that even an exponentially inflating space could impact causality in the primordial Universe.
- *Professor:* Superluminal expansion of the Universe spatial scale does not mean that it is possible to find an observer for whom the sequence of events reverses, and who can loop in time against the cosmic time arrow. Therefore, even if the superluminal scale expansion seems good, I am pretty sure that our views about the first instances of a Big-Bang will continue to evolve. Thus I believe that the topic we just discussed will be revisited many times. I know of colleagues who wish to rework this idea and interpret the homogeneous Universe in different ways.

Example 12.1

Spacelike event separation & the Lorentz-FitzGerald body contraction

Two events are observed by an inertial observer S to occur at $s_1 = (ct_1, \boldsymbol{x}_1)$ and $s_2 = (ct_2, \boldsymbol{x}_2)$ and are measured to have a spacelike separation. Demonstrate that even if $t_1 \neq t_2$ there is another inertial observer S' moving with velocity

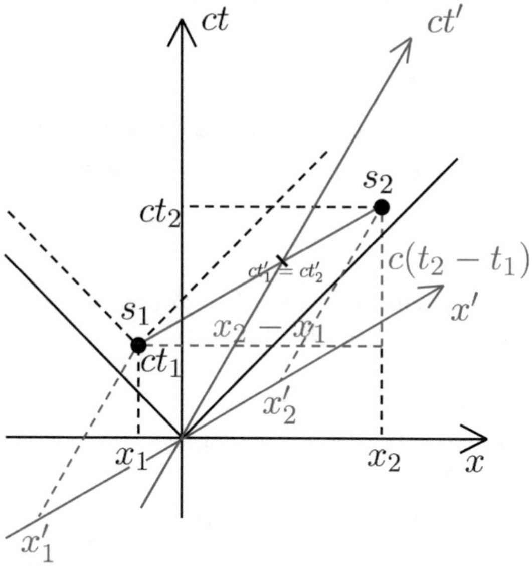

Fig. 12.2 Two events s_1 and s_2 with coordinates $ct_i, x_i,\ ct_i', x_i',\ i = 1, 2$ are shown. Their separation is spacelike since s_2 is seen outside of the light cone of s_1. For the observer S' we have $t_1' = t_2'$, which value is located on the ct' axis by the intersecting $s_1 - s_2$ line parallel to x' axis; the x'-axis traces in S the straight line $ct = \frac{v}{c}x$. For $\frac{v}{c} = \frac{ct_2 - ct_1}{x_2 - x_1}$ this line is parallel to the line connecting the s_1 and s_2 events

v who observes these events occur simultaneously as shown in Fig. 12.2. Is this also possible for timelike event separation? Obtain the spatial separation of these two events observed by $S'(v)$ and write this result solely as a function of v and $x_1 - x_2$, eliminating any dependence on t_1, t_2. Explain and interpret your finding for the case that the two events correspond to the ends of a physical body. ◀

▶ **Solution** The event separation $s^2 = (s_1 - s_2)^2$ is an invariant under Lorentz coordinate transformations, see Sect. 7.1. If S measures $s^2 < 0$ (spacelike separation Eq. (12.2)) all other observers $S'(v)$ agree that

$$\mathbf{1}\quad s^2 = s'^2 = c^2(t_1' - t_2')^2 - (\boldsymbol{x}'_1 - \boldsymbol{x}'_2)^2 < 0\,.$$

However, if S measures a timelike separation $s^2 > 0$, we cannot ever satisfy the condition $t_1' - t_2' = 0$ since in that case it follows, as seen above, that for observer S' we must have $s'^2 < 0$, which contradicts the invariance requirement $s^2 = s'^2$.

Thus only in the case of spacelike event separation can we look for a frame of reference such that $t_1' - t_2' = 0$.

$$\mathbf{2}\quad 0 = t_1' - t_2' = \gamma(t_1 - t_2 - (x_1 - x_2)v/c^2)\,,$$

where as usual we oriented the coordinate system with the x-axis being the axis of velocity v. We thus have

3 $t_1 - t_2 = (x_1 - x_2)v/c^2$,

and we keep in mind that the two events in S differ in time while they are at equal time in a reference frame S' moving with the relative speed v

4 $\frac{v}{c} = \frac{c(t_1 - t_2)}{x_1 - x_2}$.

We now compare measurements made by S' to those made by S. Observer S' reports that the spatial separation between the two events is

5 $x'_1 - x'_2 = \gamma(x_1 - x_2 - v(t_1 - t_2))$.

We use speed v to eliminate $t_1 - t_2$ and we obtain

6 $x'_1 - x'_2 = \gamma\left(1 - \frac{v^2}{c^2}\right)(x_1 - x_2) = \sqrt{1 - \frac{v^2}{c^2}}\,(x_1 - x_2) = \frac{1}{\gamma}\,(x_1 - x_2)$.

We see that the spatial separation between events recorded by the moving 'equal time observer' S' is reduced by the reciprocal of the Lorentz factor when compared to the spatial separation measured by the observer S where the events are not at equal time.

Assigning two ends of a material body to both events we can interpret this result as follows: The 'equal time observer' S' observes for a moving body, in comparison to a comoving observer S, a contracted body length. We obtained, considering the Lorentz coordinate transformations of spatial events, a result that is fully consistent with the body contraction postulated by FitzGerald and Lorentz. ◄

To conclude: We confirm the results of Chap. 8 with the added insight that a material body is spacelike, i.e., we can find an observer S' capable of observing the body ends at equal time. The here demonstrated consistency of body properties and LT is a necessary property of SR. It should be noted that this consistency is achieved by tracking carefully in which reference frame a measurement is accomplished at equal time. A commonly performed reversal of points of view between two bodies in relative motion often done without consideration of time as a part of event description leads to misunderstandings described even in introductory SR textbooks, and rampantly present on the WWW as well as in popular SR introductions.

► **Insight: 12.1 Causality and proper time** The following is a short vocabulary of concepts:

Space-time distance s between two event points, one at the origin and the other at $(ct, \boldsymbol{x})$, is

$$s^2 \equiv c^2t^2 - \boldsymbol{x}^2 .$$

This quantity is relativistically invariant. For $s^2 > 0$, these two events are said to be **timelike** since ct dominates $\boldsymbol{x}$ – in fact there is an observer S' for whom $\boldsymbol{x}' = 0$. Two such events can influence each other, respecting here the event time sequence. Conversely, when $s^2 < 0$, these two events are said to be **spacelike**. There is an observer who can claim that the time separation between these events is absent; $t' = 0$. Such events cannot communicate with each other.

Signals and their information content can propagate at a maximum speed of c and that creates the light cone $|\boldsymbol{x}| = ct$. Therefore, there is a region of space-time, the 'future' that we can influence with our actions originating at coordinate origin, and another region, the 'past', from which the coordinate origin can be influenced. Event points outside of the light cone are acausal (that is, require faster than light communication) with respect to coordinate origin.

Causality means that the time sequence of timelike events is the same for any inertial observer as long as all velocities $v \leq c$.

Quantum effects, including large distance entanglement, are compatible with relativity; there is no acausal information transmission.

Proper time of a moving body: τ in motion from origin to the space-time point $(ct, \boldsymbol{x})$ is

$$\tau^2 = s^2/c^2 \,.$$

Like s^2 it is also an invariant to which all observers agree. To address more general motion we consider the differential increment of proper time of a body given by

$$d\tau = dt\sqrt{1 - \boldsymbol{v}^2/c^2} = \sqrt{(dt)^2 - (d\boldsymbol{x})^2/c^2} \,,$$

where $\boldsymbol{v}$ is the instantaneous velocity of the body. The proper time, that is also the 'age' of the body, is then for an observer located in the laboratory and measuring time t

$$\int^t d\tau = \int^t \sqrt{1 - \boldsymbol{v}^2/c^2}\, dt \,.$$

12.2 Time Dilation Revisited

Considering how time dilation and the "twin paradox" continue to challenge many students, we want to deepen here the understanding of these concepts we first considered in Chap. 4. Imagine that an astronaut observer S_A undertakes a trip to a distant star and back. He travels in an accelerating spaceship reaching $v \approx c$, while his twin S_E remains on 'the Earth base'. S_E is an inertial coasting observer. Considering the effect of time dilation, the time in the spaceship t_A should pass at a significantly slower pace than the time in the inertial base t_E. The traveling twin will therefore be younger after the trip compared to the base twin.

However, considering the principle of relativity, one could argue (wrongly) that the Earth's base inertial system $S_{\rm E}$, which we have assumed to provide the reference frame for the relative time measurement is not any different (wrong) from a system at rest with respect to the spaceship. This hypothesis is motivated by the viewpoint that for the astronaut twin $S_{\rm A}$ on the spaceship, the spaceship is at rest, and the Earth is traveling at a large velocity, carrying the Earth's base twin $S_{\rm E}$ away. Setting up the problem in this way, we see that there must be a way to distinguish the twins, or else how can we tell which twin is younger when they reunite on Earth's base? Furthermore, we could consider another observer $S_{\rm B}$ following a different world line for whom, both the Earth's base $S_{\rm E}$, and the other spaceship, are in motion. Clearly, it cannot be that the aging of twins depends on the presence of this observer $S_{\rm B}$.

The difference between the traveling twins is established by comparing the length of their world lines. The twin who has traveled away $S_{\rm A}$, and later comes back to meet the twin $S_{\rm E}$ on Earth's inertial base, has a longer world line path in space-time since this twin was exposed to an *acceleration*. Because $S_{\rm A}$ was subject to some acceleration, however small, it is not possible to reverse the situation claiming that the Earth twin $S_{\rm E}$ traveled away. All inertial observers will agree that the accelerated body has a longer world line. This acceleration can be continuous or momentary; it does not matter. An observer comoving with an accelerated body cannot be an inertial observer for the purpose of the study of the effect of time dilation, no matter how we have applied the acceleration to the body in time.

We can cast these words in an equation by considering the respective world line length increments $dl^2 \equiv c^2dt^2 + dx^2$ in the ct, x diagram. We now look from the $S_{\rm E}$ reference frame at the difference of the world lines of both observers

$$dl_{\rm A}^2 - dl_{\rm E}^2 = (dt^2c^2 + dx^2) - dt^2c^2 = dx^2 \,, \tag{12.5}$$

where $dl_{\rm A}$ is the incremental world line length of the accelerated twin, and $dl_{\rm E}$ that of the laboratory twin. Under LT we find $dx^2 \to dx'^2 = [\gamma(dx - vdt)]^2 \geq 0$, where the increment dx'^2 vanishes for an inertial observer who at this particular instant in time is comoving with the accelerated clock, $dx/dt = v$. Except for this accidental instantaneous match, in general $dx'^2 > 0$. We therefore can conclude that the integral over the entire world line difference Eq. (12.5) is always positive and non vanishing. This means that there is an observable difference between both twins which we find considering the length difference of their world lines. As result *the accelerated twin is always younger compared to the inertial twin.*

For the purpose of clock comparison, special relativity necessitates that all travelers must meet at a common space-time point. This means that considering two twins, at least one, here $S_{\rm A}$, must accelerate and/or decelerate to meet $S_{\rm E}$ again. We have learned that the inertially coasting twin $S_{\rm E}$ will always be the oldest traveler. By how much older compared to accelerated twins we find considering the world lines.

Although it is the acceleration of the traveling twin $S_{\rm A}$ that changes his space-time path and distinguishes him from the other Earth's base twin $S_{\rm E}$, to find out who is younger we do not need to improve on special relativity nor invoke GR. This is so, since the acceleration that distinguishes the world lines of each twin can be arbitrarily small. Thus we can make the determination of the time dilation entirely using the principles of special relativity, knowing that the presence of acceleration, however small, allows the distinction between S_{A} and S_{E}.

Example 12.2

Which triplet is the youngest?

"Triplets " E, and A, B follow three different space-time paths that all start at the origin. In this example consider the period during which there is a change in velocity, i.e. when the travelers are subject to acceleration, as being negligible when compared to other travel periods.

Triplet E remains at rest at the origin defined to be Earth's inertial base. A clock comoving with E measures the reference time t.

Triplet A travels at constant velocity $v_{\mathrm{A}} < c$, starting at $t = 0$ and ending far away at $t = t^f$. We describe motion of A using the time t measured in the rest-frame E.

Triplet B travels at constant velocity $v_{\mathrm{B}} < c$ away from the origin until $t = t^f/2$, at which point she turns around and heads back to the origin meeting triplet E at time $t = t^f$.

These three world lines are shown in Fig. 12.3. Calculate and compare the proper time elapsed for each traveler between $t = 0$ and $t = t^f$ in system E, and interpret your results. Compare the relative age of the triplets for the case that A decides at some later time to return to base E. Demonstrate your qualitative argument in exact mathematical terms, presenting a precise argument why a triplet with the longest world path is the youngest. ◀

▶ **Solution** For all of the triplets we obtain the proper time in terms of the velocity of each triplet measured by the laboratory observer E.

Triplet E has speed $v = 0$; we have simply

1 $\tau_{\mathrm{E}}^f = t_{\mathrm{E}}^f \,.$

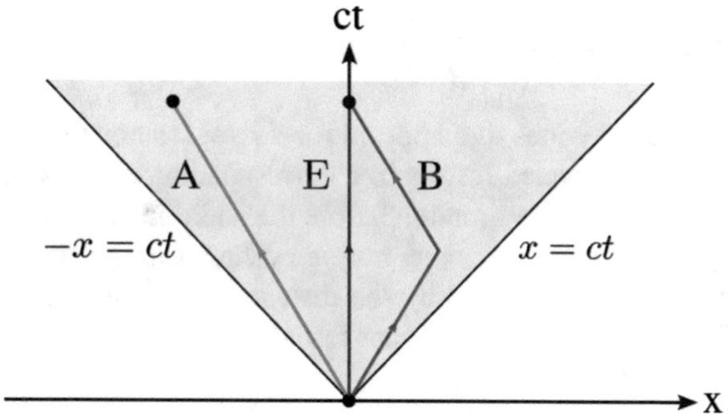

Fig. 12.3 Voyages of triplets A, B, and E. See Example 12.2

Triplet A has a constant speed v_A; hence

2 $\tau_\mathrm{A}^f = t_\mathrm{A}^f \sqrt{1 - v_\mathrm{A}^2/c^2} = t_\mathrm{A}^f/\gamma_\mathrm{A}$

describes the usual time dilation, but the triplet is still far away from Earth.

Triplet B has an effectively constant speed v_B, as we neglect the period of acceleration. Thus we have again

3 $\tau_\mathrm{B}^f = t_\mathrm{B}^f \sqrt{1 - v_\mathrm{B}^2/c^2} = t_\mathrm{B}^f/\gamma_\mathrm{B}\ .$

We note that the direction of the velocity changes allowing triplet B to return to base E.

For $v_\mathrm{A} = v_\mathrm{B}$ we have $\tau_\mathrm{A} = \tau_\mathrm{B}$ because the lengths of the paths in Fig. 12.3 are the same. Thus both triplets A and B have aged the same at time $t = t_f$. However, while triplet B is already back, triplet A needs to take the road back to join B and E (not shown in Fig. 12.3). No matter how triplet A is going to come back she has to travel at finite speed, which we can choose for simplicity to be again v_A and this part of the trip will add to time dilation of A compared to the Earth observer E and compared to the observer B who is now at rest. We learn in this example that among all travelers who were traveling with same speed the one who went farthest away will be the youngest when all reunite at the base where the non-traveling crew is in comparison the oldest.

We can characterize this situation more precisely: The incremental change in the proper time of travelers 'T' is

4 $d\tau_\mathrm{T}^2 = dt^2 - dx^2/c^2 = d\tau_\mathrm{E}^2 - dx^2/c^2\ ,$

where we inserted the relationship that the laboratory time is also the proper time of the triplet E at rest. We further note travelers T speed as measured by E

5 $v_\mathrm{T} = \dfrac{dx}{dt} = \dfrac{dx}{d\tau_\mathrm{E}}\ .$

Combining these two equations we obtain for the infinitesimal change of the location of the traveler T with $dx^2 > 0$

6 $d\tau_\mathrm{T}^2 = d\tau_\mathrm{E}^2 \left(1 - \dfrac{v_\mathrm{T}^2}{c^2}\right) \le d\tau_\mathrm{E}^2\ ,$

where the equal sign appears for $dx \to 0$ since then $v_\mathrm{T} \to 0$. Any addition to the distance traveled results in a corresponding decrease in the traveler's proper time compared to the situation before the addition of distance dx.

For unequal $v_\mathrm{A} \neq v_\mathrm{B}$ nothing changes in the above argument, since the proper time is characterized by the distance dx traveled in the base time increment dt. Hence the traveler who goes farthest away (largest dx) will be youngest. Of course, it is easy to complicate our argument in that one of the travelers makes many turns, so that the distance traveled away will be small. What matters in this case is of course the sum of the distances traveled. To operationally resolve this situation one can evaluate the path-length of the world lines considered.

Laying a thread on the world path of each traveler, we see that longest thread is associated with the most traveled, thus youngest traveler. This is so since any local extension of the thread results as demonstrated in final result in a further reduction in the proper time of the traveler. The geometric length of the world line drawn in the (t, x)-plane is a convenient measure allowing a comparison of any two observers traveling along different world lines towards the same destination. ◀

Example 12.3

Optimizing duration of travel to a distant space-time destination

We are planning a trip to a distant destination in the Universe. (a) What is the best way to execute the trip to minimize the proper time of the spaceship crew? (b) Compare the world lines of several possible space-time paths to determine a path connection with the proper time. (c) Show that the new result is consistent with prior Example 12.2. ◀

▶ **Solution** In our prior considerations, e.g. Example 12.2, we considered what can be called 'time' travel, where the travelers instead of going to a prescribed destination explore the effect of time dilation by traveling to arbitrary destinations before returning to the base and waiting for each other. We remember that in such situations the traveler with the longest world line remains the youngest. In this example all travelers aim for the same destination in space. The question how to achieve this at the least cost in proper time differs from our earlier examples.

Let us consider a near speed of light traveler 'T'. As long as we neglect the acceleration periods he practically does not age on the scale of the Earth 'E' time during his trip

1 $$d\tau_{\mathrm{T}}^2 = d\tau_{\mathrm{E}}^2\left(1 - \frac{v_{\mathrm{T}}^2}{c^2}\right) \leq d\tau_{\mathrm{E}}^2\,.$$

We now look at another traveler, who at least part of the way has been slower. Accordingly, the proper time that this traveler's clock records will be longer. This argument can be continued as often as is needed and we conclude that with certainty, given a destination in space, the fastest traveler will have aged least.

Comparing the length L_{path} of different world lines for these travelers assuming nearly constant speed of travel v, see Fig. 12.4, we find at a first sight a surprising result. With $x \to D = v\,\tau_{\mathrm{E}}$ we obtain

2 $$L_{\mathrm{path}} = \sqrt{c^2\tau_{\mathrm{E}}^2 + D^2} = D\sqrt{(c/v_{\mathrm{T}})^2 + 1}\,.$$

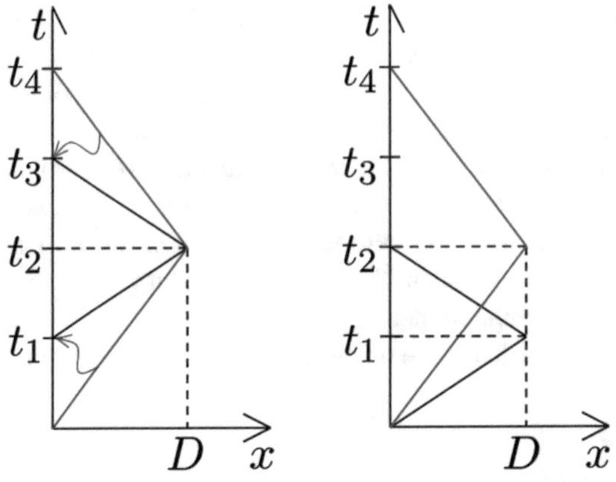

Fig. 12.4 Two travelers on the way to a prescribed spatial destination traveling at two different speeds. Case on the left: arrival at destination at the same time; and on the right: departure at the same time. In both travel cases each traveler needs the same personal proper time for the trip. The arrows on the left indicate the required path deformation aiming to overlay the paths. We see that the faster traveler has the longer path between the common events of both travelers. In this comparison the faster traveler with longer world line remains the youngest. See Example 12.3

We conclude that the shortest world line occurs for the traveler with $v_T \to c$. In the spatial destination travel case we find that the fastest traveler with the shortest proper time has the shortest world line path. This is just opposite to the result we derived in the prior Example 12.2.

However, if we include in the path length of the faster traveler the wait time at the base for the return of the slower traveler, there is no contradiction with our prior consideration. We see such a travel path on the right in Fig. 12.4, while on the left we kept the faster traveler longer at base to allow equal time arrival at the destination. The world lines of each traveler remain in both cases of equal length. However, we see on the left more clearly either by a graphic deformation or by twice using the triangle inequality that the full path, including wait time, of the faster traveler is longer. The reason is the time the faster traveler waits for the slower one. Naturally, the proper time and the space-time path length for the fastest traveler is always shortest (omitting wait times). ◄

12.3 *Essay:* Quantum Entanglement and Causality

This essay addresses questions requiring knowledge of special topics in quantum physics. It relates to the context of this book as it clarifies that long distance quantum entanglement does not conflict with causality.

The statement that the cause must precede the outcome by a time it takes a message to travel with speed of light has emerged in the context of quantum physics as not self-evident. The problem arises because 'quantum nonlocality' poses a challenge to classical local realism on which our conceptual view of the world is built and on which this book relies. To clarify what follows, proof of the 'violation of the Bell[1] inequality' means that quantum non-locality prevails over classical local realism.

We begin by citing a claim from an abstract of a prominently published research paper that there can be superluminal signaling, or in plain language, an exchange of information:[2]

> The experimental violation of Bell inequalities using spacelike separated measurements precludes the explanation of quantum correlations through causal influences propagating at subluminal speed. ... assuming the impossibility of using non-local correlations for superluminal communication, we exclude any possible explanation of quantum correlations in terms of influences propagating at any finite speed.

Let us consider S_1 as the originator of two quantum signals, and S_2, S_3 the two measuring observers. The above authors believe that it is the process of measurement at location s_2 by the observer S_2 that defines the state observed by S_3 at s_3 at a spacelike separation

$$0 > (s_2 - s_3)^2 = c^2(t_2 - t_3)^2 - (\boldsymbol{x}_2 - \boldsymbol{x}_3)^2 \,. \tag{12.6}$$

The spatial quantum nonlocality connecting spacelike distances in view of the above cited authors requires infinite speed. This topic begs for further discussion in these pages.

We are especially interested in the question: Is the instantaneous appearance of quantum information across large spacelike distances in conflict with causality? We argue now why this is not the case. It is of interest in our evaluation that neither S_3 nor S_2 know who measures first – *a priori* the observer closer to the source of the signal source S_1 measures first. However, spacelike separated observers S_2 and S_3 cannot communicate their location.

Who measures first, S_2 or S_3, is moreover a moot question in the context of relativity given Eq. (12.6). This is so since for spacelike separated events we can always find an observer S' for whom the sign of $\Delta t'$ is reversed, see Example 11.2, as compared to $\Delta t = t_2 - t_3$

$$\frac{t'_2 - t'_3}{t_2 - t_3} = \gamma\left(1 - \frac{v'}{c^2}\frac{x_2 - x_3}{t_2 - t_3}\right) < 0 \quad \text{when} \quad \frac{v'}{c} > c\,\frac{t_2 - t_3}{x_2 - x_3} \,. \tag{12.7}$$

The event sequence $t_2, \;\; t_3$ does not matter for another reason as well: The measurement outcome was set up by observer S_1 who created the entangled

[1]Named after the same John S. Bell who developed the 'Bell rocket' example extensively used in this book.

[2]J-D. Bancal, S. Pironio, A. Acin, Y-C. Liang, V. Scarani, & N. Gisin "Quantum non-locality based on finite-speed causal influences leads to superluminal signaling," Nature Physics **8**, 867 (2012).

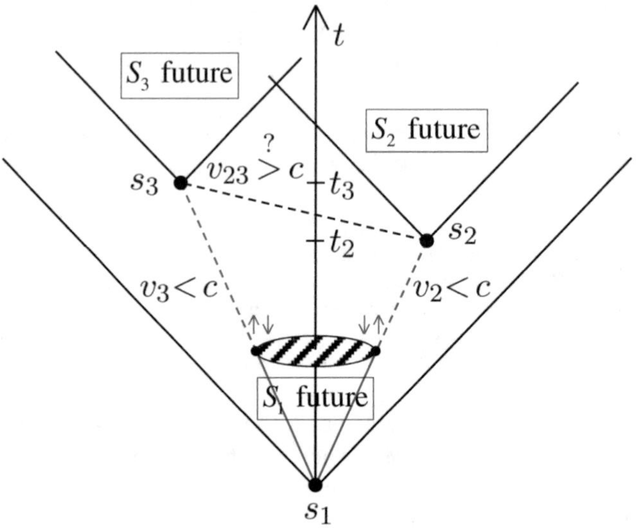

Fig. 12.5 The Minkowski diagram for a spin entangled experiment. We note the events s_2 and s_3 in the future cone of the observer S_1 who sets up the experiment. The entangled particles reach these events with speeds $v_2 < c$ and $v_3 < c$. Both s_2 and s_3 are outside the future light cone domain of the other, thus they have spacelike separation

quantum-non-local state and thus a list of probabilities for the measurement outcome. Thus the issue is that both S_2 and S_3 must be in causal sequence with S_1; the causal sequence of S_2 and S_3 does not matter at all.

The Minkowski diagram of the experiment is shown in Fig. 12.5. In the future domain of observer S_1 we see the shaded source of the spin-entangled signal observed at events s_2 and s_3. We see that by inspecting future domains of the observers S_2 and S_3 that they are not causally connected; this also means that a signal from s_2 requires superluminal speed to reach s_3. However, since Bell's inequality is violated, we cannot interpret the experiment applying classical local realism: Quantum mechanics addresses the entanglement differently, and a signal transmission is not required. The result of measurements at s_2 and s_3 has been set up by observer S_1 according to the currently understood principles of physics.

Having clarified that the presence of the observer S_1 assures that there is no conflict with relativity that quantum nonlocality entails, we could now close the discussion, especially so since there are no known conflicts between quantum physics and special relativity. However, there are a few related issues worth further discussion, for example the statement (*loc. cit.*) ... *we exclude any possible explanation of quantum correlations in terms of influences propagating at any finite speed.*

The laws of quantum physics at the microscopic scale are symmetric with regard to time arrow reversal, the so-called T-symmetry. In general, when one studies the time evolution of a quantum system to assure causal, i.e., 'arrow in time forward' evolution, we choose to consider the *forward* time arrow. The fact that we discovered direct and very weak T-breaking in behavior of certain elementary processes[3] has no immediate bearing on this argument.

A consequence of the analyticity (word used in the mathematical sense describing the behavior of complex functions) of quantum amplitudes is that a quantum particle tries to tunnel into causally forbidden domain outside the light cone, but can only go the distance consistent with the quantum uncertainty provided by the Compton wavelength $\lambda_C = h/mc$ of the particle (for an electron, $\lambda_{e\,C} = 2.4 \times 10^{-6}\,\mu$m, and for all other particles a much shorter distance[4]).

This *microscopic* quantum non-locality is a widely accepted feature of quantum physics, and addresses the behavior of one particle only; the macroscopic quantum non-locality requires two particles. For one particle a classical analogue to quantum coherence cannot be imagined (key phrase Schrodinger cat). For two and more particles one always can construct a classical model that in special situations produces an outcome that we expect for a quantum system. We keep in mind that Bell's inequality permits us to distinguish a classical simulation with hidden variables from the quantum physics case.

Let us next make the concept of quantum entanglement more precise: Source-observer S_1 sends out two particles which, by process of their creation, are 'connected' even though traveling in opposite directions. We speak of a fully correlated state when, for example, a measurement of the spin polarization of one particle predicts with certainty the behavior of the other, potentially at a very distant location. Entanglement means the non-local macroscopic presence of a quantum system, demonstrated by the fact that measurement performed 'here' implies an outcome 'there, far and away'. The conundrum is, how can a measurement outcome here influence the other location without causal time delay? This seems to touch the causality and faster than light information exchange.

Given this apparent conflict of quantum entanglement of particles with classical local realism, it would help if quantum physics non-locality were wrong – revealed by the Bell inequality as not being violated. However, we have seen definitive proof of the validity of quantum physics in a loophole proof demonstration of the violation of the Bell inequality by Hensen and colleagues[5] using entangled electron spins and detectors separated by 1.3 km. Two more such reconfirms of quantum physics

---

[3]The BABAR Collaboration: J.P. Lees et al., "Observation of Time Reversal Violation in the B0 Meson System," *Phys. Rev. Lett.* **109**, 211801 (2012).

[4]The situation with neutrinos is evolving: The three neutrinos may have masses that are much smaller – at the level of 0.01–0.1 eV which implies a quantum scale which is 5×10^6 times larger – at the scale of 10 μm. Moreover, we are with near certainty immersed in a sea of cosmic neutrinos, see for example: J. Birrell, and J. Rafelski, "Proposal for Resonant Detection of Relic Massive Neutrinos," *Eur. Phys. J. C* **75**, 91 (2015).

[5]B. Hensen et al., "Loophole-free Bell inequality violation using electron spins separated by 1.3 kilometers," *Nature Letter*. **526**, 682 (October 2015).

followed.[6] A neutrino based Bell-type experiment[7] shows quantum entanglement over the distance of 735 km.

We are certainly facing a logical challenge: How can a highly delocalized quantum system coordinate its behavior over a spacelike distance? We keep in mind that there cannot be any active communication between S_2 and S_3. All we know is that if either of the two observers S_2, S_3 measures one outcome, the other observer will measure a predictable, indeed prescribed, outcome. This does not establish faster than light communication, as we read the expected outcome off the quantum code table provided by S_1 who has set up the quantum-non-locality. It is the joint action of observers S_1 and S_2 that signals S_3, located at a timelike distance from S_1. Without the action of S_1 there is nothing to measure at S_3. Causality is assured. However, measurement by S_2 collapses the many outcomes defined by S_1 to one particular 'world'. This argument can be repeated when considering observers S_1 and S_3 signaling S_2, located at a timelike distance from S_1.

As it seems, the action of observer S_1 created virtual alternative worlds. The measurement at space-time event s_2 makes one of these real, creating the corresponding reality at the other location s_3. The analogous statement applies to the measurement at space-time event s_3 creating the reality also at s_2. Resolving the communication-causality conflict between events s_2 and s_3 we create the challenge of understanding how we collapse many possible future worlds into one reality valid everywhere. The solution to these questions awaits. At the first sight, this seems not to require any modification of relativity theory.

Even if the above discussion addressed a quantum phenomenon, it bears a resemblance to a classical physics context we now develop. This sharpens our understanding of the non-locality problem: An observer S_1 can predetermine what happens at the distant event s_2. For example, we (S_1) need to pre-program what a Mars rover at s_2 will do to assure that it can autonomously explore the distant surface for the period that signals take to make the return trip. We also call the rover observer S_2. More generally, S_1 can predetermine two events s_2 and s_3 associated with two observers S_2 and S_3 present in the future light cone of observer S_1, and being separated by spacelike distance $(s_2 - s_3)^2 < 0$. This is in particular precluding according to special relativity that S_2 and S_3 can communicate with each other.

S_1 can predetermine what happens to S_2 and S_3 in a way that is 'hidden', so that the two spacelike observers S_2 and S_3 are not necessarily aware of having been synchronized with respect to their behavior. The outcome could fool S_2 and S_3 into arguing that a superluminal information exchange has occurred between them. Here we keep in mind that the Bell inequality was invented to differentiate the quantum

[6]M. Giustina et al., "Significant-Loophole-Free Test of Bell's Theorem with Entangled Photons," *Phys. Rev. Lett.* **115**, 250401 (December 2015); L.K. Shalm et al., "Strong Loophole-Free Test of Local Realism," *Phys. Rev. Lett.* **115**, 250402 (December 2015).

[7]J.A. Formaggio, D.I. Kaiser, M.M. Murskyj, and T.E. Weiss, "Violation of the Leggett-Garg Inequality in Neutrino Oscillations," *Phys. Rev. Lett.* **117**, 050402 (2016).

world from one with such a hidden variable classical context, which we are now developing. Yet, let us proceed.

We note that additional complications arise when such a synchronization is carried out by S_1 without self-awareness of what she is up to. Now everybody involved, S_1, S_2, and S_3, can argue that faster than light communication is possible. How this happens we show by example; keep in mind we are discussing classical physics processes.

Let us assume that S_1 makes a radio broadcast. When it is received simultaneously at two opposite ends of your town, continent, planet, etc., with receiving observers S_2 and S_3 at spacelike separation $(s_2 - s_3)^2 < 0$, this does not break causality. Considering that everyone can hear the transmission of a radio station, the question of whether you turn your receiver on or not is the pivotal information content: While one listener knows what the other listener *could* hear, she does not know *if* the other observer was listening or not.

Following this line of thought, imagine the radio message instructs the spacelike separated observers S_2 and S_3 to message each other; these are events s_2 and s_3. Let there be another observer S_4 for whom the events s_2 and s_3 are in the past light cone and separated by the invariant $(s_2 - s_3)^2 < 0$ spacelike distance that is the same for all observers. S_4 reports that S_2 and S_3 were exchanging messages with each other communicating with superluminal speed. However, we know that the observers S_2 and S_3 were only executing commands introduced to them by S_1. All of this is just like the radio example; however now the observers S_2 and S_3 upon reception of instruction from S_1 appear to the observer S_4 to exchange information in superluminal fashion over spacelike distance $(s_2 - s_3)^2 < 0$.

In the final example S_1 informs both spacelike separated observers S_2 and S_3 how to incorporate a random amount of variance from certainty into the predetermined messages being exchanged, so that the messages S_2 and S_3 send each other are not fully predetermined but are the outcome of a probability-based game. Given the complexity of the game the observer S_4 is even more convinced that S_2 and S_3 are both aware of each other, and are messaging by superluminal communication.

To make the classical situation even more similar to the present day quantum entanglement experiments, let us imagine that the synchronizing communication that S_1 transmits to S_2 and S_3 was arranged by someone else (let us call this observer S_1') at the transmission station so that the actual S_1 is not even aware that the synchronization message with action instructions is sent out. To make things worse let S_4 be also S_1 but observing at a later time. Clearly, $S_4 = S_1(t + \Delta t)$ could claim she has observed superluminal communication between S_2 and S_3. She would be fooled into doing this by S_1' who may now be 'dead', and cannot rectify the claim made by S_4.

To sum up, the recognition of a possible predetermination of the measurement outcome removes in classical thinking the paradox of superluminal or even instantaneous communication in an example that mimics, but does not equal quantum entanglement. Yet this example naturally motivates in the quantum context the discussion of how S_1 by producing the entangled particles can establish a predetermined synchronization between events s_2 and s_3.

In a popular rendering of the recent results of Hensen and colleagues (*loc. cit.*) the science writer of *The Economist*[8] takes the matter much further:

> Just maybe ... all these counter intuitive findings ... were all predetermined at the Universe's birth, and all these experiments are playing out just as predetermined.

However: (a) Our example clarified that a 'grand' universal determinism is not needed. (b) Considering the Bell inequality the classical determinism in small or big is not the explanation. To date we have not found how to resolve the challenge that the two particle quantum non-locality presents. However, causality is not our problem; our understanding is challenged by the absence of local realism.

[8] *The Economist*, October 24, 2015, p. 77, "Hidden no more."

SR-Doppler Shift 13

Abstract

The relativistic Doppler light wavelength or frequency shift depends on relative velocity and yet it is reciprocal: Any two observer-sources will find a reciprocal Doppler shift which allows each of the two to identify the relative velocity vector. A distinction of motion of the light source or observer is not feasible. This effect cannot be described using time dilation, and we clarify contrary claims. We obtain the Doppler effect as a consequence of Einstein's recognition of light-wave phase being an invariant, i.e. the same for all inertial observers. We show that the aberration of the line of sight is an essential element required for the proper understanding of the Doppler shift.

13.1 Introducing the Nonrelativistic Doppler Shift

The phenomenon of motion-related change of sound frequency is a common experience today. The frequency is heard to be higher in the approach of a speeding sports car as compared to when the vehicle recedes. Fast cars, planes or trains were not available in 1842 when Doppler[1] considered the influence of binary star motion on the color of emitted light.

The classical (nonrelativistic) Doppler shift was seen as being due to a change in the time it takes for two wave crests to arrive in the detector, originating in the change in the separation between emitter and observer during this interval in time. It is easy to see, based on the wave crest idea, that when source and observer move toward each other, the frequency of wave crests increases, and the wavelength is

[1]Christian Andreas Doppler, (1803–1853), Professor of Mathematics and Practical Geometry at the Technical Institute (1841–1847) (now Czech Technical University) in Prague at the time of the Doppler shift announcement in 1842/1843.

J. Rafelski, *Modern Special Relativity*, https://doi.org/10.1007/978-3-030-54352-5_13

shorter (blue shift). Conversely when source and detector separate, the frequency of crests diminishes, and the wavelength is longer (redshift). This nonrelativistic shift thus does not exist if the relative motion of the source and observer is transverse to the line of sight; that is, the line connecting source and observer. In special relativity this shift exists and it is called the transverse, or quadratic, Doppler shift; quadratic since when expanding for small velocities one finds a leading shift proportional to $(v/c)^2$.

The original Doppler's theory applies to sound propagation in air. Air is a ponderable medium and thus sound is actually modified by the motion of the source with reference to the carrier of sound, the air. Typically either the observer or the source is at rest with respect to air. How the observer hears a moving object depends on the motion condition of the source with respect to the state of the carrier of the signal, the air.

After the development of SR and the banishment of material æther, the SR-Doppler shift of light must be understood in a different way. The light emitted by a source cannot depend on the state of motion of the source, since absolute motion is not discernible. Therefore the reader should apply the SR-Doppler shift as discussed in the following pages only to the study of light and not to the study of sound.

Consider two standard light sources, one in the laboratory, and another on a spaceship on an exploration mission in the Universe. We know, according to Chap. 12, that the spaceship bound clock is slow as compared to the time ticking in the lab. However, the individual motion of any of the two different observers, one in the lab, and the other in the spaceship, according to the theory of relativity, should not influence at the source the emitted signal. Both travelers, only aware of the standardized light signal emitted by the other body, should expect to observe the same signal subject to the same SR-Doppler shift, dependent only on relative velocity, that is what we call the relative and reciprocal SR-Doppler shift. Clearly the relative and reciprocal SR-Doppler shift is not influenced by the time dilation in one of the light sources.

To summarize, there are essential differences considering the relativistic Doppler effect of light compared with the Doppler shift of sound:

- Light travels in space in a way that must be completely independent of the state of motion of the source; there cannot be an effect of motion of the source that can be ascribed to properties of emitted light – otherwise we could tell the absolute velocity of the source.
- The SR-Doppler wavelength shift must be both **relative and reciprocal**.
- The (quadratic) SR-Doppler shift exists for motion transverse to the line of sight.

Since the SR Doppler (frequency or wavelength) shift is almost always accompanied by the effect of aberration of line of sight, we will speak of the 'Doppler effect' when referring to both the Doppler shift and the aberration effects. The Doppler shift refers to both the shift of light frequency and light wavelength.

Discussion 13.1 Doppler frequency formula

About the topic: We clarify how and why the light emitted by a distant observer is Doppler shifted by the effect of relative motion.

- *Simplicius:* I have been reading ahead in these fine pages about the Doppler shift and I wonder why Professor gave this matter so much attention.
- *Student:* The Doppler shift method allows the measurement of the speeds of stars in our galaxy, thus creating a map of motion and helping, for example, to disentangle the distribution of dark matter. The Doppler shift helps your GPS work properly. Do I need to continue to motivate your interest?
- *Simplicius:* Why not write down the good old Doppler? Just include a Lorentz-γ factor with what Doppler wrote down around 1842; this γ accounts for time dilation in the source. Basta! I think this could save some students headaches and reduce the number of pages in this book.
- *Student:* Really? Keep in mind light is always emitted just like we know it would be emitted by a hot star. It is emitted for all to see. The atoms on the star that radiate the light cannot possibly know that they are in motion with regard to the Earth observer who is going to measure them for example many thousand years later. Or, for that matter, any other observer far and away. In the absence of material æther the only thing that matters in the measurement of the signal is the relative motion state of the observer, many possible observers, each reporting a different measured frequency spectrum.
- *Simplicius:* Hearing this I feel foolish! Yes, the SR-Doppler shift must be generated in the process of measurement. However, that is not what I read in another book.
- *Student:* And yet, this is obvious: The light emitting observer sees the wave go away in his coordinate system and the light phase oscillates in space-time according to the emitter's set frequency and wavelength, that is light phase condition. However, the same phase oscillations in the other observer's frame of reference will be interpreted in a different fashion.
- *Simplicius:* This does not seem at all like the classic Doppler shift with sound waves!
- *Student:* Right, this is so since sound waves travel in air, a ponderable medium. We can incorporate in the interpretation of the signals not so much the relative motion between the loudspeaker and microphone, but the relative motion with respect to the air by either the loudspeaker or microphone or both.

(continued)

- *Simplicius:* I am still bothered by the fact that I, on this 'Earth' location, can discover the relative state of motion of a star out there.
- *Student:* This is possible if the light wave provides the information that is critical to determining the relative motion.
- *Professor:* Indeed! Einstein postulated in 1905 that the value of the phase of light is the same for all observers; today we say the phase of light is a Lorentz-invariant. An observer who sees the emitted light at its original frequency must be comoving with the light source. This means that the phase of light contains the information about source motion, which is recognized by an observer as relative velocity. This is accomplished by Lorentz-transforming the contents of the phase of light into the observer comoving system of coordinates. The phase information contents allows recognition of wavelength and frequency shifts, and a line of sight aberration. One can say that the relativistic Doppler effect is created in the measurement carried out by the observer.
- *Simplicius:* But how can an observer know there is an 'absolute' Doppler frequency or wavelength shift?
- *Professor:* This is only possible because the emitter uses a reference signal such as the spectral lines in hydrogen. There is also a remaining ambiguity in distinguishing the shift in frequency from the angle of aberration, as we will consider in the following.
- *Simplicius:* When can we ignore the aberration of the observation angle?
- *Professor:* For parallel motion only, when the line of sight is the line of relative motion.
- *Student:* To conclude, what is in your opinion the key insight and most important lesson this discussion conveys to all those preparing to teach SR?
- *Professor:* When teaching relativity many professors like to motivate students' understanding of the Doppler shift by introducing time dilation in the source. That is, however, a poor pedagogical shortcut. In my opinion it leads to pretty bad student comprehension of principles of the theory of relativity. The relativistic Doppler effect arises in the process of observation and not in the process of light emission: The source cannot have precognition about the relative motion state of the observer. Moreover, only when the aberration of the line of sight is accounted for, is the relation between source and observed frequencies reciprocal and allows consistent SR-Doppler formulas.

▶ **Insight: 13.1 SR-Doppler and the 'other' shifts** There are three *different* mechanisms influencing the observed frequency ν or, equivalently, the wavelength λ of (star)light, where $\lambda\nu = c$. These effects are due to distinct physical phenomena which are sometimes confounded:

1. **Relativistic Doppler shift:** The SR-Doppler shift that we discuss in detail in this book is due to relative motion between source and observer. The SR-Doppler shift is present within the special theory of relativity. It does not depend on either the presence of a force nor on how far light must travel. The magnitude of relative motion is established in the process of measurement by an observer, relying on the Lorentz invariance of the light wave phase.
2. **Gravitational redshift:** This author's point of view is that light emitted from within any gravitational potential well must do work to escape. This effect is in principle part of the 'general' theory of relativity; however, we describe it to a very good approximation by employing energy conservation within SR: A photon escaping from a stellar source will do work to get away and thus is always subject to a gravitational redshift. The effect has been demonstrated in a laboratory experiment[2] by R.V. Pound, and G.A. Rebka Jr. This effect is rarely confounded with the Doppler shift as it is due to photon interaction with gravity.
3. **Cosmological redshift:** A very distant source of light is in general not speeding with respect to the Universe CMB; thus we normally should expect that such light (atomic spectral lines) when emitted should be observed to be similar if not the same as the light emitted by a similar source today. However, this light travels far through an expanding Universe and in the time it takes to reach the detector on Earth, the free-streaming photons are subject to the expansion of the Universe that is 'stretching' the length and redshifting the frequency. The effect can be surprisingly large and grows with the time a photon takes to reach the observer, producing a cosmological energy redshift that grows with cosmological distance. Today it is widely agreed that this cosmological redshift is neither due to motion nor to direct action of a force. The understanding of redshift relation to source distance is today highly refined and has allowed experimental determination of the accelerated expansion of the Universe,[3] leading to the concept of repulsive gravity due to dark energy. Even so, the cosmological redshift is sometimes described in popular renditions as being due to

[2] R.V. Pound, and G.A. Rebka Jr. "Gravitational Red-Shift in Nuclear Resonance," *Physical Review Letters* **3** 439 (1959); "Apparent weight of photons," *ibid* **4** 337 (1960).

[3] The Nobel Prize in Physics 2011 was awarded to Saul Perlmutter, Brian P. Schmidt and Adam G. Riess "...for the discovery of the accelerating expansion of the Universe through observations of distant supernovae."

distant stars receding from us at increasing speed, the further away they are. Such characterization of the cosmological Doppler shift is incompatible with the wealth of experimental information.

13.2 Misunderstanding of the Relativistic Doppler Effect

The relativistic Doppler effect is frequently explained in terms of time dilation at the source and one even sees pictures of how wave crests vary due to velocity between source and the vacuum, the carrier of light. As we just argued, the magnitude of the SR-Doppler (frequency or wavelength) shift is in conceptual conflict with these concepts reminiscent of the sound Doppler shift combined with time dilation at the source, because there is no material æther. A SR-Doppler shift description in terms of two effects characteristic of the material æther is a deeply ingrained misunderstanding of SR that is found quite frequently in textbooks, online lectures, web resources etc. This point of view has even penetrated into the domain of research. Determining and explaining how we can understand the relativistic Doppler effect using SR will keep us busy for a few extra pages.

The relativistic Doppler effect originates in Einstein's postulate of Lorentz invariance of the light phase. This was a part of his 1905 SR publication, Ref. 18 in Chap. 1. In this publication on p. 915 (bottom) Einstein clarifies that all relativistic problems concerning propagation of light are to be resolved by Lorentz transforming into observer reference frame. This in particular applies to the EM-fields (called forces by Einstein), see §6, pp. 907 ff. The light wave appears in §7, pp. 910 ff under the subtitle: "Theory of the Doppler Principle and the Aberration". Einstein does not present explicit computations, but clarifies that EM-fields of a light wave must behave under Lorentz transformation in the same way as do all EM-fields. This is equivalent to the postulate of light phase invariance in the form

$$\Phi(t', x'; \nu', \lambda') = \Phi(t, x; \nu, \lambda) \,, \tag{13.1}$$

in Einstein's notation (see his equations on p. 910 bottom and p. 911 top).

We will obtain the Doppler effect using this postulate. This means that the phase of emitted light contains the information needed by any observer to determine the relative velocity with the source prevailing at the time of observation. Thus determining the Doppler effect is the light phase. This approach allows the Doppler effect to be relative and reciprocal. The time dilation at the light source is not required for the Doppler effect.

Let us begin answering the question how and why the time dilation argument entered the context of the SR-Doppler effect, see Ref. 8 in Chap. 3. We consider the hypothesis that the origin of the problem is in efforts to rush to STEM know-how in 1950–1960. Books were needed offering access to modern physics, e.g. SR and quantum physics for students at the freshman level. Quantum physics was well represented in many original sources and textbooks of the epoch in key science languages of this discovery epoch. The situation was fundamentally different with

SR. The SR sources were predominantly written in German and were older, from the pre-WWI era 1905–1915.

Among introductory SR texts the book by Max von Laue,[4] notable for its German text, was continuously updated for 50 years through 1960s. The discussion of the Doppler effect,[5] translated here according to its clearly implied scientific contents, reads as follows:

> both the Doppler γ-factor originating the new quadratic effect, and time dilation γ-factor originate in the γ-factor of the Lorentz transformation. This γ-factor is also seen in the time dilation effect.

Presumably the last comment is added to emphasize that the Lorentz transformation factor (von Laue actually refers to $1/\gamma = \sqrt{1-\beta^2}$) can be verified in different ways and thus the new 'quadratic' (in relative velocity after power law expansion) relativistic part of the Doppler shift is well established. Here one needs to remember the origin of the von Laue book in the 1910s, and Einstein's prior publication[6] in which Einstein proposes measurement of the quadratic effect as a test of his SR theory.

Von Laue is clear in his presentation of the Doppler effect. However, the text is difficult; the meaning of von Laue is only evident to a translator well versed in both SR and the German language. In many textbooks authored and coauthored by R. Resñick[7] we see a confounded rendition of the Max von Laue text: "It is instructive to note that the transverse Doppler effect has a simple time-dilation interpretation …The transverse Doppler effect is another physical example confirming the relativistic time dilation." The introduction of the words *interpretation* and *confirming time dilation* not given in the Max von Laue original is the conceptual mistake that confounded and confounds student learning especially in the US but also elsewhere.

The above is one among many examples. It is hard if not impossible to find an introduction to SR where the Doppler effect is discussed correctly and unambiguously. For example we read in Wikipedia (sourced June 20, 2020) in the opening:

> The relativistic Doppler effect is different from the non-relativistic Doppler effect as the equations include the time dilation effect of special relativity and do not involve the medium of propagation as a reference point.

---

[4]Max von Laue, *Die Relativitätstheorie* (Braunschweig: Vieweg 1960), 7 durchgesehene Auflage (7th reviewed edition).

[5]German in its original is found on pp. 101–102, Max von Laue, *loc.cit.*: "Die Wurzel $\sqrt{1-\beta^2}$ in 14.8 welche den quadratischen Doppler-Effekt bedingt, stammt, wie man an Hand der Rechnung Zurüverfolgt, aus dem Nenner $\sqrt{1-\beta^2}$ der Transformation (4.7). Da wir aus demselben Nenner in §5a auf die Zeitdilatation schlossen, finden wir auch hier den Zusammenhang zwischen dem quadratischen Effekt und der Verlangsamung des Uhrenganges."

[6]A. Einstein, "Über die Möglichkeit einer neuen Prüfung des Relativitätsprinzips (On the Possibility of a New Test of the Relativity Principle)," *Annalen der Physik* (ser. 4), **23**, 197–198 (1907).

[7]See for example: R. Resnick, *Introduction to Special Relativity* (New York: J. Wiley, 1968), pp. 90–91; D. Halliday, R. Resnick, and K. S. Krane, *Physics*, Vol. **2** (New York: J. Wiley 2002). 4th ed., pp. 897–898.

These claims do not agree with our findings:

- **On the cause of the Doppler effect** The Doppler (frequency or wavelength) shift effect follows from Einstein's postulate of the phase of light-wave invariance. The Doppler effect shares the Lorentz-γ-factor with many relativistic effects, e.g. also time dilation. However, the Doppler effect has nothing in common with time dilation. The fact that some SR experiments, see Sect. 19.5, used time dilation language does not change the circumstance.
- **Medium of propagation** In many of his writings Einstein clearly stated beginning with 1920 that the propagation of light is evidence for specific properties of the Lorentz invariant æther, as we discussed in Sect. 2.3, see also Conversation Ref. [2.2].
- **Reference point** The invariant phase of emitted light-wave is the reference point we observe that allows us to determine relative motion with respect to a distant source emitting the light long before the present day observer's relative motion to the source could be anticipated.

One can wonder how the German language Wikipedia handles the matter, considering the availability of original SR source material to the Wiki-authors. We find an article that deals both with sound and light Wikipedia-D (sourced June 20, 2020) where we read:[8]

> The finite speed of light propagation is the origin of the relativistic Doppler effect viewed as a geometrical property of space-time.

The French language (language of Poincaré and Langevin) article is even more mystifying on the subject.

13.3 SR-Aberration of Light

Light, as Maxwell realized more than 150 years ago, was a wave solution of the equations we today call Maxwell equations.[9] These are transverse waves which comprise the 'plane-wave' factor

$$W(x,t) = \cos\left(\omega t - \boldsymbol{x}\cdot\boldsymbol{k}\right) = \cos\left(2\pi\nu t - 2\pi\,\boldsymbol{x}\cdot\boldsymbol{n}/\lambda\right) \ . \tag{13.2}$$

We further introduce

$$\boldsymbol{k} = 2\pi\boldsymbol{n}/\lambda \ , \qquad \omega = 2\pi\nu \ . \tag{13.3}$$

[8]Translated by the author from the original: "Relativistische Doppler-Effekt ist darauf zurückzuführen, dass die Wellen sich mit endlicher Geschwindigkeit, nämlich der Lichtgeschwindigkeit ausbreiten. Man kann ihn als geometrischen Effekt der Raumzeit auffassen."

[9]Einstein refers in his 1905 paper to 'Maxwell-Hertz' equations, see Ref. 13 e.g. p. 29.

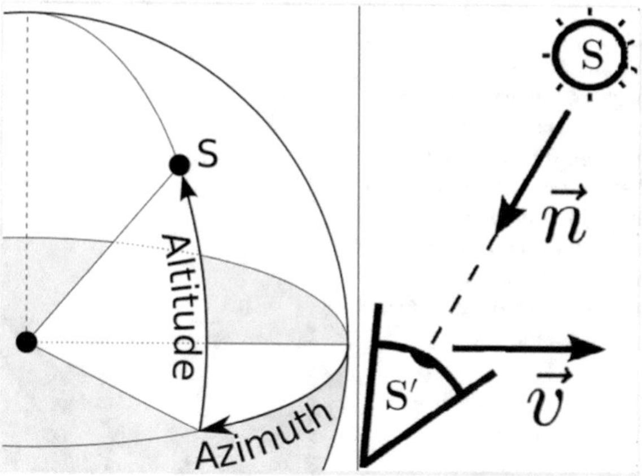

Fig. 13.1 Left: Definition of viewing altitude and azimuth angles. Right: The light emitted in the direction $\boldsymbol{n}$ from the source S is observed by S' moving at relative velocity $\boldsymbol{v}$

This is the light wave of wavelength λ emitted from a source S at rest with coordinates t, x in direction $\boldsymbol{n}$, as shown in Fig. 13.1. For any wave of light the crest of the wave $\omega t - \boldsymbol{x} \cdot \boldsymbol{k} = 0$ travels at the velocity of light since $d\boldsymbol{x}/dt = c\boldsymbol{n}$. Thus

$$\omega^2 - c^2\boldsymbol{k}^2 = 0 , \qquad |\boldsymbol{k}| = \omega/c , \rightarrow \lambda = c/\nu . \tag{13.4}$$

The observer S' in Fig. 13.1 is in motion with respect to the reference system S, the source of light. Light arrives more horizontally as discussed before in Sect. 2.3 and as shown in Fig. 13.2. Let θ be the altitude angle for $v = 0$ and θ' the actually observed altitude angle; see Fig. 13.2, then the aberration angle is

$$\alpha \equiv \theta - \theta' . \tag{13.5}$$

We determine this angle using the addition theorem of velocities, for now without the need to use the light phase invariance.

We now consider a light ray originating in a distant star. We are addressing this situation for light being a well focused incoherent light-ray created in independent atomic processes. We choose the direction of the z'-axis of observer frame of reference S' and the direction of the z-axis of the source system S without loss of generality to be parallel to the relative velocity vector $\boldsymbol{v}$. With this choice the angles θ and θ' are between the real and virtual lines of sight and the velocity vector $\boldsymbol{v}$ as seen in Fig. 13.2. In the reference frame S we choose the direction of the x-coordinate axis so that it points away from the Earth. The direction of the y-axis follows. The velocity components of the light ray emitted in S are described by the unit vector $\boldsymbol{n}$ pointing 'down' at the Earth, see Fig. 13.1. With this the velocity

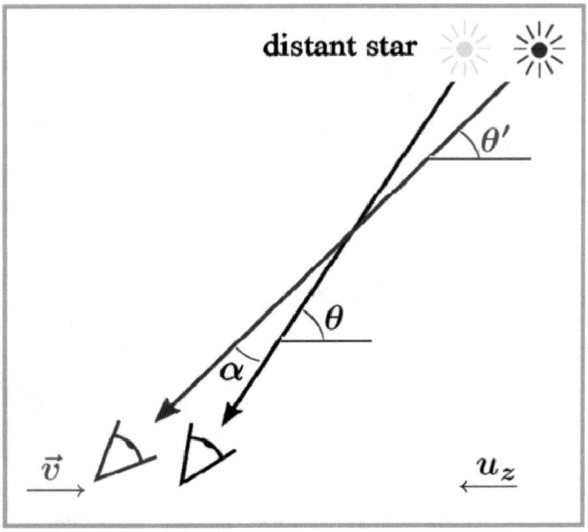

Fig. 13.2 Aberration angle α of light from a distant star: Comparison of the line of sight for an observer at rest and for another observer traveling at a constant velocity $\boldsymbol{v}$ relative to the line of sight. θ is the altitude angle for $v = 0$ while θ' is the observed virtual altitude angle. u_z is the z-component of the photon velocity

components of the incoming photons are

$$u_x = cn_x = -c\sin\theta\cos\varphi\ , \quad u_y = cn_y = -c\sin\theta\sin\varphi\ , \quad u_z = cn_z = -c\cos\theta\ . \tag{13.6}$$

φ is the azimuth angle, see Fig. 13.1. The z-component of the photon velocity points against the direction of $\boldsymbol{v}$ as we also see in Eq. (13.6) and is seen in Fig. 13.2.

We are seeking to determine the new velocity vector $\boldsymbol{u}'$ of the light ray as observed on Earth. We use the velocity addition theorem, Eq. (7.18a), where now the transformation is along the z- and not the x-axis. In the Earth's rest-frame S' the light ray velocity $\boldsymbol{u}'$ becomes:

$$\begin{aligned}
u'_x &= \frac{u_x}{\gamma(1+v'u_z/c^2)} = -\frac{c\ \sin\theta\ \cos\varphi}{\gamma(1+(v/c)\cos\theta)} \equiv -c\sin\theta'\cos\varphi' = cn'_x\ ,\\
u'_y &= \frac{u_y}{\gamma(1+v'u_z/c^2)} = -\frac{c\ \sin\theta\ \sin\varphi}{\gamma(1+(v/c)\cos\theta)} \equiv -c\sin\theta'\sin\varphi' = cn'_y\ ,\\
u'_z &= \frac{u_z+v'}{1+v'u_z/c^2} = -\frac{c\ \cos\theta+v}{1+(v/c)\cos\theta} \equiv -c\cos\theta' = cn'_z\ ,
\end{aligned} \tag{13.7}$$

where as usual $\gamma = 1/\sqrt{1-v^2/c^2} = \gamma'$, since $\boldsymbol{v}' = -\boldsymbol{v}$; it is easy to verify $u'^2_x/c^2 + u'^2_y/c^2 + u'^2_z/c^2 = 1$ as in Example 7.9. The velocity v' is that of the source S as seen by Earthbound S'. The last equality on the right hand side of Eq. (13.7) reminds us that according to Eq. (13.6) we also obtained the Lorentz-transformed components of the unit vector $\boldsymbol{n}'$ describing the virtual line of sight observed by S'.

On the right-hand side in Eq. (13.7) we have introduced the observation angles θ', φ' that an observer on Earth reports for the location of the observer S. Thus Eq. (13.7) establishes a velocity-dependent relation between the two sets of angles θ, φ and θ', φ' for the line of sight that connects the two systems: the source S and observer S' moving with relative velocity $\boldsymbol{v}$.

Comparing the two first expressions in Eq. (13.7) we see differences in $\sin\varphi$ and $\cos\varphi$ factors. Forming the ratio of the right- and left-hand sides respectively we obtain

$$\tan\varphi = \tan\varphi' \quad \rightarrow \quad \varphi = \varphi' \,, \tag{13.8}$$

and the azimuthal aberration vanishes. However, according to the last expression in Eq. (13.7) the altitude aberration α does not vanish:

$$\cos\theta' \equiv \frac{\cos\theta + (v/c)}{1 + (v/c)\cos\theta} \simeq \cos\theta + \frac{v}{c}\sin^2\theta - \left(\frac{v}{c}\right)^2 \cos\theta \sin^2\theta + \cdots \,. \tag{13.9}$$

We note that for $v > 0$ (motion of observer to right), $\cos\theta' > \cos\theta$ and thus $\theta' < \theta$ for any altitude angle $\theta \in \{0, \pi/2\}$; see Fig. 13.2. Recall that we measure the angles θ', θ against the z-axis; thus $\theta' < \theta$ means that light is seen arriving from a direction closer to the direction of motion. For $\theta \rightarrow 0$ the altitude aberration vanishes; this is the case of motion along the line of sight.

We next check the relativistic consistency of Eq. (13.9). Inverting Eq. (13.9) to obtain $\cos\theta$, we find

$$\cos\theta = \frac{\cos\theta' - (v/c)}{1 - (v/c)\cos\theta'} \qquad \Longleftrightarrow \qquad \cos\theta' = \frac{\cos\theta + (v/c)}{1 + (v/c)\cos\theta} \,. \tag{13.10}$$

The altitude aberration expressions Eqs. (13.5), (13.9), and (13.10) show the equivalence between the two observers S and S' with respect to

$$v \Leftrightarrow -v \,, \quad \theta \Leftrightarrow \theta' \quad \rightarrow \quad \alpha \Leftrightarrow -\alpha \,. \tag{13.11}$$

Equation (13.10) demonstrates that the effect of aberration for an observer S is reciprocal to that for the observer S', Eq. (13.9), see also Eq. (13.11). This reciprocity expresses the principles of relativity. In fact we could use the principle of relativity to obtain our result: By the nature of the problem the speed of light must enter into the answer, and by dimensional considerations the effect has to involve the ratio v/c. Since the nonrelativistic limit has to arise, we know that the aberration effect begins with the linear power v/c. In order to assure symmetry of the result between observers on Earth and on the Star, as described by Eq. (13.11), the functional dependence we found, Eq. (13.9), applying the velocity addition theorem, is uniquely defined alone by qualitative application of the principle of relativity, and respecting the nonrelativistic limit.

The relativistically consistent expression Eq. (13.10) allows the computation in terms of the Earth observed altitude angle θ' the 'relative rest' altitude θ provided that we know the relative velocity vector. Sometimes one speaks of θ as the 'true' altitude. However, since in SR all inertial observers are equivalent, one cannot look at the observer at relative rest as being in any way special.

We now discuss the magnitude of the effects we expect. Considering for the relative star motion the range of values $v/c < \mathcal{O}(10^{-3})$, we can neglect quadratic terms[10] in the nonrelativistic expansion shown in Eq. (13.9). In the small aberration angle limit we use

$$\cos\theta' = \cos(\theta - \alpha) = \cos\alpha\cos\theta + \sin\alpha\sin\theta \simeq \cos\theta + \alpha\sin\theta \ . \qquad (13.12)$$

The linear term in the nonrelativistic expansion shown in Eq. (13.9) provides

$$\alpha = \frac{v_\perp}{c}, \qquad v_\perp = v\sin\theta \ . \qquad (13.13)$$

This is the final aberration result in the nonrelativistic limit: The relative speed transverse to the line of sight fixes the aberration angle α. The parallel to line of sight speed $v_\parallel = v\cos\theta$ does not influence the magnitude of aberration in nonrelativistic limit.

Discussion 13.2 Light aberration and æther drag

About the topic: Aberration is a topic recurrent in the context of æther drag theories and the "relativity must be wrong" argument; i.e., it is easy to argue that the aberration effect is a measurement of the absolute speed of the æther near the Earth. Is that really the case?

- *Simplicius:* I saw a book where the author argues that since one measures the absolute speed of Earth by observing fixed star periodic aberration, one actually observes the æther motion around the Sun.

(continued)

[10]The quadratic Doppler effect was observed for the star S2 circling with up to $v \simeq 0.026c$ the central massive object in the Milky Way: GRAVITY Collaboration – R. Abuter et al. "Detection of the gravitational redshift in the orbit of the star S2 near the Galactic centre massive black hole" Astronomy & Astrophysics **615**, L15 (2018).

- *Professor:* About 150 years ago in the days of the MM experiment the motion of the stars in the sky was believed to be due to the dragged æther solar orbiting velocity.
- *Student:* Today we routinely Doppler-measure the relative velocity with many stars rather than the orbital speed of the Earth or the æther. The emphasis is on 'relative'.
- *Simplicius:* The point the author makes is that by looking at many fixed stars distributed uniformly in the sky we measure exclusively the Earth's orbital speed around the Sun. The author concludes that the measurement is that of the motion of the material æther that accompanies the Earth in the motion around the Sun.
- *Professor:* I recommend a cautious approach. It is probable that some almost 200-year-old ideas about material æther were rediscovered, without rediscovery of the arguments that countered these proposals.
- *Student:* I read on Wikipedia that the aberration effect had a big influence on the development of the understanding leading to the invention of relativity. Wikipedia points out that the hypothesis of the motion of the material æther could not account for the effect of aberration – there is even a graphic simulation to explain why. It was a winding research path so it is likely that some of these nearly 200-year-old ideas about the material æther are being rediscovered – without the discovery of arguments that led to their demise.
- *Simplicius:* I also looked at Wikipedia; that is a very theoretical web page on "aberration of light". What I see in the book I was reading are lots of experimental results addressing the aberration of fixed stars placed in all directions in the sky. When the aberration data is analyzed, the author finds the orbital speed of the Earth in each case, so he attributes the effect to the motion of the dragged æther. I am more inclined to believe the experiments. A theory could be wrong.
- *Professor:* The stars we can observe easily are in general located nearby in our Galaxy – at most a few hundred ly away. These nearby stars move along with us in the Milky Way, and do this in a manner that renders the aberration effect to be often caused mainly by the motion of the Earth around the Sun with the speed of 30 km/s. Stated in technical terms, for many nearby stars the motion transverse to the line of sight is dominated by the Earth's orbital motion around the Sun.
- *Simplicius:* What you say also means that there should be some stars for which the analysis would not give the Earth speed as 30 km/s ....
- *Professor:* ...and you wonder why this book does not mention that? In my research domain I also see good people who are inclined to reject measurements that contradict their view of the world.

(continued)

- *Student:* I see this all the time when grading labs: My students only retain results they know fit what they should find.
- *Professor:* There are people who look at experimental data selectively. Both by picking the results they obtain, and by making the experimental data they do not like have a large error. To understand how such habits form consider what would happen should a student in a lab report present a result suggesting Newton or Hooke was wrong; could she expect a good grade? I imagine students 'learn' in their introductory labs to pick out 'good' data.
- *Simplicius:* I would never do that, would I …? So the table I saw was cherry-picked data? And the measurements that did not fit were ignored? I can see now how to argue with my friend who loaned me that book. What other advice do you have?
- *Professor:* To avoid protracted argument and doubt, I would focus on experimental issues. You can argue: One should consider a series of precise annual rhythm aberration measurements choosing stars randomly. Given the wealth of data available today, one finds cases that contradict æther drag. This can be done without any additional experimental effort.
- *Simplicius:* The book shows so many data points …
- *Student:* … you do not need to find an equal number of opposite examples. It suffices to consider a few results contradicting a false hypothesis – in principle one case that strongly contradicts a hypothesis is enough. However, a few are surely more convincing.

Example 13.1

Light aberration of Gamma Draconis

Background: Today we know that the London zenith star Gamma Draconis is approaching the solar system with the speed 28.19 ± 0.36 km/s. The proper motion (angle expressed transverse speed) of Gamma Draconis is -8.48 mas/yr (milli-arc-sec/year). The observed aberration motion ± 20.2 as/yr of Gamma Draconis is an effect 2000 times larger, and must be thus dominated by the Earth motion around the Sun, just as was assumed by James Bradley, see Sect. 2.3 and Ref. 3 on p. 23.

Using the aberration effect of the London Zenith-Star Gamma Draconis obtain the speed of light. ◀

▶ **Solution** We begin stating the relevant motion components of the Earthbound observer S' in reference to the London zenith star, Gamma Draconis: For S', the velocity vector $\boldsymbol{v}$ includes the Sun's relative motion $\boldsymbol{v}_\odot$ with respect to the observed star, modulated by the annual rhythm of Earth's orbital velocity $\boldsymbol{v}_\oplus$ around the Sun, and by the daily rotation velocity $\boldsymbol{v}_{\text{rot}}$ of the Earthbound observatory:

1 $\boldsymbol{v} = \boldsymbol{v}_\odot + \boldsymbol{v}_\oplus + \boldsymbol{v}_{\text{rot}}$.

We drop Earth rotation $\boldsymbol{v}_{\text{rot}}$ ($v_{\text{rot}} \simeq 0.46\,\text{km/s}$) from any further discussion; this is possible considering that Earth's orbital speed $\boldsymbol{v}_{\oplus} \simeq 30\,\text{km/s}$ is 67 times greater. We expect that $|\boldsymbol{v}| \ll c$ and thus consider the effect of nonrelativistic aberration.

Only the transverse to line of sight relative velocity enters the nonrelativistic aberration effect, Eq. (13.13). The transverse velocity component can contain contributions from orbital Earth motion and relative star motion. However, the relative star speeds $v_{\odot}$ between the Sun and easily observed stars in our neighborhood range from essentially vanishing to barely greater than the Earth orbital speed. This is so since the nearby stars move along with the Earth around the galactic center. Some of these are good 'fixed stars', with relative motion in the direction of the line of sight. In this situation they should appear at the same sky location.

The London zenith star moves directly at us; thus this motion has a negligible impact on its angular proper motion. However, the aberration effect is due to Earth orbital motion. In fact this particular excellent fixed star was selected because Molyneux and Bradley hoped that due to its properties they could measure the distance using the parallax effect. The distance turned out to be too far. Bradley realized that the dominant transverse to line of sight motion originated in Earth orbital velocity $\boldsymbol{v}_{\oplus}$. Since Gamma Draconis is located in the zenith normal to the Earth orbital plane, only the magnitude $|\boldsymbol{v}_{\oplus}|$ is of direct relevance.

While the Earth makes a full orbit around the Sun, the Earth velocity vector $\boldsymbol{v}\,'$ produces a mirror image of the Earth's motion in the aberration angle with extreme values measured by Molyneux and Bradley to be

2 $$\delta\alpha(t) \in (-20.2'', +20.2'') \,.$$

In formula Eq. (13.13) we use radians for the angular aberration α. We recall that

3 $$1\,\text{rad} \equiv 360/2\pi = 57.2958\cdots{}^{\circ} = 57.2958^{\circ} \times 60' \times 60'' = 20.6265 \times 10^4\,\text{as} \,.$$

We now convert the observed aberration to radians and interpret the result in terms of the transverse speed, Eq. (13.13)

4 $$\frac{20.2\,\text{as}}{20.6265 \times 10^4\,\text{as}} = \frac{1}{10{,}200} = \frac{v_{\perp}}{c} \,.$$

According to our discussion $v_{\perp} \simeq |\boldsymbol{v}_{\oplus}| = 30\,\text{km/s}$, and we find $c \simeq 300{,}000\,\text{km/s}$ as was reported by Bradley, see Sect. 2.3. ◀

Example 13.2

Ultrarelativistic aberration of light.

Calculate the aberration of the observation angle for a isotropic emitting source moving with ultrarelativistic velocity $\boldsymbol{v}$ with respect to the laboratory observer. ◀

▶ **Solution** The aberration formula Eq. (13.9) is not suitable for this problem; we lost 'relativistic' sensitivity when deriving Eq. (13.9).

Fortunately, this is easily rectified. An equivalent form of Eq. (13.9) is obtained using Eq. (13.8) in one of the first two expressions in Eq. (13.7)

$$\mathbf{1}\;\; \sin\theta' = \frac{\sin\theta}{\gamma(1 + (v/c)\cos\theta)} .$$

Adding the square of the above result to the square of Eq. (13.9) and using $\sin^2\theta' + \cos^2\theta' = 1$ verifies the consistency of both results. Taking instead the ratio of these results we obtain

$$\mathbf{2}\;\; \cot\theta' = \frac{\gamma(\cos\theta + (v/c))}{\sin\theta} = \gamma\cot\theta + \frac{\gamma v/c}{\sin\theta} ,$$

a useful format when considering the case $v/c \to 1$: We now see the explicit value of the Lorentz factor $\gamma \gg 1$ which can be arbitrarily large. The limit $v \to c$ is therefore controllable.

We consider the case of light emitted by a relativistic fireball formed for example in particle collisions. The observer is at rest in the laboratory in which the fireball is formed at a speed $v_F > 0$ moving to the right. For the relative velocity where we refer to the observer as we introduced previously we thus should use $v = -|v_F|$. The fireball emits radiation isotropically in all directions in its rest-frame; i.e., uniformly as a function of θ. However, this radiation is not observed to be isotropic in the laboratory when $|v| \to c$ in consideration of the strong aberration effect. The two equivalent aberration results are

$$\mathbf{3}\;\; \sin\theta' = \frac{\sin\theta}{\gamma_F(1 - (|v_F|/c)\cos\theta)}, \qquad \cot\theta' = \gamma_F\cot\theta - \frac{\gamma_F|v_F|/c}{\sin\theta} .$$

To see how this relation works, take as an example $\theta = 45°$ such that $\sin\theta = \cos\theta = 1/\sqrt{2}$ and assume that the speed v is sufficiently ultrarelativistic to justify use of $v/c = 1$ in the first form given above. Hence we find $\sin\theta \simeq \theta = 2.4/\gamma$, thus for $\gamma = 100$ the angle of emission in the lab is $\theta' = 0.024 \times 360°/(2\pi) = 1.4°$. We recognize that the radiation is observed 'focused' into a forward cone (small θ') for large γ.

Given the experimental arrangement, in the case of a forward emission the source is catching up with the emitted particles. On the other hand, for a backward emission, the source runs away from the emitted particles. Therefore, for backward emission, the sign in the first denominator in first term above changes. We see that the aberration effect for back emission is more pronounced. ◀

13.4 SR-Doppler Shift

We adopt the conventions introduced in Sect. 13.3: We orient our coordinate systems such that the observer S' travels in the direction of the z'-axis, which is also the z-axis of the source S. The line of sight is characterized by the light we observe i.e. the unit vector $\boldsymbol{n}$, such that $\boldsymbol{k} = k\boldsymbol{n}$, Eq. (13.3) and Fig. 13.1. For simplicity and without loss of generality we study a plane wave traveling along the line of sight from source S to observer S'. This situation corresponds to the aberration situation Fig. 13.2 and the aberration of light discussion in Sect. 13.3.

However, our interest is now in the study of the relativistic Doppler shift. For this we need to explore the properties of the phase of the light wave Eq. (13.2)

$$\Phi = \omega t - \boldsymbol{x} \cdot \boldsymbol{k} = \omega t - \boldsymbol{x} \cdot \boldsymbol{n}|\boldsymbol{k}| = \omega/c(ct - \boldsymbol{x} \cdot \boldsymbol{n}) \,. \tag{13.14}$$

The magnitude of this phase Φ should not depend on the observer, see Eq. (13.1).

A different set of coordinates is attached to each observer, e.g. for S it is $t, \boldsymbol{x}$ and for S' it is $t', \boldsymbol{x}'$. We will now show that to keep the phase Φ independent of the relative motion between S and S' there must be relativistic Doppler shift in the magnitude of light frequency ω and wavelength λ. We first consider the case of motion parallel to the line of sight which leads us to the well-known relativistic Doppler shift formulas. In a second step for the case of arbitrary relative motion we find a combination of aberration with relativistic Doppler shift.

$\boldsymbol{v} \| \boldsymbol{n}$: Motion parallel to line of sight

The alignment of $\boldsymbol{n}$ and $\boldsymbol{v}$ means

$$k_x = 0 \,, \quad k_y = 0 \,, \quad k_z = -k = -\omega/c \,, \; \rightarrow \; c\Phi = \omega(ct + z) \,, \tag{13.15}$$

where we include a minus sign in consideration of the orientation of the light source coordinate system S, and the direction of emission down towards the Earth. We need to Lorentz transform in Eq. (13.15) the coordinates of the light source S into Earth observer S' comoving coordinates. A source at rest $z = 0$ in S has in S' the speed $v' = z'/t'$. Thus we are using the motion of the source observed by S'. With Lorentz transformation $z = \gamma'(z' - v't')$ we obtain

$$c\Phi = \omega(ct + z) = \omega\gamma' \left([ct' - (v'/c)\, z'] + [z' - (v'/c)\, ct']\right) = \omega\gamma'(1 - v'/c)(ct' + z') \,. \tag{13.16}$$

The phase Φ, Eq. (13.16), as we have discussed, does not change in value when measured by different observers

$$c\Phi = c\Phi' = \omega'(ct' + z') \,. \tag{13.17}$$

Comparing Eq. (13.16) with Eq. (13.17) we recognize

$$\omega' = \omega\gamma'(1 - v'/c) . \tag{13.18}$$

As noted before v' is the velocity of the source along the line of sight which is the z-axis as seen by the Earth observer S'. For $v' > 0$ the source moves away from the observer, the frequency ω is shifted to smaller value, which means that the wavelength is shifted to a higher value; this is the redshift. On the other hand, when the source approaches the observer we have $v' < 0$; the frequency ω increases; the wavelength decreases, and we have a blueshift.

Equation 13.18 is the relativistic Doppler frequency shift for motion along the line of sight where there is no aberration. However Eq. (13.18) as stated has an unusual notation:

The notation choice of primed (moving observer) and not primed (source) coordinates is made in order to relate our argument to the study of Lorentz coordinate transformation properties as presented in this book. However, in the final reference formulas describing the Doppler shift we revert in the following to the usual textbook notation: It is common to denote the frequency and the wavelength of the light source by subscript zero; that is $\nu, \lambda \to \nu_0, \lambda_0$. In addition we switch $v' \to -v$ where v is the relative motion of the observer as seen from the source. Once this is done, the prime on the observed quantities can be dropped.

In this conventional notation according to Eq. (13.18) the Doppler shifted frequency $\nu = \omega/2\pi$ of the observer moving toward the source ν_0 at the speed $v > 0$ parallel to the line of sight

$$\nu = \nu_0 \frac{1 + v/c}{\sqrt{1 - (v/c)^2}} = \nu_0 \sqrt{\frac{1 + v/c}{1 - v/c}} ; \tag{13.19}$$

that is, observed $\nu > \nu_0$. This result applies also to the wavenumber vector $\boldsymbol{k}, \ \boldsymbol{k}_0$ when the velocity vector $\boldsymbol{v}$ is parallel to direction of $\boldsymbol{k}$, since

$$\frac{k}{2\pi} = \frac{1}{\lambda} = \frac{\nu}{c} . \tag{13.20}$$

Inverting Eq. (13.19) we further find

$$\nu_0 = \nu \sqrt{\frac{1 - v/c}{1 + v/c}} = \nu \frac{1 - v/c}{\sqrt{1 - (v/c)^2}} , \tag{13.21}$$

and the same for k_0, k. This follows from Eq. (13.19) by applying transformations $\nu \Leftrightarrow \nu_0$ for $v \Leftrightarrow -v$ as required by the principle of relativity.

The observed wavelength λ, according to Eq. (13.19) and using Eq. (13.4) is

$$\lambda \equiv \frac{c}{\nu} = \lambda_0 \frac{\sqrt{1 - (v/c)^2}}{1 + v/c} = \lambda_0 \sqrt{\frac{1 - v/c}{1 + v/c}} = \lambda_0 \frac{1 + v/c}{\sqrt{1 - (v/c)^2}} . \tag{13.22}$$

We find the expected blueshifted $\lambda < \lambda_0$ wavelength for the observer and source approaching each other. Conversely, observing a redshift we conclude that the observer and source are moving apart.

General case of arbitrary $\boldsymbol{v}, \boldsymbol{n}$

We now relax the condition that the line of sight is parallel to the relative motion of the source and the observer and introduce a general directional vector $\boldsymbol{n}$ for the emitted light, see Fig. 13.1

$$\boldsymbol{k} \equiv (\omega/c)\boldsymbol{n} , \tag{13.23a}$$

where

$$k_x = -\frac{\omega}{c} \sin\theta \cos\varphi , \quad k_y = -\frac{\omega}{c} \sin\theta \sin\varphi , \quad k_z = -\frac{\omega}{c} \cos\theta . \tag{13.23b}$$

The phase of light we consider is

$$c\Phi = \omega(ct - \boldsymbol{n} \cdot \boldsymbol{x}) = \omega \left(\gamma'[ct' - (v'/c)z'] + \gamma'[z' - (v'/c)ct'] \cos\theta - \boldsymbol{x}'_\perp \cdot \boldsymbol{n}_\perp \right) , \tag{13.24a}$$

where the transverse to relative motion component is

$$-\boldsymbol{x}'_\perp \cdot \boldsymbol{n}_\perp = x' \sin\theta \cos\varphi + y' \sin\theta \sin\varphi . \tag{13.24b}$$

This is to be compared to

$$c\Phi' = \omega' \left(ct' - \boldsymbol{n}' \cdot \boldsymbol{x}' \right) = \omega' \left(ct' + z' \cos\theta' + x' \sin\theta' \cos\varphi' + y' \sin\theta' \sin\varphi' \right) . \tag{13.25}$$

Our objective is to find $\omega', \theta', \varphi'$ as functions of not primed quantities, and due to the reciprocity between a source and the observer we have $\gamma' = \gamma,\ v' = -v$. To accomplish this we compare coefficients of ct', x', y', z' in Eqs. (13.24a), (13.24b) with those in Eq. (13.25).

- **Transverse to motion direction** Since these coordinates are not subject to Lorentz transformation we find for the coefficients of x' and y' respectively

$$\omega \sin\theta \cos\varphi = \omega' \sin\theta' \cos\varphi' , \qquad \omega \sin\theta \sin\varphi = \omega' \sin\theta' \sin\varphi' . \tag{13.26}$$

Taking the ratio of right-hand sides (RHSs) and of left-hand sides (LHSs) of these relations we recognize that $\varphi = \varphi'$; the azimuth angle is not subject to aberration, see Eq. (13.8). Using this in Eq. (13.26) we find

$$\omega \sin\theta = \omega' \sin\theta' . \tag{13.27}$$

This shows clearly the strong correlation between the Doppler shift and the altitude aberration.

- **Coefficient of ct′**

$$\omega\gamma'(1 - v'/c \cos\theta) = \omega\gamma(1 + v/c \cos\theta) = \omega' . \tag{13.28}$$

Equation (13.28) describes the Doppler frequency shift.

- **Coefficient of z′**

$$\omega\gamma'(\cos\theta - v'/c) = \omega\gamma(\cos\theta + v/c) = \omega' \cos\theta' . \tag{13.29}$$

Dividing the LHSs and RHSs of Eqs. (13.29) and (13.28) respectively we find the aberration equation

$$\cos\theta' = \frac{\cos\theta + v/c}{1 + v/c\ \cos\theta} \tag{13.30}$$

in agreement with Eq. (13.9). Dividing the LHSs and RHSs of Eqs. (13.27) and (13.28) respectively we find the other format of the aberration equation we have seen in Example 13.2

$$\sin\theta' = \frac{\sin\theta}{\gamma(1 + v/c\ \cos\theta)} . \tag{13.31}$$

Thus we have obtained both the Doppler shift Eq. (13.28) and aberration from this straightforward consideration.

Written in textbook variables we have

$$\nu = \nu_0 \frac{1 + v/c\ \cos\theta_0}{\sqrt{1-(v/c)^2}} . \tag{13.32}$$

We clarified that the angle θ_0 shown is measured in the frame of the light source. This is not useful for the observer study of the source. To relate this result to the observed $\cos\theta$ line of sight of the Earthbound observer, we must insert the inverse aberration result obtained by applying the principle of relativity

$$\nu_0 \Leftrightarrow \nu\ , \quad v \Leftrightarrow -v\ , \quad \theta_0 \Leftrightarrow \theta\ .$$

The outcome is a relation between the source frequency ν_0 and both the frequency ν and the altitude angle θ as measured by an Earthbound observer whose motion is not exactly parallel or antiparallel to the line of sight

$$\nu_0 = \nu \frac{1 - v/c\ \cos\theta}{\sqrt{1-(v/c)^2}} . \tag{13.33}$$

For $\cos\theta \to 1$ we recover Eq. (13.21). Since typically ν_0 is related to a known radiation signature, this Doppler shift with aberration result creates a constraint between the direction of motion and the speed. As discussed in Sect. 2.3 the variation introduced due to the Earth circling the Sun helps to disambiguate this relation.

Example 13.3

Doppler measurement of velocities

The H_α-line of a star ($\lambda_0 = 6563$ Å) is observed to be redshifted by $\delta\lambda = 3$ Å. No lateral motion of the star is observed; thus the star must be moving along the line of sight. Discuss the direction of relative motion and determine the relative speed between the Earth and the star. ◄

► **Solution** In Sect. 13.4 we found that for the case that the observer and the star are moving apart from each other the observed frequency is smaller, thus the observed wavelength is higher, and we have redshift.

In particular, using Eq. (13.19) we have for a source moving away from us with speed v

1 $$\frac{\nu}{\nu_0} = \sqrt{\frac{1 - v/c}{1 + v/c}} .$$

As $c = \nu\lambda$, the inverse of this ratio applies for the wavelength as seen in Eq. (13.22), which we can write in the form

2 $$\frac{\lambda_0}{\lambda} = \sqrt{\frac{1 - v/c}{1 + v/c}} \, .$$

Clearly, the wavelength $\lambda > \lambda_0$ is seen by the observer. Solving for v and considering the nonrelativistic limit we obtain

3 $$v = c\frac{1 - (\lambda_0/\lambda)^2}{1 + (\lambda_0/\lambda)^2} \simeq c\,\frac{\lambda - \lambda_0}{\lambda_0} \, .$$

Using $\lambda_0 = 6563$ Å and $\delta\lambda = \lambda - \lambda_0 = 3$ Å we obtain $v = 0.000457c$. The relative velocity with the Earth is

4 $$v = 0.000457c = 0.00046 \times 300{,}000 \text{ km/sec} = 137.1 \text{ km/sec} \, .$$

This velocity is small enough for the effect to be entirely described by the non-SR-Doppler shift with $\gamma \simeq 1$ as we used above. ◀

Part VI

Mass, Energy, Momentum

In Part VI we learn about: The pivotal insight that mass and energy are connected to each other is demonstrated. The rest mass represents the energy locked into material form. Conversely, we show that the totality of energy components of a body determine its mass. We establish the relation between body energy, body momentum and body mass. The connection between the energy content of a particle at rest $E(v = 0) = E_0$, its inertial mass m, and the speed of light c: $E_0 = mc^2$ implies, given the appearance of light speed c, a profound relation of all visible material bodies to each other.

Introductory Remarks to Part VI

We show how two different observers evaluate the energy produced in the conversion of an electron-positron pair into radiation, which leads to the pivotal relation $E(v = 0) \equiv E_0 = mc^2$. We show how this relation naturally combines with the known nonrelativistic kinetic energy to a simple expression that describes the total relativistic energy $E(v)$ of a body in motion with respect to any inertial observer. We strongly discourage the consideration of $E(v)/c^2$ as a speed-dependent mass.

We introduce the relativistic momentum of a body by recalling two important insights of classical mechanics, the relation between the force inducing a change of momentum, and the work-energy theorem relating work, energy and velocity. Exploiting the analogy to coordinate LT, we introduce the transformation of energy and momentum of a body observed by different inertial observers. We show how the energy and momentum of a body that is in motion with reference to an observer arises from a LT of the rest energy of a body. Conversely, we determine the LT into a frame comoving with the body in motion. Both nonrelativistic and ultrarelativistic limits of the energy-momentum-velocity relation are explored.

Irrespective of the complex internal dynamics, a body as a whole satisfies Einstein's $E_0 = mc^2$. This result follows from the principle that the inertial state of a body always remains unaffected by internal dynamics. In an example we describe how a material body can share its energy and mass with radiation that travels within. This illuminates and confirms Einstein's consideration of how radiation carries body inertia.

We add to the understanding of rapidity introduced Sect. 7.4 by considering particle motion in terms of particle rapidity. This concept parallels the characterization of particle motion in terms of speed. We show that in the relativistic domain, the concept of additive particle rapidity is more practical than particle speed. We show how we can track particle energy and momentum observed in different reference frames by a shift in particle rapidity. Particle rapidity also allows a better characterization of ultrarelativistic particle properties, where speed loses its sensitivity.

We then turn to a detailed discussion of the origin of energy in our daily lives. We show the pivotal insight that all usable energy comes from reduction of the mass of reactants that serve in energy production: The energy always comes from the reduction of the mass of the final state as compared to the initial pre-burn state. This is irrespective of whether the power station is conventional or nuclear, is built on Earth or is naturally generated by the Sun or the Big-Bang.

Renewable energy uses solar fusion energy delivered as we read this book. Fossil energy sources employ the energy of solar fusion supplied long ago to the Earth. Nuclear energy originates from the ashes of a super-nova explosion, while future fusion reactors will exploit the energy content of the ashes of the Big-Bang nucleosynthesis process. In all cases, somewhere at some time, *the mass of matter decreased to provide the energy we use.*

Mass and Energy 14

Abstract

The pivotal relation $E(v = 0) = E_0 = mc^2$ is demonstrated by considering the consistency of the measurement of the energy content of a moving body by two different observers. We discuss the difference between the energy of a body and the mass $m = E_0/c^2$, as determined by the energy in the rest-frame of the body. We strongly discourage the consideration of $E(v)/c^2$ as a speed-dependent mass and introduce comments on this matter made by Einstein and others.

14.1 Energy of a Body at Rest

Among several spectacular publications in the year 1905, including works on the photo-electric effect and Brownian motion, Einstein formulated the Special Relativity Theory (see Ref. [18] in Chap. 1). Einstein's most famous equation, the equivalence between mass and energy,

$$E = mc^2 \,, \tag{14.1}$$

was argued in a separate, very short publication (see Ref. [20] in Chap. 1). Einstein employed the emission of radiation by an atom and thus a partial conversion of inertial mass into energy, to deduce that the kinetic energy T of a body has the form

$$T = \frac{mc^2}{\sqrt{1-(v/c)^2}} - mc^2 \,. \tag{14.2}$$

J. Rafelski, *Modern Special Relativity*, https://doi.org/10.1007/978-3-030-54352-5_14

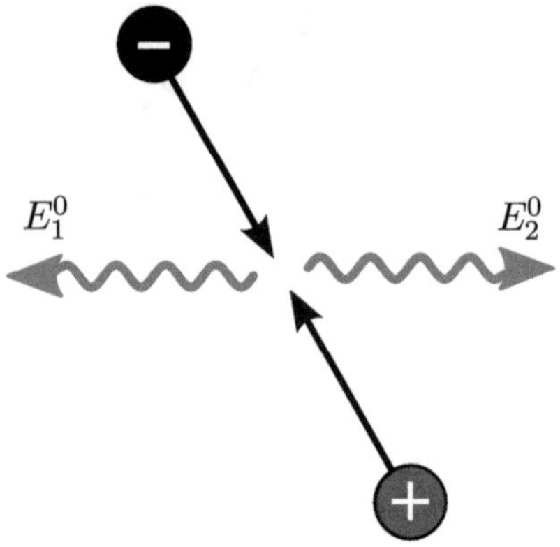

Fig. 14.1 A positron and electron of equal and opposite momenta convert into two photons

An analogous but more dramatic version of Einstein's arguments is presented in the following: We consider matter-antimatter annihilation radiation; that is, the case that all of the inertial mass of a particle-antiparticle pair is converted into the energy of emitted radiation. In this context the reader will note the appearance below of $\hbar$, the Planck quantum of action. This is not to be interpreted to mean that quantum mechanics is needed for the demonstration of Eq. (14.1). The main reason that $\hbar$ appears is that we make explicit what in Einstein's two papers is implicit, the relation between the frequency of light and the energy of the photon, which Einstein addressed earlier in another remarkable 1905 publication.[1]

Let us consider the 'meeting' of an electron e^- with its antimatter particle, the positron e^+. We can choose a frame of reference to view their encounter such that each arrives from an opposite side. A reference system in which momentum of all particles vanishes is called the center of momentum (CM) frame and will be addressed in more detail in Part 17. The two particles are shown in Fig. 14.1 approaching each other at the same speed. To simplify the discussion we further assume that each of these particles moves very slowly so that the kinetic energy is negligible in comparison to the energy locked in the rest mass.

The conservation of momentum calls for the total momentum to remain conserved in conversion of the electron-positron pair (e^+e^--pair) into photons. Since two photons are produced, if momentum was zero to begin with, these photons must be emitted in exactly opposite directions

$$\boldsymbol{p}\,_1^0 = -\boldsymbol{p}\,_2^0\,, \tag{14.3}$$

as shown in Fig. 14.1.

[1] Albert Einstein, "Über einen die Erzeugung und Verwandlung des Lichtes betreffenden heuristischen Gesichtspunkt," (translated: "On a heuristic viewpoint concerning the production and transmutation of light,") *Annalen der Physik,* **17** 132 (1905) (Received at publisher March 18, 1905).

The photoelectric effect establishes the particle nature of the photons, with their momenta given by

$$|\boldsymbol{p}\,^0_1| = \hbar k_1 = 2\pi\hbar\nu_1/c\,, \qquad |\boldsymbol{p}\,^0_2| = \hbar k_2 = 2\pi\hbar\nu_2/c\,, \tag{14.4}$$

where k_1 and k_2 are the wave numbers and ν_1 and ν_2 are the frequencies of the photons. Both photons thus have equal frequency ν_0:

$$\nu_1^0 = \nu_2^0 \equiv \nu_0. \tag{14.5}$$

Each of these photons also carries the energy

$$E_1^0 = E_2^0 \equiv E_0 = pc = 2\pi\hbar\nu_0\,. \tag{14.6}$$

Aside from momentum, energy is also conserved; thus the energy content of the $\mathrm{e}^+\mathrm{e}^-$-pair must be equal to the energy found in the two emitted photons

$$E^0_{\mathrm{e}^+\mathrm{e}^-} = 2E_0\,. \tag{14.7}$$

Here $E^0_{\mathrm{e}^+\mathrm{e}^-}$ is the intrinsic energy of the electron-positron system before annihilation.

Up to now, we have considered only the process in the reference system S in which both the electron and the positron were practically at rest. We now consider a system S' moving relative to S with a velocity $\boldsymbol{v}$, chosen here to be parallel (or antiparallel) to the direction of the emission of one of the photons.

In this frame of reference we observe a Doppler shift in which one photon's frequency increases, and the other photon's frequency decreases, as shown in Fig. 14.2. Using the SR-Doppler relationship, Eq. (13.19), for motion along the line of sight, we obtain

$$\nu^{\pm} = \nu_0 \frac{1 \pm v/c}{\sqrt{1-(v/c)^2}}\,. \tag{14.8}$$

In Eq. (14.8) we have $+v/c$ for the photon that travels antiparallel to $\boldsymbol{v}$, and $-v/c$ for the photon that travels parallel to $\boldsymbol{v}$. The energies of the photons in the moving frame of reference S' are thus also shifted and are given by

$$E^{\pm} = 2\pi\hbar\nu^{\pm} = 2\pi\hbar \frac{1 \pm v/c}{\sqrt{1-(v/c)^2}}\nu_0\,. \tag{14.9a}$$

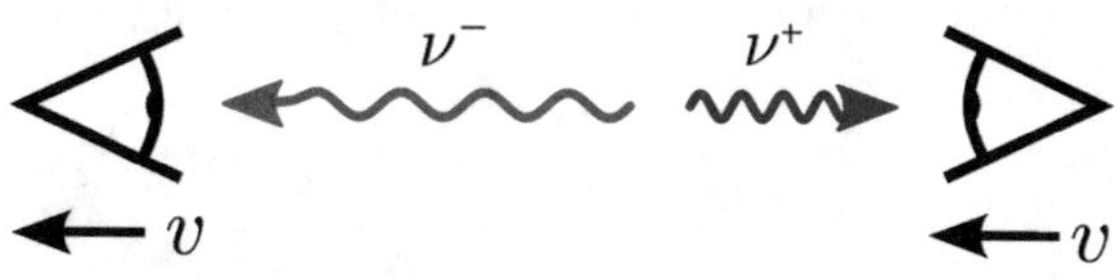

Fig. 14.2 The moving observer sees a redshift of one photon (on left) and a blueshift of the other

If we substitute in the energy from the system at rest (Eq. (14.6)), we can also write

$$E^{\pm} = E_0 \frac{1 \pm v/c}{\sqrt{1-(v/c)^2}} \ . \tag{14.9b}$$

By the principle of relativity, conservation of energy must also apply to the moving observer. This means that in system S', the combined energy of the e^+e^--pair prior to the annihilation process should be equal to the energy carried by the two photons after the annihilation. We denote by $E_{e^+e^-}(v)$ the energy of the system as observed in S', anticipating that the result must be a function of the speed v

$$E_{e^+e^-}(v) = E^+ + E^- \ . \tag{14.10a}$$

If we insert equation 14.9b, we obtain

$$E_{e^+e^-}(v) = \frac{2E_0}{\sqrt{1-(v/c)^2}} = \frac{E^0_{e^+e^-}}{\sqrt{1-(v/c)^2}} \ , \tag{14.10b}$$

where we used Eq. (14.7) to simplify the numerator.

This result differs from the result in the system at rest by the appearance of the Lorentz factor, which we now expand

$$\gamma = \frac{1}{\sqrt{1-(v/c)^2}} = 1 + \frac{1}{2}\left(\frac{v}{c}\right)^2 + \frac{3}{8}\left(\frac{v}{c}\right)^4 + \dots \ . \tag{14.11}$$

Inserting this expansion into Eq. (14.10b)

$$E_{e^+e^-}(v) = E^0_{e^+e^-} + \frac{1}{2}E^0_{e^+e^-}\left(\frac{v}{c}\right)^2 + \mathcal{O}(v^4) \ . \tag{14.12}$$

Comparing the 2nd term in Eq. (14.12) with the usual expression for the nonrelativistic kinetic energy,

$$E_{\text{kin}}(v) = \frac{1}{2}mv^2 \ , \tag{14.13}$$

we recognize a relation between the energy contained in the electron-positron system and the mass of the system

$$E^0_{e^+e^-} = m_{e^+e^-}c^2 \ . \tag{14.14}$$

We can drop the subscript 'e^+e^-' since these equations apply to any assembly of particles bound together and moving at speed v and with an inertial mass m. Rest energy

$$E_0 = mc^2 \tag{14.15}$$

is arguably the most famous equation of twentieth century. It states that any visible material body of inertial mass m has a comoving 'proper energy' content E_0. Because c^2 appears for each and every visible material particle, this relation establishes that all visible material bodies are related.[2]

Discussion 14.1 Mass energy equivalence

About the topic: In this discussion we question the meaning of rest mass and its energy content. Inevitably discussion of mass-energy equivalence leads to the question: Where does the energy we are using come from? We anticipate this topic here and we return to in Chap. 16.

- *Simplicius:* Does $E = mc^2$ mean that matter is a condensed form of energy?
- *Professor:* Of a sort, mass is one way energy can be locked up; it is one form of energy, but it is not the only form. Kinetic and potential energy are other well known forms of energy.
- *Simplicius:* Conversely, do objects with energy, therefore, always have mass?
- *Professor:* No, light can have energy even though photons most certainly do not have mass. The energy locked in light comes from a reaction where mass is transformed into energy. For example, the light from the Sun is produced in a nuclear fusion reaction in which the mass of reaction products is reduced, and in the end a large part of this mass defect becomes energy which is released as light.
- *Simplicius:* This makes sense with fusion, but does it explain chemical reactions as well? The mass of the reactants is the same as the products; how is this explained?
- *Professor:* When you have a reaction, one chemical compound converts to another. In the reaction, the binding energy of the compound is released. When an energy releasing reaction occurs, the mass of the new reaction products is reduced by a corresponding amount. This is difficult to measure directly, but many related phenomena confirm this insight. We will address this more precisely in Sect. 16.1 where we consider several examples.
- *Student:* This is true for more than nuclear or chemical reactions. This mass-energy conversion governs much of particle physics where we observe particle production, particle decays and particle reactions.

(continued)

[2]We speak here of 'visible' given the factor c^2 which indicates a relationship between all visible matter. It is possible that this equation may not apply to truly dark matter, observable solely by virtue of its gravity.

- *Professor:* Agreed. This applies to the most elementary and common process: The mass of a hydrogen atom is less than the sum of the individual parts. The mass defect corresponds to electron binding energy of 13.6 eV. The hydrogen atom has that much less mass 13.6 eV/c^2 than an electron and a proton before they bind to create atomic hydrogen (for the energy unit 'eV' see Insight *Elementary energy units* in Sect. 16.2).
- *Student:* Thus when an electron and proton bind, the mass equivalent of binding energy is released in the form of several photons, quanta of light.
- *Simplicius:* So, photons, as we discussed, have energy but certainly not mass. I still do not see how a massless object can have energy.
- *Professor:* Light holds energy in its oscillations. The higher the frequency of these oscillations, the greater the energy.
- *Simplicius:* The intrinsic energy of light seems to me to be different in qualitative terms compared to massive particles.
- *Professor:* Yes. Photons have no rest energy, as they have no mass. Thus their kinetic energy is not proportional to mass as is the case with all massive particles. Instead, like it is with waves in material bodies (sound in air, water waves) the motion energy is locked in wave oscillation. The pivotal insight that quanta of light, massless photons are acting to move energy (and thus mass) from a source to an absorber earned Einstein the Nobel Prize of 1921; the citation reads "...discovery of the law of the photoelectric effect".
- *Simplicius:* Is this mass loss of another body carried by photons very significant?
- *Professor:* Often the difference in mass can be small.
- *Student:* But fusion energy is said to run on mass-energy equivalence.
- *Professor:* Indeed! An α-particle made of two protons and two neutrons weighs slightly less, about 0.5%, than the sum of masses of these constituents.
- *Simplicius:* Such a mass defect would take nearly a pound off my weight! I would notice it on my bathroom scale.
- *Professor:* You are right; this is large enough to be measured directly. It turns out this 0.5% is a lot of energy, a million times more compared to energy produced per particle in comparable chemical processes.

Example 14.1

Lorentz transformation of photon energy and momentum

Consider corpuscular properties of a photon, its energy E_γ and momentum p_γ, and determine how E_γ and p_γ transform when evaluated by another observer moving along the line of motion of the photon. Compare your result to the behavior of coordinates time t, and position x, under Lorentz coordinate transformation. ◄

► **Solution** Our starting point is the SR-Doppler relationship, Eq. (13.19): We recast this relation into the usual transformation context replacing $\nu \to \nu'$ and $\nu_0 \to \nu$. After multiplication of the frequency ν with $2\pi\hbar$ we obtain the photon energy. Thus Eq. (13.19) reads

1 $$E'_\gamma = E_\gamma \frac{1 + v/c}{\sqrt{1 - (v/c)^2}} \,.$$

Applying this procedure to wavenumber k according to Eq. (13.20)

2 $$p'_\gamma = p_\gamma \frac{1 + v/c}{\sqrt{1 - (v/c)^2}} \,.$$

We implicitly used above the photon momentum $p_\gamma = E_\gamma/c$. We now rewrite above relations in a format reminiscent of the Lorentz coordinate transformation

3 $$E'_\gamma = \left(E_\gamma + v_x p_{x\,\gamma}\right) \frac{1}{\sqrt{1 - (v/c)^2}} \,,$$
$$p'_\gamma = \left(p_\gamma + \frac{v_x}{c^2} E_\gamma\right) \frac{1}{\sqrt{1 - (v/c)^2}} \,,$$

where we oriented the coordinate x-axis in the direction of photon motion.

This format demonstrates a deeper relation under Lorentz transformation between energy and momentum of a photon. Comparing above result with the Lorentz transformation of time, Eq. (6.21), we see that these forms are equal when we assign $ct \to E$ and $\boldsymbol{x} \to c\boldsymbol{p}$. However, unlike a Lorentz transformation of coordinates, the sign between energy and momentum on RHS in above equation is positive, a situation we will address in the following Chap. 15. ◄

14.2 Relativistic Energy of a Moving Body

Using Eq. (14.14) in Eq. (14.10b), we obtain the total energy for an inertially moving massive body:

$$E(v) = \frac{mc^2}{\sqrt{1 - (v/c)^2}} = \gamma mc^2 \,. \tag{14.16}$$

The important message in Eq. (14.16) is that all of the energy a particle carries is controlled by the particle inertia m. This energy includes the energy due to motion, the kinetic energy. However, when $v \to 0$ this expression does not vanish.

The energy content of a body at (relative) rest, see Eq. (14.15), is the rest energy. The two energy components, the rest energy and the kinetic energy, combine into a relatively simple expression, Eq. (14.16), where the Lorentz factor γ appears. To separate these components, we restate Eq. (14.16) using Eq. (14.11)

$$E(v) = mc^2 + \frac{1}{2}mv^2 + \frac{3}{8}mv^2\left(\frac{v}{c}\right)^2 + \dots . \tag{14.17}$$

The first term,

$$E_0 = mc^2 , \tag{14.18}$$

is the rest energy.

The full relativistic kinetic energy is obtained by removing the rest energy in Eq. (14.16)

$$T = E(v) - mc^2 = (\gamma - 1)mc^2 \simeq \frac{1}{2}mv^2\left[1 + \frac{3}{4}\left(\frac{v}{c}\right)^2 + \frac{5}{8}\left(\frac{v}{c}\right)^4 + \dots\right], \tag{14.19}$$

which expands to show as the first term the nonrelativistic kinetic energy followed by corrections in $(v/c)^2$ and higher order.

As already noted, the mass of a body defines the energy content of the body. Of particular interest is of course that there is energy in the limit $v \to 0$, Eq. (14.18), which is the historic insight. What we learn is best stated by a direct citation from the Einstein paper (see Ref. [20], 4th last Paragraph):

> *If a body gives off an energy δE in form of radiation, its mass diminishes by the corresponding amount $\delta E/c^2$.*

(Einstein used notation $\delta E \equiv L$ and $c^2 \equiv V^2$.) In our case all of the energy of the particle-antiparticle pair was given off, so we had obtained the complete energy equivalence between m and E.

We compare the relativistic energy Eq. (14.16) to the nonrelativistic limit, the first two terms in Eq. (14.17). Both expressions for the energy are plotted in Fig. 14.3 as a function of speed v. We scale energies with mc^2 and the speed v with c. The difference between the two results is most significant near to $v = c$. Within the special theory of relativity, the energy of a body diverges as $v \to c$: No matter how much work is done on a body, its speed never reaches the speed of light. In this sense, the speed of light is (again) recognized to be a maximum speed limit for all material bodies.

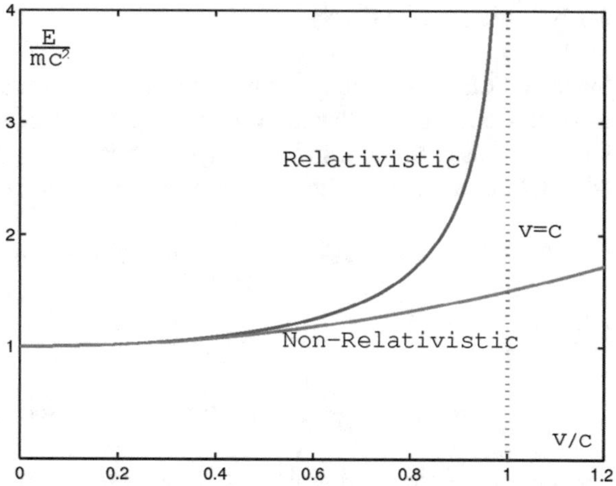

Fig. 14.3 $E(v)$ in the relativistic Eq. (14.16) and nonrelativistic cases Eq. (14.13) (which must be supplemented by Eq. (14.18))

Example 14.2

Kinetic energy and speed

Given a kinetic energy T determine the speed of a particle and compare it to the known nonrelativistic expression and ultrarelativistic limit. ◀

▶ **Solution** For a given kinetic energy T the total energy of a body is

1 $E = mc^2 + T$.

We use this in Eq. (14.16) and solve for v

2 $$mc^2 + T = \frac{mc^2}{\sqrt{1-(v/c)^2}} \Rightarrow v = c\frac{\sqrt{(mc^2+T)^2 - m^2c^4}}{mc^2+T} .$$

We recognize that for $T/mc^2 \gg 1$ the body reaches speed of light $v \to c$. We now cancel in the root the quadratic mass term and pull out a common multiplicative factor

3 $$v = \sqrt{\frac{2T}{m}}\,\frac{\sqrt{1+\frac{1}{2}(T/mc^2)}}{1+T/mc^2} .$$

The first factor is the nonrelativistic speed as a function of the kinetic energy. The second factor is the relativistic correction. This correction factor is always smaller than unity. We return to the study of particle kinetic energy in Example 15.9. ◀

14.3 Mass of a Body

Since in SR a body velocity is a relative quantity, one cannot imagine redefining the mass of a particle such that the velocity dependent factor in Eq. (14.16) is absorbed into the mass definition. In such an approach a body would also have no kinetic energy. Even so, in older literature it was common to write

$$M(v) = \frac{m}{\sqrt{1-(v/c)^2}}\,, \qquad \text{notation } \textbf{not} \text{ used in this book}\,,$$

and thus

$$M(v) = \frac{E(v)}{c^2}\,, \qquad \text{notation also } \textbf{not} \text{ used in this book.}$$

The above definition requires that we always attach to the velocity independent mass the subscript '0' to distinguish the (true rest) mass from the quantity which actually is not a mass but $E(v)/c^2$. As we already observed, $M(v)$ depends on the observer, since it can be assumed that the relative speed v between the body and the observer varies. $M(v)$ is thus not a Lorentz invariant, as it changes under the Lorentz transformation of coordinates. On the other hand the true mass m is a Lorentz invariant: Its value is the same for all observers.

It is natural to reserve the word 'mass' to refer to the Lorentz invariant quantity that describes the inertial mass of a body which has the same value for all observers. Mass is also a physical property of the body and, according to the principle of relativity, it should be independent of the inertial system in which this body is at rest. Thus, in addition to the proper time τ, there is a second Lorentz-invariant body property, the proper energy of a body, i.e. rest mass energy mc^2. We return to deepen the understanding of mass invariance in the following Chap. 15, see Eq. (15.14).

The use of body energy in motion as an effective mass has been criticized as follows:

Albert Einstein[3]

> It is not good to introduce the concept of the mass $M = m/(1-v^2/c^2)^{1/2}$ of a moving body for which no clear definition can be given. It is better to introduce no other mass concept than the 'rest mass' m.

Lev Okun[4]

> The concept of relativistic mass, which increases with velocity, is not compatible with the standard language of relativity theory and impedes the understanding and learning of the theory.

[3] In a letter of 19 June 1948 to Lincoln K. Barnett discussing the draft of *The Universe and Dr. Einstein*. Sourced by Lev Okun at Hebrew University of Jerusalem. For image of the original German language letter see: L. Okun, "The Concept of Mass," *Physics Today* June 1989, p. 31.

[4] L. Okun, "Mass versus relativistic and rest masses," *Am. J. Phys.* **77** 430–431 (May 2009). See also "The Einstein formula: $E_0 = mc^2$ 'Isn't the Lord laughing?'," arXiv:0808.0437.

We conclude that there is only one mass of a body, the rest mass m. And yet a confusion of energy with mass arises, as is seen in e.g. the Max Born's *Relativity* of 1920 vintage, or even in Richard P. Feynman's *Lectures*, but not in Landau-Lifshitz's *Course*.

There is also a misunderstanding possible using $M \to m$ while not showing the dependence on v explicitly, perhaps adding to the challenge by using natural units $c = 1$. In this book we will always distinguish energy and mass and keep c in all equations. In this book the inertial mass of a body is:

$$m \equiv \frac{E(v=0)}{c^2} \equiv \frac{E_0}{c^2} \,. \tag{14.20}$$

The energy a body of mass m moving at speed v is

$$E = \frac{mc^2}{\sqrt{1-(v/c)^2}} \,. \tag{14.21}$$

Discussion 14.2 What is mass?

About the topic: In this discussion we illuminate why it is not good to introduce a velocity dependent mass.

- *Simplicius:* The mass-energy relation is still confusing to me. As I read it, a popular science article published under a British seal of trustworthiness states that the more an object speeds up, the heavier it gets.
- *Student:* Really? Will your car become heavier or lighter just because I ride my bike near to it? See, speed is relative. That is where the name 'Relativity' comes from.
- *Professor:* This way to argue in a press article is indeed not a good way to gain reader attention. A mass cannot change when observed from different reference frames.
- *Simplicius:* I see, a particle does not get heavier by motion. However, could the force of gravity become stronger?
- *Professor:* A yet better question is: What is it that gravitates?
- *Simplicius:* When you look at Newton's laws, you would think it must be mass.
- *Student:* I heard that light does not escape black holes, hence their name. Therefore light is influenced by gravity, but it has no mass.
- *Professor:* Clearly there is more to gravity than just mass. This may be what causes the press article you mentioned to be confusing about the meaning of mass.

(continued)

- *Simplicius:* Let me guess, it is not mass, but the energy that gravitates!
- *Professor:* We are now on the right track. But before we walk down this path, we must remember there are two different questions: What is the source of gravity? And, how does gravity influence motion?
- *Student:* Do you refer here to Einstein's equivalence principle? First you distinguish between gravitational mass, and inertial mass, and then you make these always proportional to each other.
- *Simplicius:* I never understood this: We introduce the source of gravity to be gravitational mass and then we say that how the particle feels gravity force, and the inertial mass, are equivalent. Could we simply stick to Newton's $Gm_1 \cdot m_2$?
- *Professor:* In General Relativity (GR) created by Einstein a dozen years after SR, the source of gravity is the 'energy-momentum tensor'. This name implies that all motion and its energy content inherent to a body contribute to gravity.
- *Student:* This sounds as if I need to modify $Gm_1 \cdot m_2$ to read $GE_1 \cdot E_2$, introducing energy instead of mass.
- *Professor:* In spirit, you are right. In detail the situation of course is much more complicated; we stay away from this complication for now.
- *Student:* Would gravitational attraction change if I spin the two bodies around their axes?
- *Professor:* Yes, but in our daily life, the effect is so tiny we could not easily measure it. We return to discuss energy storage in rotating bodies, see Sect. 16.2.
- *Simplicius:* So perhaps rotating merging neutron stars?
- *Student:* That seems to be a good example.
- *Professor:* Let us here and now remember that spinning bodies are, indeed even if only slightly, heavier as their rotational energy has a mass equivalent, and thus more gravity is available.

Particle Momentum

15

Abstract

The product of velocity with the momentum changing force provides a relation between energy change, impulse change and velocity, called the work-energy theorem. We exploit this relation to introduce the velocity dependence of relativistic momentum. Both nonrelativistic and ultra-relativistic limits of the energy-momentum-velocity relation are considered. Considering that the linearity of the Galilean velocity addition is lost, the concept of rapidity is further advanced in the context of several examples in particle dynamics. Properties of relativistic particle beams are studied.

15.1 Relation Between Energy and Momentum

We already know that energy and momentum of a photon transform together when an observer's frame of reference is changed, see last Example 14.1. We now obtain the form of the relativistic particle momentum which is consistent with the relativistic energy form Eq. (14.21). We recall that kinetic energy is imparted to a body through the action of a force which does work on the body. We consider an infinitesimal change in the energy of a body according to the work-energy theorem

$$dE = \boldsymbol{F} \cdot d\boldsymbol{x} = \frac{d\boldsymbol{p}}{dt} \cdot d\boldsymbol{x} \tag{15.1}$$

$$= d\boldsymbol{p} \cdot \frac{d\boldsymbol{x}}{dt} = d\boldsymbol{p} \cdot \boldsymbol{v} \; ; \tag{15.2}$$

that is, the infinitesimal change in kinetic energy of a body is described by the product of velocity $\boldsymbol{v}$ with the infinitesimal change in momentum $d\boldsymbol{p}$. This equation is exactly the same as the equation one considers for nonrelativistic dynamics; one can regard Eq. (15.2) as guiding us to the definition of the momentum vector.

J. Rafelski, *Modern Special Relativity*, https://doi.org/10.1007/978-3-030-54352-5_15

Relativity enters this momentum definition in the specific form Eq. (14.21) for $E(v)$, which implies

$$dE = \frac{m\,\boldsymbol{v}\cdot d\boldsymbol{v}}{\left(1-(v/c)^2\right)^{3/2}}\ . \tag{15.3}$$

We combine Eq. (15.2) with Eq. (15.3), eliminating dE

$$\boldsymbol{v}\cdot d\boldsymbol{p} = \frac{\boldsymbol{v}\cdot d\boldsymbol{v}\,m}{\left(1-(v/c)^2\right)^{3/2}}\ . \tag{15.4}$$

We now show that the momentum component parallel to velocity is

$$\boldsymbol{p} = \frac{m\boldsymbol{v}}{\sqrt{1-(v/c)^2}} = m\boldsymbol{v}\gamma\ . \tag{15.5}$$

To prove Eq. (15.5) we compute the left-hand side of Eq. (15.4) using Eq. (15.5)

$$\boldsymbol{v}\cdot d\boldsymbol{p} = m\gamma\boldsymbol{v}\cdot d\boldsymbol{v} + mv^2 d\gamma = m\gamma^3(1-v^2/c^2)\frac{dv^2}{2} + m\gamma^3(v^2/c^2)\frac{dv^2}{2} = m\gamma^3\frac{dv^2}{2}\ ,$$

which agrees with Eq. (15.4).

For a given momentum and energy we determine the velocity of a particle from

$$\boldsymbol{\beta} = \frac{\boldsymbol{v}}{c} = \frac{c\boldsymbol{p}}{E}\ . \tag{15.6}$$

This simple result appears since both $\boldsymbol{p}$ and E are proportional to γ.

Expanding in v/c the relativistic form of momentum Eq. (15.5), we obtain

$$\boldsymbol{p} = m\boldsymbol{v} + m\boldsymbol{v}\left[\frac{1}{2}\left(\frac{v}{c}\right)^2 + \frac{3}{8}\left(\frac{v}{c}\right)^4 + \frac{5}{16}\left(\frac{v}{c}\right)^6 + \frac{35}{128}\left(\frac{v}{c}\right)^8 + \ldots\right]\ . \tag{15.7}$$

The first term is the nonrelativistic momentum. This reaffirms that the vector $\boldsymbol{p} = \gamma m\boldsymbol{v}$ is an appropriate relativistic generalization of the momentum vector: since there is no transverse to velocity vector component in nonrelativistic limit, there is none for the relativistic observer either. Thus a momentum component orthogonal to $\boldsymbol{v}$ does not exist.

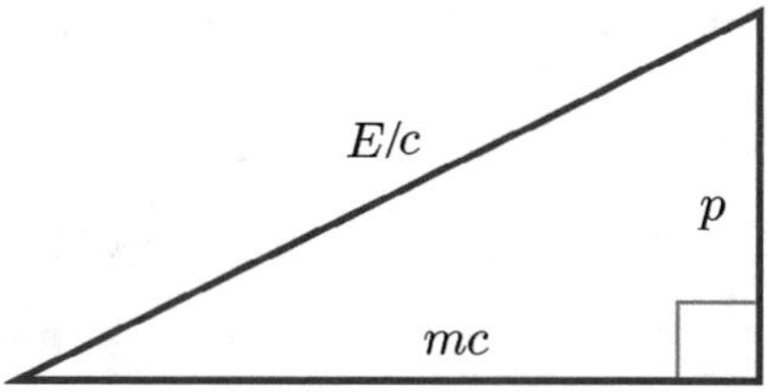

Fig. 15.1 The relationship between energy, momentum, and rest mass, as given in Eq. (15.9)

To recapitulate, the relativistic momentum of a particle given by Eq. (15.5) has been obtained from the relativistic energy of the moving body, Eq. (14.21) and the work-energy theorem Eq. (15.2), remembering the form of the non-relativistic limit Eq. (15.7).

$$E^2 - \boldsymbol{p}^2c^2 = m^2c^4\gamma^2(1 - \boldsymbol{v}^2/c^2) = m^2c^4 \,, \tag{15.8}$$

which may be conveniently written in the form

$$E/c = \sqrt{m^2c^2 + \boldsymbol{p}^2} \,. \tag{15.9}$$

It can be helpful to associate this relationship with the sides of a right triangle, as shown in Fig. 15.1. Note that the base of the triangle is the same in any reference frame, as the rest mass is an invariant quantity. However, a moving observer S' will always observe a different magnitude of momentum, allowing energy to close the triangle.

Collecting the different relativistic energy and momentum relations for a particle moving with velocity $\boldsymbol{v}$

$$\begin{aligned} mc^2 &= \sqrt{E^2 - (\boldsymbol{p}c)^2} \,, \\ E &= \gamma mc^2 = \sqrt{(mc^2)^2 + (\boldsymbol{p}c)^2} \,, \\ c\boldsymbol{p} &= \gamma mc\boldsymbol{v} = \gamma \boldsymbol{\beta} mc^2 \,, \\ c|\boldsymbol{p}| &= \sqrt{E^2 - (mc^2)^2} \,. \end{aligned} \tag{15.10}$$

We have already seen in Example 14.1 how the energy of a photon transforms for different IOs. We now extend the insights of Example 14.1 to the case of a particle of finite mass. Motivated by Eq. (15.10), we consider the following analogy between the space-time coordinates, and, momentum and energy

$$(c\tau, \, ct, \, \boldsymbol{x}) \Longleftrightarrow \left(mc^2, \, E, \, c\boldsymbol{p}\right) \,. \tag{15.11}$$

We are already familiar with the relations

$$
\begin{aligned}
c\tau &= \sqrt{(ct)^2 - (\boldsymbol{x})^2}\,,\\
ct &= \gamma c\tau = \sqrt{(c\tau)^2 + (\boldsymbol{x})^2}\,,\\
\boldsymbol{x} &= \gamma\tau\boldsymbol{v} = \gamma\boldsymbol{\beta}c\tau\,,\\
|\boldsymbol{x}| &= \sqrt{(ct)^2 - (c\tau)^2}\,,
\end{aligned}
$$

which are analogous to Eq. (15.10) for energy and momentum. Therefore we read Eq. (15.11) as a predictor of the behavior of energy and momentum under Lorentz coordinate transformation. This argument extends the already known case of massless photons, see Example 14.1, to case $m \neq 0$, and it shows that the proper time τ as well as the mass of a material body will be found to have the same value for any IO. This implies that subject to Lorentz coordinate transformation, E transforms just like ct, and $c\boldsymbol{p}$ just like $\boldsymbol{x}$. The LT for the energy E and momentum $c\boldsymbol{p}$ for the case of finite mass agrees with the relation Eq. (15.11) for massless photons

$$
\begin{aligned}
E' &= \gamma(E + cp_x\beta)\,,\\
cp'_x &= \gamma(cp_x + E\beta)\,,\\
p'_y &= p_y,\;\; p'_z = p_z\,.
\end{aligned}
\tag{15.12}
$$

The signs are chosen so that for the usual coordinate choice, a particle after a transformation acquires a motion 'to the right'.

The interesting special case is that of a body at rest with $E = mc^2$ and $\boldsymbol{p} = 0$. We determine the corresponding finite velocity v_x properties applying a boost employing Eq. (15.12) and obtain the result seen in Eq. (15.10)

$$
\begin{aligned}
E' &= \gamma\left((E = mc^2) + (cp_x = 0)\beta\right) = \gamma\, mc^2\,,\\
cp'_x &= \gamma\left((cp_x = 0) + (E = mc^2)\beta\right) = \gamma\beta\, mc^2\,,
\end{aligned}
\tag{15.13}
$$

with by choice of coordinates $p'_y = p'_z = 0$. By explicit computation we verify

$$
\begin{aligned}
E'^2 - \boldsymbol{p}'^2c^2 &= \gamma^2[(E + cp_x\beta)^2 - (cp_x + E\beta)^2] - p_y^2c^2 - p_z^2c^2\\
&= \gamma^2(1 - \beta^2)(E^2 - (cp_x)^2) - p_y^2c^2 - p_z^2c^2 = E^2 - \boldsymbol{p}^2c^2\,.
\end{aligned}
$$

$$
E'^2 - \boldsymbol{p}'^2c^2 = E^2 - \boldsymbol{p}^2c^2 = (mc^2)^2\,. \tag{15.14}
$$

We confirm that the mass value of a body reported by two observers, one at rest with respect to the body, and the other moving with any $\boldsymbol{v}$, are the same. Of course each observer records different values of energy and momentum.

We found that the body inertial mass m is invariant under LT and therefore the energy locked in the mass is the same for all IOs, including in particular the comoving observer. The realization that the intrinsic energy content of a body is the same for all IOs is an insight of great practical importance, allowing the generalization of the energy conservation law to include the intrinsic energy content of all bodies involved; see the in-depth discussion in the following Chap. 16.

Example 15.1

LT into a comoving reference frame

Determine the speed $\beta_x = v_x/c$ of the LT into a comoving frame of reference for a body with energy E and momentum p_x. State the LT transformed coordinates $t', \boldsymbol{x}'$ with reference to this observer, explicitly using energy and momentum as dynamical variables. Establish a relation of comoving coordinate time t' with the proper time τ of the body. ◀

▶ **Solution** To determine the speed of the comoving observer (variables with a prime) we set $p'_x = 0$ in the second Eq. (15.12) and obtain

1 $$\beta_x = -\frac{cp_x}{E} \,.$$

We confirm this result, computing the (rest) energy observed by the comoving observer by inserting into the first Eq. (15.12)

2 $$E'_{\text{rest}} = \gamma\,\frac{E^2 - (cp_x)^2}{E} = \frac{\gamma(mc^2)^2}{E} = mc^2 \,.$$

In the last relation we used

3 $$\gamma = \frac{E}{mc^2} \,.$$

These relations allow the presentation of the explicit form of LT into the comoving reference frame, beginning in the – non primed – laboratory frame coordinates by using the LT seen in Eq. (6.19)

4 $$x' = \gamma(x - \beta ct) \;\to\; x' = \frac{E}{mc^2}\left(x + \frac{cp_x}{E}ct\right) \,.$$

This can be further simplified

5 $$x' = x\,\frac{E}{mc^2} + ct\,\frac{cp_x}{mc^2} \,.$$

For the comoving observer the considered particle is at rest, i.e. for $x' = 0$. We can check that this agrees with our initial definition of β_x for the uniformly moving body

$$6\ \ x' = 0 \rightarrow x = -ct\,\frac{cp_x}{E}\,.$$

For the time t' in the comoving frame we obtain proceeding in the same way according to Eq. (6.20)

$$7\ \ ct' = ct\,\frac{E}{mc^2} + x\,\frac{cp_x}{mc^2}\,.$$

This can be further simplified combining last two relations to eliminate x

$$8\ \ ct' = ct\left(\frac{E}{mc^2} - \frac{c^2p_x^2}{Emc^2}\right) = ct\left(\frac{E^2 - c^2p_x^2}{Emc^2}\right) = ct\,\frac{mc^2}{E} = \frac{ct}{\gamma} \equiv c\tau\,.$$

As could be expected, the comoving time is the proper time of the body. ◀

Example 15.2

Nonrelativistic limit of energy-momentum relation

Obtain the series expansion for the energy of a nonrelativistic particle in terms of its momentum and compare to Eq. (14.19); explain the difference. ◀

▶ **Solution** We begin with the form seen in Eq. (15.10)

$$1\ \ E = \sqrt{(pc)^2 + (mc^2)^2} = mc^2\sqrt{1+x}\,, \qquad x = \frac{p^2}{(mc)^2}\,.$$

We use the well-known Taylor expansion

$$2\ \ \sqrt{1+x} = 1 + \frac{1}{2}x - \frac{1}{8}x^2 + \frac{1}{16}x^3 - \frac{5}{128}x^4 + \dots\,,$$

to obtain the expression for energy

$$3\ \ E = mc^2 + \frac{p^2}{2m} - mc^2\left[\frac{1}{8}\frac{p^4}{(mc)^4} - \frac{1}{16}\frac{p^6}{(mc)^6} + \frac{5}{128}\frac{p^8}{(mc)^8} - \dots\right]\,.$$

The coefficients of the power series we see in the velocity expansion Eq. (14.19) of energy are different from those we have obtained above. The reason is the relation between the momentum and velocity, which also is a power series, see Eq. (15.7) and which when used in our above result will reproduce the velocity series Eq. (14.19). The momentum expansion of energy we obtained in this example converges faster compared to velocity expansion Eq. (14.19). ◀

Example 15.3

Ultrarelativistic limit

Find an approximation for the energy of an ultrarelativistic particle, i.e. with kinetic energy much greater than its rest energy, in terms of its momentum. Obtain the deviation from $E - pc \to 0$ valid for massless particles in an exact form. Obtain the deviation of the speed of a relativistic particle from that of light $c - v$ in terms of the Lorentz factor γ, both as a practical approximate form, and exactly. ◄

► **Solution** Large kinetic energy implies $p^2c^2 \gg m^2c^4$ so we can begin with the expansion

$$\textbf{1}\quad E = \sqrt{p^2c^2 + m^2c^4} = pc\sqrt{1 + \left(\frac{mc}{p}\right)^2} = pc + mc^2\frac{mc}{2p} + \dots .$$

To first order $E = pc$ this is just the relationship for massless particles, e.g. photons. It is clear that the deviation $E - pc$ is a series in odd powers of mc/p. We can obtain an exact result performing a simple algebraic manipulation

$$\textbf{2}\quad E - pc = \frac{(E - pc)(E + pc)}{E + pc} = \frac{(\sqrt{p^2c^2 + m^2c^4})^2 - (pc)^2}{\sqrt{p^2c^2 + m^2c^4} + pc} = \frac{m^2c^4}{E + pc} .$$

This can be written as

$$\textbf{3}\quad E(p) = pc + mc^2\,\frac{mc/p}{\sqrt{1 + (mc/p)^2} + 1} .$$

Setting in the root the quadratically small term $(mc/p)^2 \to 0$, we recover the initial expansion term seen above.

To determine the speed using Eq. (15.6)

$$\textbf{4}\quad c - v = c\left(1 - \tfrac{v}{c}\right) = c\left(1 - \tfrac{pc}{E}\right) = c\,\frac{(E - pc)(E + pc)}{E(E + pc)} = \frac{c\,(mc^2)^2}{mc^2\gamma(mc^2\gamma + mcv\gamma)} .$$

Upon cancelling the common factors mc^2 we obtain

$$\textbf{5}\quad c - v = \frac{c^2}{c + v}\frac{1}{\gamma^2} \simeq \frac{c}{2\gamma^2} .$$

The last approximate relation for ultrarelativistic particles follows since we can simply use $v \approx c$. An exact solution for $c - v$ as a function of γ can also be obtained by replacing above $c + v \to 2c - (c - v)$ and solving for $c - v$. We obtain

6 $$c - v = \frac{c}{\gamma^2 + \gamma\sqrt{\gamma^2 - 1}} \,.$$

Multiplying this result by γmc we recover the result shown above for $E - pc$.

In Example 15.8 we will further explore the small deviation from the speed of light in terms of particle rapidity. ◄

Example 15.4

Nonrelativistic limit of LT of particle E, and $\boldsymbol{p}$

Show that the Lorentz coordinate transformation of energy E and momentum $\boldsymbol{p}$ of a particle reduce to the well-known nonrelativistic expressions taking the limit $c \to \infty$. Begin by recalling the nonrelativistic Galilean transformation. ◄

► **Solution** For the momentum the Galilean transformation, for a particle of mass m boosted with transformation velocity $\boldsymbol{v} = (v_x, 0, 0)$, reads

1 $$p'_x = p_x + mv_x \,, \qquad p'_y = p_y \,, \qquad p'_z = p_z \,.$$

Using the nonrelativistic expression for the kinetic energy we find its boosted value

2 $$T' = \frac{(\boldsymbol{p}')^2}{2m} = \frac{1}{2m}\left((p'_x)^2 + (p_y)^2 + (p_z)^2\right) = T + p_x v_x + \frac{1}{2}mv_x^2 \,.$$

For the case of Lorentz transformation we first look at the momentum in Eq. (15.12)

3 $$p'_x = \gamma\left(p_x + \frac{E}{c^2}v_x\right) \to p_x + mv_x + \mathcal{O}(1/c^2) \,.$$

Considering terms in lowest order in $1/c^2$, we have $\gamma \to 1$ and $E/c^2 \to m$. This limit is thus the Galilean limit. Following on the leading order terms all additional terms include $1/c^2$ and vanish in the limit $c \to \infty$.

We now evaluate the energy transformation in Eq. (15.12)

4 $$E' = \gamma(E + p_x v_x) \,.$$

Adopting $\gamma \to 1$ limit, the last term is c independent. However, the situation is more complicated when considering the first term γE, which will be needed to the second order in both terms. This is so since we must recover the rest energy of a particle not 'counted' in the usual form for nonrelativistic motion energy and be sure to cancel the second term above. We look back at Example 15.2 and employ the Lorentz factor expansion to obtain through order $1/c^2$

5 $$\gamma E = \left(1 + \frac{1}{2}\left(\frac{v_x}{c}\right)^2 + \ldots\right)\left(mc^2 + \frac{p^2}{2m} - \ldots\right) = mc^2 + \frac{p^2}{2m} + \frac{1}{2}mv_x^2 + \ldots .$$

Collecting terms and moving the mc^2 term to the other side of the equation, we obtain

6 $$T' \equiv E' - mc^2 = \frac{p^2}{2m} + \frac{1}{2}mv_x^2 + p_x v_x + \mathcal{O}(1/c^2)\,,$$

which agrees with the nonrelativistic Galilean transformation shown above.

It is of interest to note that in order to obtain the correct kinetic energy transformation we needed to account for the cross term in the $\mathcal{O}(1/c^2)$ expansion of both momentum and energy. Therefore one can argue that the nonrelativistic LT limit tests the LT format seen in Eq. (15.12) to order $1/c^2$. ◄

Example 15.5

"Radiation transfers inertia"

Einstein's publication on the equivalence of mass and energy ends with the phrase

> If the theory relates to nature, radiation transfers inertia between the emitting and absorbing bodies.

This example shows how that transfer of mass happens even though the photon itself is massless: As shown in Fig. 15.2 a few photons are emitted from the left end and travel toward the right end of a closed box of length L. Demonstrate the equivalence of energy and mass by considering the recoil of the box that causes a displacement D of the box but no motion of the system's center of mass. The energy and momentum of a single photon are related by $E_\gamma = p_\gamma c$. Assume here that signals travel faster within the box compared to the rigid box walls. We will study the opposite case in Example 15.6. ◄

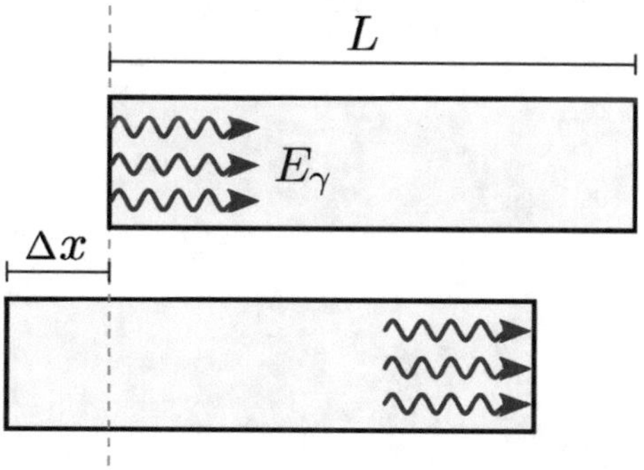

Fig. 15.2 Photons carry a fraction of box mass from the left to the right while the box recoils by Δx; see Examples 15.5 and 15.6

► **Solution** We study the box in its rest-frame. When the photon is emitted from the left end, heading to the right end of the box, it gives the box an impulse to the left. When the light is absorbed again, it releases the momentum p_γ back to the box which comes to rest. Even so, the box has moved by Δx as seen in Fig. 15.2.

Seeing this we can ask what happened to the center of mass of the box? Did it also move by Δx? However, this cannot be the case, since the center of mass of a closed system cannot change its position! The only other explanation is that the mass distribution in the body also changed. We conclude that there must have been movement of mass from the left end to the right end of the box, an increment m_{eq} transported by radiation. This then compensated the shift of the box, maintaining the location of the center of mass.

We thus see two contributions that can change the location of the center of mass of the system. There is mass m_{eq} moved by radiation by distance L to the right, and the second, the motion of the rigid box of mass M by Δx to the left. Both contributions to Δx_{cm} must compensate exactly. From this consideration we derive the condition for the final state of box

1 $$M\Delta x + m_{\text{eq}}L = 0 \quad \Rightarrow \quad m_{\text{eq}} = -M\frac{\Delta x}{L} .$$

The recoil against the emitted radiation sets the box into uniform slow motion with speed v

2 $$\Delta x = v\Delta\tilde{t} , \quad (\Delta x < 0) ,$$

where $\Delta\tilde{t}$ is the dilated travel time increment we met in the context of muon travel to the surface of the Earth. Therefore the box can travel further, since

3 $$\Delta\tilde{t} = \frac{\Delta t}{\sqrt{1 - v^2/c^2}} .$$

The laboratory time Δt follows from the distance traveled by light

4 $$L = c\Delta t .$$

Combining the last three equations we obtain

5 $$\Delta x = L\frac{v/c}{\sqrt{1 - v^2/c^2}} = L\frac{p_M}{Mc} = -L\frac{p_\gamma}{Mc} ,$$

where we used Eq. (15.5) and the momentum conservation. We use this in the condition posited at beginning for the final state of box and obtain

$$\textbf{6}\quad m_{\text{eq}} = \frac{p_\gamma}{c} = \frac{E_\gamma}{c^2}\ .$$

The proportionality factor $1/c^2$ in Eq. 6 follows with $1/c$ each from time-distance, and from energy-momentum of the photon proportionality relations.

We have determined that the energy and the transported mass of the electromagnetic wave are related by the usual Einstein form seen above. This effect also explains how the photon energy contributes to the mass of the box-photon system. When the photon is emitted, this energy must come from the mass of the box, and when it is reabsorbed (through atomic and thermal excitation), this energy must contribute again to the box mass. We will generalize this finding to other forms of energy in Chap. 16. ◀

Example 15.6

Back to radiation and inertia

We learned in Example 15.5 that light movement within a body can alter the mass distribution. From the principle that a body cannot spontaneously change its location we derived Einstein's mass-energy equivalence relation. We return to this problem in order to consider how this result holds up when the speed of signal propagation along the box material is faster than inside the box (gas-filled) chamber: Show by considering the displacement of the box that also in this situation the result of Example 15.5 holds. ◀

▶ **Solution** We address this problem in the same way we dealt with the time clock in Chap. 5: The box responds rapidly to the recoil generated by light emission and now the box wall moves towards the traveling light beam. We allow for the Lorentz-FitzGerald contraction of the moving body of the box. Considering the results seen in Eqs. (5.4), (5.5) and (5.6), we obtain the box displacement employed at the opening of Example 15.5

$$\textbf{1}\quad \Delta x = \frac{L}{c}\frac{v}{\sqrt{1-(v/c)^2}} = -L\frac{p_\gamma}{Mc} = -L\frac{E_\gamma}{Mc^2}\ .$$

We used Eq. (15.5) to replace the speed term by the recoil momentum (hence the minus sign). Our result is identical to the result obtained in Example 15.5. Thus the relevant discussion and all results of Example 15.5 remain valid.

It is beyond the scope of this book to consider further cases involving light motion within a closed physical system. It seems evident based on the here explored examples that the principles of SR assure the outcome: photons (light) within a body share in the mass, in the energy, and in the momentum of the body observed from outside; internal processes, even if the recoil is due to internal radiation processes, cannot lead to spontaneous displacement of the system. ◀

15.2 Particle Rapidity

The Lorentz coordinate transformation Eq. (15.12) for particle energy and momentum can also be written using the rapidity $y_{\rm p}$ of the Lorentz coordinate transformation introduced in Sect. 7.4. We have, using rapidity, the transformed energy and momentum for a boost in the x-direction in a format analogous to transformation of coordinates Eq. (7.39)

$$\begin{aligned} E' &= \cosh y_{\rm r}\, E + \sinh y_{\rm r}\, cp_x\,, \\ cp'_x &= \cosh y_{\rm r}\, cp_x + \sinh y_{\rm r}\, E\,, \\ p'_y &= p_y, \quad p'_z = p_z\,. \end{aligned} \tag{15.15}$$

Introducing P with dimension of momentum akin to X that had the dimension of length, this can be written using a Minkowski space boost introduced in Example 7.13 as

$$P' = \Lambda_{p_x} P\,, \qquad \Lambda_{p_x} = \begin{pmatrix} \cosh y_{\rm r} & \sinh y_{\rm r} & 0 & 0 \\ \sinh y_{\rm r} & \cosh y_{\rm r} & 0 & 0 \\ 0 & 0 & 1 & 0 \\ 0 & 0 & 0 & 1 \end{pmatrix}, \qquad P' = \begin{pmatrix} E'/c \\ p'_x \\ p'_y \\ p'_z \end{pmatrix}. \tag{15.16}$$

An interesting special case is that particle momentum in the x-direction is entirely acquired in the boost process; in other words, we start with a particle with $p_x = 0$ but in general $\boldsymbol{p}_\perp \neq 0$. We find

$$E(p_x = 0) = \sqrt{m^2c^4 + (p_y^2 + p_z^2)c^2} \equiv \sqrt{m^2c^4 + \boldsymbol{p}_\perp^2 c^2} = E_\perp\,. \tag{15.17}$$

After we carry out the transformation Eq. (15.15) the particle is observed to have a 'longitudinal' momentum $p'_x \equiv p_\parallel$ and an energy $E' \to E$, where we rename the primed values of all three momentum components.

We call the rapidity of the transformation that generates the x-directed momentum the particle rapidity[1] y_p. Then Eq. (15.15) reads

$$\begin{aligned} E &= \cosh y_\mathrm{p}\, E_\perp\,, \\ cp_\parallel &= \sinh y_\mathrm{p}\, E_\perp\,, \\ p'_y &= p_y, \quad p'_z = p_z\,. \end{aligned} \tag{15.18}$$

Equation (15.18) is a clever way to express the total energy and longitudinal momentum of a particle. We are using the transverse energy $E_\perp$ and a new variable, particle rapidity y_p related to the longitudinal motion.

How we choose a parallel ($\parallel$) directed particle is discretionary – usually the physical problem involves a preferred axis. For example, when the direction of motion of the particle defines the ($\parallel$) direction $\boldsymbol{p}_\perp = 0$, $p_\parallel = p$ and

$$\begin{aligned} E &= \cosh y_\mathrm{p}\, mc^2\,, \\ cp &= \sinh y_\mathrm{p}\, mc^2\,, \\ p'_\perp &= p_\perp = 0\,. \end{aligned} \tag{15.19}$$

We now develop further the relation of particle energy and momentum with particle rapidity that follows from Eq. (15.18): We see that the relation is consistent by evaluating

$$E^2 - c^2 p_\parallel^2 = (\cosh^2 y_\mathrm{p} - \sinh^2 y_\mathrm{p}) E_\perp^2 = E_\perp^2 \to E^2 = m^2c^4 + c^2\boldsymbol{p}_\perp^2 + c^2 p_\parallel^2\,.$$

Next we divide the two non-trivial equations Eq. (15.18) one by the other

$$\beta_\parallel \equiv \frac{cp_\parallel}{E} = \tanh y_\mathrm{p} \quad \to \quad y_\mathrm{p} = \mathrm{artanh}(cp_\parallel/E)\,. \tag{15.20}$$

The first expression in Eq. (15.20) reads

$$\beta_\parallel = \tanh y_\mathrm{p}\,, \tag{15.21}$$

which remembering that the inverse function of tanh is ln implies

[1] We recall that subscript 'p' reminds up that y_p is not a coordinate but the *particle* rapidity.

$$y_p = \frac{1}{2} \ln \left(\frac{1+\beta_\parallel}{1-\beta_\parallel} \right) . \tag{15.22}$$

More generally the last expression in Eq. (15.20) leads to

$$\begin{aligned} y_p &= \frac{1}{2} \ln \left(\frac{E+cp_\parallel}{E-cp_\parallel} \right) , \\ y_p &= \ln \frac{E+cp_\parallel}{E_\perp} , \\ y_p &= \ln \frac{E_\perp}{E-cp_\parallel} . \end{aligned} \tag{15.23}$$

The last two expressions arise by multiplying nominator and denominator with $E \pm cp_\parallel$, respectively.

Example 15.7

Properties of particle rapidity

Show that for a particle with $\boldsymbol{p}_\perp \neq 0$, $p_\parallel = 0$, we have $y_p = 0$. Show that in the nonrelativistic limit the particle rapidity is particle speed in the $\parallel$-direction divided by c. Perform an explicit Lorentz boost in the $\parallel$-direction and find the boosted particle rapidity. ◀

▶ **Solution** We see from the first form of Eq. (15.23) that $y_p \propto \ln 1 = 0$ for $p_\parallel = 0$, verifying the first assertion.

In order to see the behavior of particle rapidity y_p in the nonrelativistic limit we rewrite the first form of Eq. (15.23) to read

1 $y_p = \frac{1}{2} \ln \left(1 + \frac{2cp_\parallel}{E-cp_\parallel} \right) \to \frac{cp_\parallel}{E-cp_\parallel} \qquad |\boldsymbol{p}| \ll mc .$

The nonrelativistic limit is shown. In the denominator the rest energy dominates in this limit, hence $E - cp_\parallel \to mc^2$, and we obtain

2 $y_p \simeq \frac{cp_\parallel}{mc^2} = \beta_\parallel .$

Another way to reach the same result employs Eq. (15.21)

3 $\beta_\| \equiv \frac{cp_\|}{E} = \tanh y_\mathrm{p} \simeq y_\mathrm{p} + \mathcal{O}(y_\mathrm{p}^3)\ .$

We boost the particle in the direction of $\boldsymbol{v}$. Hence we will be referring with $\perp,\ \|$ to the direction of $\boldsymbol{v}$. We use Eq. (15.15) with $y_\mathrm{r} \to y_{\mathrm{p}'}$

$$E' = \cosh y_{\mathrm{p}'}\, E + \sinh y_{\mathrm{p}'}\, cp_\|\ ,$$

4 $$cp'_\| = \cosh y_{\mathrm{p}'}\, cp_\| + \sinh y_{\mathrm{p}'}\, E\ ,$$

$$p'_\perp = p_\perp\ .$$

Setting $\cosh y_{\mathrm{p}'} = \gamma,\ \sinh y_{\mathrm{p}'} = \beta\gamma$ we obtain Eq. (15.12).

We compute the transformed particle rapidity, inserting in Eq. (15.23) the transformation Eq. 4

5 $$y'_\mathrm{p} = \frac{1}{2}\ln\left(\frac{E' + cp'_\|}{E' - cp'_\|}\right)$$

$$= \frac{1}{2}\ln\left(\frac{\cosh y_{\mathrm{p}'}E + \sinh y_{\mathrm{p}'}cp_\| + \cosh y_{\mathrm{p}'}cp_\| + \sinh y_{\mathrm{p}'}E}{\cosh y_{\mathrm{p}'}E + \sinh y_{\mathrm{p}'}cp_\| - \cosh y_{\mathrm{p}'}cp_\| - \sinh y_{\mathrm{p}'}E}\right)\ .$$

The coefficients of E and $p_\|$ involve $\cosh y \pm \sinh y = e^{\pm y}$

6 $$y'_\mathrm{p} = \frac{1}{2}\ln\left(\frac{Ee^{y_{\mathrm{p}'}} + cp_\| e^{y_{\mathrm{p}'}}}{Ee^{-y_{\mathrm{p}'}} - cp_\| e^{-y_{\mathrm{p}'}}}\right) = \frac{1}{2}\ln\left[e^{2y_{\mathrm{p}'}}\left(\frac{E + cp_\|}{E - cp_\|}\right)\right] = y_{\mathrm{p}'} + y_\mathrm{p}\ .$$

This is a very general result called the addition theorem of (particle) rapidities.

We can perform multiple boosts in sequence: Starting with $p_\| = 0$ we have $y_\mathrm{p} = 0$ and thus we find that the first boost rapidity $y_{\mathrm{p}'}$ is the particle rapidity. In the first boost we have moved the particle from $y = 0$ to $y_{\mathrm{p}'} = y_1$. Allowing a second boost we obtain

7 $$y_{\mathrm{p}'} = y_1 + y_2\ .$$

This can continue as many times as we need: We have learned that particle rapidities behave additively under LT carried out in the same direction. On first sight the expressions in Eq. (15.18) introducing particle rapidity y_p are a bit more complex and particle rapidity additivity is not immediately apparent. However, this property is analog to LT boost rapidities. ◀

Example 15.8

The deviation $c - v$

Obtain the exact particle speed deviations $c - v$ as a function of particle rapidity y_p. Present a relativistic particle approximation. ◀

▶ **Solution** We proceed using exact relativistic equations, and only in the last step do we make the required approximation. We use Eq. (15.6) and begin in a manner similar to the second part of Example 15.3

$$\mathbf{1}\quad c - v = c(1 - v/c) = c(1 - pc/E) = c\frac{E - pc}{E} = c\frac{(mc^2)^2}{(E + pc)E}\,.$$

We now use $y_{\rm p}$, see Eq. (15.19)

$$\mathbf{2}\quad c - v = c\frac{1}{(\cosh y_{\rm p} + \sinh y_{\rm p})\cosh y_{\rm p}} = c\,\frac{2}{1 + {\rm e}^{2y_{\rm p}}} = 2\,c\,{\rm e}^{-2y_{\rm p}}\frac{1}{1 + {\rm e}^{-2y_{\rm p}}}\,.$$

This result is exact. For ultrarelativistic particles we can neglect in the denominator the exponentially small term compared to unity and obtain

$$\mathbf{3}\quad c - v \simeq 2\,c\,{\rm e}^{-2y_{\rm p}}\,.$$

Remembering that in the relativistic limit $\gamma = {\rm e}^{-y_{\rm p}}$, this is just the result seen at the end of Example 15.3.

Knowing the relativistic particle rapidity, or equivalently the Lorentz-factor γ, we can now evaluate the deviation of the relativistic particle speed from that of light with relative ease. ◀

Example 15.9

Electron and proton beams

What are the speed and rapidity of an electron and a proton, in a particle beam with kinetic energy 1 MeV? 1 GeV? 1 TeV? The electron and proton rest masses are respectively $m_e \approx 0.511\,\text{MeV/c}^2$ and $m_p \approx 938\,\text{MeV/c}^2$. Use the results of Example 14.2. ◀

▶ **Solution** We begin with the relationship Eq. (15.6), $v = pc^2/E$, which with Eq. (15.9) leads to

$$\mathbf{1}\quad \frac{v}{c} = \frac{\sqrt{E^2 - m^2c^4}}{E} = \sqrt{1 - \frac{m^2c^4}{E^2}}\,.$$

The kinetic energy T is defined in Eq. (14.19) as the difference between the total energy and the rest energy, $T = E - mc^2$, so we obtain with Eq. 1 a form suitable for small mc^2/T,

$$\mathbf{2}\quad \frac{v}{c} = \sqrt{1 - \frac{m^2c^4}{(T + mc^2)^2}} = \frac{\sqrt{1 + 2mc^2/T}}{1 + mc^2/T}\,.$$

To evaluate particle rapidity for a given kinetic energy we employ the second form of Eq. (15.23), where we insert $E = T + mc^2$ and $cp_\parallel = cp = \sqrt{(T+mc^2)^2 - (mc^2)^2} = \sqrt{T^2 + 2Tmc^2}$, thus

3 $$y_\mathrm{p} = \ln\left(1 + \frac{T}{mc^2} + \sqrt{\frac{T^2}{(mc^2)^2} + 2\frac{T}{mc^2}}\right) \to \ln\left(1 + 2\frac{T}{mc^2}\right) .$$

Here the last limit applies for $T/mc^2 > 1$. This specific form more easily provides a more precise answer since in all cases under consideration $T/mc^2 > 1$, except for the single case of proton at 1 MeV where $T/mc^2 \ll 1$. However, since all terms in the exact argument of the logarithm are positive, there is no cancellation, and this expression also works in this nonrelativistic case.

We avoid the use of Eq. (15.22) as it is singular in the limit $\beta_\parallel \to 1$. Instead, by multiplying both numerator and denominator in the argument of the logarithm in Eq. (15.22) with $1 + \beta_\parallel$, we obtain

4 $y_\mathrm{p} = \ln[\gamma(1+\beta)]$.
In the ultrarelativistic limit we can take $\beta \to 1$ and set $\gamma = 1 + T/mc^2$. We obtain the following speeds and rapidities: GeV-energy electrons and TeV-energy protons have velocities that differ from the speed of light by only a few parts in ten million.

T	T/m_ec^2	v_e	y_e	T/m_pc^2	v_p	y_p
1 MeV	1.957	$0.941c$	1.747	0.001	$0.046c$	0.046
1 GeV	1957	$0.99999987c$	8.27	1.066	$0.875c$	1.36
1 TeV	1.96×10^6	$\sim c$	15.2	1066	$0.99999956c$	7.665

This result shows that when (kinetic) energy is much larger than the rest mass of the particle it is hard to distinguish particle speed from the speed of light. On the other hand, the rapidity can characterize the relativistic particle motion quite adequately. ◀

Example 15.10

Lorentz transformation of the phase

Show that the phase $\Phi = Et - \boldsymbol{p}\cdot\boldsymbol{x}$ is invariant under LT of coordinates, and of energy and momentum. Use here the rapidity akin to a rotation, see Example 7.13 and Eq. (15.16). ◀

▶ **Solution** Our interest is founded in the Einstein postulate that the phase of light is the same for all observers, see Chap. 13. This example generalizes this concept to massive particles described in quantum physics by the de Broglie matter waves. We ask when and how

$$\mathbf{1}\ \ \Phi = E/c\, ct - \boldsymbol{p}\cdot\boldsymbol{x} = \Phi' = E'/c\, ct' - \boldsymbol{p}'\cdot\boldsymbol{x}'$$

is true. As in Example 7.13 we write

$$\mathbf{2}\ \ \Phi = P^{\mathrm{T}}\, G\, X\,, \qquad P^{\mathrm{T}} = (E/c,\ p_x,\ p_y,\ p_z)\,.$$

The matrices G, and X were introduced in Example 7.13.

We check the condition for the normalization of the momentum transformation matrix Eq. (15.16)

$$\mathbf{3}\ \ m^2c^2 = P^{\mathrm{T}}\, G\, P = P^{\mathrm{T}\prime}\, G\, P' = P^{\mathrm{T}} \Lambda_{p_x}^{\mathrm{T}}\, G\, \Lambda_{p_x} P\,.$$

Just like for boosts of coordinates we also find here

$$\mathbf{4}\ \ \Lambda_{p_x}^{\mathrm{T}}\, G\, \Lambda_{p_x} = G\,.$$

We are now ready to look at the wave phase Φ transformation: We recognize that its invariance requires

$$\mathbf{5}\ \ \Lambda_{p_x}^{\mathrm{T}}\, G\, \widetilde{\Lambda_x} = G\,,$$

where to be safe we have used a yet unknown matrix $\widetilde{\Lambda_x}$ for the transformation of X. The unique solution of these two equations is

$$\mathbf{6}\ \ \widetilde{\Lambda_x} = \Lambda_{p_x} = \begin{pmatrix} \cosh y_{\mathrm{r}} & \sinh y_{\mathrm{r}} & 0 & 0 \\ \sinh y_{\mathrm{r}} & \cosh y_{\mathrm{r}} & 0 & 0 \\ 0 & 0 & 1 & 0 \\ 0 & 0 & 0 & 1 \end{pmatrix}.$$

This result applies to both massive and massless particles, i.e. photons. When we let particles, including photons, run to the right along the x-axis, as is implied by the use of Λ_{p_x}, this is equivalent to back transformation for the LT of coordinates. These are changing with a shift of coordinate axis to the left. This agrees with our study of the Doppler effect: Light that a star was shining at the observer on Earth was oriented against the observer's reference system; see Chap. 13.

We have shown that the wave phase Φ is an invariant when the LT transformation of coordinates $ct, \boldsymbol{x}$ is chosen to be opposite to the direction of the LT transformation of $E/c, p$

$$\mathbf{7}\ \ \Lambda_{p_x} = \widetilde{\Lambda_x}(y_{\mathrm{r}}) = \Lambda_x(-y_{\mathrm{r}})\,. \blacktriangleleft$$

Example 15.11

Aberration for massive particles

This problem generalizes and expands considerably Example 7.7 and presents the aberration formula for massive particles, generalizing Example 13.2. A 'fireball' created in high energy collisions of elementary particles moves with speed $\beta_F c$, or equivalently, rapidity y_F. Let θ_p be the polar angle of an emitted particle measured from the direction transverse to the fireball motion – transverse direction is not altered when we perform a Lorentz coordinate transformation along the fireball motion axis. The polar angle is complementary to the altitude angle seen in Fig. 13.1. What is the relation of θ'_p, the polar emission angle in the fireball comoving frame to the laboratory polar angle θ_p for a particle source moving with rapidity y_F? Discuss production of particles with reference to the altitude angle $\tilde{\theta}_p$, i.e. the laboratory angle of emission measured against the direction of motion, and introduce the relativistic aberration formula for massive particles. ◀

▶ **Solution** In terms of the polar emission angle we have

1 $$p_\parallel = p \sin\theta_p, \quad p_\perp = p\cos\theta_p\,,$$

valid for both primed, and not primed reference frames. The ratio of both momentum components of an emitted particle in the comoving frame of reference is, according to Eq. (15.15)

2 $$\tan\theta'_p = \frac{p'_\parallel}{p'_\perp} = \frac{\cosh y_F\, p_\parallel - \sinh y_F (E/c)}{p_\perp} = \frac{\cosh y_F \sin\theta_p - \sinh y_F (E/cp)}{\cos\theta_p}\,.$$

Here the particle direction of motion expressed in the sign of y_F is chosen such that the particle is emitted in the direction of the observer (typically comoving with the source) velocity vector.

The inverse transform is obtained by exchanging prime and not primed quantities, and changing $y_F \to -y_F$

3 $$\tan\theta_p = \frac{p_\parallel}{p_\perp} = \frac{\cosh y_F\, p'_\parallel + \sinh y_F (E'/c)}{p'_\perp} = \frac{\cosh y_F \sin\theta'_p + \sinh y_F (E'/cp')}{\cos\theta'_p}\,.$$

We recall that for fireball motion we have $\cosh y_F = \gamma_F$, $\sinh y_F = \beta_F\gamma_F$. The particle is emitted from the comoving frame with $E' = \gamma'_p mc^2$, $cp' = \gamma'_p\beta'_p mc^2$. We obtain

4 $$\tan\theta_p = \gamma_F \frac{\sin\theta'_p + \beta_F/\beta'_p}{\cos\theta'_p}\,.$$

In the fireball (primed) comoving reference frame, particles are typically emitted in an isotropic distribution in θ'_p – there is no preferred emission direction. However, for large γ_F the only possible value of the laboratory emission angle according to the above equation is the polar angle $\theta_{\rm p} \to \pi/2$; all particles appear focused along the direction of motion.

We can refine this insight with the help of the altitude angle $\tilde{\theta}$, an angle measured in the laboratory with respect to the direction of the fireball source motion. In this case $\sin\theta'_{\rm p} \to \cos\tilde{\theta}'_{\rm p}$, $\cos\theta'_{\rm p} \to \sin\tilde{\theta}'_{\rm p}$, $\tan\theta_{\rm p} \to \cot\tilde{\theta}_{\rm p}$, resulting in

$$\mathbf{5}\quad \tan\tilde{\theta}_{\rm p} = \frac{1}{\gamma_{\rm F}}\,\frac{\sin\tilde{\theta}'_{\rm p}}{\cos\tilde{\theta}'_{\rm p} + \beta_{\rm F}/\beta'_{\rm p}}\,.$$

This result contains the (inverse of) altitude angle aberration formula obtained in Example 13.2, Eq. 2: we see the relation setting $\beta_{\rm F}/\beta'_{\rm p} \to -v_{\rm F}/c$, since the observer velocity $v = -v_{\rm F}$, and for light, the particle velocity is always the light velocity.

For relativistic motion of the fireball, and relativistic particles emitted parallel to direction of motion $\beta_{\rm F}/\beta'_{\rm p} \to 1$

$$\mathbf{6}\quad \tan\tilde{\theta}_{\rm p} \simeq \frac{1}{\gamma_{\rm F}}\tan\frac{\tilde{\theta}'_{\rm p}}{2}\,,$$

where we used the half-angle formula

$$\mathbf{7}\quad \tan(\theta/2) = \frac{\sin\theta}{\cos\theta + 1}$$

obtained remembering $\sin 2\alpha = 2\sin\alpha\,\cos\alpha$ and $\cos 2\alpha = \cos^2\alpha - \sin^2\alpha$. For large values of $\gamma_{\rm F}$ we can replace $\tan\tilde{\theta}_{\rm p} \to \theta'_{\rm p}$, showing that the laboratory altitude angle always spans a range of small values. For particles emitted isotropically in the CM system the effect of the relativistic fireball (source) motion is that all emitted particles are focused into a relatively small forward cone in the laboratory reference system.

This effect allows us to recognize the presence of one or more fireballs moving with different relativistic speeds with reference to the observer: We expect single or multiple azimuthal bunching of particles into related cones with corresponding azimuthal emission rings around the motion axis, characteristic of the different values of source $\gamma_{\rm F}$.

Further Reading: These considerations played an important role in the understanding of reaction mechanisms in reactions of strongly interacting particles[2] – the meaning of the term 'emission angle' in this publication was later clarified.[3] ◂

[2]N.M. Duller, W.D. Walker, "High-Energy Meson Production," *Phys. Rev.* **93** (1954), pp. 215–226.

[3]R. Hagedorn, "The Long Way to the Statistical Model," in *Hot Hadronic Matter: Theory and Experiment* NATO-ASI-Series B **346**, pp. 13–46, eds. J. Letessier et al., Plenum Press, New York, (1995).

Generalized Mass-Energy Equivalence

16

Abstract

A composite body with complex internal dynamics always satisfies Einstein's $E = mc^2$, irrespective of what is going on inside the body in terms of motion, chemical and nuclear transmutation reactions, and emission and absorbtion of radiation. Within a body, the origin of usable energy is explored in many examples involving kinetic energy, potential energy, and internal body energy. We show that available energy always derives from the reduction of the mass of the final state as compared to the initial state of a reacting system, irrespective of whether, for example, the power station is based on renewable, conventional or nuclear energy.

16.1 Where Does Energy Come From?

In the following we will show, case by case, that any energy we extract or deposit in a material body relates directly to the energy locked in the rest mass of the body. All the energy we use, in whatever format, ultimately comes from conversions of a small fraction of mass of matter into another form of energy that can serve a useful task.

Just before the French revolution, Antoine Lavoisier[1] demonstrated that the combustion of gases preserves the mass of the material cast in a new form. However, having recognized the enormous energy reservoir of the rest mass $E_0 = mc^2$, we now take the next step: Any heat produced in combustion originates in a reduction of the rest mass of all final reaction products. When we 'use' energy, say, drive a car, the combustion of gasoline with air creates a tiny mass defect

[1] Antoine-Laurent de Lavoisier (1743–1794), French nobleman and chemist central to the eighteenth-century chemical revolution.

J. Rafelski, *Modern Special Relativity*, https://doi.org/10.1007/978-3-030-54352-5_16

well beyond Lavoisier's experimental precision, and this mass defect originates the kinetic energy of the vehicle.

Consider as another example the conversion of energy that takes place in our Sun. Our Sun is a giant fusion reactor, burning up some of its mass. When some of the emitted radiation reaches Earth, it helps in biological processes to break CO_2 molecules, undoing the binding of carbon and oxygen. This temporarily adds to the mass of Earth as what we later release in our power stations by burning fossil fuel. The waste heat is (hopefully) radiated away, and Earth's energy balance is restored.

On the other hand, Earthbound nuclear fission, and in the future, fusion power stations, convert elements into each other, and the resultant energy leads ultimately to additionally produced heat. Assuming that the produced heat is ultimately radiated into space, the element conversion on Earth will result in a slight reduction of Earth mass.

Thus the energy we use comes from: (a) reduction of the mass of the Sun today (for 'green' energy), or yesterday (for 'fossil' energy) and, (b) when nuclear processes are employed, reduction of the mass of the Earth. In the following, we will make our argument case by case; in Sect. 17.4 we return to a more quantitative description of how a radioactive decay of a heavy particle into two lighter particles results in a net energy gain.

16.2 Examples of Mass-Energy Equivalence

Mass Equivalence for Kinetic Energy in a Gas

Let us consider a box containing a mono-atomic gas in thermal equilibrium at temperature T. When the temperature of the gas increases, the amount of thermal energy contained therein does as well. Thus, we expect the mass of the box to increase. The average energy of mono-atomic and nonrelativistic gas particles is given by the equipartition theorem,

$$\langle E\rangle = \frac{3}{2}k_bT \ , \tag{16.1}$$

where $k_b = 1.38 \times 10^{-23}$ J/K is the Boltzmann constant. As we have already seen, the mass of the system depends on the kinetic energy of its components, and the motion of each of the gas atoms contributes to the system mass

$$\delta M = \frac{\langle E\rangle}{c^2} \ . \tag{16.2}$$

As temperature is increased, the total mass of the gas increases. This is the topic of our following example.

▶ **Insight: 16.1 Elementary energy units** The use of the elementary energy unit 'eV' is advantageous in considerations that address properties of single elementary objects in (sub) atomic physics. It is widely accepted as a complement to the SI-units. We combine the SI-unit 'V=volt' with the elementary charge e to form an elementary unit of energy 'eV', i.e. the amount of kinetic energy that a particle of elementary charge e acquires when accelerated across a potential step of 1 V.

The relationship with the joule SI-unit of energy follows by definition: 1 V = 1 J/C; thus for the conversion we need the measured value of elementary charge $e = 1.602177 \times 10^{-19}\,\mathrm{C}$

$$\mathrm{eV} = 1.602177 \times 10^{-19}\,\mathrm{J}\,, \qquad 1\,\mathrm{J} = 6.241509 \times 10^{18}\,\mathrm{eV}\,. \tag{16.3}$$

Physicists often quote particle masses using as basic unit eV/c^2. Many elementary particles have an energy-mass equivalent at GeV scale (giga-eV: $= 10^9$ eV). The case of the electron is special; $m_e c^2 =$ 0.511 MeV (mega-eV $= 10^6$ eV).

To convert back to SI mass unit kg we note

$$1\,\mathrm{J} = 1\,\mathrm{kg}\frac{\mathrm{m}^2}{\mathrm{s}^2} \;\to\; 1\,\mathrm{kg} = 6.241509 \times 10^{18}\,\frac{\mathrm{eV}}{c^2} \times \frac{(299{,}792{,}458)^2\,\mathrm{m}^2/\mathrm{s}^2}{\mathrm{m}^2/\mathrm{s}^2}\,. \tag{16.4}$$

We thus obtain

$$1\,\mathrm{kg} = 5.609589 \times 10^{35}\,\frac{\mathrm{eV}}{c^2}\,, \qquad 1\,\mathrm{eV} = 1.782662 \times 10^{-36}\,\mathrm{kg}\,c^2\,. \tag{16.5}$$

The large number appearing in Eq. (16.5) arises since we multiply the number of elementary charges in one coulomb with the square of the speed of light.

Another important insight is how 1 eV energy per atom or molecule translates into joules per mol. We multiply the eV–J relation in Eq. (16.3) with the Avogadro constant N_A

$$N_A = 6.022141 \times 10^{23}/\mathrm{mol} \;\to\; \frac{\mathrm{eV}}{\mathrm{Atom}} \equiv 0.964853 \times 10^5\,\frac{\mathrm{J}}{\mathrm{mol}}\,. \tag{16.6}$$

SCALES: In chemistry the unit meV (milli-eV: $= 10^{-3}$ eV) appears: The average kinetic energy of an atom or molecule air component at room temperature is about 40 meV, also the scale of rotation energy of many molecules. The typical molecular binding energy is a few eV, and molecular vibration energies are a fraction of eV; electron binding energy in heavy atoms reaches to 100 keV (kilo-eV: $= 10^3$ eV); the nuclear binding energy is up to 15 MeV. The kinetic energy available in the highest energy elementary particle accelerator is below 10 TeV (tera-eV: $= 10^{12}$ eV). Higher exponents PeV (peta-eV: $= 10^{15}$ eV), EeV (exa-eV: $= 10^{18}$ eV), ZeV (zeta-eV: $= 10^{21}$ eV) appear, for example, when we deal with instantaneous power of ultra-pulsed lasers.

Example 16.1

Thermal mass content of a hot Zeppelin

A Super-Zeppelin is kept afloat in the sky by 100,000 kg of helium gas[2] providing a lift capability[3] similar to the Boeing 747. Over the course of a flight on a sunny day, the temperature of the gas increases from $T_1 = 270\,\mathrm{K}$ to $T_2 = 310\,\mathrm{K}$. By how much does the mass of the gas increase? ◀

▶ **Solution** The atomic mass of helium is $\mu \approx 4$ g/mol. The number of particles in the gas is given by

$$N = \frac{M}{\mu} = \frac{10^8\ \mathrm{g}}{4\ \mathrm{g/mol}} N_A \approx 1.5 \times 10^{31}\ , \tag{1}$$

where $N_A = 6.02 \times 10^{23}\ \mathrm{mol}^{-1}$ is Avogadro's number. The change in total kinetic energy due to heating is then given by

$$\Delta E = E_2 - E_1 = \frac{3}{2} N k_b (T_2 - T_1) \approx 1.2 \times 10^{10}\ \mathrm{J}\ . \tag{2}$$

[2]The *Hindenburg* destroyed by fire in 1937 got its lift from 200,000 m^3 of hydrogen with a weight of 18,000 kg.

[3]A volume filled with the inert and second lightest gas helium has nearly 93% of the lift of hydrogen: this is so since $(29 - 4)/(29 - 2) = 0.926$ comparing to the air molecular weight 29.0. Helium supplies are limited: It is believed to originate from radioactive decay in the depths of the Earth and is found mixed with natural gas, sometimes in abundance exceeding 5%. With a boiling point at 4.2 K at one atmosphere it plays an essential role in a wide range of technological and scientific applications. Therefore helium is a strategic material. The *Hindenburg* helium filling would have been 36,000 kg – the twenty-first century Super-Zeppelin would be only about 40% larger in all three directions.

This seems to be large, yet this is the power of the Sun, impacting the Super-Zeppelin within an hour. The fractional change in mass of the gas due to the change in thermal energy content, the fractional mass increase, is

3 $$\frac{\Delta M}{M} = \frac{\Delta E}{Mc^2} \approx \frac{0.14\ \text{mg}}{M} \approx 10^{-12} .$$

As the smallness of this mass-gain result suggests, the blimp is in no danger of crashing. Note that it is next to impossible to measure a fractional mass change that is so small. ◀

Rotational Energy Mass Equivalence

A special form of motion is rotation of a flywheel. We know that the kinetic energy internal to a body contributes to the body mass. Rotation is a special form of kinetic motion and thus a body that rotates acquires additional energy equivalent mass content. The rotational energy must contribute since locally it arises from the speed v_i of each of the constituents of the rotating body of mass m.

This effect of rotational energy mass equivalence can, in principle, be measured as is depicted in Fig. 16.1. The spring-scale measures the weight of a body m at rest relative to the scale, and another identical scale on the right measures the weight of an identical body in uniform rotational motion relative to the scale. The rotating body weighs in with $E_{\text{rot}}^{\text{NR}}(v)/c^2$ more mass. In 'constructing' the scale in Fig. 16.1, we tacitly used the equivalence of inertial and gravitational mass. This is known as the principle of equivalence, a consideration which leads ultimately to the gravity

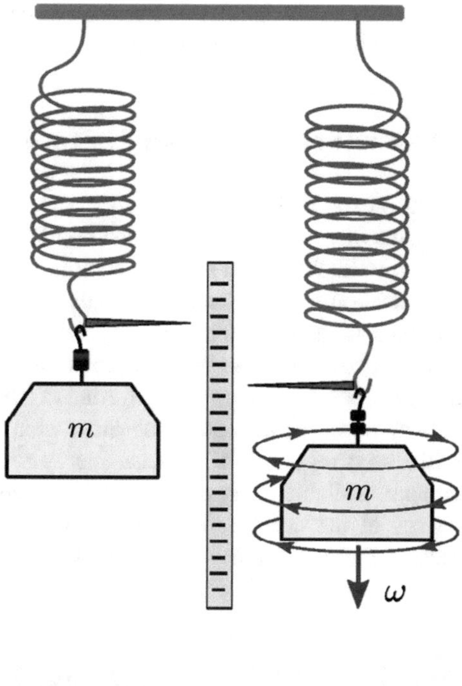

Fig. 16.1 A comparison of the mass m of a body to the mass of the same but rotating body on right. The spring-scale measures the entire energy content including the rotational kinetic energy of the body

relativity, GR. Since we needed to mention GR, we expect that in the presence of relativistic motion, the exact rotational mass equivalent can only be determined using GR.

The fractional increase in flywheel mass we achieve by speeding it up remains rather small. The best flywheels will not exceed the 10^{-13}–10^{-11} fraction we discussed in Example 16.1. This is so since the speed of the flywheel rotation must be below the speed scale of the effect of thermal motion, to assure that the centripetal forces do not break the device.

In the case $(v/c) \ll 1$ we have for the rotating body (assuming the center of mass is stationary),

$$\frac{E(v)}{c^2} = m + \frac{E_{\mathrm{rot}}^{\mathrm{NR}}}{c^2} > m\ , \tag{16.7}$$

in which $E_{\mathrm{rot}}^{\mathrm{NR}}$ is the nonrelativistic rotational kinetic energy obtained using a classic form of moments of inertia.

There are ongoing efforts to build a car kinetic energy recovery system (KERS) based on an ultra-rapid flywheel has seen use in race car design. Before taking a relativity class you would think of rotation energy, originating in rotational motion, as the energy source. A new way to think about this situation is to realize that the rotating flywheel is somewhat more massive when it stores energy. As the energy of the flywheel is consumed, it slows down and therefore a lower mass value of the rotating flywheel is reached.

We can say that the car with flywheel power-assist drive derives its energy from the variability of the mass of the flywheel, which depends on the wheel rotation. A rotation hidden within the body clearly adds to body mass. This is unlike body translation kinetic energy, which is not mass, see Sect. 14.3. The consideration of rotational energy parallels the storage of usable energy exploiting potential energy mass equivalent we discuss next.

Potential Energy Mass Equivalence

We now explore the mass of a device containing two balls of identical mass m joined by a massless spring, compressed and latched at a high potential V, as depicted in Fig. 16.2. When the spring's latch is released, the balls are thrown off. Due to conservation of momentum, the balls must have equal and opposite momentum and thus at equal mass the same speed $\pm u$ at all times as measured in the original system's rest-frame. As the extension of the spring reaches the equilibrium position, all of the potential energy is converted to the kinetic energy of both balls.

Any observer who does not know what is 'inside' the box shown in Fig. 16.2 will always report the same mass-energy equivalent for this device, the only energy content available to an observer outside the box. In other words, as the potential energy of the compressed spring is released into the kinetic energy of the two masses, there is no change in the mass equivalent of the experimental box. From the

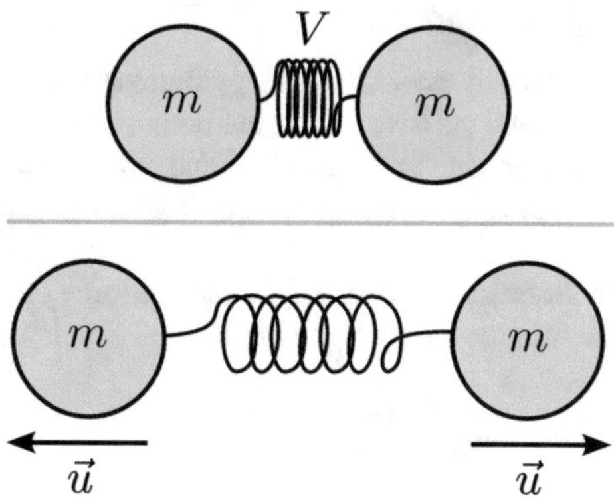

Fig. 16.2 The total energy of a device consisting of two balls attached to a spring does not depend on spring compression – this assures that the potential energy V counts as much as kinetic energy of oscillating masses m towards the total device rest energy and thus device mass

conservation of energy in the original rest-frame, considering the situation before and after spring release we see that we have

$$V(x(t)) + 2\frac{mc^2}{\sqrt{1 - u(t)^2/c^2}} = Const. \tag{16.8}$$

Once all potential energy is converted into motion of the balls we have, at this time t_m a speed u_m and

$$2\frac{mc^2}{\sqrt{1 - u_m^2/c^2}} = Mc^2 , \tag{16.9}$$

where M is the mass of the box comprising the balls, setting aside the contributions which are not changing with time. This fixes the constant in Eq. (16.8). Comparing the energy content at $t = 0$ where $u(0) = 0$ and $V(x(0)) = V_0$ with the time t_m where all potential energy is converted into motion, we have, using Eq. (16.8)

$$V_0 + 2mc^2 = 2\frac{mc^2}{\sqrt{1 - u_m^2/c^2}} = Mc^2 . \tag{16.10}$$

The merit of this consideration is now clear. When potential energy internal to a body is converted to kinetic energy we see no change in the total mass of the body. The potential energy stored in the compressed spring is recognized as contributing the amount $\delta M = V/c^2$ to the mass of the body. Conversely, after all available potential energy is converted to motion, we have exactly this mass effect now in the internal kinetic energy of components of the body which contribute to the rest mass accordingly.

Consideration of changes in mass related to potential energy provides a fundamental understanding of physics mechanisms involved in the storage of usable energy. For example the potential energy as part of mass explains where the energy stored in, e.g., a hydroelectric power generation system, comes from.

Example 16.2

Gravity potential energy mass defect

All massive bodies are bound by gravity to Earth's surface. Evaluate the mass defect due to the gravitational potential of the Earth in Newtonian approximation, that is ignoring the effects of general relativity. ◂

▸ **Solution** The magnitude of the force on a mass m located at Earth's surface and pointing towards the center of the Earth is

$$F_g = mg = \frac{GMm}{R^2} \,, \tag{1}$$

where $g = 9.80\,\text{m/s}^2$; G is the gravitational coupling constant; M is the Earth mass and $R = 6.371 \times 10^6\,\text{m}$ is the (average) Earth radius. The introduction of g allows us to sidestep the computation using large numbers.

The gravitational potential energy for a body at the surface can be written in the form

$$V_g = -\frac{GMm}{R} = -mgR \,, \tag{2}$$

and this is by how much the mass is decreased

$$\delta mc^2 = V \,, \tag{3}$$

since V is negative. The fractional gravity mass defect thus is

$$\frac{\delta m}{m} = \frac{V/c^2}{m} = -\frac{gR}{c^2} = -6.94 \times 10^{-10} \,. \tag{4}$$

This gravitational mass defect is, as we see, small but not utterly negligible in computing the magnitude, for example, for a human body: $\delta m \to 50\,\mu\text{g}$. The magnitude of gravitational mass defect on Earth is comparable to chemical mass defect, but smaller than atomic mass defect we consider next. We expect that this result remains qualitatively (i.e. up to a factor $\mathcal{O}(1)$) correct within the relativistic theory of gravity, GR, since the consideration is carried out in the Newtonian limit and does not involve relativistic motion. ◂

Atomic Mass Defect

When an atomic electron moves into an 'orbit' with lower energy, the difference in energy is carried away in the form of an emitted photon. As we have already studied in Example 15.5, a photon can carry mass away from a body, so the mass of the atom decreases with the change in binding energy of the atomic electron. In the simplest case of an electron e^- binding to a proton p forming an atom of hydrogen we have

$$e^- + p \to \mathrm{H} + \bar{Q} \,. \tag{16.11}$$

The energy released per each reaction in form of radiation after the electron reaches the ground state is $\overline{Q} = 13.6\,\mathrm{eV}$. The energy-mass balance of the reaction is

$$m_e c^2 + m_p c^2 = m_H c^2 + \bar{Q} \,. \tag{16.12}$$

The mass of the product of the reaction, here atomic hydrogen, is thus smaller than the masses of the components by the amount

$$\delta m = \frac{\bar{Q}}{c^2} = m_e + m_p - m_H \,. \tag{16.13}$$

The fractional mass defect of the most bound electron in hydrogen is

$$\frac{\delta m}{m_H} = \frac{\bar{Q}}{m_H c^2} = 1.45 \times 10^{-8} \,, \tag{16.14}$$

where we have used electron binding $\bar{Q} = 13.6\,\mathrm{eV}$ and $m_H c^2 = 0.938 \times 10^9\,\mathrm{eV}$. This atomic mass defect is at the same time small, and yet large e.g. when compared to the effect of the gravitational potential.

Chemical Energy Mass Defect

When we burn hydrocarbon fuels the source of the energy gain, as we now realize, is in reducing the mass of the reaction products as compared to the initial components. We now check how big this effect is for hydrogen burning in air, arguably the most efficient of these chemical reactions

$$\mathrm{H_2} + \frac{1}{2}\mathrm{O_2} \to \mathrm{H_2O} + Q_h \,, \tag{16.15}$$

where Q is the heat released; in this example, Q_h is heat of hydrogen combustion.

The released heat Q is usually given in Joules per mol ($Q_h = 286\,\mathrm{kJ/mol}$) but when considering elementary reactions we prefer to use units of eV per single molecule created in the reaction, see Insight 16.1: *Elementary energy units*. The conversion goes as follows: There are 6.022×10^{23} molecules in each mole ($N_A = 6.022 \times 10^{23}\,\mathrm{mol}^{-1}$ = Avogadro constant in SI units) and one $\mathrm{J} \equiv 6.24 \times 10^{18}\,\mathrm{eV}$ (the coefficient is the number of electron charges in a 'Coulomb' of electrical charge). The unit conversion factor thus is

$$R_Q = \frac{6.24 \times 10^{18}}{6.022 \times 10^{23}} = 1.03643 \times 10^{-5}\,\mathrm{eV}\,\frac{\mathrm{mol}}{\mathrm{J}}\,. \tag{16.16}$$

This leads to a molecular energy gain in each reaction of

$$\overline{Q}_h \equiv Q_h R_Q = 2.86 \times 10^5\,\frac{\mathrm{J}}{\mathrm{mol}} \times 1.03643 \times 10^{-5}\,\mathrm{eV}\,\frac{\mathrm{mol}}{\mathrm{J}} = 2.96\,\mathrm{eV}\,. \tag{16.17}$$

Many particles participate in each reaction Eq. (16.15). Since each hydrogen comprises one proton and one electron, and oxygen has 8 electrons, 8 protons and 8 neutrons, in total we have 10 electrons, 10 protons and 8 neutrons, and of course the nucleons carry practically all the mass. The energy equivalents of each of these particles in eV units, see Insight 16.1: *Elementary energy units* is

$$m_e c^2 = 0.5110\,\mathrm{MeV}\,, \quad m_p c^2 = 938.3\,\mathrm{MeV}\,, \quad m_n c^2 = 939.6\,\mathrm{MeV}\,. \tag{16.18}$$

We estimate the total mass of participating particles in Eq. (16.15) at 16,900 MeV. Each reaction Eq. (16.15) gains approx. 3 eV out of 1.7×10^{10} eV, which amounts to a fractional reduction of mass at the level of 1.8×10^{-10}

$$\left.\frac{\delta m}{m}\right|_{\mathrm{chemical}} \lesssim 1.8 \times 10^{-10}\,. \tag{16.19}$$

This is, arguably, the largest mass defect we can find among chemical reactions; even so, it is a factor 4 smaller than gravitational mass defect, see Example 16.2. This explains why it is difficult to get off the Earth surface using chemical propulsion: The gain in mass defect from a chemical reaction must compensate the gravitational potential mass defect.

Nuclear Mass Defect

In practically all nuclear reactions the mass defect energy gain per reaction is about a million times greater as compared to atomic reactions. Since this is so, nuclear reaction energy is typically measured on the scale of MeV as compared to the eV energy scale for atomic components. Here 'M' stands for Mega $= 10^6$. For example, to make a deuteron $d = pn$, one proton binds with a neutron

$$p + n \to d + \gamma + \bar{Q}_d\,. \tag{16.20}$$

The energy released is $\overline{Q}_d = 2.2245\,\mathrm{MeV}$. The difference in energy between the fusing nuclei and the fused nucleus formed up to a very small d-recoil energy is removed by (gamma) γ-radiation; that is, a very energetic photon of nuclear origin. The fractional nuclear mass defect for the reaction Eq. (16.20) in view of the masses Eq. (16.18) is

$$\frac{\delta m}{m} \simeq \frac{2.2245}{938.3 + 939.6} = 1.2 \times 10^{-3} . \tag{16.21}$$

In reaction Eq. (16.20) the energy is released in form of EM radiation. However, in general, the energy released in nuclear reactions such as the fusion of light elements or the fission of heavy elements manifests itself as the kinetic energy of released reaction fragments. In all these reactions, just as in this example, the binding mass defect is at a noticeable level of 0.1% and often even somewhat higher.

In water on Earth one out of 6,420 hydrogen atoms is actually a deuteron atom. This deuteron abundance around us is a leftover from the primordial nuclear burning period of the early Universe. Each atomic deuteron D has a nucleus d made of one proton and one neutron $d = pn$. The relatively small amount remaining testifies to the relative ease with which deuterium can serve as nuclear fuel in primordial fusion burn processes. A possible near term path to fusion energy involves deuterium reaction with tritium $t = pnn$: $dt \to \alpha + n$, likely to power the first generation fusion reactors.

In a dt fusion reaction the mass defect per nucleon is about three times as large as that given in Eq. (16.21) for the deuteron. While the energy yield is high, and the fusion reaction most easily achieved, there are other process and engineering difficulties. The produced neutrons, being neutral, are very penetrating and can travel far in their lifespan of about $\tau_n = 15\,\mathrm{min}$. No established method exists to control these high energy neutrons that are produced in first generation dt-fusion plasma devices. Moreover, there is a need to artificially create the unstable tritium nuclei required in the reaction, using these highly energetic neutrons.

The search for an acceptable method of nuclear (fusion) energy continues. An alternate path is aneutronic fusion, a controlled nuclear burn without direct production of neutrons. An aneutronic approach is the so called pB (proton-Boron) reaction induced by direct laser acceleration of reactants[4]

$$p + \mathrm{B}^{11} \to \alpha + \alpha + \alpha \,, \qquad \frac{\delta m}{m} = \frac{8.68}{938.3 + 11.007 \times 931.5} = 0.8 \times 10^{-3} . \tag{16.22}$$

[4] C. Labaune, C. Baccou, S. Depierreux, C. Goyon, G. Loisel, V. Yahia, and J. Rafelski, "Fusion reactions initiated by laser-accelerated particle beams in a laser-produced plasma," *Nature Communications* **4** 2506 (2013); and C. Labaune, C. Baccou, V. Yahia, C. Neuville, and J. Rafelski, "Laser-initiated primary and secondary nuclear reactions in Boron-Nitride," *Nature Scientific Reports* **6** 21202 (2016).

The nuclear reaction Eq. (16.22) releases 8.68 MeV, while the reference mass consists of the proton mass equivalent energy and the B^{11} nuclear mass energy equivalent written in atomic mass units, removing the mass part of electrons. We see that the mass defect is comparable to that seen in Eq. (16.21), a situation typical for many light element nuclear exothermic reactions. It should be noted that reaction Eq. (16.22) involves readily available B^{11}. And the reaction products are three inert charged α-particles; hence we call such reactions aneutronic.

Example 16.3

Energetics of deuterium content in water

Evaluate, as matter of principle, the accessible fusion energy content of one gallon of distilled water and compare this to the chemical energy content of an equal volume of gasoline, $Q_{\text{gas}} \simeq 125\,\text{MJ}$. Begin with a qualitative estimate. ◀

▶ **Solution** We consider in this example the non-metric unit gallon for the benefit of US readers. Irrespective of the question if we are considering a gallon or a liter of fuel, any potential nuclear reaction would yield a million times more energy per reactants compared to chemical energy. However since deuterons are less than a part in 6000 in water we can expect our analysis to produce a result that exceeds the gasoline energy content by about a factor 150.

Turning to more precise evaluation: One gallon of water, that is 3.785 kg, comprises 210 moles of water since a mole of H_2O weighs 18 grams. The number of deuterons in one gallon of water is

$$\mathbf{1}\quad N_d = \frac{210}{6420} 2N_A = 3.94 \times 10^{22}, \qquad N_A = 6.022 \times 10^{23}\text{mol}^{-1} ,$$

where we used the Avogadro number N_A and we recall that only 1/6240 hydrogen fraction is a deuteron. The additional factor ‘2’ reminds us that there are two hydrogen atoms in each molecule of water.

We now estimate and assume that on average one d can help release an energy of $\overline{Q}_d = 5\,\text{MeV}$. We recall from Insight 16.1: *Elementary energy units*

$$\mathbf{2}\quad 1\text{eV} \equiv 1.6 \times 10^{-19}\,\text{J} , \qquad 1\,\text{MeV} \equiv 1.6 \times 10^{-13}\,\text{J} .$$

The estimate of the total energy content of deuterium content in water is

$$\mathbf{3}\quad Q_d = N_d\, 5\,\text{MeV}\, 1.6 \times 10^{-13} \frac{\text{J}}{\text{MeV}} = N_d 8 \times 10^{-13}\,\text{J} .$$

In one gallon of water the energy content is thus

$$\mathbf{4}\quad Q_d^{\text{gallon}} = 3.16 \times 10^{10}\,\text{J} > 250\, Q_{\text{gas}} .$$

The energy content available with future technology originating in heavy hydrogen in water is more than a 100 times above the chemical energy available in an equal volume of gasoline. ◀

Discussion 16.1 Nuclear energy

About the topic: Fusion energy powers our sun, but will this last? Where does energy in nuclear fission reactors come from?

- *Simplicius:* Is this the way stars get their energy, through exploiting a small difference in binding energy of fusing nuclei?
- *Professor:* Yes, the Sun is powered in this way. All its energy comes from a nuclear mass defect when lighter particles bind together to create heavier particles.
- *Student:* But stars do not last forever.
- *Professor:* Stars burn, fusing nuclei of hydrogen, protons, which ultimately turn into the α-particle, and α-particles turn to carbon, and eventually this runs into a point where it takes more energy to bind the atomic nuclei than the reaction yields. This point is near iron, atomic number 26 in the periodic table. Many of these fusion products with atomic number below 26 are found abundantly on Earth.
- *Simplicius:* But we have on Earth many, many elements heavier than iron. Where do they come from?
- *Professor:* A massive enough star can become unstable when it runs out of light element nuclear fuel. It collapses, and then bounces and explodes. Much happens along the way, including heavy element production.
- *Student:* What keeps a star from imploding anytime?
- *Professor:* Normally a star is kept stable by the balance between the force of gravity, which wants the star to contract, and the energy of burning, the pressure that the heat generates, which wants the star to explode.
- *Student:* So when there is no longer enough energy available from nuclear fusion, the star collapses in on itself?
- *Professor:* Yes, this is so, though there are other triggers that could collapse a star. As the star collapses and bounces, new nuclear processes become possible. These can make, in the absence of equilibrium, heavier elements that are not as tightly bound as is iron.
- *Simplicius:* If these elements heavier than iron are not as tightly bound, how are these heavier elements stable?
- *Professor:* Excellent question; you are right; they are not always stable even if left to themselves in empty space. For example, uranium is not stable; we have built fission reactors to exploit this instability.
- *Simplicius:* Why then do we have these heavier elements on Earth?
- *Professor:* It is a question of time. Uranium must be sufficiently stable, that is, live for a long time after it was produced or we would not have it today. To be precise, the ^{235}U isotope has a half-life of 0.7 billion years. The more abundant ^{238}U is more stable with a half-life of 4.5 billion years,

(continued)

close to the lifespan of the Solar system. If we wait long enough very little uranium will be left, as it will all eventually decay on its own.

- *Student:* Is this decay where we get fission energy from?
- *Professor:* Yes. We accelerate or perhaps better said, induce natural decay, and extract the energy from the kinetic motion of rapidly separating fission fragments. In this way, the nuclear fission energy we gain in a nuclear reactor is taken from the ashes of a stellar explosion.
- *Simplicius:* So a nuclear reactor is just a place to put a lot of uranium?
- *Professor:* Not exactly, in a nuclear reactor we make the natural radioactive decay process go faster. It is a very complex machine. There is fear of nuclear energy as experience has shown that mistakes can be made both in the design and the operation of fission reactors.
- *Student:* There are many elements heavier than iron which seem all to have energy available, right?
- *Professor:* Right, and if one counts not only elements but also the different variants with differing numbers of neutrons, the isotopes, one can say that very few heavy element isotopes are stable; most decay, and those which do not decay naturally can be used in a element transmutation reactor to gain energy. Such accelerator driven reactors are of interest as they are less prone to human error.
- *Student:* You said some of these isotopes are stable, but how can this be? Anything beyond iron should not be stable.
- *Professor:* Stable is anything that will remain in the same form during the lifespan of the Universe. Water is stable, although its nuclear components could fuse. However, one day in the future we may master technology to help water fuse …
- *Simplicius:* Just like Mr. Fusion in the *Back to the Future* movies series?
- *Professor:* Yes. For now we only have fission reactors and are busy working on fusion of lightest element isotopes. Mr. Fusion must have traveled more than a century into the future.
- *Simplicius:* The energy we get out of nuclear reactions is rather low compared to the whole amount of energy available in reactant mass, is it?
- *Professor:* Yes, most energy remains locked in the mass of reaction products; we just transmute elements to gain a fraction of a percent in mass-equivalent energy in breaking up or fusing isotopes.
- *Student:* That is so because we are not annihilating matter.
- *Professor:* Right, to do that we would need to obtain the unobtainium, the antimatter, or learn how to control the stability of matter.

16.3 Origin of Energy, Origin of Matter

We derive usable energy in reactions that reduce the mass of the reactants. Therefore in order to understand the origin of available energy we need to know the origin of the mass of matter. This question leads us to consider the origin of energy, and eventually to the consideration of the energy balance of the Earth.

Employing relativistic nuclear (referred to as heavy ion) collisions we have established the primordial quark-gluon plasma (QGP) phase[5] as the source of matter. In the *hadronization* of QGP a considerable fraction of the primordial energy of the early Universe is locked into the mass of nearly symmetrical abundance of matter and antimatter.

Once almost all matter and antimatter annihilates we are left with a small nano-excess of matter over antimatter. We have discovered within the laws of physics an asymmetry, called CP-violation. This allows the communication, to a distant observer, that we are made of matter. It is believed that due to CP-Violation primordial energy becomes locked in the mass of the nano-excess of matter. However, why does matter surround us, and not antimatter? We do not yet know; we are still seeking to identify possible reaction mechanisms.[6]

The mass content of the Universe is evolving. After the antimatter annihilation, there were Big-Bang nuclear (BBN) fusion processes in the Universe as a whole; this is how the light elements such as helium and lithium were formed. Later, after stars formed, all the elements we are familiar with were ultimately created. Today we derive fusion energy using light elements left over in the BBN and star fusion burn. Fission energy originates in burned-up stars' collapse: The gravitational binding of the stellar remnant delivers in (super-) novae explosions an increase in the expelled heavy element mass liberated in fission process.

Our Sun has a lot of light element fuel left to burn. The solar fusion power is transmitted in the form of radiation to Earth. Only a tiny fraction, 173,000 TW, corresponding to the fraction of the surface of Earth compared to the entire spherical surface irradiated by solar radiation, hits the Earth. In addition, natural radiogenic (fission) heat flows from the Earth's interior occur at a rate of about[7] 44 ± 1 TW; about half of this heat can today be convincingly attributed to natural radioactive decay.

Practically all solar and radiogenic energy is again reradiated, assuring the stability of the Earth's climate. Human activity perturbs directly only in a negligible way this natural energy balance: The total annual world energy consumption is about 5×10^8 TJ, corresponding (upon division with 3.15×10^7 s/year) to 16 TW average power, well below the small radiogenic heat flow from the Earth's

---

[5] J. Letessier & J. Rafelski, *Hadrons and Quark-Gluon Plasma,* Cambridge University Press (2002).

[6] See for example: C.T. Yang and J. Rafelski, "Possibility of bottom-catalyzed matter genesis near to primordial QGP hadronization," arXiv:2004.06771.

[7] The KamLAND Collaboration, A. Gando, et al., "Partial radiogenic heat model for Earth revealed by geoneutrino measurements," *Nature Geoscience*, **4** 647 (2011).

interior. Therefore, harvesting a small 1 in 10,000 part of the available natural energy would satisfy all present human power needs. In this way the production of pollutants perturbing the natural energy balance and inducing climate change can be minimized.

Let us establish how the power unit, TW $= 10^{12}$ J/s, relates to equivalent mass burn, and practice the use of the eV-energy unit, see Insight 16.1 on *Elementary energy units*. We have

$$1\,\mathrm{TW} = 10^{12}\,\frac{\mathrm{CV}}{\mathrm{s}} = \frac{10^{12}\,\mathrm{eV}}{1.602177\,\times\,10^{-19}\,\mathrm{s}} = 6.2415\,\times\,10^{30}\,\frac{\mathrm{eV}}{\mathrm{s}}\,, \tag{16.23}$$

where we used for the electron entering the unit 'eV' $e = 1.602177 \times 10^{-19}$ C. One gram of hydrogen contains Avogadro's number $N_\mathrm{A} = 6.022141 \times 10^{23}\,\mathrm{mol}^{-1}$ of individual atoms

$$1\mathrm{g}c^2 = N_\mathrm{A}\,M_\mathrm{H}c^2 = 6.022 \times 10^{23}(938.2 + 0.51)10^6\,\mathrm{eV} = 5.653 \times 10^{32}\,\mathrm{eV}\,. \tag{16.24}$$

Combining Eq. (16.23) with Eq. (16.24) we obtain

$$1\,\mathrm{TW} = 1.1\,\times\,10^{-5}\,\frac{\mathrm{kg}\,c^2}{\mathrm{s}}\,. \tag{16.25}$$

Equation (16.25) connects the two SI units,[8] TW, where $W = V \times A$, with kgc^2/s. According to Eq. (16.25) and the radiation flow of 173,000 TW; that means every second the Sun delivers to Earth 1.9 kg in radiation energy equivalent; in a year that is 60,000,000 kg in radiation energy equivalent. The net amount remaining as fossil mass-energy equivalent is much smaller, since a vast majority of the solar energy is re-radiated as heat. In a stable climate condition the gain and loss in energy are in balance.

In summary: Nuclear power, be it fission or fusion, on Earth or in the Sun, provides our energy needs. The process exploits the primordial elemental abundance and element mass, created by processes operating prior to the formation of the solar system. In the context of balancing the energy today on Earth, 'renewable' means the exploitation on Earth of the solar fusion energy, and 'fossil energy' describes the solar energy of the past stocked in the last 10s million of years in coal, oil and gas underground.

[8]TW and kgc^2/s originate in two distinct areas of physics, electromagnetism and classical mechanics. Their definitions were introduced before energy and mass equivalence $E = mc^2$ was recognized. The recent 2019 revision of SI units has fixed the numerical value of relation Eq. (16.24) and other related conversions between equivalent SI units. However, this novel SI procedure does not (yet) introduce a system of well chosen SI units unifying in the process the two independently defined unit systems.

Part VII

Collisions, Particle Production and Decays

In Part VII: We explore the kinematic tools needed in the study of elastic particle collisions, in the inelastic conversion of massive particles from one form into another, and in the conversion of energy into matter, and matter into energy. All processes are governed by energy-momentum conservation. Examples involving elastic and inelastic collisions and particle decays exhibit the methods.

Introductory Remarks to Part VII

The kinematic tools needed to study the conversion of massive particles from one form into another, and the conversion of energy into matter and matter into energy are the main topics of Part VII of the book. The concept of the center of momentum (CM) reference frame is introduced. In the CM-frame by definition the sum of all considered particle momenta adds to zero. The CM-frame replaces the center of mass as the primary frame of reference of a multi-particle system; we explore the coordinate transformation into the CM-frame. A conventional center of mass reference frame is not allowed, as there cannot be a specific privileged point in space given the principles of SR.

We consider in detail the decay of a quasi-stable particle into several lighter particles, and look at the situation in several relevant frames of reference. The energy balance of a particle decay is then considered in the CM-frame. The kinematic variables, energy and momentum, products of the decaying body are characterized, among several applications explored in examples. Special attention is given to the decay that produces one of the reaction products at rest in the frame of reference considered. Another example shows that photons do not decay into two massive particles in the vacuum.

We then turn to the study of two-body collisions: In an elastic collision, the masses of particles remain unchanged and no new particles are produced. In an inelastic reaction, the masses of particles participating change, particle numbers change in conversion of particle kinetic energy into matter, or matter into kinetic energy. The main goal of this introduction is to explore the role of energy and momentum conservation constraints.

As an introduction to this domain of relativistic physics a few select key results will be described. An example of microscopic elastic collision we consider is the Compton scattering of photons on electrons. The dynamics of particles elastically back-scattered from a moving wall is of special interest: The relativistic wall motion with the Lorentz factor γ generates a reflected particle with up to $4\gamma^2$ enhanced energy-momentum. Our presentation generalizes the remarks found in the Einstein 1905 relativity publication.

Kinematic constraints present in particle production are introduced, with special attention given to production and annihilation of antimatter. We also describe the processes that occur in inelastic multi-particle production events involving collisions of relativistic nuclei.

Preferred Frame of Reference 17

Abstract

The energy and momentum conservation laws govern particle collision and decay processes. The center of momentum reference frame (CM-frame) is introduced and the Lorentz coordinate transformation into the CM-frame constructed. We study decays of particles into several lighter particles, and look at these processes in the laboratory frame and in the CM-frame of reference. The possible production of one of the decay products at rest in the laboratory is explored. Massless photons do not decay in the vacuum into two massive particles.

17.1 The Center of Momentum Reference Frame (CM-Frame)

A good choice of a frame of reference is of considerable importance in the study of the properties of any physical system. This is especially true when one or more particles are relativistic. However, the idea of the *center of mass reference system*, which is so useful in nonrelativistic mechanics, must be generalized to the relativistic domain by allowing for novel features: particles can be created (and destroyed); their masses can experience mass defect due to binding; and more generally, a transmutation of particles can occur accompanied by modification of their masses. Moreover, according to principle of relativity, a fixed point of reference does not exist. These novel insights of SR imply that the center of mass reference frame is not available to us. In the study of an assembly of relativistic particles we are led to consider instead the *center of momentum frame of reference* (CM-frame).

The existence of the CM-frame as a preferred reference system is assured by the law of the conservation of momentum, which has very deep experimental roots. In the theoretical formulation this is a result of translational spatial symmetry: We expect that the laws of physics are the same everywhere in the Universe. The relation of this natural postulate to the momentum conservation law was

J. Rafelski, *Modern Special Relativity*, https://doi.org/10.1007/978-3-030-54352-5_17

presented by E. Noether[1] as a part of the renowned Noether theorem rooted in the Lagrangian formulation of dynamics. Another important aspect of the Noether theorem is that for the laws of physics to be immutable as function of time, energy conservation must also hold in particle reactions. For this reason we rely in this book on energy and momentum conservation as the foundational principles governing particle reactions. We furthermore recognize through the Noether theorem that the laws of physics are the same everywhere in time and space.

We first consider the mass of a body in which some internal components are in motion. In fact almost any material body, when looked at in greater depth is composed of smaller constituents that are in motion with respect to each other. When considering a system of many particles we have for the total energy E and the total momentum $\boldsymbol{P}$ of the system

$$E = \sum_i E_i , \qquad \boldsymbol{P} = \sum_i \boldsymbol{p}_i . \tag{17.1}$$

A body consisting of many moving bodies has its rest-frame which is the CM-frame. This is the reference frame where all the momentum vectors of individual particles i which make up the object of interest sum to zero,

$$\bar{\boldsymbol{P}} = 0 = \sum_i \bar{\boldsymbol{p}}_i , \tag{17.2}$$

where in this book the over-bar denotes that this quantity is evaluated in the CM-frame.

Each of the particles contributing to the body also contributes to the total energy obtained by summing over the energy of all particles

$$\bar{E} = \sum_i \bar{E}_i . \tag{17.3}$$

This relation can implicitly account for effect of interactions if we understand this relation as including the potential energy: In Sect. 16.2 we have shown that the presence of an interparticle potential complements internal motion and contributes to the energy and mass content of the body. However, in this chapter we focus attention on situations in which the relativistic particle energy can be considered as the dominant if not the sole contributor. This is for example the case in collision of two or more particles interacting only at a very short distance. In this situation each

[1] Amalie Emmy Noether (1882–1935) is to this day considered the most important woman in the history of theoretical physics and mathematics. A complete biography of this remarkable scientist is available at the web pages of the School of Mathematics and Statistics, University of St Andrews, Scotland.

of the contributing components in Eq. (17.3) is due to the motion of an individual particle, while in the CM-frame the sum of all components is at rest.

Given these considerations we also know the mass of the body that comprises these dynamical particle components

$$M = \frac{\bar{E}}{c^2} > 0 \,. \tag{17.4}$$

We did not highlight by an over-bar the mass M of the composite system as being measured in the CM-frame, since the mass of a body does not depend on the frame of reference in which it is observed. However, the reader should be always aware that only in the CM-frame, in which Eq. (17.2) is satisfied, is Eq. (17.4) correct.

We can evaluate the invariant mass M in any frame of reference. This is done by introducing a Lorentz invariant expression[2] analogous to Eq. (15.14)

$$M^2c^4 = E^2 - c^2\boldsymbol{P}^2 = \left(\sum_i E_i\right)^2 - \left(c\sum_i \boldsymbol{p}_i\right)^2 \equiv s \,. \tag{17.5}$$

In Eq. (15.14) we considered m to be the intrinsic (rest) mass of a particle. M in Eq. (17.5) is the definition of the mass of a system of particles. This expression is exact for a system of freely moving particles. It is common in the latter case to denote the resultant CM-frame energy Mc^2 by the symbol $\sqrt{s}$ as indicated in Eq. (17.5). We will characterize this variable further in Sect. 18.5.

We note that Eq. (17.3) makes good sense for the 'trivial' system of a single particle. In this case there is no sum and we find that the CM-frame is the particle rest-frame where $\boldsymbol{p}_{i=1} = 0$. The mass is given by $M = E_{i=1}/c^2 = m_{i=1}$, where $E_{i=1}$ is the energy of the particle in its rest-frame.

For the case of two particles observed in the CM-frame, for their momentum to vanish, these particles must collide head-on. In much of the remainder of this chapter we look at various methods of observation which rely of specific frames of reference. This means that we mostly look at collisions where same particles are present before and after, with exception of decay processes where one particle splits into a few, typically two or three. We widen our study to include the more general case of two particle collisions including mass modifications and production of new particles in Chap. 18.

[2]In Eq. (17.5) we introduce a 'Mandelstam' variable denoted by s which is the same letter symbol as used elsewhere in this book for the invariant event separation. For historical reasons we adhere to this only in the narrow context ambiguous nomenclature, despite our goal to use unambiguous nomenclature.

17.2 The Lorentz Transformation to the CM-Frame

We are interested in understanding the available two-particle system energy and compare two cases: first, collisions involving a beam of particles hitting a target at rest in the laboratory; and second, the collider mode where the two particles approach from opposite directions, and the CM-frame is, for example, the laboratory frame.

We now determine for the case of two particle collisions the Lorentz coordinate transformation between the CM-frame and the laboratory frame. As a first step we select a coordinate system such that the momentum $\boldsymbol{P}$ of the colliding particle system in the lab frame is parallel to the x-axis, $\boldsymbol{P} = P_x\hat{x}$. We now boost to a frame moving at velocity $\boldsymbol{\beta} = \boldsymbol{v}/c$ also along the x-axis; the transformed energy and momentum are

$$\begin{aligned} \bar{E} &= \gamma(E - \beta P_x c)\,, \qquad & \bar{P}_x &= \gamma\left(P_x - \beta\tfrac{E}{c}\right)\,, \\ \bar{P}_y &= P_y = 0\,, & \bar{P}_z &= P_z = 0\,. \end{aligned} \tag{17.6}$$

For this transformation to lead to the CM-frame, the condition in Eq. (17.6) is $\bar{P}_x = 0$. Solving for β we can now determine the speed of the CM-frame system relative to the laboratory system, which we label with the subscript 'CM'

$$\beta_{\mathrm{CM}} = \frac{cP_x}{E}\,. \tag{17.7a}$$

We remember that the total momentum and energy are composed of individual contributions

$$\boldsymbol{\beta}_{\mathrm{CM}} = \frac{c\sum_i \boldsymbol{p}_i}{\sum_i E_i}\,. \tag{17.7b}$$

We obtain also γ_{CM}:

$$\gamma_{\mathrm{CM}} = \frac{1}{\sqrt{1-\beta_{\mathrm{CM}}^2}} = \frac{E}{\sqrt{E^2 - P^2c^2}}\,,$$

which simplifies using variable s, see Eq. (17.5)

$$\gamma_{\mathrm{CM}} = \frac{E}{Mc^2} = \frac{\sum_i E_i}{\sqrt{s}} \,. \tag{17.8}$$

Quantities with the subscript 'CM' describe the properties of the CM-frame of reference, in particular its motion, as observed in the reference frame in which all the other quantities are observed. Therefore we cannot characterize these with a 'bar'. By definition $\bar{\beta}_{\mathrm{CM}} = 0$ and $\bar{\gamma}_{\mathrm{CM}} = 1$.

Example 17.1

Transformation to CM-frame in two-particle collisions

Obtain the Lorentz coordinate transformation parameters from the lab system S to CM-frame S' for the CERN-LHC $E_1 = 7000\,\mathrm{GeV}$ proton beam hitting a laboratory target, and consider the physical relevance of such experiments. ◀

▶ **Solution** The $E_1 = 7000\,\mathrm{GeV}$ LHC proton beam collides with a nucleon in the target wall, thus $m_1 = m_2 = m$, with $E_2 = mc^2 = 0.94\,\mathrm{GeV}$. For now we keep the two masses $m_1,\ m_2$ in equations.

Note that

1 $p_1c = \sqrt{E_1^2 - m_1^2c^4} = \sqrt{7000^2 - 0.94^2}\,\mathrm{GeV} = 6999.999937\,\mathrm{GeV}\,.$

To a reasonable precision the energy E_1 and momentum p_1c of the beam have the same value, yet as will be seen, it is often better to keep all terms to the end when deriving equations as significant cancellation occurs.

The (high) collision system CM-frame velocity as evaluated from the lab frame is

2 $\beta_{\mathrm{CM}} = \dfrac{p_1c + p_2c}{E_1 + E_2} = \dfrac{\sqrt{E_1^2 - m_1^2c^4}}{E_1 + m_2c^2} = \sqrt{\dfrac{E_1 - mc^2}{E_1 + mc^2}}\,,$

where in the last equality we used the fact that the projectile and target have the same mass m. In this case $\sqrt{E_1 + mc^2}$ between numerator and denominator cancel. We expand in powers of m/E_1

3 $\beta_{\mathrm{CM}} = \sqrt{\dfrac{E_1 - mc^2}{E_1 + mc^2}} \simeq 1 - \dfrac{mc^2}{E_1} \simeq \dfrac{7000 - 0.94}{7000} = 0.9998657\,.$

Given the simplicity of the analytic result Eq. 2 we compute the Lorentz factor γ_{CM} directly

$$\mathbf{4}\quad \gamma_{\mathrm{CM}} = \frac{1}{\sqrt{1-\beta_{\mathrm{CM}}^2}} = \sqrt{\frac{E_1 + mc^2}{2mc^2}} = \sqrt{\frac{7000.94}{2\cdot 0.94}} \simeq 61.02\,.$$

To test for consistency, we can cross-check using Eq. (17.8) and we find for the case of collision with a particle at rest in the laboratory

$$\mathbf{5}\quad \gamma = \frac{E}{\sqrt{s}} = \frac{E_1 + m_2c^2}{\sqrt{(m_1c^2)^2 + (m_2c^2)^2 + 2m_2c^2E_1}} = \frac{7000.94}{\sqrt{2\cdot 0.94\cdot 7000.94}} = 61.02\,.$$

When the circulating LHC beam of protons in the LHC is dumped into a specially designated wall situated at rest in the laboratory, we can observe (laboratory) 'fixed' target collision events. In the specific LHC case we find, according to Eq. 4, that in the CM system we have $\gamma_{\mathrm{CM}} m_p c^2 = 57$ GeV/proton head-on collisions, thus for the pair of projectile-target nucleons we have the equivalent $\sqrt{s_{\mathrm{NN}}} = 114$ GeV, which we compare to the collider Energy 2×7000 GeV. We see that for primary physical processes such as particle production the fixed target collisions make available only $114/14{,}000 = 0.8\%$ of the energy which is available in the LHC collider mode. We note that $\sqrt{s_{\mathrm{NN}}} = 114$ GeV is in the middle of the RHIC-collider[3] energy range.

Despite this, there is interest in experiments using the LHC in fixed target mode, in consideration of the significantly longer laboratory frame lifespan attained due to the dilation factor γ_{CM}. This offers experimental advantages improving the capability for detection of short lived particles. A fixed target LHC experimental program involving a gas jet target has been tested at the LHC-B detector, and its physics case is under exploration.[4] ◀

Example 17.2

CM rapidity of colliding particles

Obtain CM-rapidity in terms of kinematic variables of two colliding particles. Your objective is to understand how CM-rapidity depends on colliding particle rapidities and masses. Consider the special case that both masses are equal. ◀

▶ **Solution** Two colliding particles which are not exactly collinear in the laboratory reference frame particles are seen in Fig. 17.1. We see that the momenta of the two individual particles have a net transverse momentum $\boldsymbol{p}_{\perp 1} + \boldsymbol{p}_{\perp 2} \neq 0$, while the longitudinal momentum components add to zero $\boldsymbol{p}_{\parallel 1} + \boldsymbol{p}_{\parallel 2} = 0$. If we choose

[3] RHIC=Relativistic Heavy-Ion Collider at Brookhaven National Laboratory, Long Island, New York, has been in operation since year 2000.

[4] See for example a special issue of: *Advances in High Energy Physics* Volume **2015** (2015), Guest Editors: J.-Ph. Lansberg, G. Cavoto, C. Hadjidakis, J. He, C. Lorcé, and B. Trzeciak.

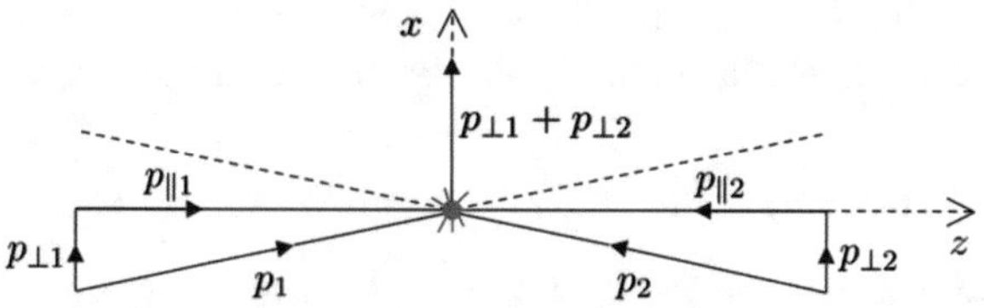

Fig. 17.1 CM-frame particle momenta in two-particle collision decomposed into transverse $p_\perp$ and longitudinal $p_\parallel$ components with reference to the collision z-axis with both particles showing a symmetric relative impact angle

a reference frame where the collision is collinear along the z-axis, the transverse momentum vanishes as well. This is per definition the CM-reference frame.

The value of CM-rapidity can be obtained using Eq. (17.7a)

1 $$\tanh y_{\mathrm{CM}} \equiv \beta_{\mathrm{CM}} = \frac{cP_\parallel}{E} = \frac{cp_{\parallel 1} + cp_{\parallel 2}}{E_1 + E_2} ,$$

which leads to

2 $$y_{\mathrm{CM}} = \tfrac{1}{2} \ln \left(\frac{E_1 + E_2 + cp_{\parallel 1} + cp_{\parallel 2}}{E_1 + E_2 - cp_{\parallel 1} - cp_{\parallel 2}} \right) .$$

We can use the particle rapidity relations Eq. (15.18) to write

3 $$E_{1,2} \pm cp_{\parallel 1,2} = E_{\perp 1,2}\, e^{\pm y_{1,2}} ,$$

which allows us to write

4 $$y_{\mathrm{CM}} = \frac{1}{2} \ln \left(\frac{E_{\perp 1}\, e^{y_1} + E_{\perp 2}\, e^{y_2}}{E_{\perp 1}\, e^{-y_1} + E_{\perp 2}\, e^{-y_2}} \right) , \quad E_{\perp i} = \sqrt{m_i^2 c^4 + p_{\perp i}^2 c^2} , \; i = 1, 2 .$$

We first look at the special case $m_1 = m_2$ which is frequently studied experimentally. We also have $\boldsymbol{p}_{\perp 1}^2 = \boldsymbol{p}_{\perp 2}^2$ and thus $E_{\perp 1} = E_{\perp 2}$

5 $$E_{\perp 1} = \sqrt{m_1^2 c^4 + \boldsymbol{p}_{\perp 1}^2 c^2} = E_{\perp 2} = \sqrt{m_2^2 c^4 + \boldsymbol{p}_{\perp 2}^2 c^2} ,$$

Now we can cancel the common energy factor in Eq. 4 and we obtain

6 $$y_{\mathrm{CM}} = \frac{1}{2} \ln \left(\frac{e^{y_1} + e^{y_2}}{e^{-y_1} + e^{-y_2}} \right) = \frac{1}{2} \ln \left(e^{y_1 + y_2} \frac{e^{-y_2} + e^{-y_1}}{e^{-y_1} + e^{-y_2}} \right) = \frac{y_1 + y_2}{2} , \quad m_1 = m_2 .$$

When $m_1 = m_2$, all mass-dependent terms vanish exactly. We recognize the average of both colliding particle rapidities as the CM-rapidity.

We now determine the rapidity of the particles in the CM-system. Since the rapidities add considering the LT along the direction of the z-axis, we obtain

$$\textbf{7}\quad \bar{y}_1 = y_1 - y_{\mathrm{CM}} = \frac{y_1 - y_2}{2}\,, \qquad \bar{y}_2 = y_2 - y_{\mathrm{CM}} = \frac{y_2 - y_1}{2}\,.$$

A special case confirms this result: For a target at rest in the laboratory; that is, $y_2 = 0$, we find $\bar{y}_1 = y_1/2$ and $\bar{y}_2 = -y_1/2$.

We now consider the general case $m_1 \neq m_2$, thus also $E_{\perp 1} \neq E_{\perp 2}$ in Eq. 4. We take in the numerator the factor e^{y_1} and in the denominator the factor e^{-y_2} out of the logarithm and obtain:

$$\textbf{8}\quad y_{\mathrm{CM}} = \frac{y_1 + y_2}{2} + \frac{1}{2}\ln\left(\frac{E_{\perp 1} + E_{\perp 2}\, e^{-(y_1 - y_2)}}{E_{\perp 2} + E_{\perp 1}\, e^{-(y_1 - y_2)}}\right).$$

We note that this expression is symmetric under renaming $1 \Leftrightarrow 2$. In most cases of interest, $y_1 - y_2 \gg 0$, including the collider mode in the collision of beams of opposite momentum. In this case the rapidity gap between two relativistic colliding particles is large, and the exponential function above is very small allowing the expansion

$$\textbf{9}\quad y_{\mathrm{CM}} \simeq \frac{y_1 + y_2}{2} + \frac{1}{2}\ln\left(\frac{E_{\perp 1}}{E_{\perp 2}}\right) + \frac{E_{\perp 2}^2 - E_{\perp 1}^2}{2E_{\perp 1}E_{\perp 2}} e^{-(y_1 - y_2)} + \cdots\,.$$

We can use the additivity of rapidity, see Sect. 7.4, to relate the laboratory and CM-frame particle rapidities

$$\textbf{10}\quad y_1 = \bar{y}_1 + y_{\mathrm{CM}}\,, \qquad y_2 = \bar{y}_2 + y_{\mathrm{CM}}\,.$$

Substituting the first form above into Eq. 8 we find

$$\textbf{11}\quad \bar{y}_1 = \frac{y_1 - y_2}{2} - \frac{1}{2}\ln\left(\frac{E_{\perp 1} + E_{\perp 2}\, e^{-(y_1 - y_2)}}{E_{\perp 2} + E_{\perp 1}\, e^{-(y_1 - y_2)}}\right).$$

Using $1 \Leftrightarrow 2$ symmetry a similar result follows for $\bar{y}_2$. For $|\boldsymbol{p}_\perp| \ll m_i c$ we can substitute $E_{\perp i}/c^2 \to m_i$. This is in general the case in collider experiments for particles approaching from opposite sides. In this case we have usually $y_2 \simeq -y_1$ since the particles move against each other, thus in general also a relatively large $|y_1 - y_2|$. Choosing particle '1' to move to the right, $y_1 > 0$ and we have $y_1 - y_2 = y_1 + |y_2| \gg 1$, so that the exponential function can be neglected

$$\textbf{12}\quad \bar{y}_1 = \frac{y_1 - y_2}{2} - \frac{1}{2}\ln\left(\frac{m_1}{m_2}\right), \quad |\boldsymbol{p}_\perp| \ll c\, m_i\,, \quad \text{for} \quad y_1 - y_2 \gg 1\,. \blacktriangleleft$$

17.3 Particle Decay in the CM-Frame

When a heavy nucleus undergoes spontaneous or induced fission, the resulting fragments carry usable kinetic energy. This is a special case of a family of reactions called particle decays. We address here the two-body decay processes, in which a mother particle of mass M decays into two daughter particles with mass m_1 and m_2, illustrated in Fig. 17.2.

We study here the decay in the rest-frame of the mother particle, which, due to conservation of momentum, must also be the CM-frame of all (two in the present example) daughter particles. Conservation of energy and momentum in this reference system can be expressed as

$$Mc^2 = \bar{E}_1 + \bar{E}_2 \,, \tag{17.9}$$

$$0 = \bar{\boldsymbol{p}}_1 + \bar{\boldsymbol{p}}_2 \,, \tag{17.10}$$

and therefore naturally also

$$|\bar{\boldsymbol{p}}_2| = |\bar{\boldsymbol{p}}_1| \,. \tag{17.11}$$

The conservation of energy as stated in Eq. (17.9) implies

$$M = \sqrt{m_1^2 + \left(\frac{\bar{\boldsymbol{p}}_1}{c}\right)^2} + \sqrt{m_2^2 + \left(\frac{\bar{\boldsymbol{p}}_1}{c}\right)^2} > m_1 + m_2 \,. \tag{17.12}$$

The inequality reminds us that for the mother particle to decay, the daughter particles must be less massive; the balance of energy is found in their motion.

We now solve Eq. (17.12) for the only unknown $|\overline{p}_1|$. These manipulations are straightforward but nontrivial. We use the trick of placing one of the roots alone on one side of the expression. In this way Eq. (17.12) yields

$$\left(M - \sqrt{m_1^2 + \left(\frac{\bar{\boldsymbol{p}}_1}{c}\right)^2}\right)^2 = m_2^2 + \left(\frac{\bar{\boldsymbol{p}}_1}{c}\right)^2 ,$$

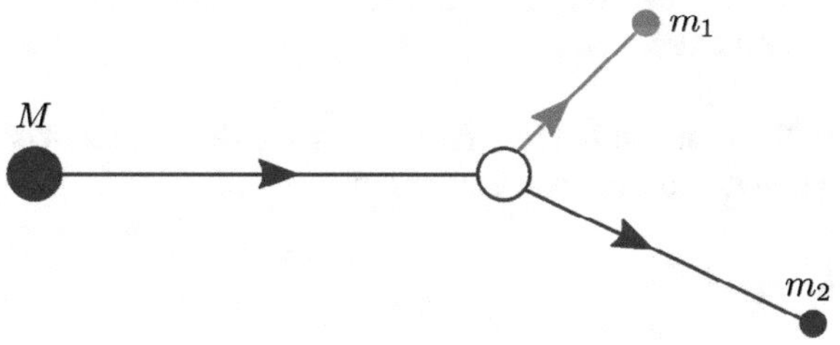

Fig. 17.2 The two-body decay

and simplifying:

$$M^2 - 2M\sqrt{m_1^2 + \left(\frac{\bar{p}_1}{c}\right)^2} + m_1^2 = m_2^2\,.$$

We obtain

$$\bar{p}_1^2 = c^2\left[\left(\frac{M^2 + m_1^2 - m_2^2}{2M}\right)^2 - m_1^2\right], \tag{17.13}$$

and so

$$\bar{p}_1^2 = c^2\left(\frac{M^4 + m_1^4 + m_2^4 - 2M^2m_1^2 - 2M^2m_2^2 - 2m_1^2m_2^2}{4M^2}\right). \tag{17.14}$$

The energy of each of the daughter particles follows from Eq. (17.13)

$$\frac{\bar{E}_1}{c^2} = \sqrt{m_1^2 + \left(\frac{\bar{p}_1}{c}\right)^2} = \frac{M^2 + m_1^2 - m_2^2}{2M}, \tag{17.15}$$

and

$$\frac{\bar{E}_2}{c^2} = \sqrt{m_2^2 + \left(\frac{\bar{p}_1}{c}\right)^2} = \frac{M^2 + m_2^2 - m_1^2}{2M}. \tag{17.16}$$

Note that the solution checks; that is, $\bar{E}_1 + \bar{E}_2 = Mc^2$, and the expressions are symmetric under the exchange of indices (particles), $1 \leftrightarrow 2$.

Example 17.3

Momentum of π^- in Λ-decay

Λ is a neutral elementary particle of mass $m_\Lambda = 1115.7\,\mathrm{MeV}/c^2$, which decays with about 2/3 probability into a proton p (mass $m_p = 938.3\,\mathrm{MeV}/c^2$) and a negatively charged pion, π^- (mass $m_\pi = 139.6\,\mathrm{MeV}/c^2$). Show that the momentum of π^- produced in the decay $\Lambda \to \pi^- + p$ is exactly $101\,\mathrm{MeV}/c$ when measured in the rest-frame of Λ, and compare this value to the momentum of the produced proton. ◀

▶ **Solution** In the rest-frame of Λ the decay products must emerge with opposite momenta so that momentum is conserved

1 $\bar{p}_{\pi^-} + \bar{p}_p = \bar{p}_\Lambda = 0\,.$

Thus the momenta of decay products will be equal in magnitude and opposite in direction. To obtain the momentum magnitude we use Eq. (17.14)

2 $$\bar{p}^2 = c^2 \frac{m_\Lambda^4 + m_p^4 + m_\pi^4 - 2m_\Lambda^2(m_p^2 + m_\pi^2) - 2m_p^2 m_\pi^2}{4m_\Lambda^2} \, .$$

When evaluating the numerical values (units MeV) caution is advised since

3 $$m_p c^2 + m_\pi c^2 = 938.3 + 139.6 = 1077.9 = m_\Lambda c^2 - 37.8\,\mathrm{MeV} \, .$$

This near balance of masses signals considerable cancellation of large numbers requiring more than a few digits precision. Another path is to seek a better reformulation of our result. A version appropriate in such a situation is

4 $$\bar{p}^2 = c^2 \frac{\left[m_\Lambda^2 - (m_p + m_\pi)^2\right]\left[m_\Lambda^2 - (m_p - m_\pi)^2\right]}{4m_\Lambda^2} \, .$$

This result is also symmetric under exchange of subscripts $p \leftrightarrow \pi$. Now, the first factor in the numerator of Eq. 4 can also be written

5 $$m_\Lambda^2 - (m_p + m_\pi)^2 = \left[m_\Lambda - (m_p + m_\pi)\right]\left[m_\Lambda + (m_p + m_\pi)\right] \, ,$$

allowing factorization with the small number, $m_\Lambda - (m_p + m_\pi) = 37.8\,\mathrm{MeV}/c^2$. After combining the two last equations a few digits precision computation succeeds yielding the stated result $\bar{p} = 101\,\mathrm{MeV}/c$. ◀

17.4 Decay Energy Balance in the CM-Frame

To see how much energy is freed in the decay we evaluate the kinetic energy $T_i = E_i - m_i c^2$ for both particles $i = 1, 2$.

$$\bar{T}_1 = \bar{E}_1 - m_1 c^2 = \frac{(M - m_1)^2 - m_2^2}{2M} c^2 \, , \tag{17.17}$$

and

$$\bar{T}_2 = \bar{E}_2 - m_2 c^2 = \frac{(M - m_2)^2 - m_1^2}{2M} c^2 \, . \tag{17.18}$$

Thus the total freed kinetic energy is:

$$\bar{T} = \bar{T}_1 + \bar{T}_2 = (M - m_1 - m_2)c^2 \, , \tag{17.19}$$

which is what we expect: The available 'motion' energy is the energy of the mother particle, minus the energy locked in the mass of daughter particles. This is also a correct statement in case the final state has more than two particles, and Eq. (17.19) can be generalized accordingly for decays observed in the rest-frame of the mother particle.

Many satellites, in particular those going far away from the Sun, use radioactive decay "batteries" based on this principle. In such an energy source, a radioactive slowly decaying isotope is employed, and the decay products are stopped e.g. by a lead mantle. The heat released in the decay process is used to both warm the electronics and to power a generator. A bit more sophisticated is the decay reaction induced in a nuclear fission reactor. Here the impact of a thermal neutron induces nuclear fission. The energy freed in the reaction is also described by Eq. (17.19).

Clearly a spontaneous (i.e., on its own) decay can only occur if the energy released is positive. Conversely, energy can be gained from the binding of several components into fewer and less massive particles; the energy gain is the mass defect; that is, the reduction of the total mass M due to binding of e.g. two components, m_1, m_2

$$(m_1 + m_2 - M)c^2 = \delta \bar{E} \ . \tag{17.20}$$

The reaction proceeds if the binding energy can be released, e.g. by radiation of a photon, or as is the case with many chemical reactions, recoil on a catalyst surface. For further details we refer to our prior discussion in Chap. 16.

17.5 Decay of a Body in Flight

The above considerations yielded properties of decay particles in the rest-frame of the mother particle. In the more general case the mother particle is in motion in the laboratory frame where it has the energy

$$E = \sqrt{M^2c^4 + \boldsymbol{P}^2c^2} \tag{17.21}$$

and momentum $\boldsymbol{P}$. The transformation from the lab frame to the CM-frame is given by Eqs. (17.7a) and (17.8). The inverse transformation from the CM-frame to the lab frame is then given here with parameters

$$\gamma = \frac{E}{Mc^2}, \quad \boldsymbol{\beta} = \frac{-\boldsymbol{P}c}{E}, \quad \gamma\boldsymbol{\beta} = \frac{-\boldsymbol{P}}{Mc} \ . \tag{17.22}$$

Decomposing the momenta of the daughter particles into components parallel (∥) and perpendicular (⊥) to $\boldsymbol{P}$, we apply the Lorentz coordinate transformation of Eq. (17.6) with Eq. (17.22) to find the lab frame properties of both the daughter particles ($i = 1, 2$),

$$E_i = \frac{1}{M}\left(\frac{E}{c^2}\bar{E}_i + P\ \bar{p}_{i\parallel}\right) , \tag{17.23}$$

$$p_{i\parallel} = \frac{1}{M}\left(\frac{E}{c^2}\bar{p}_{i\parallel} + \frac{P}{c^2}\bar{E}_i\right) , \tag{17.24}$$

$$\boldsymbol{p}_{1\perp} = -\boldsymbol{p}_{2\perp} = \boldsymbol{p}_{\perp} \, . \tag{17.25}$$

In an experiment we usually measure the laboratory momenta and energies of particles produced and would like to obtain the CM values for better understanding of the reaction process. We find this result with the help of inverse transformation or by solution of Eqs. (17.23) and (17.24); both procedures lead to the same result.

We multiply Eq. (17.23) with E and Eq. (17.24) with P and the difference of the two readings is

$$\bar{E}_i = \frac{1}{M}\left(\frac{E}{c^2}E_i - P p_{i\parallel}\right) . \tag{17.26}$$

Note that using Eqs. (17.7a) and (17.8) to replace E/M and P/M in Eq. (17.26), this equation is confirmed as being the Lorentz coordinate transformation from lab to CM-frame, replacing β by $-\beta$ in the inverse transformation.

Similarly, we can use Eq. (17.26) in Eq. (17.24) along with Eq. (17.21) and obtain

$$\bar{p}_{i\parallel} = \frac{1}{M}\left(\frac{E}{c^2}p_{i\parallel} - \frac{P}{c^2}E_i\right) . \tag{17.27}$$

This is also easily recognized as a transformation to CM-frame replacing β by $-\beta$ in the inverse transformation from lab to CM-frame.

Because of energy-momentum conservation in the lab frame we also have

$$\begin{aligned} P &= p_{1\parallel} + p_{2\parallel} , \\ E &= E_1 + E_2 , \\ E_i &= \sqrt{m_i^2 c^4 + p_{i\parallel}^2 c^2 + p_{i\perp}^2 c^2} , \quad i = 1, 2 , \\ Mc^2 &= \sqrt{E^2 - P^2 c^2} , \end{aligned} \tag{17.28}$$

which we can insert above to obtain the CM-frame daughter particle properties in terms of laboratory measured daughter particles momenta $p_{1\parallel}$, $p_{2\parallel}$.

Example 17.4

Decay of the η-particle

About 40% of η-particle decays result in the production of two γ particles. Obtain the energy and momenta of the γ particles both in the CM-frame and for an η having a momentum of 1 GeV/c in the direction of the motion of one of the decay photons. Note $M_\eta c^2 = 0.5475\,\text{GeV}$. ◀

▶ **Solution** The decay process in the CM-frame divides the η-mass into two massless photons and thus we have for each

1 $$\bar{p}_{1,2}c = \bar{E}_{1,2} = 0.5 M_\eta c^2 = 0.274\,\text{GeV}\,,$$

with the photons moving in opposite directions.

We now consider a boost to a momentum of 1 GeV/c for the η. The energy of the η is now

2 $$E_\eta = \sqrt{M_\eta^2 c^4 + P_\eta^2 c^2} = \sqrt{0.5475^2 + 1^2}\,\text{GeV} = 1.1401\,\text{GeV}\,.$$

We also have

3 $$\frac{P_\eta}{M_\eta c} = \gamma\beta = 1.8265\,, \qquad \frac{E_\eta}{M_\eta c^2} = \gamma = 2.0823\,, \qquad \frac{P_\eta c}{E_\eta} = \beta = 0.8771\,.$$

In the following we will use lower index γ to denote the decay particles, which should not be confused with the Lorentz factor. We use the above input in Eq. (17.23), which we write as:

4 $$E_{\gamma\pm} = \left(\gamma \bar{E}_\gamma \pm \gamma\beta\, \bar{p}_\gamma c\right) = 2.0823 \bar{E}_\gamma \pm 1.8265 \bar{p}_\gamma c\,,$$

where the $\pm$ refers to the photon being parallel and, respectively, antiparallel with reference to the motion of the η particle.

Remembering that each photon carries in the CM-frame half of the η energy, we obtain inserting the numerical values (units GeV)

5 $$\begin{aligned} E_{\gamma+} &= (2.0823 + 1.8265)0.274 = 1.0710\,\text{GeV}\,, \\ E_{\gamma-} &= (2.0823 - 1.8265)0.274 = 0.0701\,\text{GeV}\,. \end{aligned}$$

The energy conservation checks, $E_{\gamma+} + E_{\gamma-} = 1.0710 + 0.0701 = 1.1411\,\text{GeV}$, which is the energy of the η in the lab frame. Since the decay particles are massless, $E_{\gamma\pm} = p_{\gamma\pm}c$, thus up to the factor c the result is valid also for the momenta of the photons.

As we see the two decay photon energies are now very different; in fact one is nearly zero as measured on the scale of the total energy. We learn that choosing a suitable frame of reference of the mother particle results in one of the daughter photons being produced with very little energy in the lab frame. In the following Example 17.5 we will generalize this result. ◀

Example 17.5

The 'magic' decay momentum

Sometimes we would like to study a short-lived daughter particle at rest in the laboratory, in which case we would like to create this particle nearly at rest in the lab. The question arises how to choose the energy-momentum of the mother particle so that this can happen. It turns out that it is necessary for the mother particle to have a special momentum P_m, which we call the magic momentum. Considering momentum conservation, the second decay particle must carry all of the mother particle momentum, without a transverse momentum component. However, only energy conservation constrains the transverse momentum of daughter particles, see Eq. (17.28); $\bar{p}_\perp$ has no dependence on the momentum P of the mother particle, see Eq. (17.25). Thus $\bar{p}_\perp = 0$ can only happen with a relatively small probability.

Determine P_m in general for a two-body decay process, where we want the daughter particle with mass m_1 to be at rest in the laboratory frame. Then calculate the exact value of P_m for the example decay of a (charged) pion π

$$\pi^- \to \mu^- + \bar{\nu}, \quad M_\pi c^2 = 139.57\ \text{MeV}, \quad m_{1=\mu} c^2 = 105.66\ \text{MeV}, \quad m_{2=\bar{\nu}} c^2 \simeq 0\,,$$

where aside of the massless antineutrino $\bar{\nu}$, also the heavy electron, the muon μ, is produced at rest in the laboratory. ◀

▶ **Solution** When the mother particle has the required 'magic' momentum P_m, the daughter has $p_{1\|} = 0$. To obtain the magnitude of the magic momentum we can set $p_{1\|} = 0$ in Eq. (17.24)

1 $$\frac{E_\text{m}}{P_\text{m}} = \frac{\bar{E}_1}{\bar{p}_1}\,.$$

Squaring both sides and simplifying by using the relativistic energy relationship Eq. (15.9) we find

2 $$P_m^2 = \frac{M^2}{m_1^2}\bar{p}_1^{\,2}\,.$$

As we have already calculated the CM-frame quantities, we insert $\bar{p}_1^2$ from Eq. (17.14), yielding upon cancellation of the common M^2 factor

3 $$P_m^2 = c^2\left(\frac{M^4 + m_1^4 + m_2^4 - 2M^2m_1^2 - 2M^2m_2^2 - 2m_1^2m_2^2}{4m_1^2}\right)\,.$$

Now we can calculate the 'magic' momentum for the specific reaction mentioned above, $\pi^- \to \mu^- + \bar{\nu}$. In this case $M = m_\pi$, $m_1 = m_\mu$, and $m_2 = m_\nu = 0$. Using these relations we obtain for the magic momentum of the pion

4 $$P_m^2 = c^2 \left(\frac{m_\pi^4 + m_\mu^4 - 2m_\pi^2 m_\mu^2}{4m_\mu^2} \right) ,$$

therefore

5 $$P_m = c \left(\frac{m_\pi^2 - m_\mu^2}{2m_\mu} \right) = 39.4 \text{ MeV/c} .$$

We can also calculate explicitly the speed β of the pion and the Lorentz factor

6 $$\beta = \frac{P_m c}{\sqrt{P_m^2 c^2 + m_\pi^2 c^4}} = 0.27 , \qquad \gamma = \frac{\sqrt{P_m^2 c^2 + m_\pi^2 c^4}}{Mc^2} = 1.04 . ◂$$

Example 17.6

Are photon decay reactions possible?

Show that a single photon cannot decay into an electron-positron pair and explain how photon pair 'conversion' detectors work despite this. ◂

▶ **Solution** Assume that the reaction shown in Fig. 17.3 on the left

$$\gamma \xrightarrow{?} e^+ + e^-$$

can occur. Then the sum of the energy and momentum of the electron-positron pair must equal the energy and momentum of the photon:

1 $$E_\gamma = E_{e^+} + E_{e^-} , \qquad \boldsymbol{p}_\gamma = \boldsymbol{p}_{e^+} + \boldsymbol{p}_{e^-} .$$

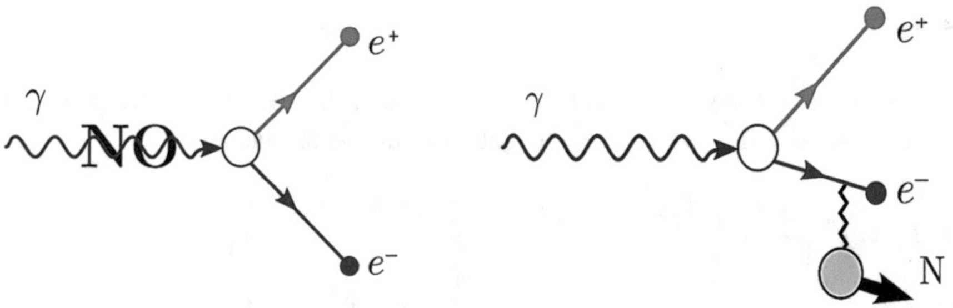

Fig. 17.3 Photon decay: forbidden decay (on the left); allowed photon conversion into a $e^- e^+$-pair on the nucleus N (on the right). See Example 17.6

We are familiar with the relation $E_\gamma = c|\boldsymbol{p}_\gamma|$. This is actually why photon decay is forbidden. We restate therefore the relativistic relation Eq. (15.9) between energy and momentum

2 $E_\gamma^2 - \boldsymbol{p}_\gamma^2 c^2 = m_\gamma^2 c^4$.

Since the photon is massless we obtain

3 $E_\gamma^2 - \boldsymbol{p}_\gamma^2 c^2 = 0$.

The energy and momentum of the photon are given by presumed decay product properties as stated above. We thus combine the above two relations and use the relativistic energy-momentum relation Eq. (15.9) for electrons and positrons to obtain after division by $2c^2$

4 $m^2c^2 + \frac{E_{e^+}E_{e^-}}{c^2} - \boldsymbol{p}_{e^+}\cdot\boldsymbol{p}_{e^-} = 0$,

where $m = m_{e^+} = m_{e^-}$ is the mass of an electron or positron. Using the relativistic energy Eq. (15.9)

5 $m^2c^2 + \sqrt{\boldsymbol{p}_{e^+}^2 + m^2c^2}\sqrt{\boldsymbol{p}_{e^-}^2 + m^2c^2} - \boldsymbol{p}_{e^+}\cdot\boldsymbol{p}_{e^-} = 0$.

Since $m^2c^2 > 0$, and with the Cauchy-Schwartz inequality, the left side above cannot equal zero. Thus our assumption must be false; the decay of a free photon in the vacuum into an electron-positron pair is not possible.

This result is easily generalized to a theorem: A particle of mass M can decay into several (at least two) particles of mass m_i if and only if $M \geq \sum_i m_i$. This we can recognize without any computation by going to the rest-frame of the particle M. There, the decaying particle has energy Mc^2 and the decay products must also have this energy considering energy conservation. The condition we stated corresponds to the 'threshold'; that is, all decay products are at rest, and have no kinetic energy. The effort we made above to resolve for the special case $M \to 0$ of a photon derives from the circumstance that there is no rest-frame for a massless particle and in this special case we need to make more explicit computation.

Reality is more complex – in practice, the photon 'converts' into a e^+e^--pair. In the conversion process illustrated on the right in Fig. 17.3 there is another body participating in the interaction. Specifically, when γ is in the proximity of a nucleus N and this nucleus participates in the reaction, the presence of the nucleus fixes a special CM-frame of reference, which is nearly the rest-frame of the nucleus. Thus the reaction we consider is the collision of γ with the nucleus

6 γ + nucleus at rest $\longrightarrow e^+ + e^- +$ (moving nucleus, momentum $= P_{\mathrm{N}}$).

Our condition Eq. 4 is now supplemented by two additional terms on the right, related to the nucleus:

$$\mathbf{7}\quad 2(\ldots) = 2\boldsymbol{P}_{\mathrm{N}} \cdot (\boldsymbol{p}_{e^-} + \boldsymbol{p}_{e^+}) + \boldsymbol{P}_{\mathrm{N}}^2 \left(1 - \frac{E_{e^+} + E_{e^-}}{M_{\mathrm{N}} c^2}\right) .$$

We see that the reaction $\gamma \rightarrow e^+ + e^-$ is no longer forbidden. The energy of the photon now measured in the rest-frame of the nucleus is found in comparison to the light e^+e^--pair; see Fig. 17.3 on the right.

Such 'catalyzed' photon decay (conversion) reactions are of practical importance, being employed in detectors. How this works is shown in following Example 17.7. ◀

Example 17.7

Formation of e^+e^--pairs

A photon of energy $E_\gamma = 1.1\,\mathrm{MeV}$ collides with a lead plate of mass $M_{\mathrm{Pb}} = 0.020\,\mathrm{kg}$. In this collision the photon converts into an electron-positron pair, which in the example maintains the motion direction of the photon. Determine the speed and the energy of the lead nucleus in the lead plate assuming that the energy of the photon is retained nearly in full and in equal part by the produced electron and positron. Begin considering the masses of the bodies involved to be able to justify this hypothesis. ◀

▶ **Solution** We have

$$\mathbf{1}\quad m_{e^\pm} = \frac{511\,\mathrm{keV}}{c^2} \equiv 9.11 \times 10^{-31}\,\mathrm{kg}\,, \qquad \frac{m_{e^\pm}}{M_{\mathrm{Pb}}} = 4.55 \times 10^{-29}\,.$$

However, only one lead nucleus of mass m_{Pb} participates in the reaction

$$\mathbf{2}\quad \frac{m_{e^\pm}}{m_{\mathrm{Pb}}} = \frac{1}{1840 \times 207} = 2.63 \times 10^{-6}\,.$$

Considering the smallness of the mass ratio of the electron to the lead nucleus we expect that the recoil energy is small, allowing us to ignore it at first. We will verify this hypothesis at the end of the problem.

The energy of the photon is shared equally by the electron and positron. We convert into SI units from the eV units using the value of the elementary charge $e = 1.602 \times 10^{-19}\,\mathrm{C}$, see Insight 16.1: *Elementary energy units.*

$$\mathbf{3}\quad \begin{aligned} E_{e^+} &= E_{e^-} = \frac{E_\gamma}{2} = 550\,\mathrm{keV}\,, \\ E_{e^+} &= E_{e^-} = 550 \times 10^3 \times 1.602 \times 10^{-19}\mathrm{J} = 8.81 \times 10^{-14}\,\mathrm{J}\,. \end{aligned}$$

We determine the magnitude of the particle momentum and convert again into the SI units

$$p_{e^\pm} = \frac{1}{c}\sqrt{E_{e^\pm}^2 - c^4 m_{e^\pm}^2} = 203.4\,\frac{\text{keV}}{c}$$

4

$$= \frac{230.4 \times 10^3 \times 1.602 \times 10^{-19}}{2.998 \times 10^8}\,\frac{\text{kg m}}{\text{s}} = 1.09 \times 10^{-22}\,\frac{\text{kg m}}{\text{s}} \,.$$

Thus the speed of produced particles is

5 $$v_{e^+} = v_{e^-} = \frac{c^2 p_{e^\pm}}{E_{e^\pm}} = c\,\frac{203.4}{550} = 0.370c = 1.109 \times 10^8\,\frac{\text{m}}{\text{s}} \,.$$

The momentum conservation requires

6 $$\boldsymbol{P}_{\text{Pb}} = \boldsymbol{p}_\gamma - (\boldsymbol{p}_{e^+} + \boldsymbol{p}_{e^-}) \,,$$

where $cp_\gamma = E_\gamma$. In the special case that the motion of the electron and positron is along the motion direction of the primary photon we have

$$P_{\text{Pb}} = \frac{E_\gamma}{c} - p_{e^+} - p_{e^-}$$

7

$$= \frac{(1100 - 2 \times 203.4) \times 10^3 \times 1.602 \times 10^{-19}}{2.998 \times 10^8}\,\frac{\text{kg m}}{\text{s}}$$

$$= 3.70 \times 10^{-22}\,\frac{\text{kg m}}{\text{s}} \,.$$

A comparison with produced particle momentum computed above shows that a large part of the primary momentum is absorbed by the lead nucleus. However, due to its large mass Eq. 2 its velocity is small

8 $$v_{\text{Pb--Nucleus}} \simeq \frac{P_{\text{Pb}}}{m_{\text{Pb}}} = \frac{3.70 \times 10^{-22}}{3.46 \times 10^{-25}}\,\frac{\text{km}}{\text{s}} = 1.07\,\frac{\text{km}}{\text{s}} \,.$$

This is five orders of magnitude smaller compared to electron speed.

With the help of momentum conservation we can obtain the speed of the entire lead plate once the nucleus stops within. Given the large number N_{Pb} of lead nuclei in the plate

9 $$N_{\text{Pb}} = \frac{M_{\text{Pb}}}{m_{\text{Pb}}} = 5.78 \times 10^{22} \,,$$

this speed of the entire plate is imperceptibly small. We now determine the energy retained in the plate due to motion of the recoiling nucleus

10

$$E_{\text{Pb-Nucleus}} = \frac{P_{\text{Pb}}^2}{2\,m_{\text{Pb}}} = \frac{13.7 \times 10^{-44}}{6.92 \times 10^{-25}}\,\text{J} = 1.97 \times 10^{-19}\,\text{J}$$
$$= \frac{1.97 \times 10^{-19}}{1.602 \times 10^{-19}}\,\text{eV} = 1.23\,\text{eV}\;,$$

where we used nonrelativistic approximation. The value is six orders of magnitude smaller compared to the primary photon energy. We thus learn that the transfer of momentum to the lead plate catalyst is important, but in the process the energy is not transferred; the energy is practically in full retained by the produced electron-positron pair. The initial assumption was justified.

The emerging electron-positron charged particle pair e^+e^- can be detected and their energy and momentum evaluated. This allows precise characterization of the incipient primary high energy photon with $E_\gamma > 2m_ec^2$. ◀

Particle Reactions 18

Abstract

In an elastic collision, e.g. the Compton scattering, the masses of particles remain unchanged and no new particles are produced. We further consider the example of elastic back scattering from a relativistic moving wall. In an inelastic reaction, such as the production and annihilation of antimatter, the masses of particles participating change and new particles can be produced. We also characterize the inelastic processes occurring in inelastic multi-particle production events involving relativistic heavy ion collisions.

18.1 Elastic Two-Body Reactions

We first consider elastic two-body collisions in which all physical characteristics of the interacting particles are unchanged. Elastic collisions are characterized by the following:

(a) the internal structure of the interacting particles remains unchanged, and
(b) no new particles are produced.

At relativistic energies elastic collisions are an exceptional situation, and it is hard to find a physical situation in relativistic scattering where the rules of elastic scattering apply. Two examples we consider in greater depth are (a) Compton scattering (photon electron scattering) in Sect. 18.2 and (b) scattering from a moving wall in Sect. 18.3.

We indicate values of particle properties after the collision with a prime; for example E_1 refers to the energy of particle '1' *before* the collision, and E_1' refers to the energy of this particle *after* the collision. The complete naming convention for the scattering process where the 'target' particle '2' is at rest in the laboratory frame is shown in Fig. 18.1.

J. Rafelski, *Modern Special Relativity*, https://doi.org/10.1007/978-3-030-54352-5_18

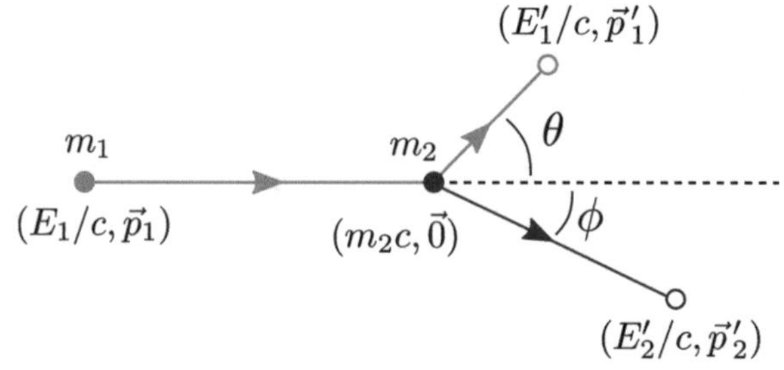

Fig. 18.1 Scattering of particle '1' with energy E_1 and momentum $\boldsymbol{p}_1$ on a stationary target particle '2' with energy $E_2 = m_2c^2$ and momentum $\boldsymbol{p}_2 = 0$; primed quantities refer to values after collision; the scattering angle of incoming particle '1' is called θ, and ϕ for the recoiling target particle '2'

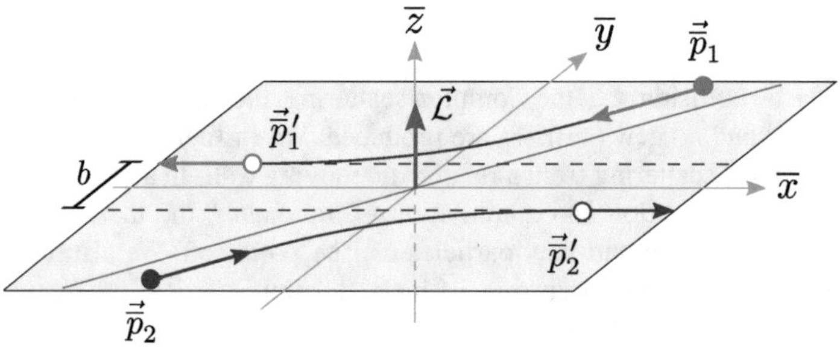

Fig. 18.2 Scattering plane of two particles colliding in the CM-frame with normal in z-direction defined by angular momentum $\boldsymbol{\mathcal{L}}$; we also see the scattering parameter b, for other variables see Fig. 18.1

As before in Sect. 17.1 we will identify values in the center of momentum system (CM-frame) through the use of an over-bar; for example $\overline{\boldsymbol{p}}_1$ refers to the momentum of particle '1' before the collision in the CM-frame, and $\bar{\boldsymbol{p}}'_1$ in the CM-frame after the collision. These variables can be seen in Fig. 18.2 showing the scattering plane normal to the angular momentum $\boldsymbol{\mathcal{L}}$ – conservation of angular momentum for central forces assures that a scattering plane exists. We note that both particles '1' and '2' are incoming from opposite sides and their scattering is, when momentum is considered, symmetric, a situation very different from the case seen in Fig. 18.1. Just as in the nonrelativistic case, in the CM-frame elastic reaction 'rotates' the momentum of the participating particles; the magnitude remains unchanged as we show in Example 18.1

$$\bar{p}' = |\bar{\boldsymbol{p}}'_1| = |\bar{\boldsymbol{p}}_1| = |\bar{\boldsymbol{p}}_2| = |\bar{\boldsymbol{p}}'_2| \equiv \bar{p}\,. \tag{18.1}$$

In scattering where 'action at a distance' occurs, such as scattering due to electromagnetic force, the impact parameter b is important in characterizing the outcome of the collision; its definition is shown in the left part of Fig. 18.2.

In elastic collisions as in all collision reactions the total energy and momentum are conserved. We consider the situation before and after collision

$$E = E_1 + E_2,\ \boldsymbol{p} = \boldsymbol{p}_1 + \boldsymbol{p}_2\,, \qquad E' = E_1' + E_2'\,,\ \boldsymbol{p}' = \boldsymbol{p}_1' + \boldsymbol{p}_2'\,, \tag{18.2}$$

which does not change

$$E = E'\,, \qquad \boldsymbol{p} = \boldsymbol{p}'\,. \tag{18.3}$$

Furthermore, by definition of elastic collision the masses of participating particles remain unchanged

$$m_1^2c^4 = E_1^2 - (cp_1)^2 = E_1'^{\,2} - (cp_1')^2\,, \tag{18.4}$$

and similarly

$$m_2^2c^4 = E_2^2 - (cp_2)^2 = E_2'^{\,2} - (cp_2')^2\,. \tag{18.5}$$

Example 18.1

'Rotating' momenta in the CM-frame

Show that the magnitude of each particle momentum in an elastic collision considered in the CM-frame does not change; illustrate graphically the meaning of this result. ◀

▶ **Solution** We consider the conservation of energy relation:

1 $\bar{E}_1 + \bar{E}_2 = \bar{E}'_1 + \bar{E}'_2\,,$

and insert the unchanged masses before and after the collision, see Eq. (18.1)

2 $\sqrt{m_1^2c^4 + \bar{p}^2c^2} + \sqrt{m_2^2c^4 + \bar{p}^2c^2} = \sqrt{m_1^2c^4 + \bar{p}'^2c^2} + \sqrt{m_2^2c^4 + \bar{p}'^2c^2}\,.$

This equation can be satisfied if and only if the magnitude of the momenta of particles remains equal,

3 $\bar{p} = \bar{p}'\,.$

This means that in the CM-frame, the elastic scattering process rotates the particle momentum vectors without changing their magnitude, as is shown in Fig. 18.3. ◀

Fig. 18.3 Elastic collision in CM-frame, see Example 18.1

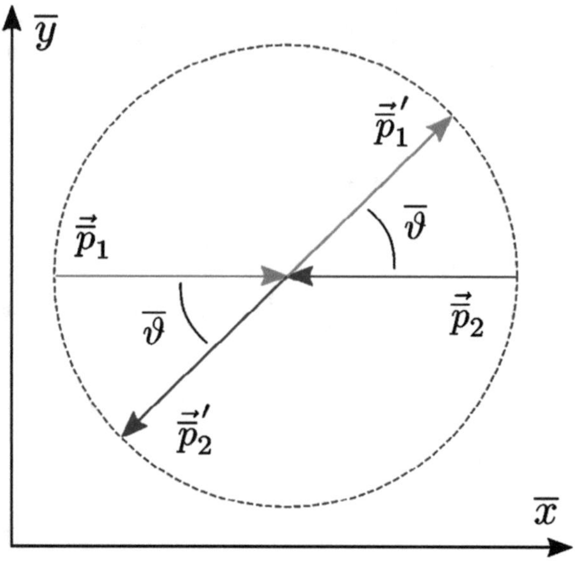

18.2 Compton Scattering

A special and very important type of elastic collision involves one massless particle, e.g. a photon, which hits a target at rest in the laboratory frame, typically an electron. This specific process is called Compton scattering.[1] In the lab frame, the target electron has no momentum; $\boldsymbol{p}_e = 0$. The photon, though massless, has momentum $|\boldsymbol{p}_\gamma| = E_\gamma/c$. Conservation of momentum requires that the initial momentum of the photon must be equal to the sum of the final momenta of the electron and photon:

$$\boldsymbol{p}_\gamma - \boldsymbol{p}'_\gamma = \boldsymbol{p}'_e \,. \tag{18.6}$$

Squaring and expanding this equation gives us

$$p_e'^2 = {p_\gamma}^2 + p_\gamma'^2 - 2p'_\gamma p_\gamma \cos\theta \,, \tag{18.7}$$

where the angle of the scattered photon is measured relative to the original axis of motion, see Fig. 18.4. Rewriting Eq. (18.7) using again $p_\gamma = E_\gamma/c$, we have

$$p_e'^{\,2} = \frac{{E_\gamma}^2}{c^2} + \frac{{E'_\gamma}^2}{c^2} - \frac{2E_\gamma E'_\gamma \cos\theta}{c^2} \,. \tag{18.8}$$

[1] Arthur H. Compton (1892–1962), American physicist and Nobel Prize Laureate 1927 for the discovery of the Compton effect.

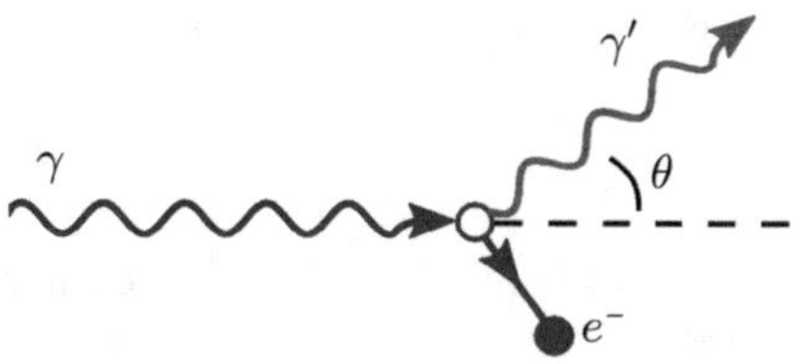

Fig. 18.4 Photon scattering off electron: Compton effect

Due to conservation of energy we have

$$E_\gamma + E_e = E'_\gamma + E'_e \,. \tag{18.9}$$

The electron has no momentum before the collision; thus

$$E_e = m_e c^2 \,. \tag{18.10}$$

After the collision, we remain in the electron's initial rest-frame, and so we must consider the momentum imparted to the electron by the collision

$$E'_e = \sqrt{p_e'^2 c^2 + m_e^2 c^4} \,. \tag{18.11}$$

Substituting Eqs. (18.10) and (18.11) into the conservation of energy equation Eq. (18.9):

$$E_\gamma + m_e c^2 = E'_\gamma + \sqrt{p_e'^2 c^2 + {m_e}^2 c^4} \,. \tag{18.12}$$

Solving for the final momentum of the electron we find

$$p_e'^2 = \frac{(E_\gamma - E'_\gamma + m_e c^2)^2 - m_e^2 c^4}{c^2} \,. \tag{18.13}$$

Setting equal the two expressions Eqs. (18.8) and (18.13) gives us

$$\frac{(E_\gamma - E'_\gamma + m_e c^2)^2 - m_e^2 c^4}{c^2} = \frac{{E_\gamma}^2}{c^2} + \frac{{E'_\gamma}^2}{c^2} - \frac{2 E_\gamma E'_\gamma \cos\theta}{c^2} \,. \tag{18.14}$$

Simplifying, we have the Compton energy formula:

$$E_\gamma E'_\gamma (1 - \cos\theta) = (E_\gamma - E'_\gamma) m_e c^2 \,, \tag{18.15}$$

which expresses how the energy of the photon changes in terms of the scattering angle. This Compton formula is usually presented in terms of the shift in wavelength of the photon. We recall that the energy of a photon E is proportional to the photon frequency ν, $E = h\nu$, with Planck constant h. This is the corpuscular property of light proposed by Einstein in 1905. Since light propagates with velocity c we further have the wavelength-frequency relation $\lambda\nu = c$. Thus, we find a relation between photon energy and wavelength:

$$E = \frac{hc}{\lambda} . \tag{18.16}$$

Substituting this into the Compton energy formula Eq. (18.15), we obtain the wavelength shift Compton condition:

$$\lambda' - \lambda = \lambda_C(1 - \cos\theta), \quad \lambda_C = \frac{h}{m_e c} = 2.42631 \times 10^{-12}\text{m} . \tag{18.17}$$

Here we introduce the Compton wavelength of the electron. The Compton wavelength is a natural length associated with any quantum wave for a particle of mass m.

The Compton wavelength shift result Eq. (18.17) agrees with our prior experience concerning the energy and momentum conservation. As the angle of deflection increases, λ' also increases, meaning that the final energy of the photon decreases. This is what we would expect from dealing with nonrelativistic collisions. The maximum possible increase in wavelength is $2\lambda_C$, corresponding to the back-scattering of the photon, ($\cos\theta = -1$). When $\lambda' - \lambda = 0$, we know that $\cos\theta = 1$. This corresponds to the photon passing straight through without colliding with an electron. This is, in fact, the most common occurrence.

As the above remark highlights, we have merely evaluated a straightforward consequence of energy-momentum conservation; that is, we found the relationship between the scattering angle and the wavelength shift. We have not touched upon the question of how often the photon-electron scattering will occur, and if it does, how often a particular scattering angle θ is to be expected.

The study of photon-electron quantum reactions is part of the field of quantum electrodynamics (QED). The result of the study of photon-electron scattering is the reaction cross-section describing the effective size of an electron in an interaction with a photon. The process of (unpolarized photon) Compton scattering is then described by the "Klein-Nishina" formula,[2]

$$\frac{d\sigma}{d\Omega} = \frac{r_e^2}{2}\frac{\lambda^2}{\lambda'^2}\left(\frac{\lambda}{\lambda'} + \frac{\lambda'}{\lambda} - \sin^2\theta\right) . \tag{18.18}$$

The quantity

$$r_e \equiv \frac{e^2}{4\pi\epsilon_0 m_e c^2} = 2.817940\,\text{fm} \tag{18.19}$$

is the 'classical' electron radius.

[2] O. Klein and Y. Nishina, "Über die Streuung von Strahlung durch freie Elektronen nach der neuen relativistischen Quantendynamik von Dirac, (On the electron-photon scattering according to relativistic Dirac quantendynamics)" *Z. Phys.* **52** p853 and p869 (1929).

The probability that the scattering reaction occurs is governed by the 'classical' radius of the electron which is of a similar magnitude as the size of the atomic nucleus. The value of r_e governing electron-photon scattering is well understood. It is of some practical importance that the effective size of an electron as 'seen' by a photon is comparable to the size of a small atomic nucleus 'seen' by an α-particle in the pivotal Rutherford scattering experiment.[3]

There is another way to understand the magnitude of Klein-Nishina scattering cross section. We rewrite Eq. (18.19) in the form

$$r_e \equiv \left(\frac{e^2}{4\pi\epsilon_0\hbar c}\right)\left(\frac{\hbar c}{m_e c^2}\right) \equiv \alpha\, \bar{\lambda}_C\,, \qquad \alpha = \frac{e^2}{4\pi\epsilon_0\hbar c} = \frac{1}{137.036}\,. \tag{18.20}$$

α, the fine-structure constant (not the α-particle), describes the strength of the electron-photon interaction in Quantum Electrodynamics. The (2π) reduced Compton quantum wave wavelength $\bar{\lambda}_C = \lambda_C/2\pi$ is therefore considered as the characteristic size scale of the quantum electron. Viewed from this quantum perspective, the small magnitude of r_e compared to the quantum electron size $\bar{\lambda}_C$ is recognized as being due to the relative smallness of the electromagnetic charge e contained within α.

Example 18.2

Thomson scattering

Determine the low energy limit of the Klein-Nishina cross section, Eq. (18.18), which is the classical Thomson limit. Does $\hbar$ appear in this result? ◀

▶ **Solution** The magnitude of the Klein-Nishina cross section, Eq. (18.18), is controlled by the classical electron radius Eq. (18.19), a quantity which is manifestly independent of $\hbar$; $\hbar$ appears only in the Compton wavelength shift Eq. (18.17). We use the Compton wavelength shift formula Eq. (18.17) to eliminate from the Klein-Nishina cross-section Eq. (18.18) the ratio λ'/λ. We find

1 $$\frac{d\sigma}{d\Omega} = \frac{r_e^2}{2}\left(\frac{1}{(1+\epsilon)^3} + \frac{1}{1+\epsilon} - \frac{\sin^2\theta}{(1+\epsilon)^2}\right), \qquad \epsilon = \frac{\lambda_C(1-\cos\theta)}{\lambda}\,.$$

Here, by assumption, we are in the low energy, high wavelength limit, so ϵ can be made as small as needed.

[3]Ernest Rutherford (1871–1937), renowned experimentalist and discoverer of the atomic nucleus by means of α-particle back-scattering, referred to as the 'father' of nuclear physics. Nobel Prize in Chemistry in 1908.

Upon integration over the scattering angle θ we obtain the Thomson limit

$$\textbf{2} \quad \sigma_{\text{Thomson}} = r_e^2 \int d\Omega \left(1 - \tfrac{1}{2}\sin^2\theta\right) = 4\pi r_e^2 \left(1 - \tfrac{1}{3}\right) = \frac{8\pi}{3} r_e^2 = 665 \text{ mb} .$$

This result is independent of $\hbar$. Since $r_e^2 \propto 1/m^2$, the lightest free particle carrying an electrical charge interacts strongest with photons, with a remarkable strength, an order of magnitude above the strong interaction scale. ◀

18.3 Elastic Bounce from a Moving Wall

The reflection of a photon from a mirror is a well-known phenomenon. In the rest-frame of the mirror the photon momentum normal to the mirror is reflected. In his landmark paper on relativity, Einstein considered the case of photon reflection from a moving mirror. He obtained the fascinating result that the reflected energy of the photon scales with γ_{mirror}^2 i.e. square of the Lorentz factor of the mirror. This is so since in the rest-frame of the mirror the incoming photon is shifted to higher energy, and after reflection in the mirror rest-frame, transformation back into the lab frame produces a second, energy-enhancing shift, since the longitudinal momentum has been reversed.

For a laboratory observer, the reflection is a 'squared' energy shift. For the case that the mirror velocity is relativistic this leads to the expectation of a very large boost in photon energy, the source of energy being the kinetic energy of the mirror, which is assumed to be so large that we can ignore the back-reaction. We show in Example 18.3 that in the nonrelativistic limit the ball bounces off much harder from such a moving wall, as the normal component picks up twice the speed of the wall.

The scattering process is illustrated in Fig. 18.5: We solve this elastic scattering problem of a particle of mass m (photons $m \to 0$) impacting with velocity $\boldsymbol{v}$ a

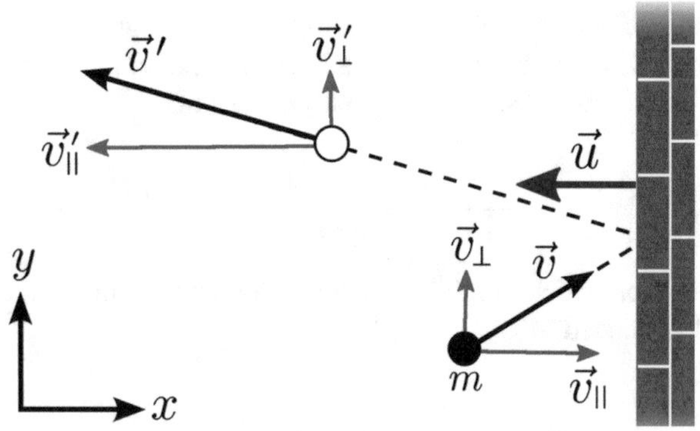

Fig. 18.5 Scattering of massive particle m, impacting a wall moving with velocity $\boldsymbol{u}$. Directions $\perp$ and $\parallel$ denote the components of particle velocity $\boldsymbol{v}$ or momentum $\boldsymbol{p}$ perpendicular and parallel to $\boldsymbol{u}$ and the coordinate system is selected such that the wall moves against the x-axis, while the y-axis points in the direction parallel to transverse motion of the particle

much heavier mirror moving with velocity $\boldsymbol{u}$. Elastic here means that the momentum normal to the wall in the rest-frame of the wall is reversed, while the transverse momentum is unchanged.

Let $\perp$ and $\|$ denote the components of particle momentum perpendicular and parallel to the direction of motion of the wall, see Fig. 18.5. These will relate to the Lorentz coordinate transformation, which we will introduce to solve the problem. The total energy of the incoming particle is given by

$$E = \sqrt{m^2c^4 + p^2c^2} = \sqrt{m^2c^4 + p_\perp^2c^2 + p_\|^2c^2} \equiv \sqrt{E_\perp^2 + p_\|^2c^2}\,, \tag{18.21}$$

where $E_\perp = \sqrt{m^2c^4 + p_\perp^2c^2}$. With the x-axis being the direction (opposite) of wall motion defining the $\|$ component, and the y-axis parallel to $\boldsymbol{p}_\perp$, we further have for the x-component $p_\| = m\gamma v_\|$, and the y-component $\boldsymbol{p}_\perp = m\gamma \boldsymbol{v}_\perp$, while the z-component of both the momentum and velocity vectors vanishes. Note that we can also write

$$E = E_\perp \cosh y_p\,, \qquad p_\| c = E_\perp \sinh y_p\,, \tag{18.22}$$

where y_p is particle rapidity, see Eq. (15.18). We will also use y_W, the wall rapidity related as usual to the speed u of the wall, see Eq. (7.31)

$$\begin{aligned} y_\text{W} &= \frac{1}{2}\ln\left(\frac{1+\beta_\text{W}}{1-\beta_\text{W}}\right), & \beta_\text{W} &= \frac{u}{c}\,, \\ \gamma_\text{W} &= \frac{1}{\sqrt{1-\beta_\text{W}^2}} = \cosh y_\text{W}\,, & \beta_\text{W}\gamma_\text{W} &= \sinh y_\text{W}\,. \end{aligned} \tag{18.23}$$

We will present the energy and momentum components as columns; the coordinate system is as shown in Fig. 18.5

$$\begin{bmatrix} E \\ cp_\| \\ cp_\perp \\ 0 \end{bmatrix} = \begin{bmatrix} E_\perp \cosh y_p \\ E_\perp \sinh y_p \\ cp_\perp \\ 0 \end{bmatrix} \to \underset{\text{wall}}{[\text{LT}]} \to \begin{bmatrix} E_\perp \cosh(y_p + y_\text{W}) \\ E_\perp \sinh(y_p + y_\text{W}) \\ cp_\perp \\ 0 \end{bmatrix}. \tag{18.24}$$

In the last step we applied an active Lorentz coordinate transformation taking us to the rest-frame of the wall. Since we have decomposed the particle momentum into the appropriate components, and the wall motion is opposite the motion of the particle, the rapidities add as we saw in Example 15.7. Seen in the rest-frame of the wall the incoming particle is always more energetic as we see explicitly in Eq. (18.24). From now on we omit the last two entries in columns as the transverse momentum does not change.

The next step is for the particle to be reflected off the wall. This of course happens that way in the frame of reference of the wall. This means that the momentum parallel to the direction of the wall's motion is reversed; the superscript 'bb' indicates 'bounced from the wall as seen in the wall rest-frame':

$$\begin{bmatrix} E^{\mathrm{bb}} \\ cp_{\parallel}^{\mathrm{bb}} \end{bmatrix} = \begin{bmatrix} E_{\perp} \cosh(y_p + y_{\mathrm{W}}) \\ -E_{\perp} \sinh(y_p + y_{\mathrm{W}}) \end{bmatrix} = \begin{bmatrix} E_{\perp} \cosh y_{\mathrm{pR}} \\ E_{\perp} \sinh y_{\mathrm{pR}} \end{bmatrix}, \tag{18.25}$$

where we introduced y_{pR} as the rapidity of the particle after the reflective bounce

$$y_{\mathrm{pR}} = -y_p - y_{\mathrm{W}} \,. \tag{18.26}$$

Note the negative sign implying that the particle is now moving in the opposite direction.

We now transform back to the laboratory frame, where the prime indicates 'bounced from wall as seen in the lab':

$$\begin{bmatrix} E' \\ cp'_{\parallel} \end{bmatrix} = \begin{bmatrix} E_{\perp} \cosh(y_{\mathrm{pR}} - y_{\mathrm{W}}) \\ E_{\perp} \sinh(y_{\mathrm{pR}} - y_{\mathrm{W}}) \end{bmatrix} = \begin{bmatrix} E_{\perp} \cosh(y_p + 2y_{\mathrm{W}}) \\ -E_{\perp} \sinh(y_p + 2y_{\mathrm{W}}) \end{bmatrix}. \tag{18.27}$$

This result shows that the reflected particle acquires in the direction of the wall motion twice the rapidity of the wall, and a changed parallel momentum sign shows it is moving in a direction opposite from the original.

This is not quite the final result; we need to revert to the original quantity $p_{\parallel}$ for the result to become more transparent. This is accomplished by employing the addition theorems for hyperbolic functions

$$\begin{aligned} \cosh(a+b) &= \cosh a \cosh b + \sinh a \sinh b \,, \\ \sinh(a+b) &= \cosh a \sinh b + \sinh a \cosh b \,. \end{aligned} \tag{18.28}$$

This leads to

$$\begin{bmatrix} E' \\ cp'_{\parallel} \end{bmatrix} = \begin{bmatrix} E_{\perp} \cosh y_p \cosh 2y_{\mathrm{W}} + E_{\perp} \sinh y_p \sinh 2y_{\mathrm{W}} \\ -E_{\perp} \cosh y_p \sinh 2y_{\mathrm{W}} - E_{\perp} \sinh y_p \cosh 2y_{\mathrm{W}} \end{bmatrix}. \tag{18.29}$$

Remembering Eq. (18.24) we can revert to the original laboratory energy and momentum of the incoming particle

$$\begin{bmatrix} E' \\ cp'_{\parallel} \end{bmatrix} = \begin{bmatrix} E \cosh 2y_{\mathrm{W}} + cp_{\parallel} \sinh 2y_{\mathrm{W}} \\ -cp_{\parallel} \cosh 2y_{\mathrm{W}} - E \sinh 2y_{\mathrm{W}} \end{bmatrix}, \tag{18.30}$$

which is the final result wherein we retain y_{W} to characterize the effect of the moving mirror attached to a wall.

For applications in nonrelativistic and ultrarelativistic wall motion we seek a result which employs velocity vectors. Using Eq. (18.28) we obtain

$$\begin{aligned}\cosh 2y_{\mathrm{W}} &= \cosh^2 y_{\mathrm{W}} + \sinh^2 y_{\mathrm{W}} = (1+\beta_{\mathrm{W}}^2)\gamma_{\mathrm{W}}^2 \to 2\gamma_{\mathrm{W}}^2\,,\\ \sinh 2y_{\mathrm{W}} &= 2\cosh y_{\mathrm{W}} \sinh y_{\mathrm{W}} = 2\beta_{\mathrm{W}}\gamma_{\mathrm{W}}^2 \to 2\gamma_{\mathrm{W}}^2\,,\end{aligned} \tag{18.31}$$

where we show the ultrarelativistic limit explicitly. We now can write Eq. (18.30) using the wall velocity and Lorentz factor γ_{W}, restoring for completeness the transverse momentum in explicit form

$$\begin{bmatrix} E' \\ cp'_{\parallel} \\ cp'_{\perp} \\ 0 \end{bmatrix} = \gamma_{\mathrm{W}}^2 \begin{bmatrix} (1+\beta_{\mathrm{W}}^2)E + 2\beta_{\mathrm{W}} cp_{\parallel} \\ -(1+\beta_{\mathrm{W}}^2)cp_{\parallel} - 2\beta_{\mathrm{W}} E \\ cp_{\perp}/\gamma_{\mathrm{W}}^2 \\ 0 \end{bmatrix} \xrightarrow{\beta_{\mathrm{W}}\to 1} 2\gamma_{\mathrm{W}}^2 \begin{bmatrix} E + cp_{\parallel} \\ -(cp_{\parallel} + E) \\ cp_{\perp}/2\gamma_{\mathrm{W}}^2 \\ 0 \end{bmatrix}, \tag{18.32}$$

which makes explicit the coefficient γ_{W}^2. The appearance of a γ_{W}^2 factor boosting the incoming energy and momentum, but not the transverse momentum, suggests that ultrarelativistic particle beams could be realized employing wall bounce.

Both Eqs. (18.30) and (18.32) remain valid considering the special case of massless particles (i.e., photons), since in the presented derivation we used the particle energy and momentum, and did not specify the value of particle mass. We recover from Eq. (18.32) the well-known Einstein limit

$$E' \simeq cp' \simeq 4\gamma_{\mathrm{W}}^2 E \tag{18.33}$$

for head-on collisions.

This remarkable result is today no longer a merely theoretical topic. A 'relativistic mirror' has been realized forming a relativistic electron cloud using an ultra short high edge laser pulse interacting with nanometer-thin carbon foil.[4] In another approach an intense electron pulse formed in a conventional linear accelerator provides the moving mirror.[5]

Not every electron cloud is a suitable mirror, considering the greatly reduced light wavelength observed in the rest-frame of the moving relativistic mirror. Such hard photons can behave and actually are X-rays and can penetrate a thin cloud. In order to achieve in a bounce a gamma energy of 1 MeV, thus, to achieve upscale by a factor 10^6, a laser beam with 1 eV photons would need, according to Eq. (18.33),

---

[4]D. Kiefer, et al. "Relativistic electron mirrors from nano scale foils for coherent frequency upshift to the extreme ultraviolet," *Nature Communications*, 1763 (2013).

[5]N.V. Zamfir, "Nuclear physics with 10 PW laser beams at ELI-NP," *EPJ-ST* **223** 1221 (2014).

to be head-on reflected from an electron cloud that has $\gamma = 500$, i.e. an electron energy of 256 MeV. In the rest-frame of this electron mirror, where the reflection occurs, the incoming 1-eV-laser beam has a Doppler 2γ-shortened wavelength and accordingly increased energy per photon, $1\,\text{eV} \times 2\gamma = 1\,\text{keV}$, corresponding to a wavelength $\lambda = hc/1\text{keV} = 12.4\,\text{Å}$. For the mirror to work, it has to be able to reflect light in the X-ray domain. Other dynamical effects, not considered here, such as the recoil of the mirror 'fragment' were investigated;[6] they also influence the achievable energy upscale.

Example 18.3

Nonrelativistic bounce limit

Obtain the velocity vector of a particle scattered from a moving wall from consideration of the nonrelativistic limit of the general result presented in Eq. (18.32). ◀

▶ **Solution** In the nonrelativistic limit we employ the leading terms

$$E = \left(1 + \frac{\boldsymbol{v}^2}{2c^2} + \ldots\right) mc^2\,,$$

1
$$p_\parallel = m v_\parallel + \ldots\,,$$
$$p_\perp = m v_\perp + \ldots\,,$$

$$\gamma_{\mathrm{W}}^2 = 1 + \frac{\boldsymbol{u}^2}{c^2} + \ldots\,,$$

in Eq. (18.32) to obtain

2
$$\begin{bmatrix} E' \\ cp'_\parallel \\ cp'_\perp \\ 0 \end{bmatrix} \to \begin{bmatrix} mc^2\left(1 + \frac{2u^2}{c^2}\right)\left(1 + \frac{\boldsymbol{v}^2}{2c^2}\right) + 2muv_\parallel \\ -mc^2\left[\left(1 + \frac{2u^2}{c^2}\right)\frac{v_\parallel}{c} - \left(1 + \frac{\boldsymbol{v}^2}{2c^2}\right)\frac{2u}{c}\right] \\ cp_\perp \\ 0 \end{bmatrix} = \begin{bmatrix} (2\boldsymbol{u} + \boldsymbol{v})^2 \frac{m}{2} + mc^2 \\ -(2u + v_\parallel)\, mc \\ v_\perp\, mc \\ 0 \end{bmatrix}.$$

This shows that the bounce off a moving wall adds twice the velocity of the wall to the reversed (bounced) velocity vector component, $v_\parallel$, that is parallel to the motion of the wall. The total velocity vector of a body after a bounce from a moving wall is thus

[6]P. Valenta, et al. "Recoil effects on reflection from relativistic mirrors in laser plasmas," *Physics of Plasmas* **27**, 032109 (2020).

3 $\boldsymbol{v}' = -2\boldsymbol{u}\left(1 + \frac{\boldsymbol{u}\cdot\boldsymbol{v}}{\boldsymbol{u}^2}\right) + \boldsymbol{v}\,.$

As a cross-check we compute the square of the velocity vector

4 $(\boldsymbol{v}')^2 = 4(\boldsymbol{u})^2 + 8\boldsymbol{u}\cdot\boldsymbol{v} + 4\frac{(\boldsymbol{u}\cdot\boldsymbol{v})^2}{\boldsymbol{u}^2} - 4\boldsymbol{u}\cdot\boldsymbol{v} - 4\frac{(\boldsymbol{u}\cdot\boldsymbol{v})^2}{\boldsymbol{u}^2} + (\boldsymbol{v})^2\,,$

which verifies the kinetic energy upshift: Upon effecting two trivial cancelations we obtain the result

5 $(\boldsymbol{v}')^2 = (2\boldsymbol{u} + \boldsymbol{v})^2\,.$

Further considerations: The bounce from a wall provides a first view of the 'swing-by' method used to boost (or reduce) the velocity of a satellite 'bouncing' due to a gravity potential, that is, not by actual impact, off a moving stellar body. ◀

18.4 Inelastic Two-Body Reaction Threshold

We turn now to inelastic reactions in which energy can be converted into particle rest mass and vice versa, as we have already seen considering decay reactions. In these cases, some of the physical characteristics of the participating particles change in the interaction. This change can be relatively minor, for example, the excitation of internal dynamics of a nucleus or an atom. However, more often in relativistic collisions we see a much more radical modification, such as a breakup of particles into more elementary constituents or the formation of new particles, such as antiprotons in proton-proton collisions, an example we will present.

A 'practical' example of an inelastic reaction is a nuclear fusion reaction in which two heavy hydrogen isotopes (deuteron $d = pn$, and triton $t = pnn$) react into an α-particle ($ppnn$) and a neutron n – here p denotes a proton:

$$d + t \to \alpha + n + 17.6\,\text{MeV}\,.$$

While this reaction is exothermic (it releases $Q = 17.6\,\text{MeV}$ kinetic energy as indicated by the '+' sign), many two-body reactions are endothermic; that is, energy needs to be supplied in order for the reaction to proceed. In such a case one often indicates the amount of threshold energy required with a '$-$' sign.

In the field of elementary particle physics the production of new particles, in general, consumes energy due to the conversion of kinetic collision energy into the rest energy of newly produced particles. As an example consider the production of two particles called 'strange', Λ and K^0:

$$m_\Lambda = 1115.68\ \text{MeV/c}^2\,, \qquad m_{K^0} = 497.65\ \text{MeV/c}^2\,.$$

The new physics particles Λ, K^0 were called 'strange' in the mid-twentieth century because of their unusual properties and the name stuck. The 'strangeness' producing reaction is strongly endothermic

$$\pi^- + p \to \Lambda + K^0 - 535.5\,\text{MeV}\,, \tag{18.34}$$

as shown by the negative sign of the Q-value, the last term in Eq. (18.34). We compute the energy threshold; that is, the required energy from the masses of the participating particles

$$Q = m_p + m_{\pi^-} - m_\Lambda - m_{K^0}\,,$$

where the incoming particle masses are:

$$m_{\pi^-} = 139.57\ \text{MeV/c}^2\,, \qquad m_p = 938.27\ \text{MeV/c}^2\,.$$

This reaction and other endothermic reactions obviously require that the reactants have a kinetic energy greater than some minimum, below which there is insufficient energy available to convert to the rest mass of the reaction products. The threshold energy shown is the energy amount needed. This is not the kinetic energy needed to make the reaction happen. Considering the need to transform into the CM-frame, the kinetic energy a projectile requires has usually a much greater value. We will now show this considering reaction Eq. (18.34).

We determine the minimum kinetic energy and momentum of the reactants as follows. In the CM-frame of reference, the mass M of the system is related to the energy $\overline{E}$ by $\overline{E} = Mc^2$. The Lorentz coordinate transformation equation out of the CM-frame to any other frame reads

$$E = \gamma \bar{E} = \gamma M c^2\,. \tag{18.35}$$

We see that minimizing the required mass-energy equivalent Mc^2 of a system minimizes its energy in any other frame of reference. We therefore consider the process Eq. (18.34) in the CM-frame, where we have

$$\bar{p}_{\pi^-} = -\bar{p}_p = \bar{p}\,. \tag{18.36}$$

The total energy of participants is therefore

$$\bar{E}_T = \bar{E}_{\pi^-} + \bar{E}_p = \sqrt{m_{\pi^-}^2 c^4 + \bar{p}^2 c^2} + \sqrt{m_p^2 c^4 + \bar{p}^2 c^2}\,. \tag{18.37}$$

After the collision, energy conservation requires that the final state particles must have the same total energy given by Eq. (18.37). Thus we have

$$\bar{E}_T = E_\Lambda + E_{K^0} = \sqrt{m_\Lambda^2 c^4 + \bar{p}_\Lambda^2 c^2} + \sqrt{m_{K^0}^2 c^4 + \bar{p}_{K^0}^2 c^2}\,. \tag{18.38}$$

The minimum energy necessary for the endothermic reaction to proceed corresponds to the state in which the reaction products are at rest in the CM-frame system, in which case we have for Eq. (18.38)

$$\bar{E}_{T\min} = m_\Lambda c^2 + m_{K^0} c^2\,. \tag{18.39}$$

From Eqs. (18.37) and (18.39) we obtain an implicit equation for the minimum CM-frame particle momentum required for the reaction,

$$\sqrt{m_{\pi^-}^2 c^4 + \bar{p}_{\min}^2 c^2} + \sqrt{m_p^2 c^4 + \bar{p}_{\min}^2 c^2} = (m_\Lambda + m_{K^0})c^2\,. \tag{18.40}$$

We can now transform $\overline{p}_{\min}$ into the laboratory frame where the unstable particle, the pion, moves towards a target proton which is at rest. The pion moves toward the proton with the minimum possible laboratory momentum, $p_{\pi^-\min}$. We find the transformation by considering how to transform the proton at rest to the CM-frame where it has momentum $\overline{p}_{\min}$:

$$\bar{p}_{\min} = \gamma\beta m_p c = \frac{p_{\pi^-\min}}{M} m_p\,, \tag{18.41}$$

where we have substituted for γ and β from Eqs. (17.7a) and (17.8). The invariant mass M of the system is given in terms of the reaction products, which are both at rest in the CM-frame:

$$M = m_\Lambda + m_{K^0}\,. \tag{18.42}$$

With this Eq. (18.41) becomes

$$\bar{p}_{min} = \frac{m_p}{m_\Lambda + m_{K^0}} p_{\pi^-\min}\,. \tag{18.43}$$

It is convenient for the following to measure the momentum of the pion $p_{\pi^-\min}$ in units of $m_{\pi^-}c$, i.e.

$$p_{\pi^-\min} = f m_{\pi^-} c\,.$$

Substituting Eq. (18.43) into Eq. (18.40) we obtain for f the following implicit equation:

$$\sqrt{f^2+\left(\frac{m_\Lambda+m_{K^0}}{m_p}\right)^2}+\sqrt{f^2+\left(\frac{m_\Lambda+m_{K^0}}{m_{\pi^-}}\right)^2}=\frac{(m_\Lambda+m_{K^0})^2}{m_p m_{\pi^-}}\,. \tag{18.44}$$

Note that Eq. (18.44) is symmetric when the names of particles in the initial and/or final state are exchanged.

We now solve Eq. (18.44) for f. Moving one of the roots to the other side, and squaring, and later isolating the remaining root and squaring again we obtain

$$4f^2=\frac{m_p^2}{m_{\pi^-}^2}+\frac{m_{\pi^-}^2}{m_p^2}+\frac{(m_\Lambda+m_{K^0})^4}{m_p^2 m_{\pi^-}^2}-2\left(\frac{(m_\Lambda+m_{K^0})^2}{m_{\pi^-}^2}+\frac{(m_\Lambda+m_{K^0})^2}{m_p^2}+1\right)\,. \tag{18.45}$$

Again we observe that this expression is symmetric under the exchange of the initial state particles π^- and p, as well as the final state particles Λ and K^0 shown in Eq. (18.34).

We then insert the numeric values for m_{π^-}, m_p, m_Λ, m_{K^0} and find $f \simeq 6.4$, which easily checks with the defining Eq. (18.44)

$$\sqrt{f^2+1.7195^2}+\sqrt{f^2+11.56^2}=19.88\,.$$

The pion must hit the stationary proton with a minimum momentum $p_{\pi^-\mathrm{min}} \approx 6.4\, m_{\pi^-}c \simeq 890\,\mathrm{MeV/c}$ for the reaction Eq. (18.34) to be possible. We determine the kinetic energy of the incoming pion which is the total kinetic energy available for the reaction

$$T \equiv E_\pi - m_\pi c^2 = m_{\pi^-}c^2\left(\sqrt{f^2+1}-1\right)=139.6\left(\sqrt{6.4^2+1}-1\right)=765\,\mathrm{MeV}\,. \tag{18.46}$$

We learn that the initial state kinetic energy is indeed significantly larger compared to the Q-value, Eq. (18.34).

Example 18.4

Laboratory frame annihilation

A positron with kinetic energy of T_e collides with an electron at rest. An annihilation takes place, from which two photons emerge:

$$e^+ + e^- \to \gamma + \gamma\,.$$

One of the photons is observed to move in the direction of motion of the incident positron, which of course is not always the case. In which direction does the other photon move? How big are the laboratory energies of the two photons? ◀

▶ **Solution** As the momentum must be conserved, by demanding that one of the photons moves in the direction of the incoming positron we know that the second photon cannot have any transverse momentum.

We therefore can focus our attention on the conservation of the momentum parallel to the direction of motion of the positron with momentum p_e. Introducing photon energy $E_{1,2} = c|\boldsymbol{p}_{1,2}|$ the conservation of momentum for massless photons means

1 $$cp_e = E_1 + \varepsilon E_2 \ .$$

For $\varepsilon = +1$ the second photon momentum is in the direction of the first, i.e. aligned with the positron motion, and for $\varepsilon = -1$ it is opposite to the first photon and the positron. We now express p_e in terms of kinetic energy term T_e using the relativistic relation between energy and momentum

2 $$E_e^2 \equiv (T_e + m_e c^2)^2 = (p_e c)^2 + (m_e c^2)^2 \ .$$

We take the p_e from this relationship and insert it into the conservation of momentum

3 $$E_1 + \varepsilon E_2 = \sqrt{(T_e + m_e c^2)^2 - (m_e c^2)^2} = \sqrt{T_e^2 + 2m_e c^2 T_e} \ .$$

We must also conserve energy in the annihilation process

4 $$E_1 + E_2 = T_e + m_e c^2 + m_e c^2 \ ,$$

where we are equating photon energy on the left with e^+e^- energy on the right. Taking the difference of these two conservation equations we obtain

5 $$E_2(1 - \varepsilon) = T_e + 2m_e c^2 - \sqrt{T_e^2 + 2m_e c^2 T_e} > 0 \ .$$

One easily checks that the right-hand side is always positive. However, for $\varepsilon = +1$ the left-hand side vanishes. Thus we must have $\varepsilon = -1$, which means that the second photon moves in the opposite direction to the first; that is, against the original motion of the incoming positron.

Adding up both conservation equations and restating above result we obtain

6 $$\begin{aligned} 2E_1 &= T_e + 2m_e c^2 + \sqrt{T_e^2 + 2m_e c^2 T_e} \ , \\ 2E_2 &= T_e + 2m_e c^2 - \sqrt{T_e^2 + 2m_e c^2 T_e} \ , \end{aligned}$$

for the energies and momenta of the two photons, where the photon of smaller energy (E_2) moves against the original momentum of the positron.

We look at an example, taking $T_e = 3\,\mathrm{MeV}$, $m_e c^2 = 0.511\,\mathrm{MeV}$, and obtain for the energies

7 $$p_1 = \frac{E_1}{c} = 3.748\,\mathrm{MeV}/c\,, \qquad p_2 = -\frac{E_2}{c} = -0.274\,\mathrm{MeV}/c\,.$$

We also can check the energy conservation:

8 $$E_1 + E_2 = (3.748 + 0.274)\,\mathrm{MeV} = 4.022\,\mathrm{MeV} = 3\,\mathrm{MeV} + 2 \times 0.511\,\mathrm{MeV}\,.$$

In the CM-frame the annihilation photons of equal energy emerge in opposite directions and thus the LT from CM to laboratory system enhances the energy of one of the photons while reducing the energy of the other. ◀

18.5 Energy Available in a Two-Particle Collision

We determine the general expression for the mass of a system of two particles observed in the lab frame S in which particle '1' collides with particle '2'. The energy equivalent Mc^2 of these two particles is:

$$Mc^2 = \sqrt{(E_1 + E_2)^2 - c^2(\boldsymbol{p}_1 + \boldsymbol{p}_2)^2}\,. \tag{18.47}$$

The physical meaning of Mc^2 is best understood by considering the reverse process; a particle of a putative mass M and in its rest-frame energy Mc^2 can decay into two particles, each with energy $E_i,\ i = 1, 2$ and momentum $\boldsymbol{p}_i,\ i = 1, 2$. We thus conclude that M is the quasi-mass that two colliding particles can create, and Mc^2 is the energy available in the frame in which M is at rest. Given the definition Eq. (18.47) we recognize the quasi-mass M as a Lorentz invariant.

The invariant energy is denoted, see Eq. (17.5), by $\sqrt{s}$

$$\sqrt{s} = Mc^2\,. \tag{18.48}$$

This is the energy available for further processes (e.g. particle production). As already mentioned, s (and t, and u, which we do not discuss in this book) are the Mandelstam variables. Note that the $\sqrt{s}$ in Eq. (18.48) should not be confused with the invariant space-time interval s^2 defined earlier; see Eq. (11.7). Unfortunately, in the common notation these two quantities must be distinguished by their context.

We can reorder in Eq. (18.47) the terms and noting that $E_i^2 - c^2 p_i^2 = m_i^2 c^4$, $i = 1, 2$, we find

$$\sqrt{s} = \sqrt{m_1^2 c^4 + m_2^2 c^4 + 2(E_1 E_2 - c^2 \boldsymbol{p}_1 \cdot \boldsymbol{p}_2)}\ . \tag{18.49}$$

Two cases are of particular importance:

1. Collision with particle '2' at rest. We have $E_2 = m_2 c^2$, $\boldsymbol{p}_2 = 0$. We find

$$\sqrt{s} = \sqrt{m_1^2 c^4 + m_2^2 c^4 + 2E_1 m_2 c^2}\ . \tag{18.50}$$

2. Collision of two particles that collide coming head-on from opposite directions: For the special case that $m_1 = m_2$ we find

$$\sqrt{s} = 2E\ , \tag{18.51}$$

since for $\boldsymbol{p}_1 = -\boldsymbol{p}_2 = \boldsymbol{p}$, and $m_1 = m_2 = m$ we have $E_1 = E_2 = E = \sqrt{m^2 c^4 + p^2 c^2}$.

The energy reach of an accelerator is often presented in terms of the magnitude of $\sqrt{s}$ obtainable in two-particle collisions. Presently, the most powerful accelerator is the Large Hadron Collider (LHC) at CERN. Here the particles (protons) collide from opposite directions with equal magnitudes of energy and momentum; thus the lab frame is also the CM-frame. Hence according to Eq. (18.51) we form the sum of the two proton beam energies, $\sqrt{s} = (7 + 7)\,\text{TeV} = 14\,\text{TeV}$, exceeding the proton mass equivalent by more than 14,000 times. A considerably greater amount of energy would be required in collisions of moving protons with protons at rest. To see this, solving Eq. (18.50) for E_1 we find

$$E_1 = \frac{s - m_1^2 c^4 - m_2^2 c^4}{2 m_2 c^2}\ . \tag{18.52}$$

We note that s and not $\sqrt{s}$ enters on the right. Consequently, we find for the LHC equivalent energy of a beam impacting a laboratory fixed target $E_{\text{LHC}} =$ 104,000 TeV! Clearly, the particle colliding in the collider concept allows the realization of an enormous economy: We use two 7 TeV beams and achieve as much as with one 104,000 TeV beam. Moreover, such a proton beam of 104 PeV (PeV = 10^{15} eV) is far beyond present day technological capabilities.

Example 18.5

Crash of relativistic rockets

Two identical rockets each of mass $m = 50{,}000$ kg are launched with missions to test special relativity. Unfortunately, after reaching their cruising speeds of $v = 0.6c$, they collide head-on. Assuming that no debris or radiation is released, and that all energy turns into mass, no pieces fly out, and nothing moves, what is the mass of the wreck? ◀

▶ **Solution** This example addresses the case of completely inelastic collisions. Using Eq. (14.21), the energies of the two colliding rockets are

1 $$E_1 = E_2 = \frac{mc^2}{\sqrt{1 - v^2/c^2}} = \frac{5}{4}mc^2 \, .$$

The conservation of energy implies that the energy of the wreck is equal to the total energy,

2 $$E = E_1 + E_2 = \frac{5}{2}mc^2 \, ,$$

and since the collision was head-on of two equal objects, the wreck is stationary; that is, the collision frame of reference is the CM-frame. In that case all energy must be contained in the compound body of crashed mass M,

3 $$E = Mc^2 \, .$$

Using these result we obtain

4 $$M = E/c^2 = \frac{5}{2}m = 125{,}000\,\text{kg} \, .$$

The mass of the wreck is 25,000 kg greater than the combined mass of the colliding rockets: 25% of additional mass is created in the materialization of the kinetic energy of two rockets. ◀

Example 18.6

Discovery of the Ψ-particle

A contemporary example of a totally inelastic collision of two bodies involves the conversion of energetic elementary matter and antimatter, electrons and positrons, by way of radiation formation, into a new particle. Assume that in collider mode beams of equal energy electrons and positrons approach each other at rapidities $\pm\bar{y}_e$ and turn into a new particle of mass $M_\Psi c^2 = 3097$ MeV at rest in laboratory. What is the value of $\bar{y}_e$? What is the value of y_e, the rapidity of a positron that hits a stationary electron in the laboratory producing a Ψ-particle? What is the energy of this laboratory positron? ◀

▶ **Solution** To produce a particle of mass M each of the beams of colliding elementary matter and antimatter must bring into the collision half of the required energy; that is

$$\textbf{1}\quad \bar{E}_e = \frac{Mc^2}{2} = 1548.5\,\text{MeV}\,.$$

We recall the particle rapidity relation to energy, Eq. (15.18), thus

$$\textbf{2}\quad \bar{y}_e = \operatorname{arcosh} M/(2m_e) = \operatorname{arcosh} 3030 = 8.71\,.$$

We see that electron and positron beams of rapidity 8.71 must collide to produce the Ψ-particle at rest in laboratory.

Applying a LT to transform the moving electron to the laboratory frame, and remembering that rapidities are additive under LT, we shift the positrons rapidity to twice the value

$$\textbf{3}\quad y_e = 2\bar{y}_e = 17.42\,,$$

which is needed to produce the Ψ-particle in fixed target mode.

To obtain the energy corresponding to the rapidity we insert it into Eq. (15.18) and obtain

$$\textbf{4}\quad E_e = m_e c^2 \cosh 2\bar{y}_e = 0.511\,\text{MeV} \times \cosh 17.42 = 9.39\,\text{TeV}\,.$$

This energy is more than an order of magnitude above the technological capabilities explored today in the planning of the International Linear Collider, ILC. Of course in a head-on collision in a collider two beams of $\simeq$1.5 GeV will suffice.

Historical note: The discovery in 1974 of the Ψ-particle by Burton Richter in e^+e^--collider collisions with parallel discovery in proton-beam on fixed target experiments by Sam Ting (Nobel Prize in 1976), recognized to be a bound state of heavy charmed quark pair $\bar{c}c$, paved the way for the development of the quark model, and later, the standard model of particle physics. ◀

Example 18.7

Production of antiprotons

In the collision of a relativistic proton with a proton at rest in the laboratory an additional proton-antiproton pair can be created.[7] How large must the minimum

[7] The discovery of the antiproton was achieved at the Bevatron at Berkeley in the collision of protons with nucleons bound in copper nuclei. This approach reduces the energy threshold of the reaction here computed due to occasional collisions of incoming protons with an 'orbiting' nucleon

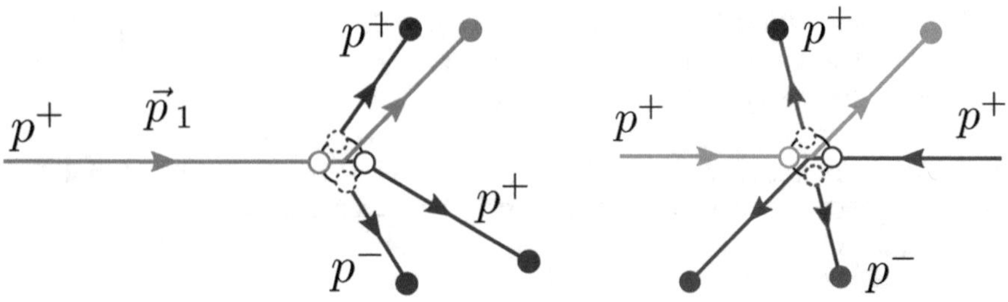

Fig. 18.6 Production of a proton-antiproton pair: (left) in the laboratory reference frame; (right) in the CM-frame. See Example 18.7

kinetic energy of the moving proton be? With what momentum and speed does the proton move? ◄

► **Solution** The situation in the laboratory system is depicted on the left in Fig. 18.6; here the momentum $\boldsymbol{p}_1$ of the incoming proton is shared by all final state particles. In the CM-frame shown on the right in Fig. 18.6, the two protons collide head-on coming from the right and the left and the new pair emerges in opposite directions. The total momentum of all particles sums to zero.

We determine the energy threshold in the CM-frame: To be able to create the additional proton-antiproton pair their mass-energy equivalent must be provided by the two colliding protons. With all four particles remaining at rest in the CM reference frame

1 $\sqrt{s_{\mathrm{Th}}} = (2+2)m_pc^2 \,.$

We enter this pair production threshold value in Eq. (18.52) in order to determine the beam energy E_1 in the laboratory frame with $m_1 = m_2 = m_p = 938.3\,\mathrm{MeV}/c^2$

2 $E_1 = \dfrac{4^2-2}{2} m_pc^2 = 7m_pc^2 = 6568\,\mathrm{MeV} \,.$

We recognize the required Lorentz factor $\gamma = 7$ and an energy difference $7m_pc^2 - 4m_pc^2 = 3m_pc^2$ as the total kinetic energy of the four particles in the laboratory frame.

In order to determine the momentum and speed of the proton we use the relativistic relations

3 $cp = \sqrt{E^2 - m^2c^4} \,, \qquad v = c\,\dfrac{cp}{E} \,.$

---

that is heading toward the incoming particle. Emilio Segré and Owen Chamberlain were awarded the Nobel Prize in Physics 1959 for the discovery of the antiproton.

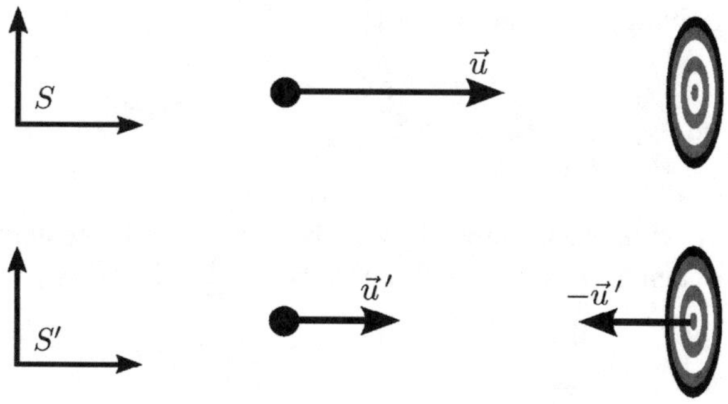

Fig. 18.7 Top: A particle collides with a target in lab frame S. Bottom: Now in a moving frame S'. See Example 18.8

Using these results we obtain

4
$$p_p = \sqrt{7^2 - 1}\, m_p c = 6.928\, m_p c = 6501\ \mathrm{MeV}/c\,,$$
$$v_p = c\,\frac{\sqrt{7^2 - 1}}{7} = c\sqrt{48/49} = c\,0.9897\,.$$ ◄

Example 18.8

Equal speed reference frame

A particle with a given Lorentz factor $\gamma = 3$ and corresponding speed $|\boldsymbol{u}|$ collides with a target particle at rest in the laboratory frame S. There is a system S' in which the velocities of the particle and the target are equal in magnitude and opposite in direction ($\boldsymbol{u}'_{\text{particle}} = -\boldsymbol{u}'_{\text{target}}$); i.e. the particles collide head-on with each other at equal velocity. Determine the magnitude of this velocity, $u' = |\boldsymbol{u}'_{\text{particle}}| = |\boldsymbol{u}'_{\text{target}}|$ and the corresponding Lorentz factor γ'. The situation in both S and S' is depicted in Fig. 18.7. ◄

► **Solution** An observer at rest in system S' sees the particle moving with velocity $+\boldsymbol{u}\,'$ and the target moving with $-\boldsymbol{u}\,'$. Since the target is at rest in the lab frame S, the velocity of frame S relative to frame S' must also be $\boldsymbol{v} = -\boldsymbol{u}\,'$. Using Eq. (7.18a) we transform the velocity of the particle in S' to S (note that here we transform from primed to 'not-primed' coordinates, while Eq. (7.18a) transforms from 'not-primed' to primed)

1
$$u = \frac{u' - v}{1 - u'v/c^2} = \frac{2u'}{1 + u'^2/c^2}\,.$$

Solving for u' yields

$$\mathbf{2}\quad \frac{u'}{c} = \frac{c}{u}\left(1 \pm \sqrt{1 - \frac{u^2}{c^2}}\right) .$$

We choose a negative sign for the root so that in the nonrelativistic limit, $u \to 0$, we obtain the nonrelativistic answer $u' = \frac{1}{2}u$. Given $1/\sqrt{1 - u^2/c^2} = \gamma$ we have

$$\mathbf{3}\quad \frac{u}{c} = \frac{\sqrt{\gamma^2 - 1}}{\gamma} .$$

For u'/c we now obtain

$$\mathbf{4}\quad \frac{u'}{c} = \frac{\gamma - 1}{\sqrt{\gamma^2 - 1}} = \sqrt{\frac{\gamma - 1}{\gamma + 1}} .$$

Now using $1 - (u'/c)^2 = 2/(\gamma + 1)$

$$\mathbf{5}\quad \gamma' = \frac{1}{\sqrt{1 - (u'/c)^2}} = \sqrt{\frac{\gamma + 1}{2}} .$$

We evaluate

$$\mathbf{6}\quad \gamma = 3\,, \to \quad \frac{u}{c} = \sqrt{\frac{8}{9}} ,$$

and in S' for the equal velocity collision

$$\mathbf{7}\quad \frac{u'}{c} = \frac{1}{\sqrt{2}} , \quad \gamma' = \sqrt{2} .$$

We compare the results of this example with those seen in Example 17.1. In particular we now also obtain for the transformation to the CM-frame from the beam dump frame the value $\gamma = 61.02$. We reconfirm the Example 17.1 result since we have $m_1 = m_2$ and thus the CM-frame, an equal momentum frame, is also an equal velocity frame as required in this example.

The physical relevance of the equal speed system is recognized when the objects colliding have different masses, but their components are the same, as is the case for two different atomic nuclei. In this situation in individual component collision the equal speed system is the collision CM reference frame as long as only such single particle reactions are relevant. This is the case in the study of the production of particles that are very massive. Their high energy production threshold is only significantly exceeded in primary collisions. ◀

18.6 Inelastic Collision and Particle Production

We consider collisions of two relativistic strongly interacting particles coming from the right and the left in the CM-frame as shown in Fig. 18.8. Typically these can be protons, or more generally, heavy ions; that is, atomic nuclei. One speaks of "Relativistic Heavy Ion" (RHI) collisions. It is known that in such highly energetic collisions many secondary particles can be created. The momentum $\bar{\boldsymbol{p}}$ of each emitted secondary particle can be decomposed as shown in Fig. 18.8 into components that are parallel and perpendicular with respect to the original collision axis

$$\bar{p}_{\parallel} = |\bar{\boldsymbol{p}}| \cos\bar{\theta} \,, \qquad |\bar{\boldsymbol{p}}_{\perp}| = |\bar{\boldsymbol{p}}| \sin\bar{\theta} \,. \tag{18.53}$$

Similarly, in the laboratory frame

$$p_{\parallel} = |\boldsymbol{p}| \cos\theta \,, \qquad |\boldsymbol{p}_{\perp}| = |\boldsymbol{p}| \sin\theta \,. \tag{18.54}$$

The distribution of particles in the azimuthal angle φ is often symmetric around the collision axis. However, this is not always the case and the distribution asymmetry in φ contains important information about the inelastic processes that lead to abundant particle production.

The particle production is measured in detectors located in the laboratory frame of reference. Once we have measured the momentum $\boldsymbol{p}$ and the angle of production θ, and we have identified the particle produced, i.e. we know its mass and energy $E_p = \sqrt{m_p^2 c^4 + \boldsymbol{p}^2 c^2}$, we can then obtain the particle rapidity by evaluating, see Eq. (15.23)

$$y_p = \frac{1}{2} \ln\left(\frac{E_p + c p_{\parallel}}{E_p - c p_{\parallel}}\right) = \frac{1}{2} \ln\left(\frac{E_p + c|\boldsymbol{p}|\cos\theta}{E_p - c|\boldsymbol{p}|\cos\theta}\right) . \tag{18.55}$$

In an experiment it is common to evaluate how many particles are produced per rapidity interval dy; that is, to form dN/dy, where for example $N \to \pi^-$ when we measure the production of negatively charged pions π^-. In the following we address π^- rapidity spectra obtained by the NA61/SHINE CERN-SPS collaboration

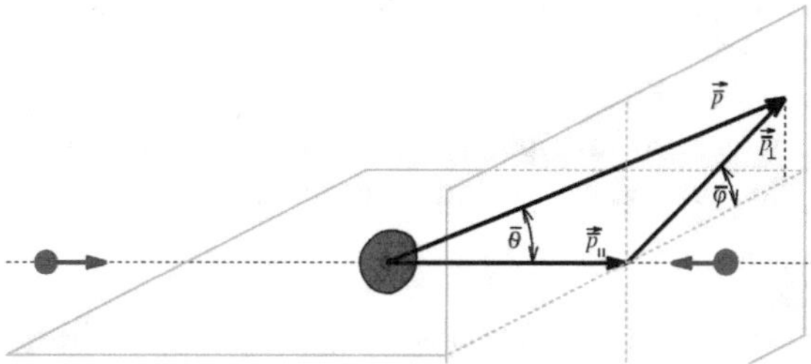

Fig. 18.8 The CM-frame momentum $\bar{\boldsymbol{p}}$ decomposition into the parallel $\bar{\boldsymbol{p}}_{\parallel}$ and perpendicular $\bar{\boldsymbol{p}}_{\perp}$ components with respect to the collision axis. Note the inclination angle θ of $\bar{\boldsymbol{p}}$ and the azimuthal angle φ of $\bar{\boldsymbol{p}}_{\perp}$

Table 18.1 Collisions of p+p (NA61/SHINE CERN-SPS collaboration, see Ref. [8]), at the beam momentum p_{beam} shown in first column, the corresponding CM-rapidity y_{CM} in 2nd column, the available p+p energy $\sqrt{s_{\text{pp}}}$ in 3rd column, the total multiplicity N_{π^-} of π^- in 4th column, and the energy cost of producing a π^- in the last column

p_{beam} [GeV/c]	y_{CM}	$\sqrt{s_{\text{pp}}}$ [GeV]	N_{π^-}	$\frac{\sqrt{s_{\text{pp}}}}{N_{\pi^-}}$ [GeV]
158	2.91	17.27	2.444 ± 0.130	7.1 ± 0.3
80	2.57	12.32	1.938 ± 0.080	6.4 ± 0.3
40	2.22	8.76	1.478 ± 0.051	5.9 ± 0.2
31	2.10	7.74	1.312 ± 0.069	5.9 ± 0.3
20	1.88	6.27	1.047 ± 0.051	6.0 ± 0.3

(SPS: super proton synchrotron).[8] These results were obtained for a stationary in laboratory frame proton target varying the momentum of the incoming proton, with momentum values shown in the first column in Table 18.1.

In order to compare inelastic production of particles for processes where the collisions in the laboratory occur at different energies, we shift the laboratory measured yields $d\pi^-/dy$ to the CM-frame by remembering that the CM-rapidity $\bar{y}$ arises from the laboratory measured particle rapidity y_p according to

$$\bar{y} = y_p - y_{\text{CM}} \, . \tag{18.56}$$

The value y_{CM} is obtained as described in Example 17.2: When both colliding particles have the same mass it is the average of the rapidities of the colliding particles.

For the target particle, here the proton p, being at rest in laboratory means that the CM-frame rapidity y_{CM} is half of the projectile rapidity $y_{\text{CM}} = y_{\text{p}}/2$. To obtain y_{p} for a given high value of incoming proton momentum $|\boldsymbol{p}_{\text{p}}| \gg m_{\text{p}}c$ we can use the second form in Eq. (15.23) simplified for $\boldsymbol{p}_{\text{p}\,\perp} = 0$ to read

$$y_{\text{p}} = \ln \frac{E_{\text{p}} + cp_{\text{p}\,\parallel}}{E_{\text{p}\,\perp}} = \ln\left(\sqrt{1 + \frac{\boldsymbol{p}_{\text{p}}^2}{m_{\text{p}}^2c^2}} + \frac{|\boldsymbol{p}_{\text{p}}|}{m_{\text{p}}c}\right) = \ln\left(\frac{2|\boldsymbol{p}_{\text{p}}|}{m_{\text{p}}c}\right) + \frac{1}{4}\frac{m_{\text{p}}^2c^2}{\boldsymbol{p}_{\text{p}}^2} + \dots \, . \tag{18.57}$$

The resulting values of y_{CM} are shown in the second column in Table 18.1.

The third and fourth column in Table 18.1 are also from Ref. [8]; the third column shows the total available p+p energy $\sqrt{s_{\text{pp}}}$, see Sect. 18.5. The fourth column shows the integrated multiplicity of produced π^- with the systematic error originating in the required extrapolation to domains of momentum $\boldsymbol{p}$ where the experiment

[8] N. Abgrall et al., "Measurement of negatively charged pion spectra in inelastic p+p interactions at $p_{\text{lab}} = 20, 31, 40, 80$ and 158 GeV/c." *Eur. Phys. J. C* **74** 2794 (2014).

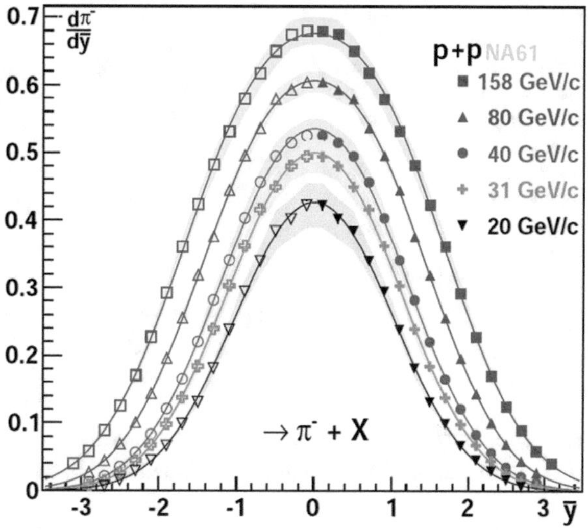

Fig. 18.9 Rapidity spectra of π^- produced in inelastic collisions of protons with momenta $p_{\rm lab} = 20, 31, 40, 80$ and 158 GeV/c, where target protons are at rest in laboratory. The other reaction products are indicated by 'X'. The statistical errors are smaller than the symbol size. The systematic uncertainties are indicated by the lightly shaded bands. Results of NA61/SHINE CERN-SPS collaboration. (Adapted from figure 20 in Ref. [8])

NA61 was not able to take data. The resultant rapidity distributions of inelastically produced π^- are shown in Fig. 18.9. Each of the distributions has been shifted from the laboratory frame to the CM-frame. The localization of observed particle yields in rapidity centered at CM-frame $\bar{y} = 0$ suggests a significant non-transparency, and a large energy loss of colliding matter. This remains an unexpected result for this elementary proton-proton collision system, albeit one must remember that these results are obtained in highly selected set of high multiplicity events.

We show in the last column in Table 18.1 the computed energy cost to produce the π^-. The growth of this value with energy indicates collision system 'transparency', with much of available energy remaining as longitudinal 'flow' of matter. It is of interest to observe that within error this value is a constant for the four lowest values of the projectile proton momentum.

The shape of the particle rapidity distribution is shown in Fig. 18.9. Such a distribution can be interpreted according to Hagedorn:[9] In a thermal model one assumes that in the average rest-frame of particle source the unknown momentum distribution of particles is only energy dependent, $\tilde{n}(\boldsymbol{p}) \rightarrow n(E)$. In the Hagedorn approach one explores a thermal Boltzmann (relativistic) distribution with a temperature T_H near to Hagedorn temperature.[10] The Hagedorn temperature describes the dissolution

[9]Rolf Hagedorn (1919–2003), a German theoretical physicist who, working at CERN, introduced the thermal model of particle production.

[10]J. Rafelski (Editor), *Melting Hadrons, Boiling Quarks: From Hagedorn Temperature to Ultra--Relativistic Heavy-Ion Collisions at CERN* (Springer Open 2016).

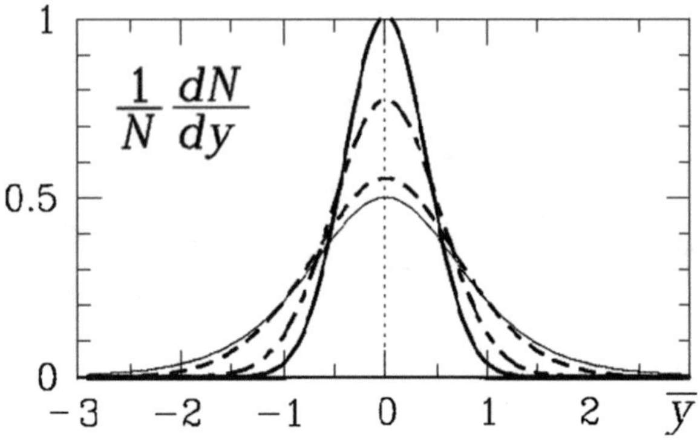

Fig. 18.10 Normalized (to unity) rapidity spectra of particles in a thermal model for $T = 160$ MeV. Thin line: massless particles; dashed line: pions ($m_\pi c^2 \simeq 140$ MeV); chain line; kaons ($m_K c^2 \simeq 500$ MeV); solid line: nucleons ($m_N c^2 \simeq 940$ MeV)

of matter into a quark-gluon plasma, and, conversely the creation of matter in the process of quark-gluon plasma 'hadronization' back into normal matter.

$$n(E) = Ae^{-E/T_H} = Ae^{-E_\perp \cosh y/T_H} , \tag{18.58}$$

where we used Eq. (15.19). The distribution parameter T_H is measured in units of energy and A is chosen to normalize the distribution to one particle. By definition we have

$$\frac{dN}{dy} = \int n(E)\frac{dp_\parallel}{dy}\, d^2p_\perp = A\int e^{-E_\perp \cosh y/T}\, E_\perp/c\, \cosh y\, d^2p_\perp , \tag{18.59}$$

where to obtain the last form we again used Eq. (15.19). Differentiating $E_\perp^2 = m^2c^4 + \boldsymbol{p}_\perp^2c^2$, we see that we can replace $d^2p_\perp \to 2\pi E_\perp dE_\perp/c^2$, where the factor 2π arises from azimuthal integration. Substituting $w = E_\perp \cosh y/T$ we obtain

$$\frac{dN}{dy} = \frac{A\,4\pi T^3}{c^3\cosh^2 y}\int_{m\cosh y/T}^{\infty} e^{-w} w^2 dw , \tag{18.60}$$

where the additional factor 2 accounts for the presence of both positive and negative $p_\parallel$.

The integral of Eq. (18.60) is elementary and the result is presented in Fig. 18.10. The π distribution (dashed line) has the desired shape fitting the lowest collision energy seen in Fig. 18.9; for higher energies this simple thermal Boltzmann distribution model does not suffice. The interacting particles and the produced secondaries retain some of the primary longitudinal momentum, which leads to the observed widening of the experimental rapidity distribution in Fig. 18.9. For further related discussion see Chapter III in Ref. 5 on page 287.

Further Study and Reading: This topic has attracted a lot of attention; a classic view on kinematic questions is offered by Hagedorn[11] himself. Thermal models of particle production are addressed by Letessier and Rafelski see Ref. [5] in Chap. 16. A dedicated YouTube channel *Creation of Matter* includes *original* and unseen footage from CERN history vault addressing the processes of quark-gluon plasma hadronization and matter formation, see https://www.youtube.com/channel/UCRYj1UAYtCfXQdLeeiENu8A. This includes in particular Hagedorn's two lectures describing his thermal SBM: Statistical Bootstrap Model and its development, presented in 1994 at a conference in *Divonne-les-Bains*, near CERN.[12]

[11] R. Hagedorn, *Relativistic Kinematics* (W. A. Benjamin, New York 1963 and 1973), reprinted by Literary Licensing, LLC (2012).

[12] For proceedings see: J.Letessier, H.H. Gutbrod, and J. Rafelski (Editors) *Hot Hadronic Matter: Theory and Experiment*, NATO-Advanced Study Institute Series B Vol. **346**, pp. 12+562, (Plenum Press 1995: Hardcover ISBN: 0-306-45008-9), (Springer 2012: eBook ISBN 978-1-4615-1945-4, Softcover ISBN 978-1-4613-5798-8).

Part VIII

SR-Tests and Open Questions

In Part VIII: We survey the status of experimental verification of SR which addresses the questions: How sure are we of the correctness and completeness of SR? Are there limits of validity regarding the principles of SR, or the results that follow? We than turn to discuss physics questions reaching beyond the SR framework with the overarching conundrum: the acceleration. We ask, in what way have these unresolved challenges already perturbed the SR edifice?

Introductory Remarks to Part VIII

Everybody expects that one day we will find a theoretical framework more inclusive than SR. In Chap. 19 we address both the status of current experimental searches for any discernible validity limit of SR. We then turn in Chap. 20 to acceleration which does not fit into the structure of SR. We ask how strong acceleration has to be in order to crack the foundations of SR.

Many experiments searching for the limits of SR have been performed. We sort these to address first the key postulates, and second the consequences and results of SR. In this context the Michelson-Morley experiment has been often presented as body-contraction confirmation. However, it is per se a test of the principle of relativity, the key postulate of SR. Another principle we look at is the constancy of the speed of light. The key results we consider are the changes related to material body: time dilation and the Lorentz-FitzGerald body contraction. We also address verification of the energy-mass equivalence.

We introduce the large family of tests of the Doppler effect. We recall that contrary to what is often stated, these efforts do not test time dilation. The Doppler effect relies within the SR primarily on Lorentz transformation properties of space-time coordinates. Beyond SR, the relativistic Doppler effect relies on the nature of light waves, and specifically, the observer independence of the light wave phase.

The cosmological history of the Universe and continued Universe expansion imply that ultimately such experiments must find a limit of validity of SR since space and time are not fully equivalent, even though the Minkowski space-time and the Lorentz coordinate transformation rely on this assumption. We therefore also offer a discussion of how time is different from space.

In Chap. 20 we consider how acceleration and SR interface. Einstein's inertial frame of reference and coordinate transformation formulation avoids the use of acceleration. Lorentz's approach to SR allows for a mechanism to transfer an observer from one inertial frame of reference to another. This requires (small) acceleration, and opens up the need to recognize acceleration in the context of SR. We recall the accelerated reference frame is a concept reaching beyond SR ideas. Specifically one may not LT into an accelerated rest-frame. We can only transform into an instantaneous comoving inertial frame.

Electromagnetic fields impart non inertial motion on charged particles. We discuss how EM forces available today compare to forces which could be considered to have an elementary unit strength. This is done by evaluating how field strength can be created from elementary quantities e, m, c; that is, the charge and mass of the particle subject to the field force, and the speed of light. Renowned examples of such an exploration of boundaries of the known domains of physics are the Planck units which we also describe. We show that current tests of SR are carried out in environments that experience acceleration many orders of magnitude below elementary 'critical' strength at which the instantaneous acceleration strength could become visible. We argue for future tests of SR in the presence of critical acceleration.

We show that acceleration of a body is defined locally: It manifests itself by the emission of radiation. For non-inertial charged particles the associated loss in energy generates through radiation emission an energy loss and a reaction force, acting just like vacuum-friction for accelerated particles. Such effects have not yet found their definitive understanding. We call the search for an extended conceptual framework of the laws of physics incorporating the vacuum friction force the 'acceleration frontier'. We describe experimental environments where this physics frontier can be explored.

Tests of Special Relativity 19

Abstract

We study the experimental verification status of two postulates on which SR was constructed: the principle of relativity and the constancy of the speed of light. We then describe tests of SR results: time dilation, body contraction, and mass-energy equivalence. We introduce the relativistic Doppler effect as a test of SR *per se*. We discuss how the fact that (cosmological) time, profoundly different from space could modify the SR framework. We close by introducing other related topics: the cosmic microwave background as a preferred 'Machian' reference frame, and the time dependence of natural constants.

19.1 Overview: Testing SR

In this book, one of the basic physics principles we adhere to is that 'absolute' inertial motion is unobservable. The MM experiment demonstrated this, and it remains the experimental test of the principle of relativity. MM type experiments continue today, setting ever better limits on our inability to observe the state of motion of an inertial apparatus. We will describe the current status of these experiments in the following Sect. 19.2.

SR further relies on the constancy of the speed of light postulate made by Einstein, see Eqs. (2.6) and (6.12): We can test if the speed of light is independent of the observer's, or the source's, state of motion. The MM type experiments rely on this hypothesis, but there is more to this principle that can be tested. For example, we can ask if the speed of light is independent of the wavelength of light. We discuss related matters in Sect. 19.3.

We then consider in Sect. 19.4 how the three key results of SR regarding properties of material bodies – time dilation, Lorentz-FitzGerald body contraction, and mass-energy equivalence ($E = mc^2$) – stand up to experimental tests.

J. Rafelski, *Modern Special Relativity*, https://doi.org/10.1007/978-3-030-54352-5_19

In Sect. 19.5 we look again at the Doppler effect. We recall that it depends on the Lorentz transformation of coordinates, and the property of light that allows it to transport the information about the state of motion of the emitter to the observer, where this information is decoded. There is no test of time dilation inherent to this process. Since the Doppler effect depends primarily on the validity of the Lorentz transformation which itself is a result that we derive from the principles that govern SR, this ultra precise experiment tests globally the SR properties and the nature of light.

In the final Sect. 19.6 we address how properties of time challenge SR: As the tests of SR advance, one day we will reach the precision that will probe the difference between space and time. This is so since the Universe had a beginning in time; however, the Universe is not bounded in space. On the cosmological space-time distance scale this directly conflicts with Minkowski space-time symmetry. We also look at the Machian idea of a preferred inertial frame of reference, defined by stars and today by cosmic microwave background (CMB). We will explain why this idea is compatible with SR. With time being different from space on the cosmological scale, we can ask if the so-called 'natural' constants which are constants in space, are also constant in time.

19.2 The Michelson-Morley Experiment Today

The Michelson-Morley experiment tests the principle of relativity by showing that absolute motion of the apparatus in the Universe is undetectable. In the past century we have seen highly significant improvements in the Michelson-Morley experiment; see the time line shown in Fig. 19.1. At the beginning this was due to the continuing efforts of Michelson himself, resulting in improved limits shown in the upper left corner in the figure. Note that this figure addresses the other postulate of SR, the constancy of the light speed, presenting analysis of the possible variation of velocity of light $\Delta c/c$, or signal frequency $\Delta\nu/\nu$ arising from the presumed 'æther drag'.

In the year 2009 incarnation (bottom right corner Fig. 19.1), the MM experiment involves an apparatus that floats on a thin cushion of air above a 1.3 metric ton granite table. It comprises two optical cavities, essentially pairs of mirrors that reflect light back and forth. They are both about 8.4 cm long and at right angles to each other. Because the cavities are slightly different in length, they have slightly different resonant frequencies.

A laser beam is split into two beams, one for each cavity. The frequencies of the beams are then tuned to resonate in their respective cavities using acousto-optic modulators. If the speed of light were different in different directions, it would affect the resonant frequencies of the two cavities, which could then be detected as a shift in the beat frequency once the apparatus is rotated and two beams are recombined.

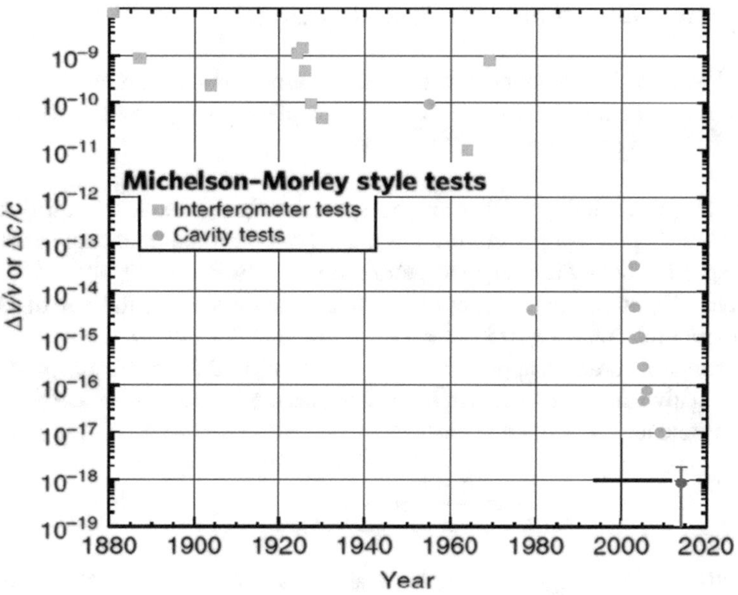

Fig. 19.1 Signal interferometry experiments of Michelson-Morley type set limits on the variation of velocity of light $\Delta c/c$, or, signal frequency $\Delta \nu/\nu$ due to the presumed 'æther drag'. The experimental progress evolution representation follows Ref. 2

This experiment, carried over a period of 13 months, shows the local isotropy of light propagation at the 10^{-17} level.[1]

An even more recent experiment[2] comprised two sapphire cylinders with their crystal axes aligned orthogonally. Resonance modes were excited in each sapphire. The difference in frequency was recorded while the apparatus was continuously rotated with a 100 s period. Evaluation of one year of data allowed the setting of a limit on the variation of frequency $\Delta \nu/\nu$ or, equivalently, light velocity $\Delta c/c$ at $9.2 \pm 10.7 \times 10^{-19}$ (95% confidence interval); this is the latest experimental result seen in bottom right in Fig. 19.1. The authors expect further improvement in the precision of this new method.

The verification that light propagation is independent of the direction of motion of the laboratory in the Universe has now shown the absence of 'æther drag' speed at the level of 3Å/s; that is, atomic size per second.

[1] C. Eisele, A. Y. Nevsky and S. Schiller, "Laboratory Test of the Isotropy of Light Propagation at the 10^{-17} Level," *Phys. Rev. Lett.*, **103**, 090401 (2009).

[2] Moritz Nagel, Stephen R. Parker, Evgeny V. Kovalchuk, Paul L. Stanwix, John G. Hartnett, Eugene N. Ivanov, Achim Peters, and Michael E. Tobar, "Direct terrestrial test of Lorentz symmetry," *Nature Comm.* **6** 8174 (2015).

19.3 How Constant Is the Speed of Light?

In Part I we introduced the speed of light measurement and we learned that the speed of light is now defined in terms of the definition of the meter, Eq. (2.2). Therefore, a test of the speed of light means that we test:

1. The frequency independence of the speed of light – explosive astrophysical events at high redshift were used to place limits[3] on the fractional variation in the speed of light with frequency, $\Delta c/c < 6.3 \times 10^{-21}$.
2. Speed of light independence of the motion of the source – for the case of a binary pulsar a limit[4] characterized by $c' = c + kv,\ k < 2 \times 10^{-9}$ was obtained.
3. Elementary particles approach speed $v \rightarrow c$ as their energy increases; for ultrarelativistic particles it is of interest to check this limit: We note the expected SR difference

$$\frac{\delta v}{c} \equiv \frac{c - v}{c} = \frac{c}{(c + v)\gamma^2} \rightarrow \frac{1}{2\gamma^2}\,. \tag{19.1}$$

Neutrinos are the lightest particles available and thus offer greatest sensitivity for testing Eq. (19.1) by providing the largest laboratory accessible Lorentz factor $\gamma = E/mc^2$ at given energy E. A laboratory experiment with neutrinos[5] obtained $(c - v)/c \simeq 0.3 \pm 1.5 \times 10^{-6}$. However, given the employed γ-factor $\simeq 10^{11}$ and given the neutrino energy available, a much better limit can be attained. Thus this type of experiment can be better carried out by observing the relative time of flight of cosmic TeV-energy neutrinos and gamma rays.[6] One expects an improvement by 13 orders of magnitude. In any case, the result of Ref. 5 presents an absolute test that the speed of neutrinos is bounded by the speed of light. In Discussion 11.1 we have considered the relevance of this measurement.

In conclusion: The speed of light c is for all frequencies, and everywhere in the visible space-time, the same and independent of the state of motion of source or observer. It provides an upper bound on the speed of all material particles, including the neutrinos. While the test of constancy of speed of light cannot be

[3] B. E. Schaefer, "Severe Limits on Variations of the Speed of Light with Frequency," *Phys. Rev. Lett.* **82**, 4964 (1999).

[4] K. Brecher, "Is the Speed of Light Independent of the Velocity of the Source?" *Phys. Rev. Lett.* **39**, 1051 (1977).

[5] P. Alvarez Sanchez et al. [BOREXINO Collaboration], "Measurement of CNGS muon neutrino speed with Borexino," *Phys. Lett. B* **716** 401 (2012).

[6] Jun-Jie Wei, Xue-Feng Wu, He Gao, and Peter Mészáros, "Limits on the neutrino velocity, Lorentz invariance, and the weak equivalence principle with TeV neutrinos from gamma-ray bursts," *JCAP*, **2016**, (August 2016).

easily disentangled from other potential violations of SR principles, the precision level reached demonstrates that the principle of relativity and the constancy of the speed of light as extremely well tested foundations of SR.

19.4 Tests of SR Material Body Properties

Test of Time Dilation

Setting aside the Doppler effect, which as we repeatedly discussed is not related to time dilation, there are several types of measurement that deserve mention:

1. The first measurement of time dilation was achieved in the study of the proper lifespan of the muon. This was historically the first test of time dilation. In 1941 Rossi and Hall (see Ref. [2]) measured a variation of the muon decay probability for two different energy groups of cosmic muons comparing their survival at different altitudes: a mountain location at 3240 m, and a plain location at 1616 m, resulting in a height distance differential of 1624 m. This differential assures a larger travel time from a muon formation point at yet higher altitude around 10 km over the Earth's surface. These were the results: in their conclusions Rossi and Hall say (note: mesotrons=muons)

   > The experiments described …(are) confirming the view that the absorption anomaly is caused by spontaneous decay of mesotrons in the atmosphere. A value of the (muon) proper lifetime …is deduced.

   and in the abstract

   > The softer (slower) group of mesotrons was found to disintegrate at a rate about three times faster …in agreement with the theoretical predictions based on the relativity change in rate of a moving clock.

2. Many experiments of similar type followed; one deserving mention was carried out in the presence of an acceleration of magnitude $10^{18} g$ due to the presence of bending magnets.[7] We will argue in Chap. 20 that the acceleration that the muon experienced is insignificant and the verification of SR in this environment is not surprising. This measurement was made at CERN, carried out with relativistic positive and negatively charged muons injected into a circular storage ring. This experiment has shown that the time dilation factor of magnitude $\gamma = 29.3$ works with a fractional error of $\simeq 10^{-3}$.
3. Time dilation can of course be tested by comparing the clock-time of a traveler to laboratory time. Such a direct time dilation experiment was carried out by Hafele and Keating in 1972. They compared clocks sent on plane trips around the Earth

[7] J. Bailey, et al., "Measurement of relativistic time dilation for positive and negative muons in a circular orbit," *Nature*, **268** 301 (1977); and "Final report on the CERN muon storage ring…," *Nucl. Phys. B* **150**, 1 (1979).

towards East and West.[8] They obtained a result in agreement with both SR and GR. GR enters the discussion, since a terrestrial time dilation measurement uses clocks operating under the effect of the Earth gravitational potential, which is greater on the Earth's surface than in flight. Both SR and GR time dilation effects are of comparable magnitude and in the experimental environment considered both effects are additive. Moreover, the SR effect is speed dependent, and the GR effect is gravitational potential dependent, in that it depends on the distance from the Earth's surface. By performing a difference experiment it is possible to isolate both effects. We discussed this measurement in Discussion 4.2.

4. The GPS satellite signal offers another opportunity to test directly time dilation, see Ref. [1] in Chap. 4. Present day laboratory experimental results provide stronger limits on SR breaking parameters when compared to GPS and thus allow us to rely on the GPS network for precise positioning.

Observation and Test of Lorentz-FitzGerald Body Contraction

We recall that the nil outcome of the MM experiment means that the transverse and longitudinal optical paths with reference to the direction of motion are equal. Within the framework of SR this is interpreted invoking the Lorentz-FitzGerald body contraction. However, there are other interpretations possible that we chose not to discuss in this book, given that SR is an established framework. The point of this remark is that the MM experiment establishes, independent of how it actually works, the principle of relativity. The MM experiment is eliminating the material æther hypothesis which involves in particular a variation in speed of light, thus variation of optical path, see also Sect. 19.2.

So we return to the question, has the Lorentz-FitzGerald body contraction ever been measured? Literature search reveals one manuscript reporting an expected Lorentz-modification of the EM field properties.[9] This result is, however, not about a change of a material body properties.

No measurement of the Lorentz-FitzGerald body contraction has been reported. This is so, since in order to compare material body of finite extent in different speed conditions involving relativistic motion we must accelerate these over a distance constrained by laboratory dimensions. This requires forces at a scale capable of inflicting a mechanical material deformation rivaling for small speeds (as compared to c) the Lorentz-FitzGerald body contraction. Thus it is difficult to determine if

[8] J.C. Hafele, and R.E. Keating, "Around-the-World Atomic Clocks: Predicted Relativistic Time Gains." *Science* **177** 166 (1972); and "Around-the-World Atomic Clocks: Observed Relativistic Time Gains." *Science* **177** 168 (1972).

[9] A. Laub, T. Doderer, S. G. Lachenmann, R. P. Huebener, and V. A. Oboznov, "Lorentz Contraction of Flux Quanta Observed in Experiments with Annular Josephson Tunnel Junctions," *Phys. Rev. Lett.* **75** 1372 (1995).

the body contraction is due to mechanical action between the involved bodies or a change in body length.

A proposal was recently made to observe the Lorentz-FitzGerald body contraction as it occurs in an intentionally destructive acceleration process,[10] where the experimental outcome depends on the presence of actual material body contraction. Consider a high intensity laser shot across a nanometer thick material foil. The lightest charged particles, electrons (possibly accompanied by some protons), emerge from the foil debris accelerated to a very high speed. Initially the spatial distribution of the foil debris retains some of the extreme effect of the Lorentz-FitzGerald body compression of the foil that has been accelerated to a relatively high rapidity. Considering the large magnitude of the Lorentz-FitzGerald body compression, one cannot interpret this in terms of inertial compression effect inflicted on the foil by the EM light wave. A natural probe of the high density electron cloud is another laser pulse impacting the foil remnant just a little later from the other side: a coherent back scattering 'mirror' reflection process can only occur for sufficiently high electron density.

Test of $E = mc^2$

Measurement of the mass M_1 before and M_2 after a nuclear transmutation induced by neutron capture with mass m_n yields the relative precision[11]

$$\frac{(M_1c^2 + m_nc^2) - (M_2c^2 + \Delta E)}{M_2c^2} = (-1.4 \pm 4.4) \times 10^{-7} , \tag{19.2}$$

where ΔE is the observed energy of binding as carried away by a gamma ray.

▶ **Insight: 19.1 Doppler effect test of SR** Many tests of relativity are carried out by verifying the Doppler shift. We recall that in SR light does not age and does not have a body size that could be contracted.

So, what is being tested? To answer this question we recall that in order to derive in Sect. 13.4 the Doppler shift formulas for both the collinear and the general form of the Doppler shift:

(i) We required that the phase of the light wave is not dependent on the state of motion of either the source or the observer, see Chap. 13. Since the observer is arbitrary we assumed that the light wave is the same for any inertial observer. This requirement implements the principle of relativity for light waves. Note that beyond the

[10] J. Rafelski, "Measurement of the Lorentz-FitzGerald Body Contraction," *Eur. Phys. J. A* **54** no.2, 29 (2018).

[11] S. Rainville, et al., "A direct test of $E = mc^2$," *Nature*, **438** 1096 (2005).

principle of relativity we were claiming that the light wave phase is not observer dependent.

(ii) We used the LT and the aberration angle obtained using LT to evaluate the changes to frequency and wavelength of the light wave needed to assure that the phase of light is the same for any inertial observer.

In particular in context of experiments where the source and observer are located near to each other in the laboratory, the Doppler shift measurement is sometimes presented as a test of time dilation. However, there is no in principle difference between such experiments and the study of distant stars. In all instances, photons moving with the speed of light have a proper time zero; i.e. being attached to the light cone, they do not age, thus there is nothing that can be time dilated. The fact that the Lorentz γ-factor seen in the Doppler shift corrects the original nonrelativistic Doppler formula does not imply that there is a test of time dilation. This factor enters practically all relativistic results and only a few relate to time dilation.

It is of course a convenient 'short cut' to explain to students that all we need to do is to correct the nonrelativistic Doppler by time dilation. A lecturer arguing in this way must, however, remember that this argument amounts to the introduction of material æther: The nonrelativistic Doppler effect relies on sound wave propagation in air.

If one day we were to discover a modification of Lorentz coordinate transformation, in order to preserve the theoretical framework that resembles SR, there would need to be also a modification of both time dilation and body contraction phenomena so that there is consistency between body properties and coordinate transformations. Even so, we need to decouple the test of coordinate transformation, clearly inherent to the Doppler effect test, from the test of body properties, see Sect. 19.4.

19.5 Doppler Effect and Tests of the Lorentz Coordinate Transformation

We present SR Doppler effect tests. As noted, these do not explore time dilation; what is foremost tested is the precise format of the relativistic coordinate transformation: Introducing the reciprocity of the SR Doppler effect in Sect. 13.4, we demonstrated that SR Doppler effect relies on the Lorentz coordinate transformation. The key logic error in the interpretation of the SR Doppler experiments as 'time dilation' tests is that the researchers forget that the experimental Doppler experiment outcome, if and when SR is valid, has nothing to do with the state of motion of the light source. For further discussion see Insight 19.1: *Doppler effect test of SR*.

Irrespective of this circumstance, the importance of the SR Doppler shift as a potential test of the relativity principle and of special relativity cannot be over-emphasized. Einstein recognized the pivotal role of the SR Doppler frequency shift, see Ref. [6] in Chap. 13, and defended his theory in the face of erroneous experimental results presenting effects 10 times larger compared to his SR predictions.

A SR Doppler effect based experiment was carried out as early as 1938 by Ives and Stilwell, and a discussion of the historical and physics background was given in a review by Ives about a decade later.[12] Today the sensitivity of Ives and Stilwell experiment is greatly enhanced by boosting between the laboratory frame and the rest-frame of the radiation absorber/emitter which in contemporary experiments is kept in a relativistic particle storage ring. This relativistic particle motion allows precision improvement compared to the stationary MM experiment. This is due to the large ratio of stored particle speed compared to the velocity of the Earth with respect to a hypothetical æther.

For this reason scions of the Ives and Stilwell experiment have seen renewed interest. The method employed consists of two complementary 'parallel' but opposite direction Doppler effects. When the two SR Doppler frequency shifts, one parallel and the other antiparallel along the line of sight, are combined in a product

$$\nu_+\nu_- = \nu_0\gamma(1+\beta)\nu_0\gamma(1-\beta) = \nu_0^2 \,, \tag{19.3}$$

according to SR the result should not depend on relative velocity at all. Therefore we call this a null-method test of Doppler.

Should the Lorentz coordinate transformation have a modified form, the cancellation in Eq. (19.3) is expected not to be satisfied in a wide range of models. This allows us to explore and set limits on modifications of special relativity where the invariant proper time is written in the form

$$d\tau = \left(1 + \alpha_{\mathrm{RMS}} V^2/c^2 + \alpha_{\mathrm{RMS},\,2} V^4/c^4 + \ldots\right)\sqrt{dt^2 - dx^2/c^2} \,, \tag{19.4}$$

where V is not only the local relative velocity but includes some unknown absolute velocity component. The parameter α_{RMS} is known as the Robertson and Mansouri, Sexl (RMS) parameter. We include a further term $\alpha_{\mathrm{RMS},\,2}$ arising from a possible velocity dependence of the RMS α-parameter.

To measure α_{RMS} a comparison of the two shifts measured at different velocities $\beta_i = v_i/c$ by the null-method was carried out. We refer to the Heidelberg[13] time dilation experiment for details. A possible deviation is extracted from the two measured frequencies obtained for the two respective values of β

$$\nu_{01} = \sqrt{\nu_+^{(1)}\nu_-^{(1)}} = 546{,}466{,}918{,}577 \pm 108\,\mathrm{kHz}, \qquad \beta_1 = 0.030, \quad v = 9000\,\mathrm{km/s}\,,$$
$$\nu_{02} = \sqrt{\nu_+^{(2)}\nu_-^{(2)}} = 546{,}466{,}918{,}493 \pm \ 98\,\mathrm{kHz}, \qquad \beta_2 = 0.064, \quad v = 19{,}200\,\mathrm{km/s}\,,$$

[12] H. Ives, "Historical Note on the Rate of a Moving Atomic Clock," *Journal of the Optical Society of America* **37** (10) 810 (1947).

[13] S. Reinhardt, et al., "Test of Relativistic Time Dilation with Fast Optical Atomic Clocks at Different Velocities," *Nature Physics* **3**, 861 (2007).

which result limits the time dilation modification

$$\alpha_{\rm RMS} = 4.8 \pm 8.5 \times 10^{-8} \,. \tag{19.5}$$

We see that this experimental result is consistent with zero at a high level of precision, confirming SR.

Yet another experiment was carried out at a more relativistic speed $\beta = 0.338$. Li^+ ions were stored in the ESR storage ring at the GSI laboratory at Darmstadt.[14] Given the value of $\beta^2 = 0.114$ employed, 30 times larger compared to the Heidelberg experiment, a stringent limit on $\alpha_{\rm RMS,\,2}$ could also be set.

$$\alpha_{\rm RMS,\,2} < 1.2 \times 10^{-7} \,. \tag{19.6}$$

More generally the result constrains the RMS parameter as a function of speed:

$$\alpha_{\rm RMS} \equiv \left| \frac{\alpha_{\rm RMS}\, V^2/c^2 + \alpha_{\rm RMS,\,2} V^4/c^4 + \ldots}{V^2/c^2} \right| \leq 2.0 \times 10^{-8} \,. \tag{19.7}$$

A later experiment[15] based on another physical method confirms the GSI result Eq. (19.7), with some further improvement

$$\alpha_{\rm RMS} < 1.1 \times 10^{-8} \,. \tag{19.8}$$

This effort is based on a light fiber connected four ultrahigh precision laser clock network located in France, Germany, and the United Kingdom.

These limits on $\alpha_{\rm RMS}$ are two orders of magnitude better compared to analysis of the GPS signal (see Ref. [1]). This means that the GPS position triangulation precision can be pushed to much higher values without concern that some not yet discovered breaking of SR will have a measurable impact.

19.6 Time

Time Is Different from Space

In special relativity theory one of the most important insights is about time, which like space, is a part of space-time, an equal partner as we have reported in the

[14] C. Novotny, et al., "Sub-Doppler Laser Spectroscopy on Relativistic Beams and Tests of Lorentz Invariance," *Phys. Rev. A* **80**, 022107 (2009); B. Botermann, et al., "Test of time Dilation Using Stored Li^+ Ions as Clocks at Relativistic Speed," *Phys. Rev. Lett.* **113**, 120405 (2014).

[15] P. Delva, et al., "Test of Special Relativity Using a Fiber Network of Optical Clocks," *Phys. Rev. Lett.* **118**, 221102 (2017).

opening of this book, see Sect. 1.2. However, there are a few subtle problems:

1. We treat time from the beginning differently: Aside from providing the additional coordinate needed to identify world events, it is a parameter that characterizes the evolution of the physical system.
2. While we can influence where we are in space, we cannot control our motion in time. The situation with time is like the circumstance of a tree; it moves along in space where the Earth motion takes it. In analogy, we coast along the time axis with the evolving Universe.
3. In an expanding Universe time and space differ and thus special relativity cannot apply at a macroscopic scale. The domain of application of special relativity is limited to the local flat static Minkowski space-time. The word local, or better, 'tangential', is to be understood in the same sense that we can use Euclidean geometry on a tangential surface on Earth.

Since time and space are united in Minkowski space-time in SR, the question is not if, but at which level of precision, the limits of validity of SR will be reached. To amplify the contrast between reality and SR let us look closer at the last of the above listed differences.

Time in Cosmology

Most physicists agree that our Universe is not static. Thus, we can measure how much time has passed since the Big-Bang. A measurement of the cosmological time, the time since the 'beginning', requires that we know a frame of reference defining the cosmic 'rest'-frame – cosmological time is defined as the proper time of our Universe. In the early Universe all the matter was on average at rest.

As the Universe expanded and cooled, the plasma of ions and electrons combined and neutral atoms formed, and thus the Universe became transparent to the cosmic background light. This light is observed today as the cosmic microwave background (CMB), a topic we return to just below. In this way, the distant-in-time essentially homogeneous matter defines the natural cosmological reference frame. This cosmic frame of reference can be determined by measurement of Earth's motion with respect to the CMB. We discussed this introducing the MM experiment, see Fig. 3.2.

The age of the Universe is determined today in a global analysis of the fluctuations of CMB photons straying away from complete spatial homogeneity. In 2012 the WMAP spacecraft collaboration reported the age of the universe to be:[16] 13.772 ± 0.059 billion years. The Planck spacecraft collaboration reported[17]

[16]C.L. Bennett, et al., "Nine-Year Wilkinson Microwave Anisotropy Probe (WMAP) Observations: Final Maps and Results," *Astrophys. J. Suppl.* **208** 20 (2013).

[17]Planck Collaboration (2015), "Planck 2015 results. XIII. Cosmological parameters (Table 4 on page 31)," *Astronomy&Astrophysics* **594**, A13 (2016).

in 2016: 13.813±0.038 billion years (revised[18] in 2018 to 13.830±0.037). All these values are obtained within the ΛCDM model of the Universe, called that way for the dark energy/Einstein constant Λ, and the letter acronym for 'Cold Dark Matter'.

We see that as a matter of principle we can tell where in time we are today. Therefore time and space are very different: We can pinpoint absolute cosmological time as an absolute coordinate. We have no way to introduce absolute spatial coordinates. Moreover, in SR time has no beginning and no end; it is the same forever. However, we now learned that time has a beginning, therefore the translation in time symmetry cannot work for arbitrarily long times.

We note that present day laboratory experiments carried out to test relativity usually last somewhat longer than a year, which is 10 orders of magnitude shorter compared to the lifespan of the Universe. Given the usual experimental precision the cosmological history of the Universe is irrelevant on a one year time scale and principles of SR apply. However, there is another context where SR may be challenged soon. The rapid advances in atomic optical clock technology may allow us to discover limits to SR generated by cosmological expansion; a first step into this technology is seen in Ref. 15. The Hubble expansion of the Universe, in SI units, is described by the Hubble parameter $H \simeq 2.2\text{–}2.3 \times 10^{-18}$/s (see Refs. 16, 17, 18); the current clock precision is less than an order of magnitude below this benchmark. Moreover, nuclear optical clocks are on the horizon which would breach the Hubble parameter limit.[19]

Cosmic Microwave Background Frame of Reference

The cosmic microwave background radiation (CMB) are the ashes of the Big-Bang in the form of radiation dating back to the hot Universe era during which atoms were formed. This radiation, discovered in 1964,[20] fills the entire present day universe with a thermal $T_{CMB} = 2.7255(6)$K microwave (cm-size wavelength) black body spectrum. If we lived outside the Earth atmosphere, we would see everywhere the primordial CMB photons that were already present when atoms were formed.

This is how it happened: the cooling of the primordial Universe due to the expansion allowed for ion-electron binding about 372,000 years after the Big-Bang. The Universe became transparent to radiation. Ongoing expansion means that the

---

[18] Planck 2018 results. VI. Cosmological parameters, (Table 2, page 15) arXiv:1807.06209.

[19] J. Thielking et al., "Laser spectroscopic characterization of the nuclear-clock isomer ^{229m}Th," *Nature* **556** no.7701, 321 (2018).

[20] Arno Penzias and Robert Woodrow Wilson were awarded the Nobel Prize for Physics 1978 for the discovery of CMB. CMB radiation was predicted in 1946, albeit at T = 50 K, by Georg Gamov (1904–1968), a Russian-American theoretical physicist, student of A. Friedman of cosmological **FLRW** model fame, best known for the explanation of nuclear alpha decay via quantum tunneling, and his work on star evolution and the early Universe, also the author of "Mr Tompkins' adventures" series of popular-scientific books.

ambient temperature today is much lower due to a 1000-fold cosmological redshift. CMB photons were originally formed at energies corresponding to temperatures $T \simeq 3000\,\mathrm{K}$ when ions and electrons filling the early Universe recombined. In the absence of free electrons the Universe became transparent to radiation.

The importance of the CMB radiation background is due to the fact that it provides a 'natural' frame of reference which can be universally recognized. A moving observer sees a Doppler-deformed, see Chap. 13, CMB radiation spectrum. This means one can recognize relative motion in the Universe with respect to the CMB rest-frame of reference. We keep in mind that the equivalence of all observers inertial with respect to the CMB rest-frame, and that our knowledge of which observer is at rest with respect to CMB does not violate the principle of relativity, equally well we could imagine measuring velocities with respect to any other inertial 'beacon-observer' in the Universe. The CMB is just a very convenient 'beacon' we can refer to.

Constancy of Natural Constants

Given that all scales of distance expand as the Universe ages we can ask how if it is possible that the principles of SR and in particular the universality of the speed of light are compatible with the changing Universe. More generally, in the above consideration of the CMB reference frame, we implicitly made the assumption that the laws of physics and thus also atomic emission lines were the same eons ago as they are today.[21]

Attempts to find time variation of natural constants continue. The limit on relative variation[22] of the fine-structure constant is

$$\alpha \equiv \frac{e^2}{4\pi\epsilon_0\hbar c} = \frac{1}{137.035399074(44)}, \quad \frac{1}{\alpha}\frac{d\alpha}{dt} = \frac{(-0.70 \pm 2.10) \times 10^{-17}}{\mathrm{yr}}, \tag{19.9}$$

and for the proton-to-electron mass ratio

$$R_{p/e} \equiv \frac{m_p}{m_e} = 1836.15267245(75), \quad \frac{1}{R_{p/e}}\frac{dR_{p/e}}{dt} = \frac{(0.2 \pm 1.1) \times 10^{-16}}{\mathrm{yr}}, \tag{19.10}$$

both obtained assuming a constant rate of change during the lifespan of the Universe.

[21] H. Fritzsch, *The Fundamental Constants: a Mystery of Physics*, World Scientific Publishing Company, Singapore (2009); J.-Ph. Uzan, "Varying Constants, Gravitation and Cosmology," *Living Rev. Relativity* **14** 2 (2011); X. Calmet, M. Keller, "Cosmological Evolution of Fundamental Constants: From Theory to Experiment" *Mod. Phys. Lett. A* **30** 1540028 (2015).

[22] M.S. Safranova, "The Search for Variation of Fundamental Constants with Clocks," *Ann. Phys. (Berlin)* **531**, 1800364 (2019).

This shows that we can proceed assuming that natural constants are constant, and consider properties of the early Universe using the physics laws determined today. We believe accepting Occam's Razor argument introduced in Discussion 2.2 that this applies also to the universal speed of light c. However, there is continued discussion regarding the question if time variation of speed of light c could in principle be observable.[23] Using key word "VSL theory" the interested reader can explore the large trail of **v**ariable **s**peed of **l**ight theories discussed predominantly in the context of cosmology.

Further reading: A comprehensive review of searches for new physics with atoms and molecules is available[24] Advances in setting limits regarding validity or violation of, *inetr alia*, Lorentz symmetry, parity violation, CPT theorem, searches for spatiotemporal variation of fundamental constants, tests of general relativity, and the equivalence principle are presented.

[23] M. J. Duff, "Comment on time-variation of fundamental constants" https://arxiv.org/pdf/hep-th/0208093 (updated November 2016).

[24] M.S. Safranova, D. Budker, D. DeMille, D.F. Jackson Kimball, A. Derevianko, and Ch.W. Clark, "Search for new physics with atoms and molecules," *Rev. Mod. Phys.* **90**, 025008 (2018).

Acceleration

20

Abstract

Accelerated observers are incompatible with the foundations of SR. Nevertheless, we need to address forces acting on a material body. We discuss how and why SR without a consistently introduced acceleration is probably an incomplete theory. We define 'strong' acceleration, and consider the in-principle theoretical and experimental challenges that its study brings to the SR context. This is the 'acceleration frontier' domain of modern physics. We describe several experimental environments allowing the search for new principles related to acceleration.

20.1 Accelerated Motion

Acceleration is a concept that reaches beyond the scope of SR; Einstein's 1905 foundation of SR avoids any mention of acceleration. However, without an acceleration that takes material bodies from one inertial frame of reference to another, many arguments in this book, e.g. made in the context of the Lorentz-Bell formulation of SR, would make little sense. Even if we believe all material bodies are subject to forces and thus acceleration, how do we know that there is acceleration at all? How and why can we distinguish inertial and accelerated systems? As we will argue, objectively acceleration exists. Therefore we should one day find a way to properly incorporate it into an extended SR framework.

All SR observers are inertial; they are in uniform motion. Within SR all these observers are equivalent. On the other hand, an acceleration marks a body in a specific way. For example, in an (de)accelerated vehicle we note the presence of inertial forces; and in general an accelerated electrically charged body radiates. The inertial observer reporting on this situation does not. Even though the acceleration-related effects are sometimes weak in the instant and thus difficult to observe

J. Rafelski, *Modern Special Relativity*, https://doi.org/10.1007/978-3-030-54352-5_20

experimentally, compounded acceleration effects mark a body in a distinct way as we noted discussing the time dilation of a traveler.

In order to account for forces, in SR acceleration is introduced in a way that resembles the treatment of velocity. However, as we just have seen, the idea that acceleration is relative, just like velocity, is untenable. We must not apply to acceleration the SR treatment of velocity, i.e. exchanging the role of an (accelerated) body with an (inertial) observer. We conclude that application of the principle of relativity to acceleration in the realm of SR is inconsistent with the relativity principle.

We can always tell 'who' is accelerated. We already noted that a charged body experiences in general emission of radiation. This means an accelerated particle radiates; that is, emits energy, and thus experiences radiation friction. As the strength of acceleration increases, it would seem that such a body will lose its energy of motion. This is clearly impossible as we are not able to define absolute motion of a body. What is possible is that the radiation reaction force can become as strong as the applied force that generates the acceleration. Such a force balance model theory does not, however, exist today. We conclude: A fully consistent theory of electromagnetic force awaits discovery.

In order to minimize confusion it is important not to confound different types of radiation processes. For example, a quantum jump by an inertial atom emitting radiation is possible when it 'transmutes' from one stationary state into another. Such a jump is not related to an acceleration process; it is akin to the radioactive decay of a metastable nucleus.

Many will have heard how the cutting edge of new physics discovery opportunities is now divided into 'energy' and 'intensity' frontiers, according to the question whether the available particle beam excels in its energy content, or, in the number of particles that can be used. The discussion we present leads to the recognition of a third frontier offering foundational physics discovery potential, the 'acceleration frontier'.

In the following we will discuss challenges related to the acceleration frontier. We discuss if acceleration exists and when we enter regime of strong acceleration. We discuss why understanding the physics of acceleration is an important element in characterizing particle motion.

20.2 The (Missing) Acceleration in SR

We recognized that an accelerated material body differs from any inertial reference observer; thus we could not extend the relativity principle to accelerated motion. We now recapitulate a few insights we have presented in this book that rely on acceleration:

- It is the traveling-accelerated 'twin' who, upon rejoining her base, notices a shorter increment of proper time measured by her clock compared to the proper time ticking in the base station;

- It is the train departing a station that will, upon reaching relativistic speed, fit into the usually not long enough tunnel.

Some body properties, for example time dilation, are associated with the process of acceleration. Thus the complete physics contents of physical processes is acceleration dependent. The Lorentz coordinate transformation cannot address the relative clock time of both observer and traveler. The clock time is unique and depends on acceleration history which marks the traveling body, see Sect. 12.2. In agreement with these considerations, clock experiments show that only the traveling twin stays younger. The theoretical discussion offered by Langevin, see Ref. [18] in Chap. 2, and time dilation experiments clearly confirm that there is no relativity (reciprocity) of time dilation.

We recall that the spatial separation between two events defining the two ends of a body is consistent with the Lorentz-FitzGerald body contraction for an observer measuring in her reference frame at the same time, see Chap. 9. We can imagine that the contraction-memory device we described in Chap. 10 is present in each material body. This allows us to record the history of contraction effects and to infer that the accelerated body is contracted. Given the memory about the acceleration process, the reversal along the line: "I see you contracted, so you see me contracted" is not allowed. We emphasize that akin to time dilation effect for train travelers our contraction-memory device will present the same outcome; the accelerated train accumulates in the contraction-memory device the memory of its acceleration history.

The compatibility of body properties and Lorentz coordinate transformations is assured, thus in the words of John Bell: "*Einstein approach is perfectly sound, and very elegant and powerful*". However, only with the knowledge of the history of motion can we assure which of the two inertial bodies is contracted. Thus the material bodies are marked by acceleration both for purpose of time dilation as well as body contraction.

While different states of inertial motion can be accommodated in SR without mentioning how acceleration allows to transit between these two states, we need to introduce as described in Chap. 10 the gentle acceleration to assure that our understanding of time dilation and the Lorentz-FitzGerald body contraction is complete. Once the history of acceleration is incorporated as an integral part of the extended SR framework there is no space for errors, omissions, and paradoxes.

The present day SR does not yet incorporate explicitly the notion of accelerated material bodies or the history of acceleration. However, every book on SR written today incorporates the EM Lorentz force, see the following Chap. 21 in this book. Introducing such a force and forgetting to account for the history of acceleration that the particle experiences cannot lead to a consistent theory: It suffices to inject a metastable muon into an EM field domain to see the challenge of understanding properly the modification of distance that the muon can travel subject to EM force.

On the other hand, the introduction of a force and thus acceleration is on first sight superfluous in the context of the two preeminent theoretical frameworks, general relativity (GR) and quantum mechanics (QM), which both are foundational theories without acceleration. Here is how this happens:

GR: By recognizing the equivalence of inertial and gravitational mass, Einstein could describe the force of gravity as inertial motion in a curved space time. A free-falling body in a gravity field does not experience any force. In a second step, Einstein's GR equations introduce space-time deformation by the action of the energy-momentum of all particles, allowing this back-reaction modification of the space-time geometry.

The free-falling particles follow trajectories in a curved space-time showing the properties we associate with motion executed according to Newton's equation including the effect of gravity force: The so-called geodesic motion equation allows us to interpret the particle motion as occurring in the presence of the (quasi-)force of gravity originating in the deformation of space time. This quasi acceleration is coordinate dependent. However, in a locally Cartesian-flat space coordinates we recover the physics of gravity force with appropriate extension to relativity (GR).

Important: In the classical point-particle model of matter all particles are free-falling and there is no (real) gravitational force. The force of gravity we feel is due to the surface normal force that keeps us from free-falling into the Earth.

QM: Quantum mechanics (QM) describes particles on a (sub) atomic length scale. The electron in a quantum orbit is stationary, and not accelerated; hence there is no continuous emission of radiation; an electron does not spiral into the nucleus. A modification needed to assure the stability of atoms against acceleration-induced radiation motivated the discovery of the atomic quantum theory in the first place. To clarify, when an excited atomic state decays, emitting a photon, this situation is like an unstable particle undergoing a radioactive decay; the emission of such radiation is not a result of accelerated particle motion.

These two examples describing flagship theories suggest that acceleration should perhaps not exist as a matter of principle. In this author's opinion this was the predominant opinion in the past. However, one can take the point of view that our pride in GR and QM is founded in the fact that these conceptual frameworks are consistent, while we have yet to account for acceleration in for example EM theory, and more generally, whenever mechanical devices with acceleration capability are introduced.

While in the past searches for theoretical extension of physical laws diminished the need for the existence of forces (an account of these efforts would fill more than this book to the rim), we take here firmly a different attitude: Acceleration exists as a rocket engine that relies on *actio* = *reactio* principle demonstrates. The issues are: (a) How to incorporate acceleration in a consistent fashion at an elementary level of foundational physics; and (b) How to stage experimental work that would help to discover and prove the appropriate theoretical formulation.

20.3 For the Existence of Acceleration

With GR and QM we developed fundamental theories without acceleration as a physical feature and yet our daily experience contradicts this. Thus we are justified in asking if there is acceleration at all – and this is synonymous with the question – *how do we know if and when a body is accelerated?* This is not a new question. Newton explored the meaning of acceleration in his widely quoted water bucket experiment and more than 300 years later this is still a good place to start. Consider a suspended rotating water bucket: At first the water in the bucket does not rotate with the bucket. When the water begins to rotate, the surface of the water becomes concave.

Newton concluded that by observing the curving of water surface in a rotating bucket, one can determine that the bucket is being rotated and not the rest of the world around the bucket. Anyone can check Newton's conclusion with this experiment: Place a water bucket in the immobile center of a merry-go-round and while enjoying a ride see if the immobile bucket water surface turns concave just because you are circling around; surely it will not. However, some will still claim that this demonstration is flawed since the experiment was not carried out in an inertial frame in empty space.

Luckily Newton supplemented his bucket experiment with what we today call 'thought experiments'. He considered two connected weights in empty space, far from everything. When they rotate around, these weights will create an outward force pulling the rope tight. We expect the rope will be subject to a measurable tension. In empty space there is nothing fixing a reference for the rotation nor determining the rope tension, except, as Newton argues, for space itself. These ideas have provided the foundation for Newton to propose the existence of absolute space.

This was the foundation of Mechanics, when about 150 years ago, Ernst Mach returned to the question raised by Newton. Mach proposed, see Ref. [4], the 'Universe at rest', defined by the fixed stars, to be Newton's absolute space. We can always measure acceleration against some known inertial frame, and the rest-frame of all fixed stars is such an inertial reference frame. Since all inertial frames form an equivalent set, this is a suitable choice.

Reference to a particular inertial frame proposed by Mach is not in conflict with special relativity. This differs from a preferred origin in Newton's absolute space required in the heliocentric worldview from the age of Newton, a notion that is manifestly incompatible with SR: This singles out a point in space and thus violates the principle of relativity. The choice of an inertial frame made by Mach makes good sense for the merry-go-round experiment in space, far and away from anything that could interfere with motion. By looking at starlight we can determine who is rotating and who is not. Now the entire Universe dictates the answer to who is accelerated.

However, this also prompts us to consider the following question: How does Newton's water bucket obtain this information from the universal inertial reference frame defined by light-years distant stars? In a book where we often check if a causal sequence of events is guaranteed, it seems doubtful that this information is transferred from distant locations of fixed stars, assuring that the just set in motion water bucket will curve its surface.

We thus infer that the information about who or what is accelerated has to be available locally. Within the laws of classical physics this is a difficult if not impossible to fulfill requirement. However, within the realm of quantum physics we have accepted that the empty space has quantum structure; we refer to the structured quantum vacuum. The structured quantum vacuum[1] provides a realization of ideas of Langevin and of Einstein about the relativistic æther described in the essay Sect. 2.3.

This topic that is now of importance transcends this book's scope; even so, we want to briefly address the pivotal insight. One can look at the situation as follows: The same quantum vacuum state properties that assure quark confinement, and predominantly define the inertial mass of matter, also define and provide locally the inertial frame of reference, allowing the recognition of acceleration at microscopic scale.

The quantum vacuum permits the introduction of the concept of absolute acceleration, in the sense proposed by Langevin, Ref. [18] in Chap. 2, in the context of his consideration of time dilation. Considering which experiment shows this unequivocally, we turn without much ado to the radiation. Emission of radiation by charged particles shall be the experimental evidence for accelerated motion.

Our insights about acceleration can be stated as follows:

1. Acceleration can be measured against any inertial reference frame. The cosmic reference frames, such as Mach's fixed star frame of reference, or, equivalently, the cosmic microwave background (CMB) frame – i.e. the frame in which the CMB spectrum is isotropic – are usually considered.
2. There is no relativity of acceleration between accelerated bodies and inertial observers. Their roles must never be exchanged (at least as long as a more fundamental theoretical framework accounting for acceleration is missing).
3. Acceleration is defined at the body location in space-time.
4. The structured quantum vacuum frame provides the required local point of reference and is generally accessible to all material bodies.
5. Accelerated charged bodies can emit radiation, unlike inertially moving bodies.

To conclude: Acceleration is a real physical quality, which we explore in the following pages.

[1]For a generally accessible discussion see J. Rafelski and B. Müller, *The Structured Vacuum: Thinking about Nothing*, Harri Deutsch, (Thun 1985); hard copy edition out of print, E-republication in 2006.

20.4 Small and Large Acceleration

We have acknowledged the presence of acceleration when we considered the Bell-rocket example, Chap. 10, and often when discussing relativistic effects that impact the property of a body. We considered this acceleration with reference to the inertial observer's frame of reference and introduced it in order to be able to move a body from one to another inertial system in small steps. The magnitude of the acceleration is assumed to be as small as needed; yet crucially, we have always been able to tell which of the compared bodies has been accelerated.

This distinction is needed since the accelerated and inertial observers are not equivalent in the context of SR. The principle of relativity does not apply to their relative motion. The concept of acceleration is not relative – unlike the case of two inertial observers where the measured velocity is relative. In this book we have argued that we can tell which body is inertial and which body is accelerated.

In order to tell what 'small acceleration' means, we need to have a reference benchmark. We look for natural scales of length and time to construct a ratio that describes the acceleration. The situation with a material body is different from that of an elementary particle as in the latter case the choice of a scale is usually not in question, and is related either to the mass or, if available, some internal structure parameter. Once we agree which is the 'natural' size L, the large 'critical' acceleration is by dimensional counting

$$a_{\mathrm{cr}} \equiv \frac{c^2}{L} \,. \tag{20.1}$$

We first consider a material object where we choose as the natural length scale L the macroscopic thickness in direction of acceleration. For an object with about $L^{\mathrm{matter}} = 10^{-3}$ m, the top laboratory acceleration that can be created today[2] is $a \simeq 10^{11}\,\mathrm{m/s^2}$. Considering Eq. (20.1), this is $10^{-9} a_{\mathrm{cr}}^{\mathrm{matter}}$. It is possible to make the reference critical acceleration $a_{\mathrm{cr}}^{\mathrm{matter}}$ bigger by choosing a smaller value of L^{matter} in Eq. (20.1). This will lead to an even smaller value of achieved acceleration compared to the reference critical value. Within the limits of present technology we have

$$a^{\mathrm{matter}} < 10^{-9} a_{\mathrm{cr}}^{\mathrm{matter}} \,. \tag{20.2}$$

Turning to elementary particles, the natural scale of the length is the (2π-reduced) Compton wavelength $\lambdabar_C = \lambda_C/2\pi$ and the natural unit of time is τ_C, the time needed by light to travel across this distance $\lambdabar_C$ where

$$\lambdabar_C = \frac{\hbar}{mc} = 386.16\,\mathrm{fm}, \qquad \tau_C = \frac{\lambdabar_C}{c} = 1.288 \times 10^{-21}\,\mathrm{s} \,. \tag{20.3}$$

[2] R.W. Lemke, M.D. Knudson, J.-P. Davis, "Magnetically driven hyper-velocity launch capability at the Sandia Z accelerator," *Int. J. of Impact Eng.* **38** 480 (2011).

The above values refer to an electron. For a particle of mass m we adopt

$$a_{\mathrm{cr}} \equiv \frac{c^2}{\lambda_{\mathrm{C}}} = mc^2\frac{c}{\hbar} \; . \tag{20.4}$$

For an electron,

$$a^e_{\mathrm{cr}} = 2.327 \times 10^{29}\ \mathrm{m/s^2} \; . \tag{20.5}$$

An electron moving with speed of light in an accelerator across a bending magnetic field of 4.4 T, a typical value, experiences, according to the Lorentz EM force we will study in the following, an acceleration

$$a^{\mathrm{particle}} = \frac{e}{m} c\mathcal{B} < 10^{-9} a^{\mathrm{particle}}_{\mathrm{cr}} \; . \tag{20.6}$$

This confirms that most extreme conditions we encounter in daily laboratory life correspond to a rather small elementary 'nano'-acceleration.

To summarize: The most extreme acceleration that can be achieved today in a laboratory is at least eight orders of magnitude below acceleration that one could call 'strong'. Furthermore, extremely small acceleration is present when testing and exploring SR. For example, SR is tested in the Earth's surface reference frame, where we are 28 orders of magnitude (or more; for heavier than an electron particles, see Eq. (20.4)) below what one could argue is 'strong' acceleration.

Equipped with this insight we now understand why it is possible to achieve very high precision verification of SR in non-inertial environment of Earth's surface. We return to this 'small' acceleration context in the following Chap. 21 where we consider the relativistic version of Newton's law and the EM Lorentz force.

▶ **Insight: 20.1 SI and Gauss EM units** The SI-unit of the electrical field $\boldsymbol{\mathcal{E}}$ is 'volt/meter'; and of the magnetic field is $\boldsymbol{\mathcal{B}}$: T = tesla = volt sec/meter^2. According to Eq. (21.10) a charged particle of charge q experiences in the presence of a field $\mathcal{E} = 1\ \mathrm{V/m}$ the same strength EM-force as when it moves with the speed $v = 1$ m/s in the presence of a magnetic field of strength $\mathcal{B} = 1$ T. For relativistic particles $v \to c$, the $\mathcal{B} = 1$ T produces a force equivalent to a force caused by an electrical field of magnitude $\mathcal{E} = 3 \times 10^8\ \mathrm{V/m}$.

This factor 300 million is introduced on purpose in the SI-unit system. The Lorentz force law Eq. (21.10) is written more naturally with the dimensionless factor $\boldsymbol{v} \to \boldsymbol{v}/c$. The c denominator can be traced back to the inclusion of c in $t \to ct$. In the cgs+Gauss-unit system one keeps the c in the force definition and the asymmetry in force units in presence of relativistic motion disappears. In the SI-unit system this factor c is absorbed into the product $e\mathcal{B}$. We thus discover the first two of the following rules of transcription of equations between cgs+Gauss and SI, and a third relation based on the way we write the Coulomb force is also stated below

$$e\mathcal{E}|_{\rm cgs} \to e\mathcal{E}|_{\rm SI}\,, \quad e\mathcal{B}|_{\rm cgs} \to c\, e\mathcal{B}|_{\rm SI}\,, \quad e^2|_{\rm cgs} \to \left.\frac{e^2}{4\pi\epsilon_0}\right|_{\rm SI}\,. \tag{20.7}$$

In the SI-system the Maxwell equations are written exactly in the same format as seen in Maxwell's work. The SI-unit system introduces dimensioned vacuum properties: the vacuum permittivity ϵ_0 and vacuum permeability μ_0 (ignoring anisotropic field induced response)

$$\mathcal{D} = \epsilon_0 \mathcal{E}\,, \qquad \mathcal{B} = \mu_0 \mathcal{H}\,, \qquad \epsilon_0\mu_0 = \frac{1}{c^2}\,. \tag{20.8}$$

The last relation defines the propagation speed of Maxwell waves. To some (including Maxwell), Eq. (20.8) implies that the vacuum is another form of matter. However, as Einstein explained, the vacuum is not material in an ordinary way – see Essay in Sect. 2.3. Therefore ϵ_0 and μ_0 do not appear in the cgs+Gauss unit system which made an effort to remove the æther from Maxwell equations, note here Ref. [1] in Chap. 21.

To obtain physical results we will use electron properties as follows

$$\frac{e^2}{4\pi\epsilon_0 m_e c^2} \equiv r_e\,, \quad m_e c^2 = 0.510999\,\text{MeV}$$
$$\to \frac{e^2}{4\pi\epsilon_0} = 1.4403 \times 10^{-9}\,\text{eV m}\,, \tag{20.9}$$

see Insight 16.1 on *Elementary energy units* for the electron classical radius r_e, and for the energy unit 'MeV,' in which we present the electron rest energy. In this book ϵ_0 always appears in the form shown in Eq. (20.9); thus we will neither need to know its numerical value, nor that of $\mu_0 = 1/\epsilon_0 c^2$. This shows that ϵ_0 (and similarly μ_0) is not an independently known natural constant; it appears due to the need to reconcile independent choices made for the elementary units, a situation similar to the speed of light, c, see Sect. 2.1. $\epsilon_0 \neq 1$ and $\mu_0 \neq 1$ does not imply that the EM field vacuum has material properties.

20.5 Achieving Strong Acceleration

In Sect. 20.4 we introduced 'small acceleration' into SR comparing to elementary strong acceleration Eq. (20.4). It is natural to ask when and how can such acceleration arise, beginning with Newton's gravitational force. Since this force increases as distance decreases we need some very short and elementary unit of length. The shortest known elementary length is the Planck length

$$\ell_{\mathrm{P}} = \sqrt{\frac{\hbar G_{\mathrm{N}}}{c^3}} \equiv 1.6162\,10^{-35}\,\mathrm{m}\,, \tag{20.10}$$

twenty orders of magnitude below the proton size. Today it is thought that near ℓ_{P}, gravity connects with quantum physics. Planck noted introducing ℓ_{P} in an appendix[3] that such a new elementary scale is of interest as it arises in consequence of the introduction of the quantum of radiation $\hbar$ and connects with Newton's Gravity.

We consider the Newton gravity acceleration by any particle of mass m at the Planck distance ℓ_{P}:

$$a_{\mathrm{P}} = \frac{G_{\mathrm{N}} m}{\ell_{\mathrm{P}}^2} = mc^2 \frac{c}{\hbar} \equiv a_{\mathrm{cr}}\,. \tag{20.11}$$

We note the cancellation of Newton's constant G_{N} and recognize a_{P} to be the natural unit of acceleration a_{cr} introduced in Eq. (20.4). Thus where quantum and gravity phenomena are expected to meet, we find that acceleration at the critical strength is present. In this sense this natural acceleration unit is connected to Planck's natural scales but it does not require G_{N}. Moreover it is in the realm of the possible that when we achieve natural unit strength acceleration, some new phenomena appear, reminding us of the deeper connection that Eq. (20.11) brings to mind.

Thus there is a profound foundational interest in achieving unit-1 acceleration experimentally. We recall that the EM force on a charged particle is proportional to the applied electric field. An electrical field required to generate this critical acceleration for the lightest elementary particle, the electron, would have the so-called 'Schwinger critical' field strength

$$E_{\mathrm{cr}} = \frac{m a_{\mathrm{cr}}}{e} = \frac{m^2 c^3}{e\hbar} = \frac{mc^2}{e\lambda\!\!\!^{-}_{\mathrm{C}}} = 1.323 \times 10^{18}\,\frac{\mathrm{V}}{\mathrm{m}}\,. \tag{20.12}$$

To evaluate the value E_{cr} in SI units, i.e. V/m, we insert for the elementary constants in Eq. (20.12) the SI values m[kg], c[m/s], e[C], $\hbar$[Js] and compute (note 'C' stands here for 'Coulomb'). However it is by far simpler to remember that

[3] M. Planck,"Über irreversible Strahlungsvorgänge," (translated: "On irreversibility of radiation processes") *Sitzungsberichte der Königlich Preußischen Akademie der Wissenschaften zu Berlin* **5** 440 (1899), see added comment (see added page 480).

$mc^2 = 0.511\,\text{MeV}$ and to cancel the e between numerator and denominator. The numerical coefficient in Eq. (20.12) is recognized as 511/386, the last number being $\lambda\!\!\!^{-}_C$ in fm, see Eq. (20.3), while the power comes from remembering that 'M' stands for mega = 10^6 and that fm = 10^{-15}m. The corresponding critical magnetic field is obtained dividing the SI value of E_{cr} by $c = 3\ 10^8$ m/s to find the SI value of the magnetic field, see the Insight 20.1: *SI and Gauss EM units*

$$B_{\text{cr}} \equiv \frac{ma_{\text{cr}}}{ce} = \frac{m^2c^2}{e\hbar} = 4.414 \times 10^9\text{T}\,. \tag{20.13}$$

The Schwinger critical electric Eq. (20.12) field was introduced as the condition when the spontaneous conversion of applied electric field into electron-positron pairs (more generally of particle pairs of mass m) is so abundant that the field is rapidly neutralized. Said differently, the field energy 'materializes' – achieving such an electric field in static condition in the laboratory is thus, in principle, not possible. However the equivalent static critical magnetic field can be in principle created and maintained on a long time scale because it does not lead to pair production.

At this point we note that there is actually a hierarchy of critical accelerations. For example we could ask what force do we need to apply to rip apart a hydrogen atom? The benchmark for atomic size is the Bohr radius, larger than the Compton wavelength $\lambda\!\!\!^{-}_C$, the quantum electron size, by a factor corresponding to the inverse fine-structure constant, $\alpha^{-1} \simeq 137.036$. We use Eq. (20.1) with the Bohr radius to obtain the acceleration, and hence the EM field required to 'critically' accelerate atoms. This field is of the magnitude

$$E_{\text{cr}}^{\text{Atom}} \equiv \frac{\alpha\, ma_{\text{cr}}}{e} = \frac{\alpha\, m^2c^3}{e\hbar} = \frac{mc^2}{e(\alpha^{-1}\lambda\!\!\!^{-}_C)} = 0.966 \times 10^{16}\,\frac{\text{V}}{\text{m}}\,, \tag{20.14}$$

and

$$B_{\text{cr}}^{\text{Atom}} \equiv \frac{m\alpha a_{\text{cr}}}{ce} = \frac{\alpha m^2c^2}{e\hbar} = 3.221 \times 10^7\text{T}\,. \tag{20.15}$$

On the other hand, fields that can rip apart an elementary particle must address distances at the scale of classical electron radius r_e, a length scale that is by the same factor $\alpha^{-1} \simeq 137.036$ smaller compared to $\lambda\!\!\!^{-}_C$, requiring stronger fields

$$E_{\text{cr}}^{\text{Part}} = \frac{m\alpha^{-1}a_{\text{cr}}}{e} = 1.813 \times 10^{20}\frac{\text{V}}{\text{m}}\,, \qquad B_{\text{cr}}^{\text{Part}} = 6.05 \times 10^{11}\text{T}\,. \tag{20.16}$$

A magnetized neutron star, a 'magnetar', could offer an astrophysical laboratory where such magnetic fields Eq. (20.16) are possibly present, and thus protons and neutrons are ripped apart into quarks and gluons. Our estimate provides the magnitude of fields required to create in relativistic heavy ion collisions the primordial phase of matter, the quark-gluon plasma, made of locally mobile constituents of protons and neutrons. Such strong fields are certainly available according to our understanding of strong interactions in terms of quantum chromodynamics.

The laboratory acceleration achieved today is unobservable viewed from the perspective of the scale considerations above: the largest bending magnetic field in accelerators is typically only a few Tesla; compare Eq. (20.6) for resultant acceleration. In the foreseeable future, static fields of 10–20 Tesla can be envisaged. The laboratory bending magnetic field is thus 8 orders of magnitude below the critical acceleration field Eq. (20.13). The only reason that we can measure and actually in certain circumstances use the emitted radiation is that we accumulate on an elementary scale a tiny effect over a macroscopic time and distance scale.

We now mention a few experimental environments in which it is perhaps possible to achieve and study critical acceleration both at present and in the near future:

Supercharged (quasi) nuclei The value $E_{\rm cr}$ Eq. (20.12) is actually small compared to fields available at the nuclear surface. However, atomic electrons are not near the nuclear surface and are delocalized. Furthermore the Schwinger critical field at the nuclear surface is found only in a small volume. Clearly, just reaching $E_{\rm cr}$ is not a key criterion for the appearance of new physics phenomena. We ask the question: What nuclear charge Ze is needed so that at distance $\lambda\!\!\!^{-}_{\rm C}$ the field becomes critical? This condition can be also written as

$$\frac{Ze}{\lambda\!\!\!^{-2}_{\rm C}} = E_{\rm cr} = \frac{m^2c^3}{e\hbar} \to Z\alpha = 1 \,, \quad \alpha = \frac{e^2}{\hbar c} = \frac{1}{137.036} \,. \tag{20.17}$$

Thus if we can assemble a nuclear charge of $Z > 137$ we could test in an atomic physics experiment the condition of critical acceleration. This situation has been thoroughly explored and described in the context of relativistic quantum mechanics.[4] Since the nuclei have finite size, the actual value of 'critical charge' $Z_{\rm cr} \simeq 171$ is recognized. At this point the quantum vacuum, the ground state, becomes charged. The experimental realization of this situation arises in the study of slow large nuclei, called 'heavy ion', $v \ll c$ collisions – but these collisions need to be fast enough to allow mutual approach to within a few nuclear diameters for $Z_1 + Z_2 > Z_{\rm cr}$. The experimental study of positron production in such heavy ion collisions carried out in the 1980s was hampered by significant nuclear reaction backgrounds. Present day perspective on these results is available.[5]

[4] W. Greiner, B. Müller and J. Rafelski, *Quantum Electrodynamics of Strong Fields*, Springer, (Heidelberg, New York 1985, 2015).

[5] J. Rafelski, J. Kirsch, B. Müller, J. Reinhardt, W. Greiner, "Probing QED Vacuum with Heavy Ions," *FIAS Interdisc. Sci. Ser.: New Horizons in Fundamental Physics* 211 (2017).

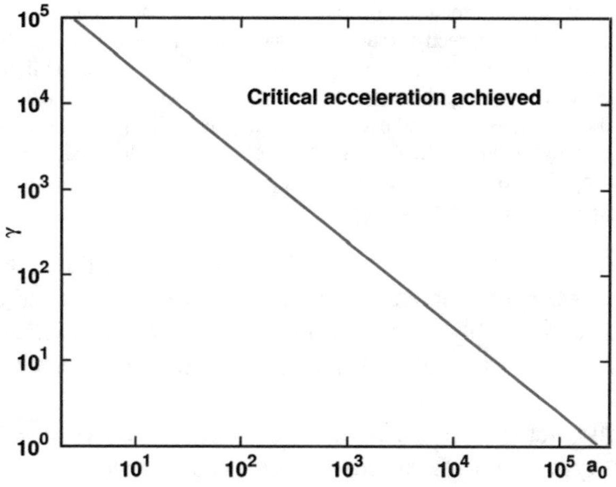

Fig. 20.1 Domain of critical acceleration in collision of $\gamma = E/mc^2$ electrons with an intense laser pulse with normalized amplitude a_0. (Adapted from Ref. 6, see text)

Ultra-Intense laser pulse collision with relativistic electrons It has been noted that when a relativistic electron is hit by a laser pulse, the required characteristics of supercritical force can occur. We need a combination of: (a) a relatively large electron Lorentz factor γ, with (b) intense laser pulses with a large normalized amplitude factor a_0, see Chap. 22. In Fig. 20.1 we consider the collision of a high energy $E = m_e c^2 \gamma$ electron with an intense laser pulse with normalized amplitude a_0 and frequency $\hbar\omega = 4.8\,\mathrm{eV}$. Above the line, the electron experiences critical acceleration.[6]

The central domain of Fig. 20.1 is where present day technology makes experiments possible: What comes handy is that such intense laser pulses are capable of accelerating electrons[7] to energies in the GeV range ($\gamma > 1000$), as we will discuss in Chap. 22. Thus one can envisage in an experimental environment the collisions of laser pulses with laser pulse generated electron pulses. Several laser pulse laboratories where the required conditions and experimental beam lines are available within the context of ELI infrastructure:[8]

> The Extreme Light Infrastructure (ELI) is a Research Infrastructure of Pan-European interest for experiments on extreme light-matter interactions at the highest intensities,

[6]Adapted from: J. Rafelski and L. Labun, "Critical Acceleration and Quantum Vacuum," *Modern Phys. Lett. A* **28** 1340014 (2013).

[7]Review Article: Tae Moon Jeong and Jongmin Lee "Femtosecond petawatt laser," *Ann. Phys.* (Berlin) **526** 157 (2014).

[8]p71 of *European Strategy Forum Report 2016 on Research Infrastructures*, prepared by the StR-ESFRI project.

shortest time scales and broadest spectral range. ELI is based on three sites (known as pillars, located in the Czech Republic, Hungary and Romania) with complementary scientific profiles, and the possible implementation of a fourth pillar, the highest intensity pillar, dependent on on-going laser technology development and validation. The fourth pillar laser power is expected to exceed that of the current ELI pillars by another order of magnitude, allowing for an extended scientific program in particle physics, nuclear physics, gravitational physics, nonlinear field theory, ultrahigh-pressure physics, astrophysics and cosmology (generating intensities exceeding 10^{23} W/cm^2).

At these facilities experimental study of a radiation reaction effect should be possible – a charged particle interacting with strong EM-fields experiences 'vacuum friction', dissipating rapidly its energy in the form of radiation.[9] A foundational consistent description of this novel physics frontier has not been achieved.

Ultrarelativistic heavy ion collisions When big nuclei collide head-on in a relativistic heavy ion collider experiment, the duration time of the collision in the laboratory frame is $\Delta t = R/c$, where R is the nuclear radius. This is the maximum time; the actual increment of time could be a fraction of Δt, but we want to be conservative in estimating the value of acceleration a

$$a \simeq \frac{1}{m}\frac{\Delta p}{\Delta t} \simeq \frac{c\Delta \sinh y_p}{\Delta t} \equiv a_{\mathrm{RHIC}}\, \Delta(\sinh y_p) \ , \tag{20.18}$$

where $\Delta(\sinh y_p)$ is the shift in rapidity of a particle within the stopping time. The scale on which we measure acceleration is

$$a_{\mathrm{RHIC}} \equiv \frac{c}{\Delta t} = \frac{c^2}{R} \simeq 2\times 10^{31}\,\frac{\mathrm{m}}{\mathrm{s}^2} \ . \tag{20.19}$$

This result is by a factor 100 greater than the values shown in Eq. (20.5) and corresponds to fields needed to rip nucleons, Eq. (20.16). As these considerations suggest, when nucleons from the incoming nucleus are stopped so that rapidity shifts significantly from a value prior to collision, critical acceleration must have been achieved, ripping nucleons apart. In such process there is creation of a large particle multiplicity. This finding coincides with the formation of quark-gluon plasma (QGP) at CERN in the year 2000, see Fig. 20.2. The abundant particle production accompanying achievement of critical acceleration could be the source of the unimpeded formation of this new state of matter.

In the above discussion the origin of large rapidity shift, that is, the stopping of colliding matter, was not discussed as it is an experimental fact. In general this

[9] Y. Hadad, L. Labun, J. Rafelski, N. Elkina, C. Klier and H. Ruhl, "Effects of Radiation-Reaction in Relativistic Laser Acceleration," *Phys. Rev. D* **82**, 096012 (2010).

Fig. 20.2 CERN announces the discovery of quark-gluon plasma in a press-release on 10 February 2000

is assumed to be caused by strong interactions governing the motion of quarks in nucleons. However, the EM forces of two relativistic colliding large nuclear charge nuclei have a rivaling strength, and the advantage of not being confined as strong forces are. The origin of large stopping of colliding quarks is one of the research frontiers in the field of relativistic heavy ion collisions.

Part IX

Lorentz Force and Particle Motion

In Part IX we explore: The Lorentz force describing the interaction of charged particles with EM fields is based on experimental experience. Therefore we study examples of the relativistic Lorentz force dynamics of charged particles in (1) constant magnetic (2) constant electric fields (3) the case of the 'Coulomb' radial force, including orbital precession and (4) the case of an electron riding an electromagnetic laser plane-wave field. We introduce the variational principle for the Lorentz force, obtain the Hamiltonian, and discuss conservation laws.

Introductory Remarks to Part IX

When exploring in the following the dynamical motion of charged particles subject to the Lorentz-force we continue to adhere to SI units. The method of conversion from Gauss units often used in classic texts and the reasoning that prompts use of these units are described.

We introduce force and acceleration such that the well-known properties of relativistic momentum and energy are respected. Upon generalization of Newton's force we explore many unexpected properties of relativistic force and acceleration. We show new physics content that the effective inertial resistance of a particle depends on its state of motion parallel or perpendicular to the applied force.

The (Lorentz-)force law describing the interaction of charged particles with electromagnetic (EM) fields is based primarily on experimental experience. The Lorentz force is most often presented in terms of electric and magnetic fields. We consider, example by example, how the well-known nonrelativistic Lorentz force dynamics generalizes to relativistic motion.

Solutions for particle motion for the four cases of general interest dominate the contents: (1) constant magnetic, (2) constant electric fields, (3) the case of the 'Coulomb' radial force, and (4) the case of an electron riding an electromagnetic plane-wave, a problem motivated by current interest in laser acceleration of charged particles.

The first and simplest example is the relativistic motion of a charged particle in a constant magnetic field, followed with the more complex case of motion in a constant electric field. The difficulty that arises in the study of electrical fields is that even if we choose initial particle motion normal to the electrical field, particle velocity turns into the field direction resulting in a catenary particle path.

We introduce a relativistic least action principle, finding a conserved canonical momentum. We identify another conserved quantity, the Hamiltonian (energy) showing a constraint that bears the signature of a 4-dimensional energy-momentum invariant. We exploit the canonical momentum and energy conservation in the study of charged particle orbits in the presence of the Coulomb potential of an atomic nucleus. We point out parallels with known relativistic quantum energies.

In order to prepare the study of particle motion in intense laser fields we introduce in more detail the electromagnetic plane wave and set up the case of an electron surfing on the electromagnetic plane wave. We show how conserved quantities constrain particle motion. The analogy with laser particle acceleration prompts consideration of the unit-language used in the high intensity laser pulse community.

We then solve for particle motion in the presence of laser fields of arbitrary strength, explicitly considering both linear and circular wave polarization. We demonstrate a significant change in particle dynamics when a threshold intensity of the plane wave is exceeded. This opens up the realm of new science domain of *relativistic optics*.

Acceleration and the Lorentz Force 21

Abstract

We introduce the concept of acceleration and force, and the force law describing the interaction of charged particles with EM fields. Example by example, we explore the well-known nonrelativistic Lorentz force dynamics generalized to relativistic motion. We consider charged particle in a constant magnetic field, and in a constant electric field. We solve the classical Coulomb problem in a radial electric field.

21.1 Newton's Second Law

The introductory form of relativistic Newton's second Law relies often on a formal generalization to the 4-dimensional notation not introduced in this volume. This step prompts a simultaneous relativistic reformulation of the Lorentz force in terms of the field tensor. However, these steps are not necessary, and as we have discussed in Chap. 20, may be incomplete: Even in its most beautiful 4-dimensional, well-accepted format, the Lorentz force does not account for some of the phenomena inherent to the interaction between particles and EM fields, such as the loss of energy due to radiation by an accelerated charged particle, mentioned often in Chap. 20.

Therefore, in this volume we always restrict the discussion of Newton's second law to the usual 3-dimensional form, and we continue with the usual well-tested 3-dimensional format introducing a minimal extension to relativity. We recall the key equation of Newtonian dynamics,

$$\frac{d\boldsymbol{p}_{\mathrm{I}}}{dt} = m\frac{d\boldsymbol{v}}{dt} = m\boldsymbol{a} = \boldsymbol{F}_{\mathrm{nr}} \,.$$

J. Rafelski, *Modern Special Relativity*, https://doi.org/10.1007/978-3-030-54352-5_21

A natural generalization of this equation is the use of relativistic momentum. In this way we obtain the relativistic form of the Newton force

$$\frac{d\boldsymbol{p}_{\mathrm{I}}}{dt} = \frac{d(m\gamma\boldsymbol{v})}{dt} \equiv \boldsymbol{F} \,. \tag{21.1}$$

The subscript 'I' (for inertial, or kinetic) reminds us that the momentum entering the Lorentz force is not the canonical momentum which we will introduce in Sect. 21.3 and which is used without indexing in this book, as is common, compare e.g. all quantum physics books.

The generalization Eq. (21.1) assures in the context of SR the validity of the work-energy theorem. According to Sect. 15.1, Eqs. (15.3), (15.4), and (15.5) we have

$$dE = d\boldsymbol{p}_{\mathrm{I}} \cdot \boldsymbol{v} = \boldsymbol{F}\, dt \cdot \boldsymbol{v} = \boldsymbol{F} \cdot d\boldsymbol{x} \,. \tag{21.2}$$

The presentation of force Eq. (21.1) without the burden of the four-dimensional space-time notation is a reminder that we are still searching for a fundamental and consistent formulation which should not reduce in the presence of strong acceleration to the traditional Lorentz force Eq. (21.1) we consider here, as it should include the radiation reaction force.

For $dm/dt = 0$, only $v(t)$ and $\gamma(t)$ contribute on the right-hand side of Eq. (21.1), and we obtain two terms, where the second arising from differentiation of the Lorentz factor γ is new:

$$m\left(\frac{\boldsymbol{a}}{\sqrt{1-\boldsymbol{v}^2/c^2}} + \frac{1}{c^2}\frac{\boldsymbol{v}\,(\boldsymbol{v}\cdot\boldsymbol{a})}{(1-\boldsymbol{v}^2/c^2)^{3/2}}\right) = \boldsymbol{F} \,, \tag{21.3}$$

and in simplified notation

$$\frac{d\boldsymbol{p}_{\mathrm{I}}}{dt} = m\gamma\boldsymbol{a} + m\gamma^3\frac{1}{c^2}\boldsymbol{v}\,(\boldsymbol{v}\cdot\boldsymbol{a}) = \boldsymbol{F} \,, \qquad m = Const. \tag{21.4}$$

The velocity of a body is a preferred direction with regard to which we usually consider all dynamics. For this reason we decompose all vectors into their parallel and orthogonal components with respect to $\boldsymbol{v}$, using the unit vectors $\boldsymbol{e}_\parallel$ and $\boldsymbol{e}_\perp$

$$\boldsymbol{v} = v\boldsymbol{e}_\parallel, \qquad \boldsymbol{a} = a_\perp \boldsymbol{e}_\perp + a_\parallel \boldsymbol{e}_\parallel . \tag{21.5}$$

In the next step we rewrite Eq. (21.4) using Eq. (21.5)

$$\frac{d\boldsymbol{p}_\mathrm{I}}{dt} = m\gamma a_\perp \boldsymbol{e}_\perp + m\gamma a_\parallel \boldsymbol{e}_\parallel + m\gamma^3 \frac{\boldsymbol{v}^2}{c^2} a_\parallel \boldsymbol{e}_\parallel = \boldsymbol{F} . \tag{21.6}$$

Introducing a common denominator for the two $\boldsymbol{e}_\parallel$ terms we obtain the much simplified equation

$$\frac{d\boldsymbol{p}_\mathrm{I}}{dt} = m\gamma a_\perp \boldsymbol{e}_\perp + m\gamma^3 a_\parallel \boldsymbol{e}_\parallel = \boldsymbol{F} . \tag{21.7}$$

We see in Eq. (21.7) that a particle reaching relativistic motion $\gamma > 1$ responds to an applied force in very different manner. The force component transverse to the velocity is γ^2 more effective at accelerating the particle as compared to the force component parallel to the velocity. To make this explicit some authors call $m\gamma$ the transverse mass and $m\gamma^3$ the parallel mass. However, there is just one body mass, so this distinction is better made by using the phrases 'transverse inertial resistance $m\gamma$' and 'parallel inertial resistance $m\gamma^3$'.

As an example recall that the force due to a magnetic field is always transverse to the velocity vector. Therefore this force goes along with the first term in Eq. (21.7). Any other force component parallel to the velocity vector acting together with the magnetic field has for ultrarelativistic particle dynamics to overcome a much greater, by factor γ^2, effective inertia, and as result is significantly less effective in producing a change in body motion. One concludes that to guide relativistic particles, magnetic fields are the natural choice.

We next consider the nonrelativistic limit of the relativistic force $\boldsymbol{F}$. We rewrite Eq. (21.7)

$$\boldsymbol{F} = m\gamma \boldsymbol{a} + m(\gamma^3 - \gamma) a_\parallel \boldsymbol{e}_\parallel , \tag{21.8}$$

which clarifies the coefficients of the post-Newtonian limit expansion

$$\frac{d\boldsymbol{p}_\mathrm{I}}{dt} = m\left(1 + \frac{1}{2}\frac{\boldsymbol{v}^2}{c^2} + \ldots\right)\boldsymbol{a} + m\left(\frac{\boldsymbol{v}^2}{c^2} + \ldots\right) a_\parallel \boldsymbol{e}_\parallel = \boldsymbol{F} . \tag{21.9}$$

In Eq. (21.9) there are two $\mathcal{O}(\boldsymbol{v}^2/c^2)$ relativistic corrections. The second of these two terms is parallel to $\boldsymbol{v}$ and thus $\boldsymbol{a}$ and $\boldsymbol{F}$ are not parallel, as illustrated in Fig. 21.1. When the force $\boldsymbol{F}$ is either exactly parallel or exactly perpendicular to the instantaneous direction of motion, $\boldsymbol{F}$ and $\boldsymbol{a}$ are parallel.

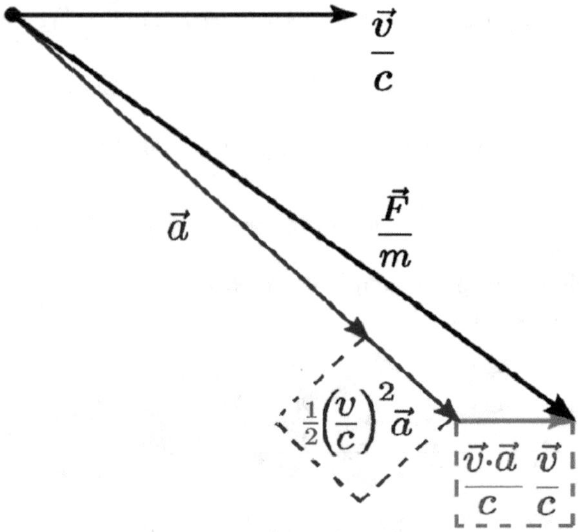

Fig. 21.1 A particle moving with velocity $\boldsymbol{v}$ subject to a non-parallel force $\boldsymbol{F}$ yields a resultant acceleration vector $\boldsymbol{a}$ shown with the two relativistic corrections, $\frac{1}{2}(\frac{v}{c})^2\boldsymbol{a}$ and $(\frac{\boldsymbol{v}\cdot\boldsymbol{a}}{c})\frac{\boldsymbol{v}}{c}$, see Eq. (21.9)

In the general case the vector describing acceleration of the body deviates in the first post-Newtonian correction from the direction of the force. This situation is familiar to sailors, given that the sailboat subject to the force $\boldsymbol{F}$ of wind in some direction is also subject to the drag force of water acting against the direction of boat velocity $\boldsymbol{v}$.

However, this comparison has limited validity. In our case, the drag is only present when the particle is accelerated; the drag also vanishes when the velocity is normal to the direction of acceleration as one can also see in Fig. 21.1. The requirement of accelerated motion reminds us of the discussion of the æther as the medium for EM waves, see Essay, Sect. 2.3. There we saw arguments of Langevin and Einstein that æther can only be perceived by an accelerated observer.

Example 21.1

Acceleration vector $\boldsymbol{a}$ in terms of the force

Cast Eqs. (21.4) and (21.6) into a form that presents the acceleration vector $\boldsymbol{a}$ directly as a function of the force $\boldsymbol{F}$ and the momentary particle velocity $\boldsymbol{v}$. ◄

► **Solution** We multiply Eq. (21.4) with $\boldsymbol{v}$ to obtain

1 $$m\gamma^3\left((1-\boldsymbol{v}^2/c^2)+\boldsymbol{v}^2/c^2\right)\boldsymbol{v}\cdot\boldsymbol{a}=\boldsymbol{v}\cdot\boldsymbol{F}\,.$$

Combining with Eq. (21.4) we find

2 $$m\gamma\boldsymbol{a}=\boldsymbol{F}-\frac{\boldsymbol{v}\,(\boldsymbol{v}\cdot\boldsymbol{F})}{c^2}\,.$$

We can analyze this result further: Decomposing the force $\boldsymbol{F}$ into the components parallel and perpendicular with reference to the direction of motion, and comparing with Eq. (21.5), we obtain

$$3\quad m\gamma \boldsymbol{a} = \boldsymbol{e}_\perp F_\perp + \boldsymbol{e}_\parallel F_\parallel - \frac{\boldsymbol{v}^2}{c^2} \boldsymbol{e}_\parallel F_\parallel \, .$$

This can be written in a manner analogous to Eq. (21.7)

$$4\quad m\gamma \boldsymbol{a} = \boldsymbol{e}_\perp F_\perp + \boldsymbol{e}_\parallel \frac{F_\parallel}{\gamma^2} \, .$$

We note that Eq. 4 has exactly the same form and content as does Eq. (21.7). In fact both vector formats look the same when written in component form:

$$5\quad \begin{aligned} m\gamma(\boldsymbol{a}\cdot\boldsymbol{e}_\perp) &= (\boldsymbol{F}\cdot\boldsymbol{e}_\perp)\,, \quad &\Leftrightarrow \quad m\gamma \boldsymbol{a}_\perp &= \boldsymbol{F}_\perp\,, \\ m\gamma^3(\boldsymbol{a}\cdot\boldsymbol{e}_\parallel) &= (\boldsymbol{F}\cdot\boldsymbol{e}_\parallel)\,, \quad &\Leftrightarrow \quad m\gamma^3 a_\parallel &= F_\parallel \, . \end{aligned}$$

When solving a problem involving relativistic motion one must remember that since the velocity vector $\boldsymbol{v}$ undergoes a continuous change in time, the unit directional vectors $\boldsymbol{e}_\perp$, $\boldsymbol{e}_\parallel$ with respect to $\boldsymbol{v}$ are in general time-dependent. ◀

21.2 Motion in Magnetic and Electric Fields

The well-known Lorentz force $\boldsymbol{F}_\mathrm{L}$, the force acting on an electron, is

$$\frac{d\boldsymbol{p}_\mathrm{I}}{dt} = q\,(\boldsymbol{\mathcal{E}} + \boldsymbol{v}\times\boldsymbol{\mathcal{B}}) \equiv \boldsymbol{F}_\mathrm{L}\,, \qquad \boldsymbol{p}_\mathrm{I} = m\gamma\boldsymbol{v} = \frac{m\boldsymbol{v}}{\sqrt{1-\boldsymbol{v}^2/c^2}}\,. \tag{21.10}$$

We use here the SI-units and thus the EM-fields $\mathcal{E}$, $\mathcal{B}$ are measured respectively, in V(olt)/m and T(esla).[1] We will typically look at the dynamics of an electron. The

[1] In a book written for students we adhere to the SI-units. However, we do this reluctantly; note that J.D. Jackson in the 3rd edition of *Classical Electrodynamics* (John Wiley & Sons, New York 1999), on page 514 switches units: "Beginning with Chapter 11 (Special Theory of Relativity) we employ Gaussian units instead of SI units...." The SI-units follow Maxwell's thinking and thus bring back æther. The reasons 'how and why' are spelled out by Richard Becker in his reissue of Max Abraham's *Electricity and Magnetism* text. In the introduction Becker says (words from the introduction to English language edition by J.Dougall, (Blackie & Son, London 1932)): "In the choice of units I have followed Abraham's last edition (which differs from earlier editions, JR) in every detail. The system used throughout is the Gaussian system,... . It does not seem possible at

electron charge is $q \to q_e \equiv e = -|e| = -1.602\,177 \times 10^{-19}$C. The modification from the nonrelativistic dynamics is seen in the relativistic form of the kinetic or, as we also say in this book, inertial momentum $\boldsymbol{p}_{\rm I}$.

Both $\boldsymbol{\mathcal{E}}$ and $\boldsymbol{\mathcal{B}}$ fields are additive; that is, they can be superposed. However, except for the case of EM-plane-wave we explore in the following examples the dynamics described in the simplified case that either $\boldsymbol{\mathcal{E}}$ or $\boldsymbol{\mathcal{B}}$ are non vanishing:

Example 21.2: The motion of a charged particle in a homogeneous time-independent magnetic field $\boldsymbol{\mathcal{B}}$, with vanishing electrical field $\boldsymbol{\mathcal{E}} = 0$. In this example the force is always normal to the direction of motion. Thus we find dynamics with which we are familiar from classic nonrelativistic electromagnetism, up to the detail that it is the energy content of the particle, and not its mass, that provides the inertia that the force must overcome, see Eq. (21.7).

Example 21.3: The motion of a charged particle in a homogeneous time independent electric field $\boldsymbol{\mathcal{E}}$ with vanishing magnetic field $\boldsymbol{\mathcal{B}} = 0$, choosing the initial momentum transverse to the direction of the field. We find that the force acts to align the particle with the force field. One can say a particle is trying to cross an 'electric river', and the flow will turn its motion into the direction of the field. Unlike the nonrelativistic case, the inertial resistance to the force depends on the velocity of the particle, which continuously changes in magnitude and direction under the influence of the electrical field.

Further below in **Example 21.4** we explore the motion and orbits of a relativistic electron in the presence of a radial Coulomb field of a point-like atomic nucleus $q\boldsymbol{\mathcal{E}} = \hat{e}_r Ze^2/r^2$. The dynamical equations differ from the nonrelativistic mechanics analogue, allowing the aphelion precession. Comparing with the well-known GR test result we determine that SR predicts 1/6 of the residual Mercury precession.

In **Chap. 22** we will study the motion in the electromagnetic field of a plane wave i.e. time-dependent waves with orthogonal fields $\boldsymbol{\mathcal{E}} \neq 0$ and $\boldsymbol{\mathcal{B}} \neq 0$.

---

present to set up a system of units which will satisfy the electrical engineer and the physicist alike. ...(This is) not a matter of notation merely but of principle. The technical view adheres much more strictly ... to the original Faraday-Maxwell theory. The engineer looks upon the vectors $\boldsymbol{\mathcal{E}}$ and $\boldsymbol{\mathcal{D}}$ – even in vacuum – as magnitudes of quite different kinds, related to each other more or less like tension and extension in the theory of elasticity.... On the other hand, **(this) distinction in principle between $\boldsymbol{\mathcal{D}}$ and $\boldsymbol{\mathcal{E}}$ which is closely connected with the mechanical theory of the æther, has been absolutely abandoned... .** (bolding by JR) The numerical identity of $\boldsymbol{\mathcal{E}}$ and $\boldsymbol{\mathcal{D}}$ – for empty space – is, in the Gaussian system of units, expression of the fact that ... (they) are actually the same thing. The introduction by an engineer of a dielectric constant and permeability not equal to 1 in a vacuum seems to the physicist to be ... an artifice... ."

Example 21.2

Relativistic motion of an electron in a constant magnetic field $\boldsymbol{\mathcal{B}}$

Examine the relativistic movement of an electron of charge $Q = -|e| = e$ in a constant homogeneous magnetic field $\boldsymbol{\mathcal{B}}$ and compare with the nonrelativistic case. ◀

▶ **Solution** Let e be the (negative) electron charge. The force on an electron in magnetic field $\boldsymbol{\mathcal{B}}$ is known to be

1 $$\frac{d\boldsymbol{p}_{\mathrm{I}}}{dt} = \boldsymbol{F} = e\boldsymbol{v} \times \boldsymbol{\mathcal{B}} .$$

We multiply with $\boldsymbol{p}_{\mathrm{I}} = m\gamma\boldsymbol{v}$ and find

2 $$\boldsymbol{p}_{\mathrm{I}} \cdot \frac{d\boldsymbol{p}_{\mathrm{I}}}{dt} = m\gamma e\boldsymbol{v} \cdot (\boldsymbol{v} \times \boldsymbol{\mathcal{B}}) = 0 .$$

This leads us to

3 $$\frac{d\boldsymbol{p}_{\mathrm{I}}^2}{dt} = 0 , \to m^2 \frac{\boldsymbol{v}^2}{1 - \boldsymbol{v}^2/c^2} = Const. , \to \boldsymbol{v}^2 = Const. , \to \gamma = Const.$$

The magnetic field $\boldsymbol{\mathcal{B}}$ is only capable of altering the direction of motion, but not the speed of a particle. This is clearly true for both relativistic and nonrelativistic particle motion as we have just seen: In the nonrelativistic case the particle mass m enters into dynamics; in the relativistic case we recognize that this is instead the energy $E/c^2 = \gamma m$.

We decompose $\boldsymbol{v}$ into components normal ($\boldsymbol{v}_\perp$) and parallel ($\boldsymbol{v}_\parallel$) to the $\boldsymbol{\mathcal{B}}$ field, so that

4 $$\boldsymbol{v} = \boldsymbol{v}_\parallel + \boldsymbol{v}_\perp , \qquad \boldsymbol{v}^2 = \boldsymbol{v}_\parallel^2 + \boldsymbol{v}_\perp^2 .$$

Inserting the first form in force equation and keeping in mind $\gamma = Const.$ as determined, we find

5 $$m\gamma \frac{d\,(\boldsymbol{v}_\parallel + \boldsymbol{v}_\perp)}{dt} = e(\boldsymbol{v}_\parallel + \boldsymbol{v}_\perp) \times \boldsymbol{\mathcal{B}} ,$$

and keeping in mind that by definition $\boldsymbol{v}_\parallel \times \boldsymbol{\mathcal{B}} = 0$, we see that

6 $$m\gamma \frac{d\boldsymbol{v}_\parallel}{dt} = 0 \to \boldsymbol{v}_\parallel = Const. , \to |\boldsymbol{v}_\perp| = Const.$$

The last condition follows since we know that the speed of the particle is constant. The motion of the particle normal to the $\boldsymbol{\mathcal{B}}$-field is according to Eq. 5 described by

7 $$m\gamma \frac{d\boldsymbol{v}_\perp}{dt} = e(\boldsymbol{v}_\perp \times \boldsymbol{\mathcal{B}}) .$$

We recognize that when a charged particle is entering a magnetic field domain, it experiences a force normal to the field and to its instantaneous velocity vector.

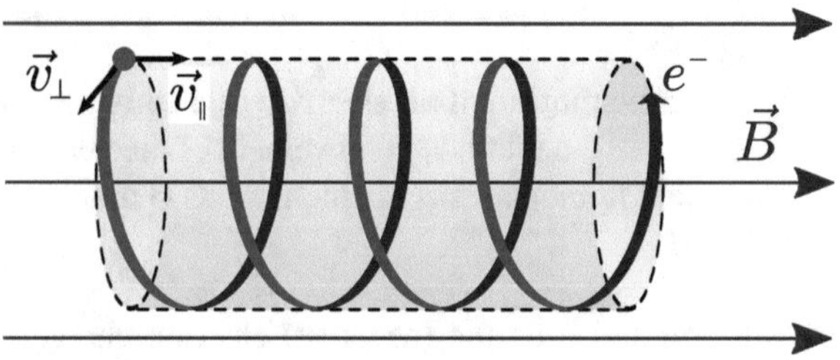

Fig. 21.2 The helical movement of an electron in a constant magnetic field

Therefore the motion is a superposition of linear motion along the field lines, Eq. 6, not affected by the magnetic field, with the motion normal to the field subject to deflection, but preserving the speed.

We choose a coordinate system such that the magnetic field is aligned with the z-axis. It follows from last relation that we must solve the equations

8 $$\frac{dv_x}{dt} = +\omega_c v_y\ , \qquad \frac{dv_y}{dt} = -\omega_c v_x\ , \qquad \frac{dv_z}{dt} = 0\ .$$

The characteristic 'cyclotron' frequency ω_c is defined by

9 $$\omega_c = \frac{eB}{\gamma m}\ .$$

Inserting one of the component equations of $d\boldsymbol{v}/dt$ in Eq. 8 into the other we obtain the second order differential equations

10 $$\frac{d^2 v_i}{dt^2} = -\omega_c^2 v_i\ , \qquad i = x, y\ .$$

We recognize the harmonic oscillator differential equation, hence the solution is

11 $$v_x = v_0 \cos\omega_c t\ , \qquad v_y = -\,v_0 \sin\omega_c t\ , \qquad v_z = v_{z\,0}\ , \qquad |\boldsymbol{v}_\perp(t=0)| \equiv v_0\ ,$$

where we chose the x-axis aligned with the initial transverse tangential speed $\boldsymbol{v}_\perp(t=0)$ pointing out of the plane of Fig. 21.2.

Time integration now provides

12 $$x = \frac{v_0}{\omega_c} \sin\omega_c t\ , \qquad y = \frac{v_0}{\omega_c} \cos\omega_c t\ , \qquad z = z_0 + v_{z\,0} t\ ,$$

showing that the cyclotron frequency governs the particle orbital rotation with radius

$$\textbf{13}\quad \rho \equiv \sqrt{x^2 + y^2} = \frac{v_0}{\omega_c}\,, \qquad \rho\omega_c = v_0\,.$$

This is a helical motion shown in Fig. 21.2 where we chose the initial condition such that the particle is at position $x_0 \equiv x(t=0) = 0$, $y_0 \equiv y(t=0) = v_0/\omega_c$. The electron path projected onto a plane orthogonal to $\mathcal{B}$ corresponds to a motion on a circle with radius ρ, as shown in Fig. 21.3. The path of motion shown in Fig. 21.2 is called a helix, and the motion called 'helical'.

Of interest is the magnitude of acceleration which we obtain differentiating Eq. 11

$$\textbf{14}\quad |\boldsymbol{a}| \equiv a = \sqrt{a_x^2 + a_y^2} = v_0\,\omega_c\,.$$

Combining two last results we can write a in several applicable forms using two out of the three variables v_0, ω_c, and ρ

$$\textbf{15}\quad a = \frac{v_0^2}{\rho} = \rho\omega_c^2 = v_0\omega_c\,.$$

Only the cyclotron frequency ω_c and the orbital velocity v_0 are experimental parameters; ρ_0 follows from Eq. 13. There is no limit on acceleration that an electron can experience with an increasing strength of a magnetic field which drives the magnitude of cyclotron frequency.

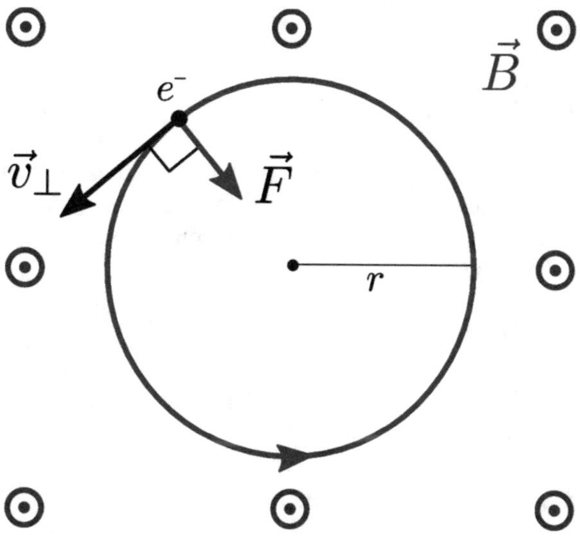

Fig. 21.3 The projection of the helical path from Fig. 21.2 onto a plane orthogonal to $\mathcal{B}$

We see that when the electron speed entering a magnetic field increases, the cyclotron frequency ω_c decreases and the radius of helical motion increases. This means that a charged relativistic particle traversing a magnetic field curves its path as if the mass of the particle is larger by a factor of γ. In fact what we found is that it is not the mass but the laboratory energy of charged particles that determines their path in a laboratory magnetic field. Particle detectors take advantage of the path curvature as a measure of particle energy.

The helical motion we obtained is practically the same as in the case of nonrelativistic dynamics. The difference is the appearance of the Lorentz factor γ in the definition of cyclotron frequency ω_c, an insight with considerable influence in the domains of plasma physics, accelerator physics, and radiation emission. ◀

Example 21.3

Relativistic motion of an electron in a constant electric field $\boldsymbol{\mathcal{E}}$

An electron moves in a space and time-independent (i.e. constant) electric field $\boldsymbol{\mathcal{E}} = \mathcal{E}\hat{e_x}$ where we oriented the x-coordinate along the field direction. Obtain the velocity $\boldsymbol{v}$, and the Lorentz factor γ of the electron as a function of the laboratory time t. Obtain the proper time τ as a function of laboratory time t and study the limits for small and large t. Determine the laboratory path $x(y)$ of the electron. Discuss all these results in quantitative terms considering the initial values $p_{0\,x} = 0$, $p_{0\,y} = mc$. As a cross-check, obtain the nonrelativistic limit for the path $x(y)$ and compare with a direct nonrelativistic solution of Newton's force equation. ◀

▶ **Solution** The force on the electron due to the electrical field $\boldsymbol{\mathcal{E}}$ is known to be $\boldsymbol{F} = e\boldsymbol{\mathcal{E}}$

$$\mathbf{1} \quad \frac{d\boldsymbol{p}_{\mathrm{I}}}{dt} = \frac{d(m\gamma\boldsymbol{v})}{dt} = e\boldsymbol{\mathcal{E}}\,, \qquad \gamma = \frac{1}{\sqrt{1-(\boldsymbol{v}/c)^2}}\,.$$

Without restriction of generality we do not need to consider motion in the third z-coordinate as we have oriented the coordinated system conveniently. We decompose Newton's force equation into the x-component parallel to $e\boldsymbol{\mathcal{E}}$, and the orthogonal y-component

$$\mathbf{2} \quad \frac{d(m\gamma v_x)}{dt} = e\mathcal{E}, \qquad \frac{d(m\gamma v_y)}{dt} = 0\,.$$

We integrate both equations with respect to t and obtain (we omit subscript 'I' from all momentum components)

$$\mathbf{3} \quad p_x = m\gamma v_x = e\mathcal{E}t + p_{0\,x}, \qquad p_y = m\gamma v_y = p_{0\,y}\,.$$

Since we know the momentum of the particle as a function of time, we also have determined its energy

4 $$E(t) = \sqrt{m^2c^4 + p_x^2c^2 + p_y^2c^2} = \sqrt{m^2c^4 + (e\mathcal{E}t + p_{0\,x})^2\,c^2 + p_{0\,y}^2c^2}\,.$$

We can now determine the velocity vector of the particle

5 $$\frac{v_x}{c} = \frac{cp_x}{E} = \frac{e\mathcal{E}t + p_{0\,x}}{\sqrt{(mc)^2 + (e\mathcal{E}t + p_{0\,x})^2 + p_{0\,y}^2}}\,,\qquad \frac{v_y}{c} = \frac{cp_y}{E} = \frac{p_{0\,y}}{\sqrt{(mc)^2 + (e\mathcal{E}t + p_{0\,x})^2 + p_{0\,y}^2}}\,.$$

We obtain the Lorentz $\gamma(t)$ factor from Eq. 4

6 $$\frac{E}{mc^2} \equiv \gamma = \sqrt{1 + \frac{(e\mathcal{E}t + p_{0\,x})^2 + p_{0\,y}^2}{(mc)^2}} = \sqrt{\gamma_0^2 + \frac{(e\mathcal{E}t)^2 + 2p_{0x}e\mathcal{E}t}{(mc)^2}}\,,$$

which increases with time. We find for the speed v

7 $$v = \sqrt{v_x^2 + v_y^2} = c\sqrt{\frac{(e\mathcal{E}t + p_{0\,x})^2 + p_{0\,y}^2}{(mc)^2 + (e\mathcal{E}t + p_{0\,x})^2 + p_{0\,y}^2}}\,,$$

For the initial values $p_{0\,x} = 0,\ p_{0\,y} = mc,\ p_{0\,z} = 0$ solutions v_x, v_y and v are depicted in Fig. 21.4. The total speed v (red, solid) monotonically increases, as does the component of the velocity in direction of the field v_x (blue, short dashed). However, given that $\gamma(t)$ also increases, the component of the velocity v_y (green, long dashed) transverse to field monotonically decreases from its initial value so that the transverse momentum not altered by the action of the electric field, see Eq. 3, can remain constant. This decrease of the transverse to the field velocity component v_y when the initial value is non-zero is an important insight about relativistic electron dynamics in constant $\mathcal{E}$ fields – this behavior is specific to relativistic dynamics; when $\gamma \to 1$ the transverse velocity remains constant.

The increment of the proper time of the electron observed to be moving at speed v in laboratory is

8 $$d\tau = dt\sqrt{1 - v^2/c^2} = dt\,\frac{1}{\gamma}\,.$$

Given that the Lorentz factor γ is a function of time we need to integrate to obtain the proper time as a function of laboratory time. We employ the formula

9 $$\int_0^t \frac{dx}{\sqrt{R}} = \frac{1}{\sqrt{k}}\ \ln[2\sqrt{kR} + 2kx + b]\,\Big|_0^t\,,\qquad R = a + bx + kx^2\,,\qquad k > 0\,,$$

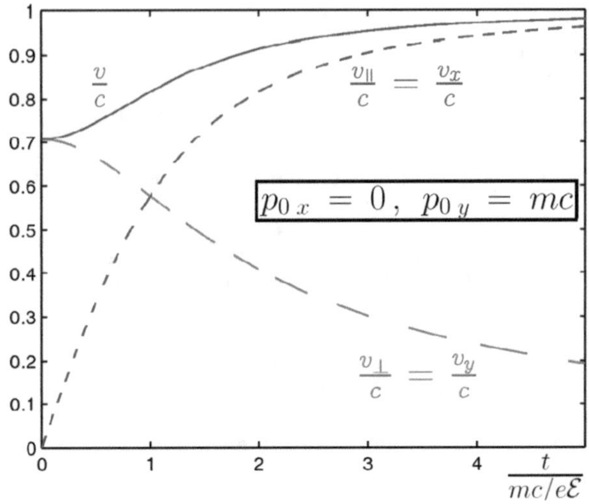

Fig. 21.4 The velocity as a function of normalized laboratory time $t/(mc/e\mathcal{E})$ for initial $p_{0\,x} = 0,\ p_{0\,y} = mc,\ p_{0\,z} = 0$: component in direction of the field $v_{\|} = v_x$ (short dashed, blue); the component normal to field $v_{\perp} = v_y$ (long dashed, green); and the total speed of the particle v/c (solid, red)

verified easily by differentiation. We have

10 $a = \gamma_0^2\,, \quad b = 2e\mathcal{E}p_{0\,x}/(mc)^2\,, \quad k = (e\mathcal{E}/mc)^2\,.$

Thus the exact solution takes the implicit form

11
$$\tau = \frac{mc}{|e\mathcal{E}|}\ln\left(\frac{2\sqrt{ka+kbt+(kt)^2}+2kt+b}{2\sqrt{ka}+b}\right)$$
$$= \frac{mc}{|e\mathcal{E}|}\ln\left(\frac{\sqrt{1+t\,b/a+t^2k/a}+t\sqrt{k/a}+b/\sqrt{4ka}}{1+b/\sqrt{4ka}}\right)\,.$$

We verify this result considering it for small t. Expanding the root within the logarithm, dropping terms in power t^2 and higher one finds after a computation

12
$$\tau \underset{t\to 0}{\longrightarrow} \frac{mc}{|e\mathcal{E}|}\ln\left(1+t\sqrt{k/a}+\mathcal{O}(t^2)\right)$$
$$= \frac{mc}{|e\mathcal{E}|}\ln\left(1+\frac{|e\mathcal{E}|t}{\gamma_0 mc}+\mathcal{O}(t^2)\right) \to \frac{t}{\gamma_0}\,.$$

This result is consistent with the proper time evaluated at $t = t_0$.

For large values of t according Eq. 5 in the limit $v_{0\,x} \to 0$ and

$$\mathbf{13} \quad \tau \underset{t\to\infty}{\longrightarrow} \frac{mc}{|e\mathcal{E}|} \ln \frac{2|e\mathcal{E}|t}{mc} .$$

Though there is a similarity to behavior we saw for $t \to 0$, the exponential relation between laboratory and proper time is a new insight.

Turning our attention to the shape of the path of the electron we introduce $v_x = dx/dt$ and $v_y = dy/dt$ in Eq. 5 allowing us to integrate with respect to t. Setting at $t = 0$, $x = 0$ and $y = 0$, the results are

$$\mathbf{14} \quad x = \frac{\sqrt{m^2c^4 + p_{0\,y}^2c^2 + (ce\mathcal{E}t + cp_{0\,x})^2} - E_0}{e\mathcal{E}} ,$$

where

$$\mathbf{15} \quad E(t = 0) \equiv E_0 = \sqrt{m^2c^4 + p_{0\,y}^2c^2 + p_{0\,x}^2c^2} .$$

The other integral is

$$y = \frac{p_{0\,y}c}{e\mathcal{E}} \left[\operatorname{arsinh}\left(\frac{e\mathcal{E}t + p_{0\,x}}{\sqrt{(mc)^2 + p_{0\,y}^2}} \right) - A \right] ,$$

16

$$A = \operatorname{arsinh}\left(\frac{p_{0\,x}}{\sqrt{(mc)^2 + p_{0\,y}^2}} \right) = \operatorname{arcosh}\left(\frac{E_0/c}{\sqrt{(mc)^2 + p_{0\,y}^2}} \right) .$$

For the case $p_{0\,x} = 0,\ p_{0\,y} = mc$, the distances traveled are shown in Fig. 21.5. We see that the distance traveled in the transverse direction dominates at first but the electron soon goes further in direction of the field lines.

We now are interested in finding the path $x(y)$, and thus we use Eq. 16 to eliminate t in Eq. 14. We find

$$\mathbf{17} \quad \sinh^2\!\left(\frac{y\,e\mathcal{E}}{p_{0\,y}c} + A \right) = \frac{(e\mathcal{E}t + p_{0\,x})^2}{(mc)^2 + p_{0\,y}^2} , \quad \frac{x\,e\mathcal{E} + E_0}{\sqrt{m^2c^4 + p_{0\,y}^2c^2}} = \sqrt{1 + \frac{(e\mathcal{E}t + p_{0\,x})^2}{(mc)^2 + p_{0\,y}^2}} ,$$

to yield

$$\mathbf{18} \quad \frac{x\,e\mathcal{E} + E_0}{\sqrt{m^2c^4 + p_{0\,y}^2c^2}} = \cosh\left(\frac{y\,e\mathcal{E}}{p_{0\,y}c} + A \right) ,$$

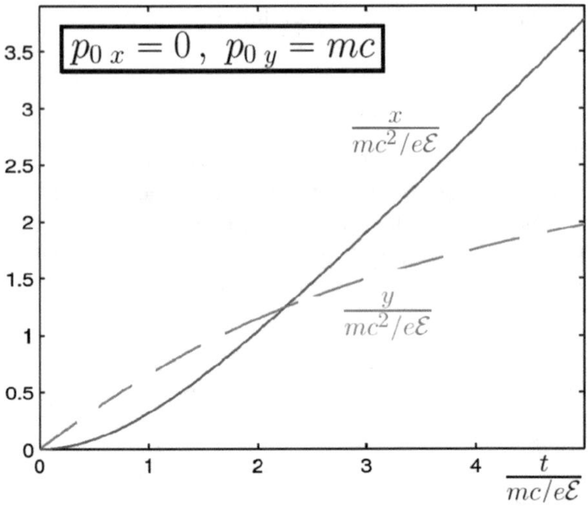

Fig. 21.5 The distance (in units of $mc^2/e\mathcal{E}$) traveled by a particle for the initial momentum $p_{0\,x} = 0\,,\ p_{0\,y} = mc\,,\ p_{0\,z} = 0$ as a function of time $t/(mc/e\mathcal{E})$: in direction of the field $x(t)$ (solid, blue); normal to the field $y(t)$ (dashed, green)

which defines the catenary path. The word catenary is derived from the Latin word *catena* = chain. A hanging chain is described by the 'cosh' function with $x \Leftrightarrow y$ compared to our form.

For the special case $p_{0\,x} = 0$ we have $E_0/c = \sqrt{(mc)^2 + p_{0\,y}^2}$, and Eq. 18 simplifies to

$$\mathbf{19}\quad x = \frac{E_0}{e\mathcal{E}}\left(\cosh\left(\frac{y\,e\mathcal{E}}{p_{0\,y}c}\right) - 1\right) \underset{c\to\infty}{\longrightarrow} x = \lim_{c\to\infty} \frac{1}{2}\frac{E_0}{p_{0\,y}^2\,c^2}\,e\mathcal{E}y^2 = \frac{1}{2}\frac{e\mathcal{E}}{m v_{0\,y}^2}\,y^2\,.$$

The path is depicted in Fig. 21.6, and the nonrelativistic limit is shown on right in Eq. 19. We see in the nonrelativistic limit that the electron follows a parabolic path. To cross-check this result we integrate the nonrelativistic equations of motion with the boundary conditions $v_{0\,x} = 0$, $x_0 = 0 = y_0$

$$\mathbf{20}\quad \ddot{x} = \frac{e\mathcal{E}}{m},\ \rightarrow\ \dot{x} = \frac{e\mathcal{E}}{m}t\ \rightarrow\ x = \frac{e\mathcal{E}}{2m}t^2\,,$$

$$\ddot{y} = 0\quad \rightarrow\ \dot{y} = v_{0\,y}\ \rightarrow\ y = v_{0\,y}\,t\,.$$

Solving for t as a (simple) function of y and inserting into the solution for x we obtain the last form in Eq. 19. ◀

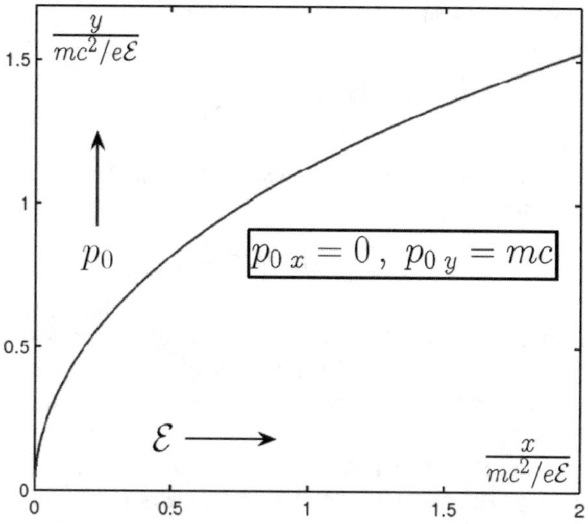

Fig. 21.6 The catenary path of an electron $y(x)$ in an electric field $\mathcal{E}\boldsymbol{e}_x$, with initial momentum $p_{0\,x} = 0\,,\ p_{0\,y} = mc\,,\ p_{0\,z} = 0$ normal to the field

21.3 Variational Principle

In classical nonrelativistic mechanics we learn that it is possible to obtain Newton's force law by considering a variation of the action I along a path a particle takes in space as a function of time. The Lagrange function L integrated along the path between events P_1 and P_2 provides the action

$$I = \int_{P_1}^{P_2} L(\boldsymbol{r}, \dot{\boldsymbol{r}})\, dt\ , \qquad L = T - U\ . \tag{21.11}$$

As indicated, the Lagrange function L and hence the action I depends on the space coordinate $\boldsymbol{r}$ where at time t the considered particle can be found, and, on the particle velocity $\dot{\boldsymbol{r}} = \boldsymbol{v}$, an independent dynamical property of the particle. Only after we have obtained dynamical equations of motion can $\dot{\boldsymbol{r}}$ be related to $\boldsymbol{r}$, as it is the case considering Newton's force law.

The search for the stationary point (in mathematical language called critical point) of the action Eq. (21.11) establishes these dynamical equations that can be solved for $\boldsymbol{r}(t)$ and $\dot{\boldsymbol{r}}(t)$. When seeking to find the best path $\boldsymbol{r}(t)$ for which the action I is smallest we require that the particle passes through the same initial P_1 and final P_2 event coordinates. To find the best path we perform a variation

$$\boldsymbol{r}(t) \to \boldsymbol{r}(t) + \delta\boldsymbol{r}\ , \qquad \delta\boldsymbol{r}|_{P_1} = 0\ , \quad \delta\boldsymbol{r}|_{P_2} = 0\ , \tag{21.12}$$

which results in

$$I = \int_{P_1}^{P_2} L(\boldsymbol{r}, \dot{\boldsymbol{r}})\, dt \to I + \delta I\ , \quad \delta I = \int_{P_1}^{P_2} dt\ \delta\boldsymbol{r} \cdot \boldsymbol{\mathcal{S}}(\boldsymbol{r}, \dot{\boldsymbol{r}})\ . \tag{21.13}$$

Thus if $\delta\boldsymbol{r}$ is arbitrary, the 'least action' path $\boldsymbol{r}(t)$ is found solving the dynamical equation

$$\boldsymbol{\mathcal{S}}(\boldsymbol{r}, \dot{\boldsymbol{r}}) = 0\ . \tag{21.14}$$

In this procedure borrowed from nonrelativistic formulation, t is a parameter of motion. Searching for the minimum of I we do not incorporate a change in elapsed time for each different path considered. Thus the unity of space and time is not (yet) implemented.

The action I includes as indicated in Eq. (21.11) both kinetic T and potential U energy. The motion related T is typically only a function of $\dot{\boldsymbol{r}}$ and not of $\boldsymbol{r}$. The action-potential U is seen as an 'external' field generated by all other particles of the system. The presumption of this approach is that the one particle we study does not perturb any of the external particles, and for that matter, neither does it perturb the field configuration significantly. In principle one can attempt to incorporate the 'back-reaction' effect; that is, to modify U by the dynamics of the considered particle. However, causality can be compromised by attempting such improvements in an ad-hoc way.[2]

To obtain δI Eq. (21.13) we consider

$$\begin{aligned}\delta I &= \int_{P_1}^{P_2} \left[L\left(\boldsymbol{r} + \delta\boldsymbol{r}, \frac{d(\boldsymbol{r} + \delta\boldsymbol{r})}{dt}\right) - L\left(\boldsymbol{r}, \dot{\boldsymbol{r}}; t\right) \right] dt \\ &= \int_{P_1}^{P_2} \left(\delta\boldsymbol{r} \frac{\partial L}{\partial \boldsymbol{r}} + \left(\frac{d\delta\boldsymbol{r}}{dt}\right) \frac{\partial L}{\partial \dot{\boldsymbol{r}}} \right) dt\ .\end{aligned} \tag{21.15}$$

We integrate by parts the second term to obtain

$$\delta I = \int_{P_1}^{P_2} \delta\boldsymbol{r} \cdot \left(\frac{\partial L}{\partial \boldsymbol{r}} - \frac{d}{dt}\frac{\partial L}{\partial \dot{\boldsymbol{r}}} \right) dt + \left(\delta\boldsymbol{r} \cdot \frac{\partial L}{\partial \dot{\boldsymbol{r}}} \right)\Bigg|_{P_1}^{P_2}\ . \tag{21.16}$$

Since according to Eq. (21.12) we consider arbitrary variations $\delta\boldsymbol{r}$ vanishing at the end points of the integration domain, we identify the usual Lagrange equation of motion

$$\frac{\partial L}{\partial \boldsymbol{r}} - \frac{d}{dt}\frac{\partial L}{\partial \dot{\boldsymbol{r}}} = 0\ . \tag{21.17}$$

[2] See for example: G. Kortemeyer, W. Bauer, K. Haglin, J. Murray, S. Pratt, "Causality Violations in Cascade Models of Nuclear Collisions," *Phys. Rev. C* **52** 2714 (1995).

The differentiation with respect to $\boldsymbol{r}$ has, for a function that depends on $\boldsymbol{r}$, the same meaning as $\boldsymbol{\nabla}$. Since the Lagrange equation is fixing the velocity of a particle $\dot{\boldsymbol{r}} = \boldsymbol{v}$, from this point forward we use $\boldsymbol{v}$ or $|\boldsymbol{v}| = v$ instead of $\dot{\boldsymbol{r}}$.

The generalization of the kinetic term $T = mv^2/2$ to the relativistic formulation must produce the modification of the $m \to m\gamma$ on the left-hand side of Newton's equations, Eq. (21.1). When used in the second term in Eq. (21.17), this must produce the inertial part of Newton's equation. We identify the relativistic generalization of $T \to T_r$ as

$$T = m\frac{v^2}{2} \to T_r = -mc^2\sqrt{1 - v^2/c^2} = -mc^2 + m\frac{v^2}{2} + \dots, \qquad (21.18)$$

where the constant term $-mc^2$ does not impact the result of variation. Some may find the format of the inertial term in the action, T_r seen in Eq. (21.18), aberrant as we learn in nonrelativistic formulation of mechanics that $L = T - U$. Hence, the relativistic form $T = mc^2\gamma$ could be expected. However, the better way to think here is that the action consists of two terms, the inertia term and the potential term. The form of these terms is determined by the already known dynamics. The nonrelativistic limit of T prescribes the form of relativistic inertia term Eq. (21.18) and we should not anticipate that the form of relativistic inertia term corresponds to the relativistic kinetic energy. The form of the inertia term given in Eq. (21.18) is the only one leading to the correct form of the relativistic inertial term in the Lorentz force

$$\frac{d}{dt}\frac{\partial T_r}{\partial \boldsymbol{v}} = \frac{d\boldsymbol{p}_{\mathrm{I}}}{dt}, \qquad \boldsymbol{p}_{\mathrm{I}} \equiv \frac{\partial T_r}{\partial \boldsymbol{v}} = \frac{m\boldsymbol{v}}{\sqrt{1 - v^2/c^2}}. \qquad (21.19)$$

In this step we encounter as before the inertial momentum $\boldsymbol{p}_{\mathrm{I}}$.

The potential term U leading to the Lorentz force in the nonrelativistic case is well known and has the form

$$U = e\,(V - \boldsymbol{v} \cdot \boldsymbol{A}), \qquad (21.20)$$

where we introduce the EM 'scalar' potential V and the EM 'vector' potential $\boldsymbol{A}$ from which the electromagnetic fields $\boldsymbol{\mathcal{E}}$ and $\boldsymbol{\mathcal{B}}$ follow as we will show below. We have placed the words scalar and vector in quotes as these quantities together form a vector in four-dimensional space and thus the use of language can be misleading.

We begin with

$$L = T_r - U \; . \tag{21.21}$$

Since the potential term U has explicit velocity dependence we recognize that there is a difference between the inertial momentum $\boldsymbol{p}_\mathrm{I}$ and the canonical momentum $\boldsymbol{p}$

$$\boldsymbol{p} \equiv \frac{\partial L}{\partial \boldsymbol{v}} = \frac{\partial T_r}{\partial \boldsymbol{v}} - \frac{\partial U}{\partial \boldsymbol{v}} = \boldsymbol{p}_\mathrm{I} + e\boldsymbol{A} \; . \tag{21.22}$$

We thus remember that in the following we will use as needed

$$\boldsymbol{p}_\mathrm{I} = \boldsymbol{p} - e\boldsymbol{A} \; . \tag{21.23}$$

Using Eq. (21.17) with Eq. (21.21) we obtain the resultant form for the EM force

$$\frac{d\boldsymbol{p}_\mathrm{I}}{dt} = \frac{d}{dt}\frac{\partial U}{\partial \boldsymbol{v}} - \frac{\partial U}{\partial \boldsymbol{r}} = -e\frac{d\boldsymbol{A}}{dt} - \boldsymbol{\nabla}\left(eV - \boldsymbol{v}\cdot e\boldsymbol{A}\right) \; . \tag{21.24}$$

The total derivative with respect to time contains two terms,

$$\frac{d\boldsymbol{A}}{dt} = \frac{\partial \boldsymbol{A}}{\partial t} + \frac{d\boldsymbol{r}}{dt}\frac{\partial \boldsymbol{A}}{\partial \boldsymbol{r}} = \frac{\partial \boldsymbol{A}}{\partial t} + \boldsymbol{v}\cdot\boldsymbol{\nabla}\boldsymbol{A} \; . \tag{21.25}$$

The last term arises from the motion of the particle along a trajectory $\boldsymbol{r}(t)$ and thus the coefficient is particle velocity $\boldsymbol{v}$ as indicated. We insert Eq. (21.25) in Eq. (21.24) and combine the terms as follows

$$\frac{d\boldsymbol{p}_\mathrm{I}}{dt} = e\left(-\boldsymbol{\nabla}V - \frac{\partial \boldsymbol{A}}{\partial t}\right) + e\left(\boldsymbol{\nabla}(\boldsymbol{v}\cdot\boldsymbol{A}) - \boldsymbol{v}\cdot\boldsymbol{\nabla}\boldsymbol{A}\right) \; . \tag{21.26}$$

We recall that the EM fields in terms of potentials are given by

$$\boldsymbol{\mathcal{E}} = -\boldsymbol{\nabla}V - \frac{\partial \boldsymbol{A}}{\partial t} \, , \qquad \boldsymbol{\mathcal{B}} = \boldsymbol{\nabla}\times\boldsymbol{A} \; . \tag{21.27}$$

With the help of BAC-CAB rule $\boldsymbol{B}(\boldsymbol{A} \cdot \boldsymbol{C}) - (\boldsymbol{A} \cdot \boldsymbol{B})\boldsymbol{C} = \boldsymbol{A} \times (\boldsymbol{B} \times \boldsymbol{C})$ we obtain

$$\boldsymbol{v} \times \boldsymbol{\mathcal{B}} = \boldsymbol{v} \times (\nabla \times \boldsymbol{A}) = \nabla(\boldsymbol{v} \cdot \boldsymbol{A}) - \boldsymbol{v} \cdot \nabla \boldsymbol{A} \tag{21.28}$$

Using Eqs. (21.27) and (21.28) in Eq. (21.26) we obtain the Lorentz force $\boldsymbol{F}_{\mathrm{L}}$ in the format given in Eq. (21.10). This confirms the choice Eq. (21.20) for the potential term in the action also in case of the relativistic dynamics.

In summary: We have found that the relativistic particle dynamics in the presence of electromagnetic fields Eq. (21.27) is described by the Lagrange function

$$L = -mc^2\sqrt{1 - v^2/c^2} - U \,, \qquad U = eV - \boldsymbol{v} \cdot e\boldsymbol{A} \,, \tag{21.29}$$

which leads upon variation to the Lorentz force, Eq. (21.10). We further note that the canonical momentum satisfies

$$\frac{d\boldsymbol{p}}{dt} = \frac{d}{dt}\left(\boldsymbol{p}_{\mathrm{I}} - \frac{\partial U}{\partial \boldsymbol{v}}\right) = -\nabla\left(eV - \boldsymbol{v} \cdot e\boldsymbol{A}\right) = -\nabla U \,. \tag{21.30}$$

This relation is useful in study of dynamical problems where one or more components of ∇U on the right-hand side in Eq. (21.30) vanish. This implies that the corresponding components of the canonical momentum are conserved.

We now explore the relationship between momentum and energy by considering the Hamiltonian $H(\boldsymbol{p}, \boldsymbol{r}; t)$. We recall

$$H(p_i, r_i; t) \equiv \sum_i \dot{r}_i p_i - L(\boldsymbol{r}_i, \dot{\boldsymbol{r}}_i; t) \,, \qquad p_i = \frac{\partial L}{\partial \dot{\boldsymbol{r}}_i} \,. \tag{21.31}$$

It is not our intent to develop Hamiltonian mechanics, e.g. to demonstrate that the Legendre transform that defines H, (see Eq. (21.31)), implies the variable dependence as shown.[3] However, we obtain Hamilton's dynamical equations from a variational principle varying the action

$$I = \int dt\, L(r_i, \dot{r}_i; t) = \int dt \left(\sum_i \dot{r}_i p_i - H(p_i, r_i; t)\right) \,. \tag{21.32}$$

Note that the subscript 'i' now indicates all independent components; these can be the vector components of one or many particles. The variation of I, Eq. (21.32), leads to, see Ref. 3

[3] See for example H. Goldstein, *Classical Mechanics*, Addison-Wesley; 3 edition (New York 2001).

$$\delta I = \int dt \sum_i \left\{ \delta p_i \left(\dot{r}_i - \frac{\partial H}{\partial p_i} \right) + \delta r_i \left(-\dot{p}_i - \frac{\partial H}{\partial r_i} \right) \right\} = 0 . \tag{21.33}$$

Since the variations are all independent we obtain the canonical equations of motion

$$\dot{r}_i = \frac{\partial H}{\partial p_i} , \qquad -\dot{p}_i = \frac{\partial H}{\partial r_i} . \tag{21.34}$$

Therefore

$$\frac{dH(p_i, r_i; t)}{dt} = \sum_i \left(\dot{r}_i \frac{\partial H}{\partial r_i} + \dot{p}_i \frac{\partial H}{\partial p_i} \right) + \frac{\partial H}{\partial t} = \frac{\partial H}{\partial t} , \tag{21.35}$$

since in consideration of the canonical equations of motion Eq. (21.34) the parentheses vanish. We thus find, Eq. (21.35), that the total system energy changes only when the action potential is explicitly time-dependent. It is well understood in nonrelativistic context that the conserved quantity associated with H is energy. Setting $H(p, r; t) \to E$, with Eqs. (21.31), (21.35), and using Eq. (21.29)

$$\frac{dE}{dt} = -\frac{\partial L}{\partial t} = \frac{\partial U}{\partial t} . \tag{21.36}$$

We now evaluate explicitly the Hamiltonian Eq. (21.31) for one particle placed in external EM potential. We use Eq. (21.22) to compute $\boldsymbol{p} \cdot \dot{\boldsymbol{r}}$ and thus we obtain for the Hamiltonian

$$H = mc^2 \frac{v^2/c^2}{\sqrt{1 - v^2/c^2}} + \boldsymbol{v} \cdot e\boldsymbol{A} + mc^2 \sqrt{1 - v^2/c^2} + e\,(V - \boldsymbol{v} \cdot \boldsymbol{A}) . \tag{21.37}$$

We see that the second and the last term cancel and we combine the first and third term to obtain

$$H - eV = \frac{mc^2}{\sqrt{1 - v^2/c^2}} . \tag{21.38}$$

We need to replace the speed in Eq. (21.38) by canonical momentum. In Eq. (21.22) we move $e\boldsymbol{A}$ to the left and square

$$(\boldsymbol{p} - e\boldsymbol{A})^2 = \boldsymbol{p}_{\mathrm{I}}^2 = \frac{m^2c^2(v^2/c^2 - 1 + 1)}{1 - v^2/c^2} = -m^2c^2 + \frac{m^2c^2}{1 - v^2/c^2} , \tag{21.39}$$

which when combined with Eq. (21.38) yields the relativistic Hamiltonian

$$H = eV + \sqrt{(mc^2)^2 + c^2(\boldsymbol{p} - e\boldsymbol{A})^2}\,. \tag{21.40}$$

We keep in mind that H is the energy of the particle; $\boldsymbol{p}$ is the canonical momentum not simply related to the measured velocity $\boldsymbol{v}$ of a particle, in that it also includes the potential term $\boldsymbol{A}$. Finally, we write Eq. (21.40) in the suggestive relativistic format

$$(H - eV)^2 - c^2(\boldsymbol{p} - e\boldsymbol{A})^2 = (mc^2)^2\,, \tag{21.41}$$

showing the relativistic relationship between all quantities explicitly. This is the format in which classical relativistic dynamics is generalized to relativistic quantum dynamics in form known as the 'Klein-Gordon Equation'.

21.4 Electron Coulomb Orbits

In the following two examples we study the classical orbital motion of an electron in the Coulomb $1/r^2$ radial electric field typical for a very heavy nucleus. This allows us to assume that the source of the force does not recoil. We are, however, primarily interested in understanding how relativity (not GR) modifies the Kepler orbits and to recognize apsidal precession in the context of the Lorentz force.

For the case of nonrelativistic motion gravity and EM forces differ only by force constants and are indistinguishable in terms of their force equations. These examples permit the comparison of these two different physical systems in the context of relativistic dynamics. We recall that Mercury precession was one of key evidences for the GR framework of gravitation force within Einstein's GR. The fact that the magnitude of apsidal precession for classical Lorentz force orbits is relatively small compared to the GR case is the backdrop against which this GR signature stands out resolving the otherwise inexplicable Mercury perihelion motion. We also discuss the SR and GR apsidal precession of more distant planets, Venus and Earth.

Example 21.4

Relativistic electron motion in the Coulomb $1/r^2$ radial electric field

Obtain periodic orbits of a relativistic particle in an attractive $1/r^2$ radial electric field and determine: (a) the angular relativistic orbits $r(\phi)$; (b) the orbit energy; and (c) the orbit size parameter, as a function of angular momentum and orbit eccentricity. Using the Bohr quantization condition adapt your solution to the special case of an electron in the Coulomb field of a point nucleus of charge $|e|Z$. ◀

▶ **Solution** We consider particle motion in a radial force field according to Eq. (21.1). The Lorentz force simplifies to the Coulomb force law

1 $$\frac{d\,m\gamma \boldsymbol{v}}{dt} = e\mathcal{E}_r(r)\hat{r}\ ,$$

where $\hat{r}$ is the unit vector in the radial direction. For the motion of an electron in the Coulomb field of a point nucleus

2 $$e\mathcal{E}_{r\,\mathrm{C}}(r) = -\frac{e^2}{4\pi\epsilon_0}\frac{Z}{r^2}\ .$$

We thus obtain the Coulomb potential eV

3 $$e\boldsymbol{\mathcal{E}} = -\boldsymbol{\nabla} eV \ \rightarrow\ e\mathcal{E}_{r\,\mathrm{C}} = -\frac{\partial eV_{\mathrm{C}}}{\partial r} \ \rightarrow\ eV_{\mathrm{C}} = -\frac{e^2}{4\pi\epsilon_0}\frac{Z}{r}\ .$$

Note that the Coulomb potential eV_{C} is negative (attractive), since the unit charges associated with a proton and an electron are of opposite sign.

The (relativistic) angular momentum

4 $$\boldsymbol{\mathcal{L}} \equiv \boldsymbol{r} \times m\gamma\boldsymbol{v}\ ,$$

is a constant of motion:

5 $$\frac{d\boldsymbol{\mathcal{L}}}{dt} = \boldsymbol{v} \times m\gamma\boldsymbol{v} + \boldsymbol{r} \times \frac{d\,m\gamma\boldsymbol{v}}{dt} = 0 + \boldsymbol{r} \times eE_r(r)\hat{r} = 0\ .$$

The constancy of $\boldsymbol{l}$ assures motion in the plane normal to $\boldsymbol{l}$, allowing use of polar coordinates (r, ϕ), see also Fig. 18.1. Since $v_\phi = \omega r$, and $\omega = d\phi/dt$, we have

6 $$|\boldsymbol{\mathcal{L}}| \equiv l = m\gamma r^2 \frac{d\phi}{dt}\ .$$

The conserved magnitude of angular momentum creates a useful relation between several dynamical variables.

A second constant of motion is recognized inspecting Eq. (21.40), which we write in the form

7 $$\left(E + \frac{e^2}{4\pi\epsilon_0}\frac{Z}{r}\right)^2 = c^2\boldsymbol{p}^2 + m^2c^4\ .$$

The conventional vector algebra can be used to find for the momentum $\boldsymbol{p}$ (canonical and inertial momentum being the same we omit subscript 'I')

$$\textbf{8}\quad \boldsymbol{p} \equiv \hat{r}\, p_r + \hat{\phi}\, p_\phi = m\gamma \left(\hat{r}\frac{dr}{dt} + \hat{\phi}\, r\,\frac{d\phi}{dt}\right) = \hat{r}\, m\gamma\frac{dr}{dt} + \hat{\phi}\,\frac{l}{r}\,,$$

where we used the constancy of the magnitude of angular momentum l in the last equality. Thus we identify

$$\textbf{9}\quad p_r \equiv m\gamma\frac{dr}{dt}\,, \qquad p_\phi \equiv \frac{l}{r}\,, \qquad p^2 = p_r^2 + p_\phi^2\,.$$

This allows us to write for Eq. 7

$$\textbf{10}\quad m^2c^4\left(\frac{E}{mc^2} + \frac{Zr_e}{r}\right)^2 = m^2c^4 + \frac{l^2c^2}{r^2} + p_r^2c^2\,,$$

where to shorthand the notation we have introduced the classical electron radius r_e, see Eqs. (18.19) and (20.9).

We are interested in the orbit equation; that is, to determine $r(\phi)$, and this also allows us to recast the last term in terms of angular momentum. The procedure is akin to the nonrelativistic case since the relativistic Lorentz factors cancel. We consider

$$\textbf{11}\quad \frac{dr}{d\phi} = \frac{dr}{dt}\,\frac{1}{d\phi/dt} = m\gamma\frac{dr}{dt}\,\frac{1}{m\gamma\, d\phi/dt} = p_r\frac{r}{p_\phi}\,,$$

$$\Rightarrow p_r = \frac{dr}{d\phi}\,\frac{1}{r}\,p_\phi = \frac{l}{r^2}\,\frac{dr}{d\phi}\,.$$

Combination of these two above results produces the relativistic orbit equation

$$\textbf{12}\quad \left(\frac{E}{mc^2} + \frac{Zr_e}{r}\right)^2 = 1 + \left(\frac{l}{mcr}\right)^2\left[1 + \left(\frac{1}{r}\,\frac{dr}{d\phi}\right)^2\right],$$

which can be solved for $r(\phi)$.

We simplify this result by introducing the orbit variable s

$$\textbf{13}\quad s = \frac{\tilde{a}}{r}\,, \qquad s' \equiv \frac{ds}{d\phi} = -\frac{\tilde{a}}{r^2}\,\frac{dr}{d\phi}\,,$$

where the scaling factor $\tilde{a}$ will be chosen shortly. We find sequencing the terms in powers of s and s'

14 $$\left(\frac{E}{mc^2}\right)^2 - 1 = -2\frac{Zr_e}{\tilde{a}}\frac{E}{mc^2}s + \left[\left(\frac{l}{mc\tilde{a}}\right)^2 - \left(\frac{Zr_e}{\tilde{a}}\right)^2\right]s^2 + \left(\frac{l}{mc\tilde{a}}\right)^2 s'^2 .$$

Like in the nonrelativistic case the simplest way to find a solving function is to differentiate with respect to ϕ. After dividing by $2s'(l/mc\tilde{a})^2$ we obtain

15 $$s'' + \left[1 - \left(\frac{Zmcr_e}{l}\right)^2\right]s = \frac{ZEm\tilde{a}r_e}{l^2} .$$

The choice of scaling factor $\tilde{a}$ is now made such that

16 $$\frac{ZEm\tilde{a}r_e}{l^2} \equiv \left[1 - \left(\frac{Zmcr_e}{l}\right)^2\right] \to \tilde{a} = \frac{Zmc^2r_e}{E}\left[\left(\frac{l}{Zmcr_e}\right)^2 - 1\right] .$$

This allows us to write for Eq. 15

17 $$(s-1)'' + \left[1 - \left(\frac{Zmcr_e}{l}\right)^2\right](s-1) = 0 ,$$

and hence we find the general periodic i.e. orbiting particle solution $s - 1 = \epsilon_x \cos(\kappa\phi + \delta)$, which implies

18 $$\frac{\tilde{a}}{r} \equiv s \to r = \frac{\tilde{a}}{1 + \epsilon_x \cos(\kappa\phi + \delta)} , \quad |\epsilon_x| < 1$$

with

$$0 < \kappa^2 = 1 - \left(\frac{Zmcr_e}{l}\right)^2 < 1 .$$

The choice of initial value $\delta \to 0$ is common. However, we can absorb the sign of the eccentricity ϵ_x by choice $\delta \to \pi$. This shows that orbits with positive and negative ϵ_x are identical.

Like in nonrelativistic mechanics, ϵ_x describes the deviation of the orbit from a circular shape. Together with energy E and magnitude of angular momentum l, the parameter ϵ_x is determined for each orbit observed. Note that in the nonrelativistic limit $c \to \infty$ since $mc^2 r_e = Const.$ we obtain $\kappa = 1$. In this limit we recover the elliptical Kepler conic section orbits. For relativistic orbits this is not the case, the rosette quasi elliptical orbits with $0 < \kappa < 1$ do not close; the orbit processes. This effect is called apsidal precession, described in Example 21.5. In this computation it will be useful to recall the orbit size parameter $\tilde{a}$, Eq. 16, written as a function of κ^2

$$\textbf{19} \quad \tilde{a} = \frac{Zmc^2 r_e}{E}\left[\left(\frac{l}{Zmcr_e}\right)^2 - 1\right] = \frac{Zmc^2 r_e}{E}\frac{\kappa^2}{1-\kappa^2} .$$

The physical meaning of the parameter $\tilde{a}$ is similar but not exactly the same as we have in the context of a nonrelativistic elliptical orbit: the mean value R of the radial distance maximum R_+ (perihelion) and minimum R_- (aphelion)

$$\textbf{20} \quad R_+ = \frac{\tilde{a}}{1-\epsilon_x} , \qquad R_- = \frac{\tilde{a}}{1+\epsilon_x} ,$$

produces the equivalent to the semi-major axis a

$$\textbf{21} \quad a \to R = \tfrac{1}{2}(R_- + R_+) = \frac{\tilde{a}}{1-\epsilon_x^2} .$$

We further note that by (implicit) definition of $\tilde{a}$, Eq. 16, and in view of Eq. 19 the orbital energy depends on $\tilde{a}$, angular momentum l and interaction strength Z

$$\textbf{22} \quad \frac{E}{mc^2} = Z\frac{r_e}{\tilde{a}}\left[\left(\frac{l}{Zmcr_e}\right)^2 - 1\right] = Z\frac{r_e}{\tilde{a}}\frac{\kappa^2}{1-\kappa^2} ,$$

but not eccentricity ϵ_x.

It is possible to eliminate $\tilde{a}$ in favor of eccentricity ϵ_x. Doing this we not only check the math, but also obtain constraints between the three parameters E, l, ϵ_x characterizing the orbit. We begin differentiating the orbit solution Eq. 18 with $\delta = 0$

$$\textbf{23} \quad \frac{\tilde{a}}{r}\frac{1}{r}\frac{dr}{d\phi} = \kappa\epsilon_x \sin\kappa\phi , \to \quad \frac{1}{r}\frac{dr}{d\phi} = \frac{\kappa\epsilon_x \sin\kappa\phi}{1+\epsilon_x\cos\kappa\phi} .$$

To shorten notation we hence will use

$$\epsilon_x \cos\kappa\phi \equiv c_\kappa \,, \qquad \epsilon_x \sin\kappa\phi \equiv s_\kappa \,, \qquad s_\kappa^2 + c_\kappa^2 = \epsilon_x^2 \,. \tag{24}$$

Inserting the r from Eq. 18 in Eq. 12 and using Eq. 23 as well as the short forms Eq. 24 we obtain

$$\left(\frac{E}{mc^2} + \frac{Zr_e}{\tilde{a}}(1+c_\kappa)\right)^2 = 1 + \left(\frac{l}{mc\tilde{a}}\right)^2 \left[(1+c_\kappa)^2 + \kappa^2 s_\kappa^2\right] \,. \tag{25}$$

On the left we regroup terms and on the right we use κ^2, Eq. 18, and simplify using also the last relation in Eq. 24

$$\begin{aligned}\left[\left(\frac{E}{mc^2} + \frac{Zr_e}{\tilde{a}}\right) + \frac{Zr_e}{\tilde{a}} c_\kappa\right]^2 = 1+ &\left(\frac{l}{mc\tilde{a}}\right)^2 (1 + \epsilon_x^2 + 2c_\kappa) \\ &- \left(\frac{l}{mc\tilde{a}} \frac{Zmcr_e}{l}\right)^2 s_\kappa^2 \,,\end{aligned} \tag{26}$$

We now group odd terms in c_κ on the left-hand side (LHS), and even on the right-hand side (RHS)

$$\text{LHS} = 2\left[\left(\frac{E}{mc^2} + \frac{Zr_e}{\tilde{a}}\right)\frac{Zr_e}{\tilde{a}} - \left(\frac{l}{mc\tilde{a}}\right)^2\right] c_\kappa \,, \tag{27}$$

$$\text{RHS} = 1 + (1+\epsilon_x^2)\left(\frac{l}{mc\tilde{a}}\right)^2 - \left(\frac{Zr_e}{\tilde{a}}\right)^2 (s_\kappa^2 + c_\kappa^2) - \left(\frac{E}{mc^2} + \frac{Zr_e}{\tilde{a}}\right)^2 \,. \tag{28}$$

Since there is no ϕ-dependence in RHS, both sides of the equation must independently vanish: LHS $= 0$, and RHS $= 0$. The condition LHS $= 0$ is equivalent to Eq. 16 and does not provide additional conditions: we see this solving Eq. 16 for E/mc^2

$$\frac{E}{mc^2} = \frac{Zr_e}{\tilde{a}}\left[\left(\frac{l}{Zmcr_e}\right)^2 - 1\right] , \tag{29}$$

and inserting into LHS. On the left-hand side we note that since $1/\tilde{a} \propto E$, see Eq. 16, the factor E^2 cancels and only l remains from among the three orbital parameters, thus this term is zero as a trivial identity. This is verified easily using the relation following from Eq. 16.

There is thus one single constraint, RHS = 0, which can be written as

30 $$0 = \text{RHS} = 1 + (1+\epsilon_x^2)\left(\frac{Zr_e}{\tilde{a}}\frac{l}{Zmcr_e}\right)^2 - \epsilon_x^2\left(\frac{Zr_e}{\tilde{a}}\right)^2 - \left(\frac{E}{mc^2} + \frac{Zr_e}{\tilde{a}}\right)^2 .$$

We use again Eq. 16 in the form

31 $$\frac{Zr_e}{\tilde{a}} = \frac{E}{mc^2}\frac{1}{(l/Zmcr_e)^2 - 1} = \frac{E}{mc^2}\frac{\tilde{Z}^2}{1-\tilde{Z}^2}, \qquad \tilde{Z} \equiv \frac{Zmcr_e}{l} .$$

Combining these last two results and dividing by the common factor $(E/mc^2)^2$ we obtain the simplified form

32 $$0 = \left(\frac{mc^2}{E}\right)^2 + (1+\epsilon_x^2)\frac{\tilde{Z}^2}{(1-\tilde{Z}^2)^2} - \epsilon_x^2\frac{\tilde{Z}^4}{(1-\tilde{Z}^2)^2} - \frac{1}{(1-\tilde{Z}^2)^2} .$$

This algebraic equation has as solution for the energy of the orbit as a function of l and ϵ_x^2

33 $$E = mc^2\sqrt{\frac{1-\tilde{Z}^2}{1-\epsilon_x^2\tilde{Z}^2}}, \qquad \tilde{Z} \equiv \frac{Zmcr_e}{l} .$$

In order to make contact with relativistic quantum mechanics we now introduce Bohr's quantization condition for the magnitude of the angular momentum l

34 $$l = n\hbar ,$$

and we write equations using the quantum-related natural constants

35 $$\lambda\!\!\!^{-}_{\text{C}} = \frac{\hbar c}{mc^2},$$
$$\alpha = \frac{r_e}{\lambda\!\!\!^{-}_{\text{C}}} = \frac{e^2}{4\pi\epsilon_0 mc^2}\frac{mc^2}{\hbar c} = \frac{e^2}{4\pi\epsilon_0\hbar c} = \frac{1}{137.035999} .$$

Checking the definition of $\tilde{Z}$ and the orbit parameter κ, Eq. 18, we find

36 $$\tilde{Z} \equiv \frac{Zmcr_e}{l} = \frac{Z\alpha}{n}, \qquad \kappa = \sqrt{1 - \left(\frac{Z\alpha}{n}\right)^2}.$$

This leads to

37 $$E = mc^2 \sqrt{\frac{1 - (Z\alpha/n)^2}{1 - (\epsilon_x Z\alpha/n)^2}}.$$

The energy of the bound particle assumes in the special case of a circular orbit $\epsilon_x = 0$ the value

38 $$E = mc^2 \sqrt{1 - \left(\frac{Z\alpha}{n}\right)^2}, \underset{Z\to 0}{\longrightarrow} mc^2 \left(1 - \frac{1}{2}\left(\frac{Z\alpha}{n}\right)^2 + \ldots\right).$$

The small-$Z\alpha$ expansion produces the known Schrodinger equation spectrum. For strong coupling $Z\alpha \to 1$ the singular behavior of the classical binding of electrons ($m_e = 0.511\,\mathrm{MeV}$) is depicted for $n = 1, 2$ as a function of Z in Fig. 21.7. We see that for $Z\alpha/n \to 1$ the total energy of the bound state vanishes: The binding energy compensates in full the energy equivalent of the electron mass.

For the value of the orbit size parameter $\tilde{a}$ we obtain

39 $$\tilde{a} = \frac{lc\tilde{Z}}{E}\left(\frac{1}{\tilde{Z}^2} - 1\right)$$
$$= \frac{l}{mc}\sqrt{\frac{1 - (\epsilon_x \tilde{Z})^2}{1 - \tilde{Z}^2}}\frac{1 - \tilde{Z}^2}{\tilde{Z}} = \frac{l}{mc\tilde{Z}}\sqrt{(1 - \tilde{Z}^2)(1 - (\epsilon_x \tilde{Z})^2)}.$$

In the Bohr quantization case we have again $l \to \hbar n$, $\tilde{Z} \to Z\alpha/n$ and we find

40 $$\tilde{a} = a_{\mathrm{I}}\frac{n^2}{Z}\sqrt{\left(1 - \left(\frac{Z\alpha}{n}\right)^2\right)\left(1 - \left(\frac{\epsilon_x Z\alpha}{n}\right)^2\right)}, \qquad a_{\mathrm{I}} = \frac{r_e}{\alpha^2} = \frac{\lambda_{\mathrm{C}}}{\alpha},$$

where we have introduced the Bohr radius $a_{\mathrm{I}} = 0.529177211 \times 10^{-10}$ m. We note the scaling with n^2/Z of the orbit radius parameter $\tilde{a}$, familiar from quantum mechanics.

This completes the solution of the relativistic Coulomb problem: We have obtained the energy of the orbit and the orbit shape as a function of the magnitude

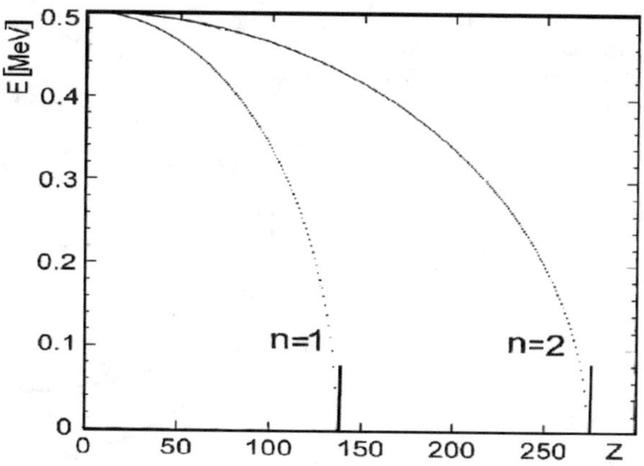

Fig. 21.7 The classical relativistic energy of electrons bound by $V = -Z\alpha/r$ Coulomb potential as a function of Z, for a circular orbit with angular momentum $l = n\hbar, n = 1, 2$

of angular momentum and orbit eccentricity. We have shown that for increasing coupling strength Z the binding consumes the entire energy equivalent mc^2 of the electron. We therefore addressed here solely the case $\tilde{Z} \equiv Z\alpha/n < 1$, i.e. $\kappa^2 > 0$ for which the periodic quantum physics solutions also exist, see 4 in Chap. 20.

Further reading: Though a relativistic solution presenting the energy in terms of action variables has been available for a long time,[4]our simple form Eq. 37, which relates to the orbit eccentricity, may have escaped attention. Another study reminds us of a century-old Darwin's spiraling-in solutions[5]for $Z\alpha/n > 1$. Lacking a viable quantum analogue this unphysical solution should be looked at with caution. ◀

Example 21.5

Apsidal precession

The apsidal precession is a famous GR test. Compare the magnitude of apsidal precession obtained for the SR-Coulomb $1/r^2$-radial electric field with the known GR gravity force effect. ◀

▶ **Solution** We recall that the $0 < \varepsilon < 1$ quasi-elliptical orbits, see Eq. 18 in Example 21.4, do not close exactly. We will study cases that deviate little from the exact elliptical shape, with an apsidal precession $\Delta\phi$ much smaller than illustrated in Fig. 21.8. The deviation of the apsidal precession of the orbit $r(\phi)$ is shown in Fig. 21.8. The aphelion (opposite of perihelion) is the farthest point from the location of the Sun in an elliptic orbit. Note that in Fig. 21.8 we show the aphelion precession more distinctly for the choice of illustration parameters.

[4]J.D. Garcia, "Quantum solutions and classical limits for strong Coulomb fields", *Phys. Rev. A* **34**, 4396 (1986), see cited Ref. [3].

[5]T.H. Boyer, "Unfamiliar trajectories for a relativistic particle in a Kepler or Coulomb potential", *Am. J. Phys.* **72**, 992 (2004), see cited Ref. [2].

Fig. 21.8 Apsidal precession: shown is the precession of the aphelion by $\Delta\phi$

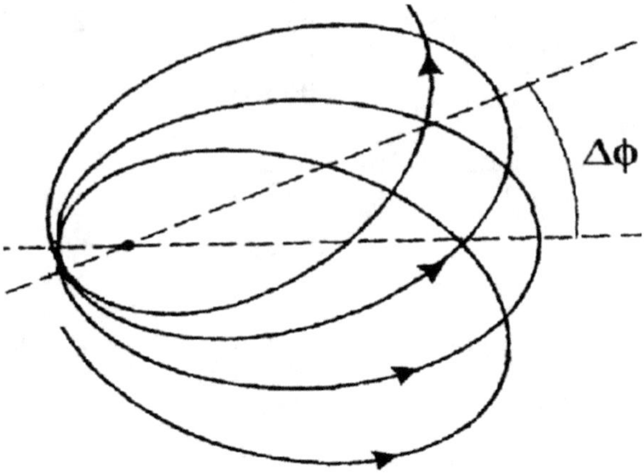

To describe the small nearly-elliptical orbit rotation we consider the cycling value of r for $\kappa\phi_C = 2\pi n, n = 1, 2, 3\ldots$. Since according to Eq. 18 in Example 21.4 $\kappa < 1$, we have $\phi_C > 2\pi, n = 1$. We find

$$1\quad \Delta\phi_C \equiv \phi_C - 2\pi = 2\pi\frac{1-\kappa}{\kappa} = 2\pi\frac{1-\kappa^2}{\kappa(1+\kappa)} = \frac{2\pi\kappa}{1+\kappa}\frac{1-\kappa^2}{\kappa^2}\,.$$

Using the value of $\tilde{a}$ from Eq. 19 in Example 21.4 we determine the value of $\Delta\phi_C$, the apsidal precession, measured in the observation of the aphelion and the perihelion precession of Mercury

$$2\quad \Delta\phi_C = \frac{2\pi\kappa}{1+\kappa}\frac{Zmc^2 r_e}{E}\frac{1}{\tilde{a}}\,.$$

In the case of Mercury, all known nonrelativistic perturbations (e.g. the influence of other planets, Sun asphericity) combine to an aphelion precession of 532"/century, with a discrepancy of 43"/century compared to the experimental value 574"/century (values rounded). Explanation of this discrepancy was of pivotal importance in establishing Einstein's GR.[6] Our SR Coulomb bound system cannot be an equivalent model of relativistic effects in planetary motion in a gravitational field.

The apsidal precession motion of Mercury according to SR is made explicit by introducing the equivalent gravitational coupling

$$3\quad \frac{Ze^2}{4\pi\epsilon_0} = mc^2(Zr_e) \to \frac{mc^2}{2}\frac{2GM}{c^2} = mc^2\frac{R_S}{2}\,, \qquad R_S = 2.953\,\text{km}\,,$$

[6] A. Einstein, "Erklärung der Perihelbewegung des Merkur aus der allgemeinen Relativitätstheorie," (translated: Explanation of the perihelion motion of Mercury from the General Theory of Relativity) *Preussische Akademie der Wissenschaften, Sitzungsberichte,* **XLVII** 831 (1915).

where R_S is the solar Schwartzschild radius, and m planetary mass. Given a typical planetary speed, $10^{-5}c$, and eccentricity $\epsilon^2 \leq 0.05$, it is appropriate in a first evaluation to use in Eq. 2 $\kappa \simeq 1$, $E = mc^2$. In consideration of Eqs. 19 , 20, and 21 in Example 21.4, we have

$$\mathbf{4} \quad \Delta\phi_C \simeq \frac{\pi}{2}\frac{R_S}{R}\,, \qquad \frac{2}{R} \equiv \frac{1}{R_-} + \frac{1}{R_+}\,.$$

In the last relation the orbit parameter R value in terms of aphelion (R_-) and perihelion (R_+) of the orbit enters. It is important to note the $1/R$-dependence is showing the largest effect for the innermost planet, Mercury. Comparing to our SR result we note that Weinberg[7] finds using post-Newtonian characterization of GR a result that is 6 times greater

$$\mathbf{5} \quad \Delta\phi_{GR} = 3\pi\,\frac{R_S}{R}\,.$$

Weinberg's result agrees with Einstein's, presented in Ref. 6 and repeated at the end of the pivotal GR manuscript.[8]

A disagreement by a factor 6 between SR-Coulomb and GR demonstrates that in specific features SR can be a poor model for the effect of GR. However, the slow decrease with orbital size ($1/R$) is specific to relativity, both to SR and GR. This $1/R$-behavior is unlike any other effect considered for apsidal precession e.g. due to planetary perturbations or solar deformation. Therefore the observation of the scaling with $1/R$ is evidence for relativity, SR or GR, and the larger apsidal precession effect is characteristic of GR.

We note that we obtained here the effect of the apsidal precession per considered planet orbit while experimental data is usually presented per 100 revolutions of the Earth (per century); for Mercury there are 415 revolutions per Earth century and thus the aphelion precession effect reported in this way is amplified by a factor 4.15. Similarly the effect on Venus is amplified by a factor 1.49. The $1/R$-dependence predicts for Venus and Earth compared to Mercury expressed per 100 Earth years as the ratio

$$\mathbf{6} \quad \frac{\Delta\phi_{Merc}}{\Delta\phi_{Venus}} = \frac{4.15}{1.49}\frac{R_{Venus}}{R_{Merc}} = 5.43\,, \qquad \frac{\Delta\phi_{Merc}}{\Delta\phi_{\oplus}} = 4.15\frac{R_{\oplus}}{R_{Merc}} = 11.2\,.$$

Here we used $R_{Mercury} = 5.55 \times 10^7$ km, $R_{Venus} = 1.08 \times 10^8$ km, and $R_{\oplus} = 1.50 \times 10^8$ km.

This specific scaling with $1/R$ (R orbit size) is observed confirming the relativistic origin of the unexplained part of the planetary apsidal precession. ◀

[7] See Eq. (9.5.17) in: S. Weinberg, *Gravitation and Cosmology* John Wiley & Sons, Inc (New York 1972).

[8] A. Einstein, "Die Grundlage der allgemeinen Relativitätstheorie," (translated: The Foundation of the General Theory of Relativity), *Annalen der Physik* **354** 769 (1916).

22 Electrons Riding a Plane Wave

Abstract

The analytical solution of Lorentz force motion of an electron in the presence of EM fields of a light (laser) plane wave are presented. The conservation laws are both emphasized and used in obtaining the solution for the electron motion. It is shown that the character of motion changes drastically with the increasing amplitude of the light wave. When the wave intensity reaches the relativistic regime, the wave force thrusts electrons forward.

22.1 Fields and Potentials for a Plane Wave

The study of the dynamics of electron motion in the electromagnetic field of an externally provided light plane wave lays the foundation for principles that lead to the understanding of relativistic particle dynamics in the presence of ultra-intense laser pulses. In the approach we present we will be exploiting the conservation laws we have established in Sect. 21.3 in order to simplify significantly the solution path.

We will be working in the transverse gauge, also called the radiation gauge, and in a different context, the Coulomb gauge. In the study of light the word transverse relates to the vector potential $\boldsymbol{A}$ being a vector transverse to the direction of propagation of the light wave. $\boldsymbol{A}$ in the transverse gauge satisfies

$$\boldsymbol{\nabla} \cdot \boldsymbol{A} = 0\,. \tag{22.1}$$

This constraint leads to the Poisson equation for the scalar potential $-\nabla^2 \epsilon_0 V = \rho$ providing the Coulomb force law, which explains the frequent naming of Eq. (22.1)

J. Rafelski, *Modern Special Relativity*, https://doi.org/10.1007/978-3-030-54352-5_22

as the Coulomb gauge. In absence of a charged source the scalar potential vanishes. Therefore in our considerations we have

$$\boldsymbol{\mathcal{E}} = -\frac{\partial \boldsymbol{A}}{\partial t}\,, \qquad \boldsymbol{\mathcal{B}} = \boldsymbol{\nabla} \times \boldsymbol{A}\,. \tag{22.2}$$

We do not allow the physical fields $\boldsymbol{\mathcal{E}}$, $\boldsymbol{\mathcal{B}}$ to be complex quantities as is sometimes done, an approach that requires for each situation a prescription how to obtain physical quantities. Since we will rely on a complex waveform, we will need to compute as appropriate the real part (symbol $\Re$) and the imaginary part (symbol $\Im$).

The first step is to gain familiarity with the electromagnetic plane waves describing the propagation of light, a solution of Maxwell's equations without sources. For a light plane wave the $\boldsymbol{\mathcal{E}}$, $\boldsymbol{\mathcal{B}}$-fields are time-dependent and transverse to direction of propagation. We choose in the transverse gauge

$$\boldsymbol{A} = \Re\left[\boldsymbol{A}_{|0}\, e^{i\chi}\right], \quad \chi = \boldsymbol{k}\cdot\boldsymbol{r} - \omega t\,, \qquad V = 0\,, \tag{22.3}$$

where $\boldsymbol{A}_{|0}$ is a constant amplitude. In order to satisfy the transverse gauge condition we must have

$$0 = \boldsymbol{\nabla}\cdot\boldsymbol{A} = \Re\left[i\boldsymbol{k}\cdot\boldsymbol{A}_{|0}\, e^{i\chi}\right] \to \boldsymbol{k}\cdot\boldsymbol{A}_{|0} = 0\,. \tag{22.4}$$

Thus we find that the 'polarization vector'

$$\boldsymbol{e}_A \equiv \frac{\boldsymbol{A}_{|0}}{A_{|0}}\,, \qquad A_{|0} = \sqrt{A^*_{|0} A_{|0}} \equiv \frac{mc}{e}\, a_o\,, \tag{22.5}$$

(where A^* is complex conjugate of A) must be always orthogonal to the direction of propagation $\boldsymbol{k}$ of the wave. We introduced here the dimensionless amplitude measure a_0. The reader should distinguish the polarization vector $\boldsymbol{e}_A$ from the elementary charge e. For a plane light-wave propagating along the light front with the light speed $c = |d\boldsymbol{r}|/dt = r/t$, we have, according to Eq. (22.3)

$$k = \frac{\omega}{c} = \frac{2\pi}{\lambda}\,, \tag{22.6}$$

and we note the last relation to the wavelength λ.

Differentiating the 3-vector $\boldsymbol{A}$ we obtain

$$\begin{aligned} \boldsymbol{\mathcal{E}} &= -\frac{\partial \boldsymbol{A}}{\partial t} = \omega\Im\left[\boldsymbol{A}_{|0}\, e^{i\chi}\right], \\ \boldsymbol{\mathcal{B}} &= \boldsymbol{\nabla}\times\boldsymbol{A} = \boldsymbol{k}\times\Im\left[\boldsymbol{A}_{|0}\, e^{i\chi}\right]. \end{aligned} \tag{22.7}$$

In view of Eq. (22.6) we have $|\boldsymbol{\mathcal{E}}| = c|\boldsymbol{\mathcal{B}}|$, and considering Eq. (22.4) we have mutual orthogonality $\boldsymbol{\mathcal{B}}\perp\boldsymbol{\mathcal{E}}\perp\boldsymbol{k}\perp\boldsymbol{\mathcal{B}}$.

If the polarization vector $\boldsymbol{e}_A$, Eq. (22.5), is real, it does not interfere with the phase χ of the wave. In this case the direction of $\boldsymbol{\mathcal{E}}\perp\boldsymbol{\mathcal{B}}$ remains fixed in space, and we have a 'linearly polarized' (LP) light plane wave. We can choose for $\boldsymbol{e}_A$ any orientation transverse to the direction of wave propagation $\boldsymbol{k} \parallel \boldsymbol{e}_3$

$$\boldsymbol{e}_A = \cos\delta\, \boldsymbol{e}_1 + \sin\delta\, \boldsymbol{e}_2\,, \quad \boldsymbol{e}_A^* \cdot \boldsymbol{e}_A = 1\,, \qquad \underline{\text{LP}}\,. \tag{22.8}$$

Here δ is a (fixed) phase angle.

To obtain circularly polarized (CP) light, we choose a complex (and constant) polarization vector

$$\boldsymbol{e}_A = \frac{\boldsymbol{e}_1 \pm i\boldsymbol{e}_2}{\sqrt{2}}\,, \quad \boldsymbol{e}_A^* \cdot \boldsymbol{e}_A = 1\,, \qquad \underline{\text{CP}}\,, \tag{22.9}$$

where $\boldsymbol{k}\perp\boldsymbol{e}_1\perp\boldsymbol{e}_2\perp\boldsymbol{k}$ assures Eq. (22.4) and the two signs represent two possible circular polarizations. From Eq. (22.7) we obtain the unit vectors and magnitudes of the fields $\boldsymbol{\mathcal{B}}\perp\boldsymbol{\mathcal{E}}$

$$\begin{aligned} \boldsymbol{e}_{\mathcal{E}} &= \frac{\boldsymbol{e}_1 \sin\chi \pm \boldsymbol{e}_2 \cos\chi}{\sqrt{2}}\,, \qquad \mathcal{E} = \omega A_{|0}\,, \qquad \underline{\text{CP}}\,, \\ \boldsymbol{e}_{\mathcal{B}} &= \boldsymbol{e}_k \times \frac{\boldsymbol{e}_1 \sin\chi \pm \boldsymbol{e}_2 \cos\chi}{\sqrt{2}}\,, \qquad \mathcal{B} = k A_{|0}\,, \qquad \underline{\text{CP}}\,, \end{aligned} \tag{22.10}$$

thus observing $\boldsymbol{e}_{\mathcal{E}}$, $\boldsymbol{e}_{\mathcal{B}}$ at the same position as a function of time; that is, choosing $\chi = -\omega t$, we see that these unit vectors rotate along the unit circle. Even though $\boldsymbol{e}_A$ does not rotate, due to the formation of the real part Eq. (22.3), the vector potential $\boldsymbol{A}$ does

$$\boldsymbol{A} = A_{|0}\frac{1}{\sqrt{2}}\left(\boldsymbol{e}_1 \cos\chi \mp \boldsymbol{e}_2 \sin\chi\right)\,. \tag{22.11}$$

Here χ is the time- and space-dependent plane wave phase. At a fixed point in space $z = 0$ we have $\chi = -\omega t$ and thus

$$\boldsymbol{A}(z=0,t) = A_{|0}\frac{1}{\sqrt{2}}\left(\boldsymbol{e}_1 \cos\omega t \pm \boldsymbol{e}_2 \sin\omega t\right)\,, \quad (\boldsymbol{A})^2 = \left(\frac{1}{\sqrt{2}}A_{|0}\right)^2\,, \qquad \underline{\text{CP}}\,, \tag{22.12}$$

is rotating in a mathematically positive sense for the (+) sign. For the right-handed coordinate system the plane wave moves out of the plane of this book towards you.

We will be interested in evaluating the intensity of the plane wave, which by definition is the magnitude of the Poynting[1] vector, which can be written in several

[1] John Henry Poynting (1852–1914), a collaborator of Maxwell and an eminent physics educator.

equivalent ways in the vacuum[2]

$$\boldsymbol{S} = c^2\epsilon_0 \boldsymbol{\mathcal{E}} \times \boldsymbol{\mathcal{B}} \equiv c^2 \boldsymbol{\mathcal{D}} \times \boldsymbol{\mathcal{B}} \equiv \frac{1}{\mu_0} \boldsymbol{\mathcal{E}} \times \boldsymbol{\mathcal{B}} \equiv \boldsymbol{\mathcal{E}} \times \boldsymbol{\mathcal{H}} \,. \tag{22.13}$$

We study $\boldsymbol{S}$ for the plane wave. The cross product $\boldsymbol{\mathcal{E}} \times \boldsymbol{\mathcal{B}}$ points, according to Eq. (22.7), in the direction of the plane wave vector $\boldsymbol{k}$. We obtain

$$\boldsymbol{S} = \frac{\boldsymbol{k}\omega}{4\pi} \frac{4\pi\epsilon_0}{e^2} \left(\Im \left[c\, e \boldsymbol{A}_{|0}\, e^{i\chi} \right] \right)^2 . \tag{22.14}$$

Note the appearance of c as a prefactor of $\boldsymbol{A}_{|0}$ in Eq. (22.14). In Gauss-units this factor c is absent as it has the same origin as the factor c in $\mathcal{B}$ in Eq. (20.7); in addition the factor $4\pi\epsilon_0/e^2$ is $1/e^2$ in Gauss-units. This is consistent with the last rule in Eq. (20.7).

We further have

$$\Im \left[e\boldsymbol{A}_{|0}\, e^{i\chi} \right] = \begin{cases} eA_{|0}\boldsymbol{e}_A \sin\chi & \underline{\text{LP}}\,, \\ \dfrac{1}{\sqrt{2}} eA_{|0}\, (\boldsymbol{e}_1 \sin\chi \pm \boldsymbol{e}_2 \cos\chi) & \underline{\text{CP}}\,, \end{cases} \tag{22.15}$$

and hence

$$\left(\Im \left[e\boldsymbol{A}_{|0}\, e^{i\chi} \right] \right)^2 = \frac{e^2 A_{|0}^2}{2} \begin{cases} 2\sin^2\chi & \underline{\text{LP}}\,, \\ 1 & \underline{\text{CP}}\,. \end{cases} \tag{22.16}$$

In consideration of the fact that we have for time and/or space average

$$2\langle \sin^2\chi \rangle = 1\,, \tag{22.17}$$

the final result for the averaged Poynting vector of a plane wave is

$$\langle \boldsymbol{S} \rangle = \frac{\boldsymbol{k}\omega}{8\pi} (ceA_{|0})^2 \frac{4\pi\epsilon_0}{e^2} = \hat{k} \frac{c\pi}{2\lambda^2} (ceA_{|0})^2 \frac{4\pi\epsilon_0}{e^2} \,. \tag{22.18}$$

[2]For discussion of historical issues, see R. N. C. Pfeifer, T. A. Nieminen, N. R. Heckenberg, and H. Rubinsztein-Dunlop, "Colloquium: Momentum of an electromagnetic wave in dielectric media," *Rev. Mod. Phys.* **79**, 1197 (2007); Erratum ibid. **81**, 443 (2009).

The last equality follows in view of the Eq. (22.6). Since our unit vectors were normalized the same way, see Eqs. (22.8) and (22.9), the result Eq. (22.18) applies equally to the two polarization cases we considered.

Example 22.1

Light waves of relativistic strength

In order to measure the strength of the extreme strength light waves it is common to consider units adapted to the relativistic motion of particles that ride the wave, typically electrons. One thus considers a dimensionless measure, see Eqs. (22.5) and (22.7)

$$\mathbf{1} \quad a_0 \equiv \frac{ceA_{|0}}{mc^2} ,$$

where the product of the light wave amplitude $A_{|0}$ with light speed c and the wave riding particle charge e is divided by the energy equivalent of the accelerated particle. For $a_0 > 1$ we speak of relativistic optics because particles can be accelerated to relativistic speeds with a Lorentz factor $\gamma \propto a_0^2$. Today laser laboratories work with $a_0 > 10$ routinely and research projects envisage $a_0 > 1000$. Note furtherthat a_0 scales the electrical field, and thus the EM-force

$$\mathbf{2} \quad e\mathcal{E} = a_0\hbar\omega \frac{mc^2}{\hbar c} \Im\left[\boldsymbol{e}_A \, e^{i\chi}\right] .$$

Explain why it makes good sense to introduce a_0. Connect the plane wave amplitude $A_{|0}$ and related field $\mathcal{E}$, $\mathcal{B}$ presented in terms of the dimensionless amplitude a_0 to relevant values in SI units. For the case of fields study the case of photons of energy 1 eV that as a typical value. Then obtain the value a_0 leading to critical field strength discussed in Sect. 20.5. ◀

▶ **Solution** The potential $\boldsymbol{A}$ multiplied with a unit charge e and speed of light c that we see in Eq. (22.14) and in reduced form in Eq. (22.18) has in the SI-unit system the unit of energy. In the context of a particle that rides the wave shown in Sect. 22.1 it would be appropriate to use as reference the rest energy of that particle to define the unit wave height. This explains why we consider the magnitude of the plane wave amplitude $A_{|0}$ in terms of dimensionless potential amplitude a_0 in Eq. 1.

In the following we chose to consider an electron as the reference particle, and the appropriate ratio of masses allows the scaling of these expressions for other elementary particles. We thus study the benchmark case $a_0 = 1$ for electrons. This corresponds to the value

$$\mathbf{3}\quad cA_{|0}\Big|_1 \equiv cA_{|0}(a_0=1) = \frac{mc^2}{e} = 0.5110 \times 10^6\ \mathrm{V}\,,$$

where we canceled the e from the MeV energy unit and made the 'mega' visible.

Let us look at a light wave with a quantum of light energy

$$\mathbf{4}\quad \hbar\omega = 1\mathrm{eV} = \hbar ck\,, \qquad \lambda = \frac{hc}{1\,\mathrm{eV}} = 1.2398\,\mu\mathrm{m}\,.$$

The corresponding electrical field strength according to Eq. 2 is

$$\mathbf{5}\quad \mathcal{E}(a_0=1, \hbar\omega = 1\,\mathrm{eV}) = \frac{\mathrm{V}}{\lambda\!\!\!^{-}_{\mathrm{C}}}\,\Im\left[\boldsymbol{e}_A\, e^{i\chi}\right]\,,$$

$$\lambda\!\!\!^{-}_{\mathrm{C}} = \frac{\lambda_{\mathrm{C}}}{2\pi} = \frac{\hbar c}{mc^2} = 386.16\,\mathrm{fm}\,,$$

where the reduced Compton wavelength appears, Eq. (20.3). Written in SI-units the magnitudes of electromagnetic fields we are now considering is

$$\mathbf{6}\quad \begin{aligned} \mathcal{E}(a_0=1, \hbar\omega = 1\,\mathrm{eV}) &= \frac{\mathrm{V}}{\lambda\!\!\!^{-}_{\mathrm{C}}} = 2.590 \times 10^{12}\,\frac{\mathrm{V}}{\mathrm{m}}\,,\\ \mathcal{B}(a_0=1, \hbar\omega = 1\,\mathrm{eV}) &= 8.638 \times 10^3\,\mathrm{T}\,. \end{aligned}$$

Below Eq. (22.7) we noted that the magnitude of the $|c\boldsymbol{\mathcal{B}}|$-field of a light wave is equal to that of the $|\boldsymbol{\mathcal{E}}|$-field, but for the coefficients c in front of the $\mathcal{B}$-field arising from the choice of SI units. Thus the last expression in Eq. 6 arises from division of the $\mathcal{E}$-field by the light speed factor 3×10^8 in SI units.

The magnitude of fields seen in Eq. 6 are hundreds of times larger than fields one can reach with EM equipment in the laboratory. It is therefore not a surprise that when these fields are exceeded highly relativistic electrons can be produced. However, we will find that the achievable electron energy depends less on field strength than on the (dimensionless) wave amplitude a_0.

Even though the fields seen in Eq. 6 must be seen as being 'strong', they lack elementary strength. This is so since we arbitrarily chose to study the case of 1 eV photons. The elementary value of electrical field strength requires a natural frequency ω_e, that is choosing $\hbar\omega_e = mc^2$, and we obtain according to Eq. 2

$$\mathbf{7}\quad \mathcal{E}_{\mathrm{cr}} \equiv \frac{mc^2}{\hbar c}\,\frac{mc^2}{e} = \frac{m^2c^3}{e\hbar}\,,$$

corresponding to the critical field E_{cr}, see Sect. 20.5 and Eq. (20.12). This field strength is in the following sense a field of natural unit strength: It is described alone in terms of elementary quantities e, m, c. In order to achieve this critical field using visible light with the typical energy of $\hbar\omega = 1\,\mathrm{eV}$ we need to compensate the low photon energy by a large value of laser wave amplitude with, $a_0^{\mathrm{cr}}|_{\hbar\omega=1\,\mathrm{eV}} \equiv mc^2/\hbar\omega = 511{,}000$. ◀

22.2 Role of Conservation Laws

A particle riding the field of a plane wave is subject to conservation laws arising from several symmetries of the problem. These allow us to constrain the particle dynamics considerably, creating an opportunity to verify that particle motion is subject to the effect of collective action of the coherent field rather than individual particle scattering processes.

For plane waves we have $V = 0$ and in transverse gauge the Lagrangian is

$$L(\boldsymbol{v}, z - ct) = -mc^2\sqrt{1 - v^2/c^2} + \boldsymbol{v} \cdot e\boldsymbol{A}(z - ct) \,, \tag{22.19}$$

where for a wave propagating along the z-axis we made explicitly visible the dependence on the 'light cone', coordinate $z - ct$, see Example 11.1.

$$e\boldsymbol{A} = mca_0\Re[\boldsymbol{e}_A e^{i\chi}] \,, \quad \chi = k(z - ct) \,, \tag{22.20}$$

where the dimensionless variable a_0 characterizes the strength of the 3-vector potential, see Example 22.1.

Two conservation laws emerge considering the explicit format Eq. (22.19):

I: The Lagrangian is independent of $\boldsymbol{r}_\perp$ and thus we have

$$0 = \frac{\partial L}{\partial \boldsymbol{r}_\perp} \to \frac{d}{dt}\frac{\partial L}{\partial \boldsymbol{v}_\perp} = 0 \quad \to \quad \frac{\partial L}{\partial \boldsymbol{v}_\perp} = Const. \,, \tag{22.21}$$

and thus

$$mc\frac{\boldsymbol{v}_\perp/c}{\sqrt{1 - v^2/c^2}} + e\boldsymbol{A}_\perp = \boldsymbol{C} \,. \tag{22.22}$$

II: For the change of energy of a particle we have according to Eq. (21.36) and using Eq. (21.11)

$$\frac{dE}{dt} = -\frac{\partial L}{\partial t} = c\frac{\partial L}{\partial z} = c\frac{d}{dt}\frac{\partial L}{\partial v_z} \quad \to \quad E = c\frac{\partial L}{\partial v_z} + Const. \tag{22.23}$$

Since in transverse gauge $A_z = 0$, the only contribution to the last term is from the kinetic energy term T_r. The result on RHS is obtained integrating over time and thus

$$E - mc^2 \frac{v_z/c}{\sqrt{1 - v^2/c^2}} = mc^2 . \tag{22.24}$$

Here the integration constant was chosen for a particle initially at rest, picking up a ride on the plane wave.

We use the two conservation laws to study constraints on particle dynamics. We use Eq. (22.24) to write

$$\begin{aligned}(E - mc^2)^2 &= m^2c^4 \frac{v_z^2/c^2 + v_\perp^2/c^2 - v_\perp^2/c^2}{1 - v^2/c^2} \\ &= m^2c^4 \frac{v^2/c^2 - 1 + 1}{1 - v^2/c^2} - m^2c^4 \frac{v_\perp^2/c^2}{1 - v^2/c^2} ,\end{aligned} \tag{22.25}$$

and substitute $E^2 = m^2c^4/(1 - v^2/c^2)$

$$E^2 - 2Emc^2 + m^2c^4 = -m^2c^4 + E^2 - m^2c^4 \frac{v_\perp^2/c^2}{1 - v^2/c^2} . \tag{22.26}$$

Solving for E we obtain

$$E = mc^2 + \frac{mc^2}{2} \frac{\boldsymbol{v}_\perp^2/c^2}{1 - v^2/c^2} = mc^2 + \frac{(e\boldsymbol{A}_\perp - \boldsymbol{C})^2}{2m} , \tag{22.27}$$

where the last equality follows by Eq. (22.22). According to the last equality in Eq. (22.27) the magnitude of the Lorentz factor for the electron riding the wave can be presented in terms of the dimensionless amplitude a_0 introduced earlier in Example 22.1

$$\frac{E}{mc^2} = \gamma \to 1 + \frac{1}{2} a_0^2 \, k , \qquad a_0 \equiv \frac{ceA_{|0}}{mc^2} . \tag{22.28}$$

We note the important factor a_0^2 governing the energy of the accelerated particle. We return to this equation in Sect. 22.3 where it will be demonstrated. The arrow in Eq. (22.28) emphasizes the qualitative nature of the result, where k is a numerical coefficient $\mathcal{O}(1)$.

We can relate the longitudinal-to-transverse velocity combining Eq. (22.24) with Eq. (22.27)

$$\frac{v_z/c}{\sqrt{1-v^2/c^2}} = \frac{1}{2}\frac{\boldsymbol{v}_\perp^2/c^2}{1-v^2/c^2}\,. \tag{22.29}$$

Condition Eq. (22.29) is valid for an electron initially at rest capturing a ride on the plane wave. We now show that this condition amounts to a relation between the angle θ of particle motion with reference to the wave axis and particle total energy. We have

$$v_\perp = v\sin\theta\,, \qquad v_z = v\cos\theta\,, \tag{22.30}$$

and thus

$$\tan^2\theta = \left(\frac{v_\perp}{v_z}\right)^2 = 2\frac{1-v^2/c^2}{v_z^2}\frac{(v_z/c)c^2}{\sqrt{1-v^2/c^2}} = 2\frac{\sqrt{1-v^2/c^2}}{v_z/c}\,. \tag{22.31}$$

Using the conservation law Eq. (22.24) we obtain

$$\tan^2\theta = \frac{2mc^2}{E-mc^2} = \frac{2}{\gamma-1}\,. \tag{22.32}$$

This quantitative statement was tested experimentally, as shown in Fig. 22.1. The results of measuring the energy of the accelerated electron and the angle θ of electron observation against the direction defined by the propagation of the laser pulse are shown.[3] We recall that in a realistic situation one obtains a relativistic strength 'plane wave' by focusing a relatively homogeneous laser pulse and capturing the particle in the focus. Thus this wave defocuses, leaving behind the surfing electron which can be later observed as shown in Fig. 22.1.

[3] C. I. Moore, J. P. Knauer, and D. D. Meyerhofer, "Observation of the Transition from Thomson to Compton Scattering in Multiphoton Interactions with Low-Energy Electrons", *Phys. Rev. Lett.* **74**, 2439, (1995).

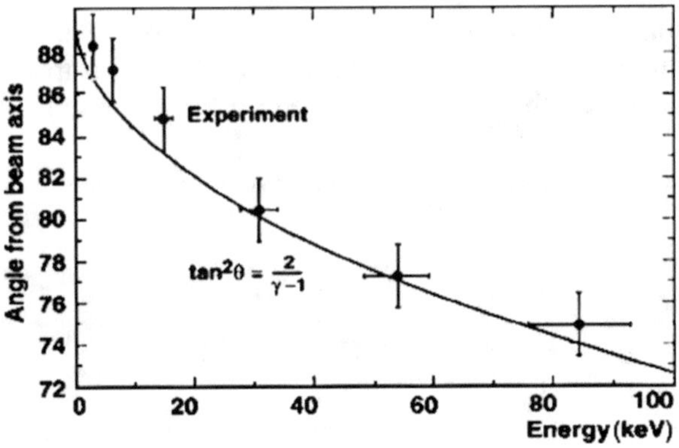

Fig. 22.1 Experimental verification of Eq. (22.32). (Adapted from Ref. [3])

Example 22.2

Laser community units

Connect the dimensionless plane wave unit a_0, see Example 22.1, to the following laser community units: power units GW and 1000 GW = TW (tera W) and W/cm^2. ◀

▶ **Solution** The intensity I of a laser is by definition measured in terms of the magnitude of the Poynting vector, see Eq. (22.18) and Eq. 1 in Example 22.1

$$\mathbf{1}\quad I \equiv |\langle \boldsymbol{S}\rangle| = \left|c^2\epsilon_0\langle \boldsymbol{\mathcal{E}} \times \boldsymbol{\mathcal{B}}\rangle\ \right| = \frac{|\boldsymbol{k}|\omega}{8\pi}\,(c\,e A_{|0})^2\frac{4\pi\epsilon_0}{e^2} = \frac{\pi}{2}\,\frac{a_0^2 c}{\lambda^2}\,(mc^2)^2\,\frac{4\pi\epsilon_0}{e^2}\,.$$

We note that intensity scales with a_0^2. The achievable power of a laser is obtained multiplying the intensity by the minimal geometric focal surface

$$\mathbf{2}\quad P_{\text{Laser}} \equiv \pi\lambda^2 I = \frac{\pi^2}{2}a_0^2\,8.7100\times 10^9\text{W} = a_0^2\ 42.982\ \text{GW}\,.$$

The SI-unit coefficient we see in Eq. 2 arises using mass-energy equivalent mc^2 of the electron, charge of the electron e, and classical electron radius r_e, Eq. (20.9)

$$\mathbf{3}\quad \frac{mc^2}{e} = 0.5110\times 10^6\text{V}\,,\qquad e = 1.602177\times 10^{-19}\text{C}\,,\qquad mc^2\frac{4\pi\epsilon_0}{e^2} = \frac{1}{r_e}\,.$$

We obtain for the prefactor in Eq. 1

$$\mathbf{4}\quad c(mc^2)^2\frac{4\pi\epsilon_0}{e^2} = \left(2.998\times 10^8\,\frac{\text{m}}{\text{s}}\right)\left(1.6022\times 10^{-19}\text{C}\right)\times\left(0.5110\times 10^6\,\text{V}\right)\frac{10^{15}}{2.8179\,\text{m}}\,.$$

Canceling the unit m(eter) and using C/s = A, AV = W, we obtain

5 $c(mc^2)^2 \frac{4\pi\epsilon_0}{e^2} = \frac{2.998 \times 1.6022 \times 5.110}{2.8179} 10^9\,\mathrm{W} = 8.7100\,\mathrm{GW}$,

the numerical result stated in Eq. 2. Thus a $P_{\mathrm{Laser}} = \mathrm{TW}$ (10^{12} W) power implies a relativistic regime with $a_0 = 4.82$. A petawatt (PW, 10^{15} W) scale laser can be focused to reach $a_0 > 150$. At the time of writing exawatt (EW, 10^{18} W) is on the visible horizon of technology, permitting $a_0 \to 5000$. We further note that when the energy in the pulse is a joule and the pulse length $\Delta t = 10^{-12}$ s, then when focused to geometric limit the power of TW is achieved as a time and space average.

Another popular laser power unit arises writing

6 $I = a_0^2\, 1.386 \times 10^{17} \frac{\mathrm{W}}{\mathrm{cm}^2} \left(\frac{\pi\,[\mu\mathrm{m}]}{\lambda}\right)^2$,

using as reference the surface of a cm^2 we obtain as seen in Eq. 6 an additional multiplicative factor 10^8 compared to Eq. 2. Omitting the last factor $\pi\,[\mu\mathrm{m}]/\lambda$ one sees how in literature statements about laser intensity I in units of $\mathrm{W/cm^2}$ can be presented. We recognize that a laser above peak intensity $10^{18}\,\mathrm{W/cm^2}$ requires $a_0 > 1$. To see this we invert the relation Eq. 6 to state the value of a_0 in terms of physical qualities of the laser pulse

7 $a_0 = 0.855 \times 10^{-9} \lambda\,[\mu\mathrm{m}] \sqrt{I[\mathrm{W/cm^2}]}$. ◀

22.3 Surfing the Plane Wave

We now solve the Lorentz force equations of motion for an electron in the field of an electromagnetic plane wave. We use the results of Sect. 22.2 to set up the dynamical equations as follows: Eq. (22.22) can be written in the form

$$\gamma \boldsymbol{v}_\perp / c = \gamma \frac{d\boldsymbol{r}_\perp}{d\,ct} = \frac{d\boldsymbol{r}_\perp}{d\,c\tau} = \boldsymbol{a}_0 + \boldsymbol{C}' \,, \tag{22.33}$$

where (possible confusion with acceleration should be avoided)

$$\boldsymbol{a}_0 \equiv \frac{-e\boldsymbol{A}}{mc} \tag{22.34}$$

and $\boldsymbol{C}' = \boldsymbol{C}/(mc)$ is an integration constant. Equation (22.33) becomes, using Eq. (22.29)

$$\gamma v_z/c = \gamma \frac{dz}{d\,ct} = \frac{dz}{d\,c\tau} = \frac{1}{2}\left(\boldsymbol{a}_0 + \boldsymbol{C}'\right)^2 , \tag{22.35}$$

where we used the proper time increment of the particle $d\tau = \gamma^{-1}dt$ to substitute for the laboratory time increment dt.

As noted in the opening of Sect. 22.2, the argument of $\boldsymbol{A}$ is $ct - z$. This variable satisfies

$$\frac{d(ct - z(\tau))}{d\,c\tau} = \gamma - \frac{dz}{d\,c\tau} = \gamma - \frac{1}{2}\left(\boldsymbol{a}_0 + \boldsymbol{C}'\right)^2 , \tag{22.36}$$

where the last relation follows using Eq. (22.35). For γ we obtain using the last form in Eq. (22.27)

$$\gamma = 1 + \frac{1}{2}\left(\boldsymbol{a}_0 + \boldsymbol{C}'\right)^2 , \tag{22.37}$$

which shows that

$$\frac{d(ct - z)}{d\,c\tau} = 1 , \Rightarrow ct - z = c\tau , \tag{22.38}$$

where we chose the coordinate system origin coinciding with the location of the particle for $\tau = 0$. Eq. (22.38) is the key result allowing direct integration of the equations of motion since the exponential phase of the plane wave is, according to Eq. (22.38), only a function of τ, the particle proper time. This implies that we will obtain the solution to the problem in terms of proper time τ of the particle riding the wave and not in terms of the laboratory time t.

We recapitulate the insights about the particle motion. According to Eq. (22.37) we have

$$\frac{cdt}{d\tau} = c\gamma = c + \frac{c}{2}\left(\boldsymbol{a}_0 + \boldsymbol{C}'\right)^2 , \tag{22.39}$$

and according to Eqs. (22.35) and (22.33) we have

$$\frac{dz}{d\tau} = \gamma v_z = \frac{c}{2}\left(\boldsymbol{a}_0 + \boldsymbol{C}'\right)^2 , \qquad \frac{d\boldsymbol{x}_\perp}{d\tau} = \gamma \boldsymbol{v}_\perp = c\boldsymbol{a}_0 + c\boldsymbol{C}' , \tag{22.40}$$

and we recall Eq. (22.34). We verify the consistency of Eqs. (22.39) and (22.40)

$$\begin{aligned} c^2 &= \left(\frac{c\,dt}{d\tau}\right)^2 - \left(\frac{dy}{d\tau}\right)^2 - \left(\frac{dx}{d\tau}\right)^2 - \left(\frac{dz}{d\tau}\right)^2 \\ &= c^2\left(1 + \frac{1}{2}\left(\boldsymbol{a}_0 + \boldsymbol{C}'\right)^2\right)^2 - c^2\left(\boldsymbol{a}_0 + \boldsymbol{C}'\right)^2 - \left(\frac{c}{2}\left(\boldsymbol{a}_0 + \boldsymbol{C}'\right)^2\right)^2 . \end{aligned} \tag{22.41}$$

We now consider in turn the motion of a charged particle riding linear (LP) and circular (CP) polarized plane waves.

LP: For a *linear polarized* plane wave with the polarization vector oriented in the *y-direction* we obtain, using Eqs. (22.33) and (22.20)

$$\frac{1}{c}\frac{dy}{d\tau} = -a_0 \cos\omega\tau + C' \,, \Rightarrow \frac{dy}{d\tau} = c\, a_0\,(1 - \cos\omega\tau) \simeq \frac{c}{2}\, a_0\, \omega^2\tau^2 \,, \tag{22.42}$$

where $C' = a_0$ is chosen for the case that the particle is initially at rest, and the start-up of motion for $\tau \simeq 0$ is shown explicitly. The particle transverse position

$$y = ƛ\, a_0\,(\omega\tau - \sin\omega\tau) \,, \qquad ƛ = \frac{c}{\omega} \,, \qquad \text{(LP)} \tag{22.43}$$

drifts with time to a large y. Here we used the 2π reduced wavelength $ƛ$ as the unit of length.

The motion along the propagation direction follows from Eq. (22.35)

$$\frac{1}{c}\frac{dz}{d\tau} = \frac{1}{2}(-a_0 \cos\omega\tau + C')^2 \,, \Rightarrow \frac{dz}{d\tau} = \frac{c a_0^2}{2}\,(1 - \cos\omega\tau)^2 \,, \tag{22.44}$$

where for $\tau = 0$ the particle is assumed to be at rest with respect to the wave. We note that the speed achieved is growing with a_0^2, and that the onset of motion along the propagation direction of the wave is highly nonlinear

$$\frac{dz}{d\tau} \underset{\tau\to 0}{\longrightarrow} \frac{c\, a_0^2}{8}\omega^4\tau^4 \,. \tag{22.45}$$

The solutions we presented for transverse motion $dy/d\tau$ Eq. (22.42), and parallel motion $dz/d\tau$ Eq. (22.44), are constrained by Eq. (22.29)

$$2c\frac{dz}{d\tau} = \left(\frac{dy}{d\tau}\right)^2 \Leftrightarrow \frac{2c\, v_z}{\sqrt{1-(v_z^2+v_y^2)/c^2}} = \left(\frac{v_y}{\sqrt{1-(v_z^2+v_y^2)/c^2}}\right)^2 . \tag{22.46}$$

This relation is manifestly satisfied by our solutions Eqs. (22.42) and (22.45).

A further integration yields

$$z = \lambda\!\!\!\bar{}\,\frac{a_0^2}{2}\left(\frac{3\omega\tau}{2} - 2\sin\omega\tau + \frac{1}{4}\sin 2\omega\tau\right), \qquad \lambda\!\!\!\bar{} = \frac{c}{\omega}, \quad \text{(LP)} \tag{22.47}$$

where we use the reduced light wave $\lambda\!\!\!\bar{} = \lambda/2\pi$ as the unit of length. The integration constant is chosen so that the coordinate origin is with the particle at $\tau = 0$.

According to Eq. (22.47) the particle moves forward but initially this is an exceedingly small effect $\propto \tau^5$. However, since the motion in the direction of wave propagation is proportional to a_0^2, for relativistic plane waves with $a_0 > 1$ the motion along with the wave wins over the transverse motion. Particles are pushed forward, but retain an asymmetry moving off the original axis in the polarization direction; we do not have azimuthal symmetry. We observe that for $a_0 < 1$ the transverse motion dominates. This clarifies the importance of achieving $a_0 > 1$ for the purpose of direct acceleration of electrons in the plane wave field.

The shape of the path of motion of the particle riding a linearly polarized wave according to Eqs. (22.43) and (22.47) is shown in Fig. 22.2a. The unit of length of the z-axis is the product of $\lambda\!\!\!\bar{}$ with a_0^2 and the unit of y-axis is the product of $\lambda\!\!\!\bar{}$ with a_0. Eliminating in both Eqs. (22.43) and (22.47) the terms growing linearly with particle proper time, we recognize in Fig. 22.2b a Lissajous figure. For $a_0 \gg 1$ the particle travels predominantly along the z-direction with path shape shown in Fig. 22.2c obtained omitting the term linear in proper time in Eq. (22.43). For the case when $a_0 \ll 1$ the path is shown in Fig. 22.2d obtained omitting term linear in τ in Eq. (22.47).

According to Eq. (22.37) we also have

$$\frac{dt}{d\tau} = \gamma = \frac{E}{mc^2} = 1 + \frac{a_0^2}{2}(1-\cos\omega\tau)^2 , \qquad \text{(LP)} \tag{22.48}$$

which shows that the energy of a particle riding a linearly polarized plane wave is bounded by

$$mc^2 \le E \le mc^2(1 + 2a_0^2) \,, \tag{22.49a}$$

while the average energy of a particle emerging from a diverging wave is obtained performing a proper time averaging of the energy ($\overline{\sin^4 x} = 3/8$)

$$\overline{E} = mc^2 \left(1 + 2\,a_0^2\, \overline{\sin^4(\omega\tau/2)}\right) = mc^2 \left(1 + a_0^2 \,\frac{3}{4}\right) . \tag{22.49b}$$

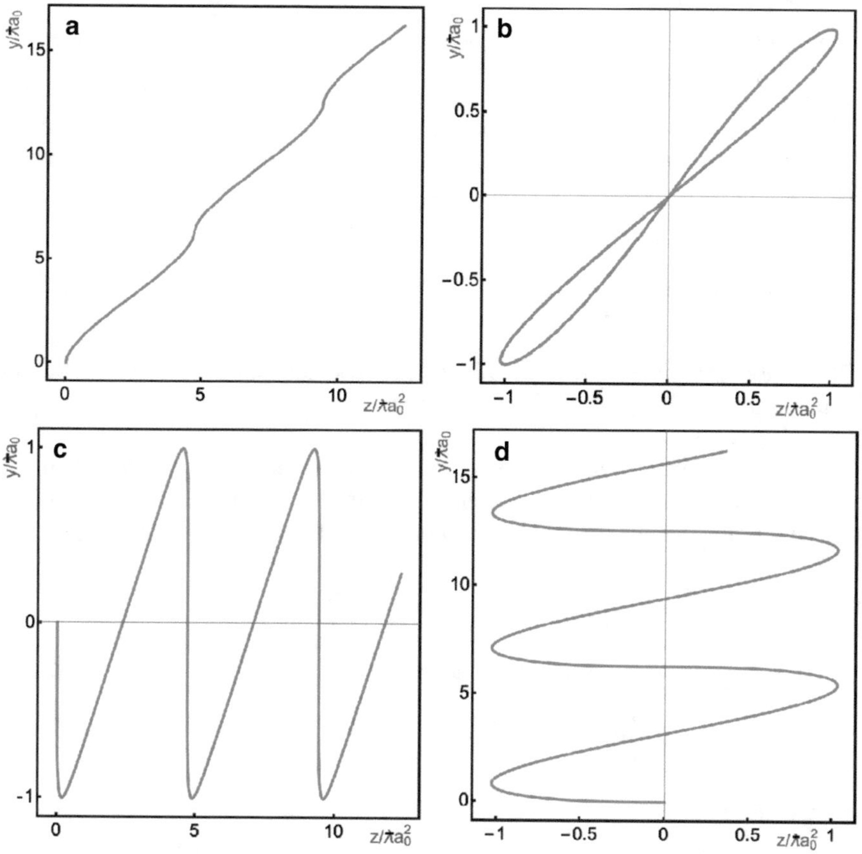

Fig. 22.2 The path of a particle in the y-z-plane according to Eqs. (22.43) and (22.47). Here a particle is riding a y-direction polarized light wave, z is the direction of wave propagation. The unit of the z-axis is $ƛa_0^2$; the unit of the y-axis is $ƛa_0$. In (**a**) we see the actual motion. In (**b**) we removed in Eqs. (22.43) and (22.47) the drift motion terms that grow linearly with proper time, which yields the Lissajous shape. In (**c**) we removed the drift in the y-direction, which shows the appearance of the path of motion for $a_0 \gg 1$. In (**d**) we removed the drift in the z-direction; this is the appearance of the particle motion for $a_0 \ll 1$

CP: For the *circular polarized* plane wave we return to Eq. (22.12) in Sect. 22.1, noting that now $\boldsymbol{A}$ is simply a function of τ

$$\begin{aligned}
\frac{1}{c}\frac{dx}{d\tau} &= -\frac{a_0}{\sqrt{2}}\cos\omega\tau + C'_x = \frac{a_0}{\sqrt{2}}(1-\cos\omega\tau)\,, \Rightarrow \omega x = c\,\frac{a_0}{\sqrt{2}}(\omega\tau - \sin\omega\tau)\,,\\
\frac{1}{c}\frac{dy}{d\tau} &= \pm\frac{a_0}{\sqrt{2}}\sin\omega\tau + C'_y = \pm\frac{a_0}{\sqrt{2}}\sin\omega\tau\,, \Rightarrow \omega y = \mp c\,\frac{a_0}{\sqrt{2}}\cos\omega\tau\,,\\
\frac{1}{c}\frac{dz}{d\tau} &= \frac{1}{2}\left(\frac{a_0^2}{2} + C'^2_x - 2\frac{a_0}{\sqrt{2}}C'_x\cos\omega\tau\right) + C'^2_z\\
&= \frac{a_0^2}{2}(1-\cos\omega\tau)\,, \Rightarrow \omega z = c\,\frac{a_0^2}{2}(\omega\tau - \sin\omega\tau)\,, \qquad \text{(CP)} \qquad (22.50)
\end{aligned}$$

where C'_x, C'_y and C'_z are chosen such that the particle is initially at rest.

According to Eq. (22.37) we also have

$$\frac{dt}{d\tau} = \gamma = \frac{E}{mc^2} = 1 + \frac{a_0^2}{2}(1-\cos\omega\tau)\,, \qquad \text{(CP)} \qquad (22.51\text{a})$$

while the upper limit and average of the electron energy is

$$E < mc^2(1+a_0^2)\,. \qquad \overline{E} = mc^2\left(1+\frac{a_0^2}{2}\right)\,. \qquad (22.51\text{b})$$

The motion described by Eqs. (22.50) and (22.51a) can be checked by subtracting from the square of Eq. (22.51a) the sum of squares from Eq. (22.50), as in the first line of Eq. (22.41).

We now consider an interesting special case where the velocity along with the wave direction is a constant in time and in the transverse direction there is only a circular motion. Such a specific CP solution emerges setting $\boldsymbol{C}' = 0$; that is,

$$\begin{aligned}
\frac{1}{c}\frac{dx}{d\tau} &= -\frac{a_0}{\sqrt{2}}\cos\omega\tau \Rightarrow \omega x = -c\,\frac{a_0}{\sqrt{2}}\sin\omega\tau\,,\\
\frac{1}{c}\frac{dy}{d\tau} &= \pm\frac{a_0}{\sqrt{2}}\sin\omega\tau \Rightarrow \omega y = \mp c\,\frac{a_0}{\sqrt{2}}\cos\omega\tau\,, \qquad (22.52)\\
\frac{1}{c}\frac{dz}{d\tau} &= \frac{a_0^2}{4} \Rightarrow z = \frac{a_0^2}{4}c\tau\,.
\end{aligned}$$

We also have

$$\frac{dt}{d\tau} = \gamma = \frac{E}{mc^2} = 1 + \frac{a_0^2}{4}(\cos^2\omega\tau + \sin^2\omega\tau) = 1 + \frac{a_0^2}{4}\,, \qquad (22.53)$$

which is constant. Since $\gamma = Const.$, see Eq. (22.48) we have

$$\tau = \frac{t}{\gamma} \,. \tag{22.54}$$

Using Eq. (22.54) in Eq. (22.52) allows us to describe the motion in laboratory variables alone. The position and longitudinal velocity of the particle as a function of laboratory time thus are

$$z = \frac{a_0^2/4}{1+a_0^2/4}\,ct \;\Rightarrow\; v_z \equiv \frac{dz}{dt} = c\frac{a_0^2/4}{1+a_0^2/4} < c \,. \tag{22.55}$$

Since the particle speed along the z-axis is ultrarelativistic for values of $a_0 \to 1000$ available in the foreseeable future, we evaluate longitudinal particle rapidity y_p, Eq. (15.22)

$$y_p = \frac{1}{2}\ln\left(1+\frac{a_0^2}{2}\right) \,. \tag{22.56}$$

According to Eq. (22.52) and using also Eq. (22.54)

$$\left(\frac{dx}{dt}\right)^2 + \left(\frac{dy}{dt}\right)^2 = c^2\left(\frac{a_0/\sqrt{2}}{1+a_0^2/4}\right)^2 , \qquad \rho \equiv \sqrt{x^2+y^2} = \lambda\frac{a_0}{\sqrt{2}} \,, \tag{22.57}$$

the motion type we found is helical as shown before in Fig. 21.2, displaying the motion in a longitudinal magnetic field.

The radius of circular motion is now governed by the light wavelength, and thus a comparison with motion in a constant magnetic field finds its limit: According to Eq. 13 in Example 21.2 if we set up using a magnetic field a rotation with $\omega = \omega_c$ the orbital speed is

$$v_0 = \rho\omega_c \;\to\; v_0^{\mathrm{eq}} = \rho\omega = \lambda\frac{a_0}{\sqrt{2}}\frac{c}{\lambda} = c\frac{a_0}{\sqrt{2}} \,. \tag{22.58}$$

We see that for $a_0 > \sqrt{2}$, in the relativistic optics domain, we would need a speed faster than light. Thus the helical particle motion in the CP plane wave becomes too extreme to be compared to particle motion in a constant magnetic field of any strength. It is further of interest to consider the strength of the magnetic field that could simulate the frequency of circular motion, see Example 21.2

$$\omega = \omega_c \;\to\; \frac{c}{\lambda} = \frac{e\mathcal{B}^{\mathrm{eq}}}{\gamma m} \,. \tag{22.59}$$

We find using for γ the value for the particle motion in the laser field, Eq. (22.53)

$$\mathcal{B}^{\rm eq} = \left(1+\frac{a_0^2}{4}\right)\frac{mc}{e\lambda\!\!\!^{-}} = \frac{m^2c^2}{e\hbar}\left(1+\frac{a_0^2}{4}\right)\frac{\hbar/mc}{\lambda\!\!\!^{-}} = B_{cr}\left(1+\frac{a_0^2}{4}\right)\frac{\lambda\!\!\!^{-}_C}{\lambda\!\!\!^{-}}, \tag{22.60}$$

where we introduced the Compton wavelength $\lambda\!\!\!^{-}_C$ and the critical field strength $B_{cr} = 4.414 \times 10^9\,$T, see Eq. (20.13). For the optical wavelength $\lambda = 1\,\mu$m we have $\lambda\!\!\!^{-}_C/\lambda\!\!\!^{-} = 2.425 \times 10^{-6}$. This means that for $a_0 \simeq 1.3 \times 10^3$ the CP laser plane wave induces particle motion features that are comparable to those generated by critical magnetic field.

To close this extensive discussion of relativistic charged particle motion in the presence of relativistic strength $a_0 > 1$ light wave, we should remember the tacit requirements permitting the use of the above idealized solutions:

(i) The growing with a_0 particle orbit means that the plane wave should not taper off in the transverse directions over a distance $a_0\lambda\!\!\!^{-}$ from the focal point.
(ii) In the longitudinal direction for the linear polarized wave the particle's primary acceleration occurs over the distance $a_0^2\lambda/(2\pi)$ around its mean location and thus the laser pulse that imitates a plane wave needs to be an approximate plane wave over this pulse time-length. For circular polarization a very similar result follows since we need to make a full circle which requires $\tau_0 = 2\pi/\omega$. Thus the pulse length in laboratory coordinate should be $\Delta z \geq a_0^2\lambda/4$.
(iii) Due to the strong acceleration accompanying the extreme conditions we described, the inclusion of radiation friction phenomena is necessary.

The first two remarks clash with laser pulse reality: To achieve a high value of a_0 one must both compress in time the pulse to as few as possible wavelengths, and simultaneously focuses towards geometric limit in the focal spot where the condition $a_0 \gg 1$ is achieved.

Clearly the solution we described cannot be pushed to these conditions. Moreover, we did not consider the process of how a particle encountering the pulse joins, or better, 'catches' or 'jumps', onto the wave. Therefore the study of motion in plane wave fields we present is but a rough orientation that must be considered with some caution when comparing to real life situations. The connection with experimental results is in general obtained in terms of numerical simulation of charged particle dynamics riding realistic light pulses. Because of the transforming impact of laser driven particle acceleration this is a very active present day research domain.

We have shown that from the point of view of relativistic physics the quantity that matters foremost in the study of particle dynamical problems is the value of a_0. One refers to lasers that achieve $a_0 > 1$ as operating in the relativistic regime.[4] The

[4] G. A. Mourou, Toshiki Tajima, and S. V. Bulanov, "Optics in the relativistic regime," *Rev. Mod. Phys.* **78** 309 (2006).

word "relativistic" reflects the profound change in particle dynamics when $a_0 > 1$. As we have seen in Eq. (22.28) in this condition a particle catching a laser wave can be accelerated and achieves relativistic dynamics.

It is of interest to the reader to know that the technology that allows relativistic optics with $a_0 > 1$ relies on non-monochromatic light waves: Superposition of nearly monochromatic light waves allows formation of finite length in time wave-trains of light. Using optical devices to change the distribution of wavelengths one can define a long wave train for the purpose of amplification, and later compress the wave train for the purpose of achieving high intensity. This method introduced for this purpose to optics by Strickland and Mourou[5] is called chirped pulse amplification (CPA).[6]

[5] D. Strickland and G. A. Mourou, "Compression of amplified chirped optical pulses," *Opt. Commun.* **56** 219 (1985).

[6] Gérard Mourou and Donna Strickland were recognized for this achievement with the 2018 Nobel prize.

Part X
Space Travel

In Part X: We describe interstellar travel for exploration of our galaxy, the Milky Way. We investigate the relativistic journey: describe the distance traveled, and consider the time dilation effect. We generalize our approach and show how one can address accelerated bodies in SR. We investigate how the conservation of energy and momentum determine the rocket equation. We show how the parameters of a star drive propulsion system should be chosen to maximize the payload fraction of the spaceship.

Introductory Remarks to Part X

Part X is dedicated to the reader dreaming about exploration of the Milky Way. Although this seems impossible today, it is not in contradiction with the basic laws of physics. Should we one day find a way (consider here $E = mc^2$) to provide for the energy needs of this expedition, we show how an exploration of the Milky Way, and perhaps one day even a greater region of the Universe, will become a possibility.

To see this we consider as an example a spaceship (subscript 's') capable of maintaining a constant proper acceleration, as example a comfortable $a_s = 1g = 9.81\,\mathrm{m/s^2}$ will be discussed. We will show that such a spaceship can travel beyond our galaxy.

We find that during a single lifespan (measured in the proper spaceship reference frame) the spaceship can travel $D \propto 100$ s million lyr. However, as result of time dilation, we expect a considerable difference between the spaceship time and the base, i.e. Earth, time: the time that passes in the inertial basis reference system is in essence the same as the distance traveled at speed of light. The time dilation factor is larger than a million.

For any expedition this will therefore be a one-way ticket. The time dilation effect separates explorers from the home base in time. Thus one of our objectives is to understand the effect of the time dilation given the desired distance D we want to travel.

In our exploration of the accelerated spaceship motion we come across inconsistencies related to LT being inapplicable when the body is accelerated. We therefore look in depth at the problem of accelerated body motion subject to arbitrary variable acceleration and determine that the gross inconsistencies disappear when the body motion is tracked in terms of light cone variables. This insight could be very useful in other domains of SR.

We then turn to study the rocket equation. We review the nonrelativistic case and develop the rocket equation that includes relativistic momentum and energy conservation laws dynamics and allows for the combustion of mass into energy. As the propellant mass is ejected, mass is also consumed, accounting for the ejected kinetic energy. The relativistic momentum and energy conservation laws constrain the rocket motion and energy evolution. The rocket retains, considering its kinetic motion, half of the original (rest) energy, but only a very tiny fraction of this energy is in the remaining rest mass (payload) of the rocket.

We use these results in a study aiming to optimize the star drive; that is, the relativistic propulsion engine, such that the payload is the largest possible fraction of the initial spaceship mass. We show that the engine exhaust beam must reach well into the relativistic domain in order to assure a large payload fraction compared to the initial total weight.

We close the book with a final student discussion concerning space travel in the coming century.

Travel in the Milky Way

23

Abstract

We study spaceship motion powered in the rest-frame with one g acceleration. We find that in a year of spaceship time, the expedition achieves travel close to speed of light. We evaluate the distance traveled as a function of the time that passes on the spaceship and the time that passes on Earth. Maintaining such powered flight for more than a decade of spaceship time allows the exploration of the entire Milky Way. We evaluate the time dilation between the traveler and the home base, finding an enormous time dilation effect. At the end we generalize our results to the case of arbitrarily accelerated body motion in SR.

23.1 Space Travel with Constant Acceleration

We explore now the constraints governing space travel at constant acceleration. Let us introduce the convention that quantities such as proper acceleration a_s with a subscript 's' refer to the frame of reference moving along with the spaceship, and quantities without a subscript refer to the inertial observer remaining at the Earth base.

We define a characteristic time related to the spaceship proper acceleration a_s to be

$$\tau_a \equiv \frac{c}{a_\mathrm{s}}\,, \qquad \text{for} \qquad a_\mathrm{s} = g \;\rightarrow\; \tau_a = 353.8 \text{ days}\,. \tag{23.1}$$

This simple computation shows that after accelerating for one year with $a_\mathrm{s} = g$, the explorer spaceship should be close to the speed of light.

J. Rafelski, *Modern Special Relativity*, https://doi.org/10.1007/978-3-030-54352-5_23

The spaceship subject to proper acceleration experiences in its frame of reference a differential velocity increase

$$a_s \equiv \frac{dv_s}{dt_s}\,, \qquad \to \qquad a_s dt_s \equiv dv_s\,. \tag{23.2}$$

As this equation actually defines the meaning of a_s, there can be no question posed regarding its validity. The increment of velocity dv_s is measured by an instantaneous inertial observer comoving with accelerated spaceship, not an inertial system. This will become relevant later on when we discuss behavior of a strongly accelerated body (spaceship), see Eq. (23.22), and the solution to this problem offered in box below Eq. (23.31).

In the rest-frame of the spaceship, the direction of proper acceleration is the preferred direction. In order to keep the present analysis at a reasonable level of complexity our spaceship keeps its prescribed linear course, and thus the increment of velocity in Eq. (23.2) always points in the same direction. For the observers remaining on Earth, the velocity increase dv must be calculated using the addition theorem for velocities Eq. (7.14)

$$v + dv = \frac{v + dv_s}{1 + v dv_s/c^2} = \frac{v + a_s dt_s}{1 + v a_s dt_s/c^2}\,. \tag{23.3}$$

$a_s dt_s$ is naturally very small compared to c, so we expand terms, taking $(a_s^2 dt_s^2) \to 0$ to obtain

$$v + dv \simeq v + a_s dt_s \left(1 - \frac{v^2}{c^2}\right)\,, \tag{23.4}$$

and therefore

$$a_s dt_s = \frac{dv}{1 - v^2/c^2} \equiv \frac{a\,dt}{1 - v^2/c^2}\,. \tag{23.5}$$

From Eq. (23.5) we learn how the acceleration a_s, experienced on the spaceship, is observed on Earth, given the Earth measurement of the speed of the spaceship. The validity of Eq. (23.5) is verifiable in the laboratory, for example when we explore the motion of charged particles subject to EM acceleration by the Lorentz force, Eq. (21.10).

We integrate Eq. (23.5) and obtain

$$a_s t_s = c \ \mathrm{artanh}\,\frac{v}{c}\,. \tag{23.6}$$

We have set the integration constant to zero since we require $v = 0$ at $t = t_s = 0$. The velocity of the spaceship as a function of the ship's proper time follows directly from Eq. (23.6)

$$v = c \tanh \frac{t_s}{\tau_a} , \tag{23.7}$$

where we used Eq. (23.1). From Eq. (23.7) we obtain the following algebraic identities we will find useful

$$\frac{1}{\sqrt{1 - (v/c)^2}} = \gamma \equiv \cosh y = \cosh \frac{t_s}{\tau_a} , \tag{23.8}$$

$$\frac{v/c}{\sqrt{1 - (v/c)^2}} = \frac{v}{c}\gamma \equiv \sinh y = \sinh \frac{t_s}{\tau_a} . \tag{23.9}$$

We recognize, using Eqs. (23.8) and (23.9), that Eq. (23.7) establishes a relation to the rapidity y of the spaceship, see Sect. 7.4. Considering Eq. (7.35) we have

$$y - y_0 = \frac{t_s - t_0}{\tau_a} = \frac{a_s}{c} (t_s - t_0) . \tag{23.10}$$

Note that in keeping with our nomenclature we omit an index for variables observed from the Earth base; a possible confusion with space coordinate y will not occur here. Thus in Eq. (23.10) y is the spaceship rapidity as measured by the Earth observer, while a_s is the spaceship acceleration measured in the spaceship frame of reference, and t_s is the proper spaceship time.

We obtain from Eq. (23.10)

$$\tau_a dy = dt_s , \qquad c dy = a_s dt_s . \tag{23.11}$$

Equation (23.11) describes the increase in spaceship rapidity, given the intrinsic-to-spaceship acceleration inherent in the factor τ_a, Eq. (23.1), and spaceship time. We could have obtained Eq. (23.11) directly from Eq. (23.5) by considering in Eq. (23.8) the incremental change

$$\frac{v dv}{(1 - (v/c)^2)^{3/2}} \equiv c^2 \sinh y \, dy \quad \rightarrow \quad \frac{dv}{1 - (v/c)^2} \equiv c \, dy . \tag{23.12}$$

We used here the definition Eq. (23.9). Given Eq. (23.5), we find Eq. (23.12).

23.2 The Effect of Time Dilation

The spaceship time t_s is related to Earth time t by the Lorentz coordinate transformation. We consider short increments of time and space

$$dt_s = \frac{dt - (v/c^2)dx}{\sqrt{1-(v/c)^2}} . \tag{23.13}$$

We rearrange terms and square to obtain

$$dt_s^2(1-(v/c)^2) = \left(1-(v/c^2)\frac{dx}{dt}\right)^2 dt^2 . \tag{23.14}$$

But we know that $\frac{dx}{dt} = v$, therefore

$$dt_s = \sqrt{1-(v/c)^2}dt, \qquad dt = \frac{dt_s}{\sqrt{1-(v/c)^2}} . \tag{23.15}$$

This is the usual time dilation formula: The time on the spaceship t_s advances slower than time on Earth t. We have re-derived this relation to show that for a small acceleration the rules of SR coordinate transformations and time dilation remain.

Using Eq. (23.8) we now obtain a relation between the time increments dt_s and dt, independent of the other variables

$$dt = dt_s \cosh\frac{t_s}{\tau_a} . \tag{23.16}$$

Integration yields

$$t = \tau_a \sinh\frac{t_s}{\tau_a} . \tag{23.17}$$

For a travel time short compared to τ_a, we see in Eq. (23.17) a linear relationship $t \simeq t_s$. However, when the spaceship proper travel time lasts longer compared to τ_a, the time on Earth t grows exponentially as a function of τ_a, leading to surprising outcomes.

For example, while for the spaceship crew 10 years pass, the Earth observer aged nearly 15,000 years. Since observed from Earth the spaceship traveled almost the entire time with speed of light, this is also in terms of ly the distance traveled. More precisely, an expedition capable of a steady one g acceleration will cross the Milky Way in 11.5 years travel time; after 20 years of travel time, the Earth base is more than 450 million years older. In this 20-year spaceship travel time the Earth could experience multiple mass extinctions, while the position of the solar system within the Milky Way changed drastically, hardly allowing the spaceship crew to find a path back to Earth. We will return to review this situation more precisely in Examples 23.1 and 23.2.

23.3 How Far Can We Travel?

We would like to refine the qualitative result we just presented. We seek to determine the travel distance D as a function of the spaceship time t_s and Earth time t. We note that the right-hand side of Eq. (23.17), with the help of Eq. (23.9), yields a relation between t and $v = dx/dt$

$$t = \tau_a \frac{v/c}{\sqrt{1-(v/c)^2}} . \tag{23.18}$$

Solving this algebraic equation for $v(t)$ results in

$$\frac{v}{c} = \frac{t/\tau_a}{\sqrt{1+(t/\tau_a)^2}} . \tag{23.19}$$

We now integrate $dx = vdt$, to obtain the distance D traveled by the ship as a function of Earth time t

$$D = \int_0^x dx = \int_0^t v\,dt = c\tau_a \int_0^t \frac{t/\tau_a}{\sqrt{1+(t/\tau_a)^2}} \frac{dt}{\tau_a} = c\tau_a \left(\sqrt{1+\left(\frac{t}{\tau_a}\right)^2} - 1 \right) . \tag{23.20}$$

For the case of a long Earth travel time $t \gg \tau_a$ we see that Eq. (23.20) agrees well with the estimate we made at the end of Sect. 23.2: $D \to ct$. We further note that for a short travel time the power law expansion of the root function leads to the classic result $D = a_s t^2/2$.

To find the travel distance as a function of spaceship time, i.e. $D(t_s)$, we substitute for t in Eq. (23.20) using Eq. (23.17)

$$D = c\tau_a \left(\cosh \frac{t_s}{\tau_a} - 1 \right) . \tag{23.21}$$

The two relations, Eqs. (23.20) and (23.21), describe the same world line traced from an origin on Earth to the spaceship location $x = D$ reached at respective time values. We see that we parameterized this world line (x, t) using the Earth time t in Eq. (23.20); and for (x, t_s) using the spaceship time t_s in Eq. (23.21).

The motion of the uniformly accelerated spaceship as observed by the Earth-bound observer appears light-like. To show this we evaluate the world line deviation from the light cone as a function of spaceship time. We combine Eq. (23.21) with Eq. (23.17) and use $(\cosh x - 1)/2 = \sinh^2(x/2)$

$$c^2t^2 - D^2 = \left(ct_s \frac{\sinh(t_s/2\tau_a)}{(t_s/2\tau_a)}\right)^2 \underset{a_s \to 0}{\longrightarrow} c^2t_s^2 . \tag{23.22}$$

The limit of negligible acceleration, $a_s \to 0$; i.e., $\tau_a \to \infty$, generates the usual proper time definition for a cruising spaceship; this is an expected result agreeing with the Lorentz coordinate transformation.

Given the exact nature of the result Eq. (23.22) we can also look at the opposite limit, $a_s \to \infty$; that is, $\tau_a \to 0$. We see here an unexpected behavior: The motion of the spaceship deviates exponentially from the light cone. This result calls for a deeper reconsideration as it is an indication that some principles of SR are inapplicable to strongly accelerated motion. We address this matter at the end of this chapter.

It is of considerable interest to consider the travel time to a prescribed target a distance D away. We thus solve Eqs. (23.20) and (23.21) for the variables t (Earth base time) and t_s (spaceship proper time)

$$t = \tau_a \sqrt{\left(\frac{D}{c\tau_a} + 1\right)^2 - 1} , \qquad t_s = \tau_a \operatorname{arcosh}\left(\frac{D}{c\tau_a} + 1\right) . \tag{23.23}$$

The asymptotic limits for $D \gg c\tau_a$ are

$$t \to \frac{D}{c} , \qquad t_s \to \tau_a \ln\left(\frac{2D}{c\tau_a}\right) . \tag{23.24}$$

Before moving on, let us remark that for the observer on Earth the travel time to a distant location is just about what the light would take, so a target 1000 ly away will be reached after 1000 years. However, the time in the spaceship advances slowly. Since $c/a_s \simeq 1$y, the travel time would be $t_s \simeq \ln 2000$ years; that is, about 7.5 years. By the time the expedition returns to home base, two millenniums have passed, while time practically stood still in the spaceship; the traveler is just 15 years older. This becomes even more extreme considering the time needed to cross the Milky Way: In the reference frame of Earth this will take about $D = 100{,}000$ ly; however only $t_s \simeq \ln 100{,}000 \simeq 11.5$ years will elapse for the traveler for the one-way trip.

The here described possible travel into Galactic interstellar domain creates a new enigma. Enrico Fermi[1] posed a famous question: "Where is everybody?" referring to absence of a contact with more advanced civilizations. Through this question Fermi was expressing the thought

> ... if technological life existed anywhere else, we would see evidence of its visits to Earth – and since we do not, such life does not exist, or some special explanation is needed.

This question was posed during a luncheon conversation with Emil Konopinski, Edward Teller, and Herbert York in the summer of 1950. Fermi's companions on that day have provided accounts of the incident.[2] This book's author likes E. Teller's response (*loc. cit.*) "... as far as our galaxy is concerned, we are living somewhere in the sticks, far removed from the metropolitan area of the galactic center." There are other interesting discussions of this so-called Fermi Paradox,[3] and a modern exploration of the Fermi question has been presented.[4]

We will further discuss space travel in Chap. 24, characterizing constraints that arise from the consideration of energy and momentum conservation for the powered flight. This will allow a first consideration of technological constraints, setting limits to our travel ambitions this century.

Example 23.1

Exploration of the Vega Star system

Consider the duration of the trip to Vega, the brightest star in the constellation Lyra, located 25 ly away from the Earth. In order to arrive in minimum time with maximum comfort we assume that the spaceship accelerates at $a = 1g$, and decelerates at the same rate after the mid-point of the trip. We allow for six months of research at Vega, and assume that the spaceship returns to Earth using the same sequence of acceleration and deceleration. ◀

▶ **Solution** The ship's space-time path, from the perspective of the Earthbound observer, is given by Eq. (23.23) and is shown in Fig. 23.1. The spaceship travels to

[1]Enrico Fermi (1901–1953), Italian-American physicist, an accomplished theorist, experimenter and inventor. The first nuclear reactor was designed and built by Fermi, and we call quantum particles that obey the Pauli principle fermions. Nobel Prize 1938.

[2]E.M. Jones, " 'Where is Everybody?' An Account of Fermi's Question," Los Alamos Report LA-10311-MS DE85 011898 https://doi.org/10.2172/5746675.

[3]R.H. Gray, "The Fermi Paradox Is Neither Fermi's Nor a Paradox," *Astrobiology* **15** 195 (2015).

[4]S. Webb, *If the Universe Is Teeming with Aliens ... Where Is Everybody?: Seventy-Five Solutions to the Fermi Paradox and the Problem of Extraterrestrial Life* (Springer Science and Fiction, 2nd ed. 2015); M. Livio and J. Silk, "Where are they?" *Physics Today* **70**, (3) 50 (2017).

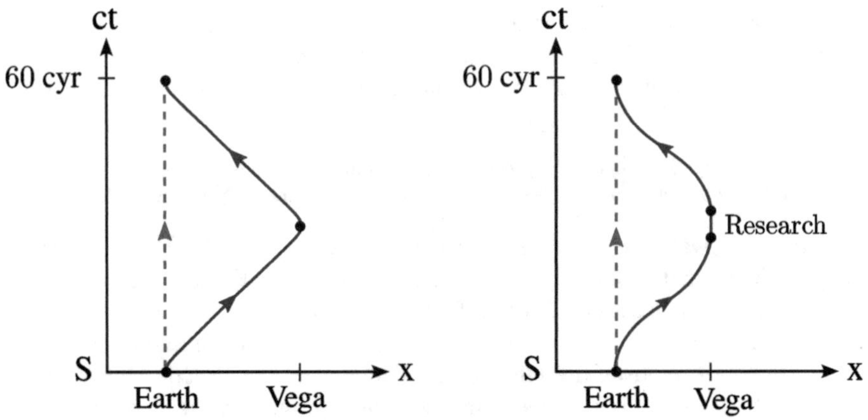

Fig. 23.1 Space-time diagram of a spaceship's voyage to Vega and back (curved line, blue) and an observer remaining on Earth (straight line, red). The actual path is shown on the left. On the right, the acceleration and research periods at Vega have been emphasized for clarity. See Example 23.1

Vega in two stages of $D = 12.5$ ly. We insert therefore the value 12.5 ly for D and we use 1 ly $= c \cdot 3.16 \times 10^7$ s, see also Eq. (2.3):

$$t = \frac{299{,}800\,\text{km/s}}{9.81\,\text{m/s}^2}\sqrt{\left(\frac{9.81\,\text{m/s}^2 \cdot 12.5 \cdot 3.16 \cdot 10^7\,\text{s} \cdot c}{299{,}800\,\text{km/s} \cdot c} + 1\right)^2 - 1} = 13.4 \text{ years}\,,$$

$$t_s = \frac{299{,}800\,\text{km/s}}{9.81\,\text{m/s}^2}\,\text{arcosh}\left(\frac{9.81\,\text{m/s}^2 \cdot 12.5 \cdot 3.16 \cdot 10^7\text{s} \cdot c}{299{,}800\,\text{km/s} \cdot c} + 1\right) = 3.22 \text{ years}\,.$$

We form the entire trip from four such increments. Including the half-year of research time we obtain the total round trip time

$$\begin{aligned}(\text{Earth})\ t_E &= 4t\ + 0.5 \text{ years} \simeq 54.1 \text{ years}\,,\\ (\text{traveler})\ t_{s\,T} &= 4t_s + 0.5 \text{ years} \simeq 13.4 \text{ years}\,.\end{aligned}$$

The main point of this example is that from the perspective of the spaceship crew, Vega, a very interesting (scientifically) star is reachable within a reasonable expedition schedule. The 'only' problem is that everyone at the base has aged by additional 40 years compared to members of the expedition. We do not need to worry too much about that as the realization of a constant 1 g-acceleration travel including a decade fuel supply is not on the current technology horizon. ◀

23.4 Variable Acceleration

We now allow for a variable $a_s(\tau)$ acceleration of a body, generalizing the discussion seen in Sect. 23.1. The quantity τ is the body proper time which we called previously t_s; the latter quantity is now reserved for the end-point of proper time evolution. In the following we will use light cone coordinates introduced in Example 11.1, a review of this example is recommended.

As before, the motion of the body is on a straight line. The instantaneous acceleration is, as before, defined by Eq. (23.2), since it does not matter here that it is not constant. This remark also applies to the incremental body speed relation Eq. (23.5), describing change in speed as seen in the laboratory frame and by the body instantaneous inertial comoving observer. We thus begin with

$$\frac{dv}{1-(v/c)^2} = a_s(\tau)d\tau \ . \tag{23.25}$$

We integrate this allowing for a variable proper body acceleration $a_s(\tau)$. This introduces a change to Eq. (23.6) and its inverse Eq. (23.7) as follows

$$\int_0^{t_s} a_s(\tau)d\tau = c \ \text{artanh}\, \frac{v(t_s)}{c} \ , \qquad \to \qquad v(t_s) = c \ \tanh \frac{1}{c}\int_0^{t_s} a_s(\tau)d\tau \ . \tag{23.26}$$

The relations seen in Eq. (23.26) describe how the laboratory observer measured body speed $v(t)$ and the variable proper acceleration $a_s(\tau)$ are related. The laboratory time t and the proper time of the body t_s describe the same instantaneous situation of the body motion; dt is the increment of the laboratory time and dt_s is the increment of the body time.

We insert the result Eq. (23.26) on the right-hand side of Eq. (23.15) and obtain

$$dt = \frac{dt_s}{\sqrt{1-(v/c)^2}} \qquad \to \qquad dt = dt_s \cosh \frac{1}{c}\int_0^{t_s} a_s(\tau)d\tau \ . \tag{23.27}$$

Considering the prior case $a_s = Const.$, we can integrate Eq. (23.27) to obtain Eq. (23.17). Inspection of Eq. (23.10) allows us to recognize the argument of cosh-function in Eq. (23.27) as the body rapidity $y_s(t_s)$, written in terms of body proper time at any given instant of the acceleration process. We obtain

$$dt = dt_s \cosh\left(y_s(t_s)\right) \ . \tag{23.28}$$

We now determine the distance dx, which the body travels during the time increments dt. We use here the right hand side of Eq. (23.26) with $v = dx/dt$ and replace dt using Eq. (23.28)

$$dx = cdt\frac{v}{c} = cdt_s \sinh \frac{1}{c}\int_0^{t_s} a_s(\tau)d\tau = cdt_s \sinh\left(y_s(t_s)\right) \ . \tag{23.29}$$

Considering the prior case $a_s = Const.$, we can integrate the argument of sinh-function in the middle term of Eq. (23.29) to obtain $a_s t_s$. Final integration of both sides leads to the earlier form Eq. (23.21).

We now form the two linear combinations of Eqs. (23.28) and (23.29) with the result

$$dx_{\pm} \equiv d(ct \pm x) = cdt_s e^{\pm y_s(t_s)} \ . \tag{23.30}$$

We realize that this relation corresponds exactly to the Lorentz transformation of the light cone coordinates with $dx_s = 0$; see Example 11.1. Acceleration is not seen as influencing Eq. (23.30).

We learn that acceleration does not affect incremental values of light cone coordinates, Eq. (23.30), which can be found by performing a Lorentz coordinate transformation into a comoving frame of reference. We note the related exact result

$$dct^2 - dx^2 = dx_+ dx_- = c^2 dt_s^2 \ . \tag{23.31}$$

Since Eq. (23.30) is identical to the result we obtain using an instantaneous Lorentz transformation, see Example 11.1, we can use LT of light cone variables $dx_+ = (dct + x)$ and $dx_- = d(ct - x)$ to track the non-inertial body motion, integrating in continuous fashion along the world path. This remark does not apply to the usual space-time increments dct and dx separately.

As just noted, even though we can obtain Eq. (23.30) with the help of a Lorentz coordinate transformation to the instantaneous comoving inertial frame of an accelerated body, this is not true for the integrated values of the body coordinates: each of the two light cone variables needs to be integrated individually. We have

$$ct = \frac{x_+ + x_-}{2} = \frac{1}{2}\left(\int_0^{t_s} cd\tau\, e^{y_s(\tau)} + \int_0^{t_s} cd\tau\, e^{-y_s(\tau)}\right) , \tag{23.32}$$

$$x = \frac{x_+ - x_-}{2} = \frac{1}{2}\left(\int_0^{t_s} cd\tau\, e^{y_s(\tau)} - \int_0^{t_s} cd\tau\, e^{-y_s(\tau)}\right) . \tag{23.33}$$

This yields a modified relation of the body light cone position to its proper time

$$(ct)^2 - x^2 = x_+ x_- = c^2 t_s^2 \, \frac{\int_0^{t_s} cd\tau \; e^{y_s(\tau)}}{\int_0^{t_s} cd\tau} \, \frac{\int_0^{t_s} cd\tau \; e^{-y_s(\tau)}}{\int_0^{t_s} cd\tau} \,. \tag{23.34}$$

Any final value of the acceleration integrated over history of the body motion accumulates in this expression.

The limit of a constant acceleration verifies Eqs. (23.32) and (23.33), and thus Eq. (23.34)

$$ct \pm x = \pm \frac{c^2}{a_s} \left(e^{\pm t_s a_s / c} - 1 \right) . \tag{23.35}$$

We form the product of both factors $ct \pm x$ to obtain, using $(\cosh z - 1)/2 = \sinh^2(z/2)$, the desired result Eq. (23.22). This result Eq. (23.35) confirms that we have achieved a general solution even though the form of Eq. (23.34) is different from the one we have seen in Eq. (23.22).

Further reading: In connection to GR a coordinate set that follows from Eq. (23.35) is called Rindler coordinates[5]appearing under heading "uniformly accelerated observer". These coordinates are discussed in many modern GR textbooks. We note that our presentation is not related to the GR context and applies to arbitrarily accelerated material bodies.

Example 23.2

Aging in space travel

Two spaceships leave Earth with the assignment of investigating habitable planets at distant stars and then returning to Earth 100 years later Earth time. One ship flies to star x_1 and then returns to Earth; the other ship flies in the opposite direction to stars x_2 and then x_3 before returning. (a) Draw a space-time diagram of the expedition. (b) What is the relative age of both crews at the time of return? (c) How can each of the crews track Earth time so that they can be back on time? ◀

▶ **Solution**

(a) The expeditions' world lines are depicted in Fig. 23.2. We see, akin to prior Example 23.1, that both expeditions rapidly achieve proximity of the speed of light; that is, the world lines are for relatively long periods of time nearly parallel to the light cone. We further see that the expedition that made two stops could not go as far as the other, otherwise it would not be back on time.
(b) The spaceship crews age with their respective proper times

[5]W. Rindler, "Hyperbolic Motion in Curved Space Time," *Phys. Rev.* **119** 2082 (1960).

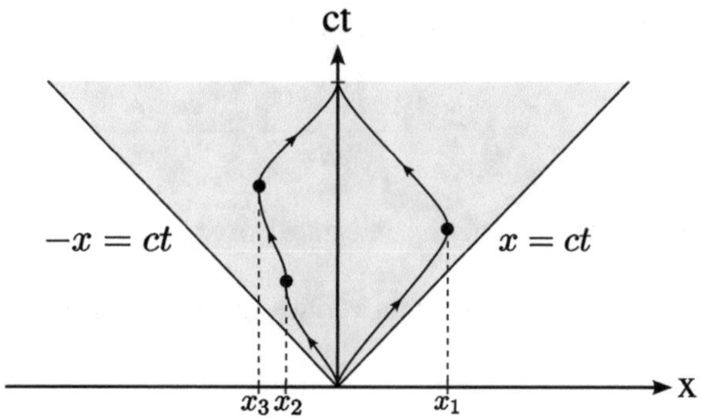

Fig. 23.2 Space-time diagram of voyages to stars at locations x_1 (on right), and x_2, x_3 (on left). See Example 23.2

$$\mathbf{1}\quad \text{AGE} \equiv \int_0^T dt_\text{s} = T\ .$$

Considering the relative proper age we note that the second expedition could not travel as far as the first. For this reason the length of the left side world line in Fig. 23.2 is shorter compared to the world line on right. This means that time dilation effect is smaller; see Example 12.2. In consequence, the crew of this expedition is older compared to the first expedition seen on the right in Fig. 23.2, which could travel further, and which has therefore a longer world line.

(c) We integrate Eq. (23.28) and obtain the Earth time t as a function of the proper time of the spaceship

$$\mathbf{2}\quad 100\,\text{years} = t = \int_0^{T_i} \cosh\left[\frac{1}{c}\int_0^{t_\text{s}} a_{i,\,\text{s}}(\tau)d\tau\right] dt_\text{s}\ , \qquad i = 1, 2\ .$$

Given the proper acceleration $a_{i,\,\text{s}}$ and the Earth time $t = 100$ years the spaceship crews can determine the desired return event T_i, $i = 1, 2$, which depends on the entire history of their respective expeditions.

When using this result one must also consider the question what happens during (a) unaccelerated periods, and (b) the braking phase. When the space ship coasts, the time dilation continues to accumulate, governed by the acquired relative velocity. Moreover, we note that the sign of a_s does not survive the even nature of the cosh-function. We further see that the integrand in the square bracket vanishes for vanishing spaceship relative speed. This means that in this 'return to base case' no additional increase in time dilation occurs; that is, $dt = dt_\text{s}$. The earlier accumulated time dilation is inherent to the external second integral. Since $\cosh z \geq 1$ we also recognize that $t \geq t_\text{s}$, the equality applies to $a_\text{s}(\tau) = 0$. The spaceship's proper time is for any travel route always shorter compared to base time. ◀

Relativistic Rocket Equation

24

Abstract

Relativistic rocket equations are derived based on energy-momentum conservation that constrains the motion of a relativistic rocket where a mass increment Δm separates at a prescribed exhaust speed from the main rocket body $m \gg \Delta m$. In response the rocket is accelerated forwards. The key new element as compared to nonrelativistic formulation is the conservation of energy, which includes the rest energy of both the rocket and ejected material. We obtain the relativistic form of the rocket equation, describing the rocket motion and corresponding evolution in time of the rocket mass.

24.1 Nonrelativistic Rocket Equation

We recapitulate the momentum conservation argument that leads to the rocket equation in nonrelativistic physics.[1] The key physics concept here is that in order to change the momentum of the payload one must eject some mass into the direction opposite to the desired change of the payload momentum. We begin by balancing momentum before and after the increment Δm mass emission.

Immediately prior to the ejection of mass increment

$$\Delta m = -dm \,, \tag{24.1}$$

[1]This insight is named after Konstantin Tsiolkovsky, though the work of Tsiolkovsky on the rocket equation occurred in parallel to a mathematical study by Ivan V. Meshchersky. Both were preceded by the work of William Moore of 1813, see: W. Johnson "Contents and commentary on William Moore's: A treatise on the motion of rockets and an essay on naval gunnery," *International Journal of Impact Engineering* **16** (3) 499 (1995). However, only Tsiolkovsky worked out many practical results for space travel that follow from a consistent application of this study and he is widely considered the grandfather of the space age.

J. Rafelski, *Modern Special Relativity*, https://doi.org/10.1007/978-3-030-54352-5_24

the magnitude of momentum of the nonrelativistic rocket is

$$p_1 = (m + \Delta m)v \,. \tag{24.2}$$

The introduction of $-dm$ in Eq. (24.1) reminds us that while Δm is a positive quantity, it is actually a negative of the negative change; i.e. $-(-\Delta m)$, in the mass of the rocket m.

After the ejection of the material we combine two contributions, the rocket and its exhaust, to obtain the new momentum p_2 of the system

$$p_2 = m(v + \Delta v) + \Delta m\, v_{\text{ex}} \,, \tag{24.3}$$

where the speed of exhausted mass is

$$v_{\text{ex}} = v - \mathcal{V} \,. \tag{24.4}$$

Here the propellant speed $\mathcal{V}$ is in the rest-frame of the rocket and derives from characteristics of the engine. We assumed that the exhaust velocity $\boldsymbol{\mathcal{V}}$ is pointing opposite to the velocity of the rocket $\boldsymbol{v}$.

Requiring momentum conservation $p_1 = p_2$, and combining the above equations, after a few simple steps we find

$$\Delta m \mathcal{V} = m\, \Delta v \,. \tag{24.5}$$

On the left we see the momentum carried out by the exhausted matter Δm and on the right the change in momentum of the rocket. Note that all dependence on the speed of the rocket v with reference to an arbitrary inertial observer has canceled. This is so since result Eq. (24.5) is independent of the choice of an observer. In particular, we could have started in the rocket rest-frame, obtaining Eq. (24.5) directly.

After we use Eq. (24.1) in Eq. (24.5) the integral form of Eq. (24.5) follows

$$\int dv + \mathcal{V} \int d \ln m = 0 \,, \tag{24.6}$$

yielding the 'Tsiolkovsky' speed-mass relationship,

$$v - v_0 = \mathcal{V} \ln \left(\frac{m_0}{m(v)} \right), \qquad m(v) = m_0 \exp \left(-\frac{v - v_0}{\mathcal{V}} \right) . \tag{24.7}$$

We see that for a required speed, e.g. escape speed to orbit, the payload m decreases exponentially from the initial mass m_0 for a given $\mathcal{V} \ll v$. Considering the required speed v, any increase in $\mathcal{V}$ leads to a major change in payload mass-fraction that can be retained. The speeds that need to be achieved to depart from Earth are at the scale $v \simeq 10\,\text{km/s}$, while for the best chemical rockets, the exhaust speed is $\mathcal{V} \lesssim 3\,\text{km/s}$.

Novel accelerator based ion propulsion technology developed in recent decades increases the value of $\mathcal{V}$ significantly, enabling the navigation between different spatial objects. The NASA spacecraft DAWN, which in 2011–2016 visited two asteroids, Vesta and Ceres, utilizes an electrically driven ion accelerator in order to achieve $\mathcal{V} = 10\,\text{km/h}$. However, at maximum 'throttle', it would take DAWN four days to accelerate from 0 to 100 km/h. On the other hand, the ion propulsion system was set to operate for thousands of days.

24.2 Relativistic Rocket Equation

To implement correctly relativistic generalization of the rocket equation we must use both momentum and energy conservation, and we must allow that use of energy $\delta E = \delta mc^2$ is required in order to eject mass increment Δm: we consume δm of the rocket mass ejecting Δm. This is so since we learned, see Sect. 16.1, that any energy required to drive an engine must come from the conversion of rest mass into energy. It is obvious that if the fuel ejection speed is relativistic, $\delta m > \Delta m$ when the ejected particle kinetic energy is larger compared to rest energy.

In order to not complicate our considerations too much we assume that the only mass-energy loss or gain experienced by the spaceship is due to the energy carried away by particles propelling it forwards. Thus Eq. (24.1) is modified to read

$$\delta m + \Delta m = -dm\ . \tag{24.8}$$

To account for momentum conservation we substitute in Eq. (24.2) for v/c the relativistic $\gamma\beta = \sinh y$, where y (no confusion here with the space coordinate) is the rapidity achieved by the spaceship rocket as seen by the base observer. Just before mass ejection we have

$$p_1 = (m + \Delta m)\, c\, \sinh y\ . \tag{24.9}$$

After mass ejection, we combine the two contributions just like in Eq. (24.3) and

$$p_2 = \widetilde{m}\, c\, \sinh(y + \Delta y) + \Delta m\, c\, \sinh y_{\text{s ex}}\ . \tag{24.10}$$

We introduced the lab observer measured rapidity of exhausted mass $\sinh y_{\text{s ex}}$. Given the need to 'burn' an additional mass δm of the spaceship to run the engine that ejected the mass Δm we have $\widetilde{m} < m$. The combustion mass defect

$$\delta m = m - \widetilde{m}\ , \tag{24.11}$$

appears immediately *after* the acceleration step is complete.

For the special case that the exhaust material is ejected against the direction of motion we can use additivity of rapidity,

$$y_{\mathrm{s\,ex}} = y - y_{\mathrm{s}\,\mathcal{V}} \,. \tag{24.12}$$

Here $y_{\mathrm{s}\,\mathcal{V}}$ is the propellant rapidity in the rocket rest-frame, where

$$\cosh y_{\mathrm{s}\,\mathcal{V}} \equiv \gamma_{\mathcal{V}} = \frac{1}{\sqrt{1-(\mathcal{V}/c)^2}}\,, \qquad \sinh y_{\mathrm{s}\,\mathcal{V}} = \gamma_{\mathcal{V}}\mathcal{V}/c\,, \qquad \tanh y_{\mathrm{s}\,\mathcal{V}} = \frac{\mathcal{V}}{c}\,. \tag{24.13}$$

We proceed to consider the energy conservation. Just before mass ejection we have

$$E_1/c^2 = (m + \Delta m)\cosh y\,. \tag{24.14}$$

After mass ejection, we combine two contributions to the energy, the rocket with increased rapidity and the residual exhaust matter

$$E_2/c^2 = \widetilde{m}\cosh(y + \Delta y) + \Delta m\, \cosh y_{\mathrm{s\,ex}}\,. \tag{24.15}$$

The conservation of momentum requires $p_2 = p_1$ and the conservation of energy $E_2 = E_1$; thus we find

$$p_2 = p_1 \quad \rightarrow \quad \widetilde{m}\sinh(y + \Delta y) + \Delta m\, \sinh y_{\mathrm{s\,ex}} = (m + \Delta m)\sinh y\,, \tag{24.16}$$

$$E_2 = E_1 \quad \rightarrow \quad \widetilde{m}\cosh(y + \Delta y) + \Delta m\, \cosh y_{\mathrm{s\,ex}} = (m + \Delta m)\cosh y\,. \tag{24.17}$$

Considering that $M^2 = E^2/c^4 - p^2/c^2$ we proceed to square both equations and evaluate the difference, which results in the exact expression for $\widetilde{m}$

$$\widetilde{m}^2 + \Delta m^2 + 2\widetilde{m}\Delta m\cosh(y_{\mathrm{s}\,\mathcal{V}} + \Delta y) = (m + \Delta m)^2\,. \tag{24.18}$$

This condition is, as it must be, exactly independent of the rapidity of the rocket; that is, independent of the choice of the reference frame of the observer.

We now can evaluate the relationship between the small quantities that enter our considerations. Using Eqs. (24.8) and (24.11) in Eq. (24.18) and expanding and keeping terms up to linear order in small deviations Δy, δm, Δm, we obtain

$$-dm = \delta m + \Delta m = \Delta m \cosh y_{\mathrm{s}\,\mathcal{V}}\,. \tag{24.19}$$

We recall that $|dm|$ is the reduction in the mass of the rocket by ejection of the material amount Δm where $\cosh y_{\mathrm{s}\,\mathcal{V}}$ is the Lorentz-γ-factor of the ejected mass. Clearly Eq. (24.19) makes good sense. Moreover, we gain additional insight when considering the nonrelativistic limit of Eq. (24.19). We use

$$\cosh y_{\mathrm{s}\,\mathcal{V}} = \gamma_{\mathcal{V}} = \frac{1}{\sqrt{1-\mathcal{V}^2/c^2}} = 1 + \frac{\mathcal{V}^2}{2c^2} + \dots \,. \tag{24.20}$$

We find the relation Eq. (24.19)

$$\delta mc^2 = \Delta m \frac{\mathcal{V}^2}{2}\,; \tag{24.21}$$

i.e., the additional mass-energy defect δmc^2 of the rocket due to relative motion of the propellant, is as expected contained within the kinetic energy of the expelled material. This mass defect δmc^2 is not incorporated in the nonrelativistic rocket equation.

To obtain a relation showing the momentum conservation in the rocket frame of reference we can combine Eq. (24.16) with Eq. (24.17) to eliminate $\widetilde{m}$ by multiplying with $\cosh(y+\Delta y)$ and $\sinh(y+\Delta y)$, respectively, and by taking the difference. This leads to the exact relation

$$\Delta m \sinh(\Delta y + y_{\mathrm{s}\,\mathcal{V}}) = (m+\Delta m)\sinh \Delta y\,, \tag{24.22}$$

which provides for small values of $\Delta m,\ \Delta y \to dy$, the relativistic analogue of the nonrelativistic rocket equation Eq. (24.5)

$$\Delta mc \sinh y_{\mathrm{s}\,\mathcal{V}} = mc\,dy\,. \tag{24.23}$$

Note that Eq. (24.23) is valid for any inertial observer since the rapidity is additive under coordinate transformations and therefore the increment of rapidity is the same in all inertial reference systems.

Thus we can equally well write Eq. (24.23) as

$$\Delta mc \sinh y_{\mathrm{s}\,\mathcal{V}} = mc\,dy_{\mathrm{s}}\,, \tag{24.24}$$

where we used increment dy_{s} with reference to an observer comoving with the space ship. This allows us to create an exact parallel behavior to the nonrelativistic rocket equation where the rocket velocity is an additive variable.

Specifically, on the left in Eq. (24.24), we see the momentum carried away by the expelled mass Δm, which is equal to the momentum increase imparted on the rocket. Note that the speed increment (in units of c) in the nonrelativistic Eq. (24.5) is replaced by a rapidity increment. In this comparison we recognize on the left the momentum carried away by the expelled mass Δm, and with the nonrelativistic factor $\mathcal{V}$ being replaced by the relativistic form $\mathcal{V} \to c\ \sinh y_{\mathrm{s}\,\mathcal{V}} = \mathcal{V}\gamma_{\mathcal{V}}$, see Eq. (24.13). The relativistic generalization introduces an additional factor $\gamma_{\mathcal{V}}$, which accounts for the change to relativistic momentum carried by the incremental mass ejection having relativistic content. The greater value of $\sinh y_{\mathrm{s}\,\mathcal{V}}$ the engine achieves, the less mass needs to be ejected in order to achieve an incremental increase in rapidity dy of the rocket.

24.3 Energy of Relativistic Rocket

In the nonrelativistic case, see Eq. (24.7), we tracked the mass of the nonrelativistic rocket. To obtain the equivalent relativistic relation, we consider the total energy of the rocket as observed from the base

$$E = m(t)c^2 \cosh y(t)\ . \tag{24.25}$$

E comprises the remaining payload and its relativistic kinetic energy. The relative change in energy of the rocket therefore contains two terms

$$dE = dmc^2 \cosh y + mc^2 dy\ \sinh y\ , \tag{24.26}$$

that is

$$\frac{dE}{E} = \frac{dm}{m} + dy\ \tanh y\ . \tag{24.27}$$

We now use Eq. (24.19) to obtain

$$\frac{dE}{E} = -\frac{\Delta m}{m} \cosh y_{\mathrm{s}\,\mathcal{V}} + dy \tanh y\ . \tag{24.28}$$

We see that the exhausted mass Δm is multiplied with $\gamma_{\mathcal{V}}$, see Eq. (24.13); thus it describes the entire energy decrease due to the operation of the rocket exhaust, while there is an incremental increase in kinetic energy of the rocket described by the last term.

We next use Eq. (24.23) to eliminate the quantity Δm in Eq. (24.28)

$$\frac{dE}{E} = dy(-\coth y_{\mathrm{s}\,\mathcal{V}} + \tanh y) = -\frac{dy}{\mathcal{V}/c} + \tanh y\, dy\ , \tag{24.29}$$

where we also used Eq. (24.13). Since $dy \tanh y = d(\ln\cosh y)$ Eq. (24.29) can be integrated to yield

$$\ln\left(\frac{E}{\cosh y}\bigg/\frac{E_0}{\cosh y_0}\right) = -\frac{y - y_0}{\mathcal{V}/c}\,. \tag{24.30}$$

Choosing $y_0 = 0$, thus $E_0 = m(y = 0)c^2 \equiv m_0c^2$, where m_0 is the mass of the fully loaded and fueled rocket at departure, we obtain

$$\frac{E/c^2}{\cosh y} \equiv m(y) = m_0 \exp\left(-\frac{y}{\mathcal{V}/c}\right). \tag{24.31}$$

Note that upon division by $\cosh y = \gamma$ in Eq. (24.30) we study the residual rest energy (or rest mass) content of the rocket in its proper frame of reference.

We recognize Eq. (24.31) as a natural generalization of the nonrelativistic result Eq. (24.7): The relativistic generalization of the rocket equation requires that we replace the nonrelativistic v/c by rapidity y. However, the exhaust speed enters as it had in the nonrelativistic case Eq. (24.7). Therefore even when $\mathcal{V}/c \to 1$, when rocket exhaust has almost light velocity, for a rocket reaching high rapidity the residual payload just like in the nonrelativistic case has an exponentially small payload.

Result Eq. (24.31) could be guessed as follows: We consider a change from mass to energy that a rocket retains and starting in the frame of an observer with $y_0 = 0$ and with the initial energy content m_0c^2 the energy is found by multiplying with the Lorentz factor $\gamma = \cosh y$. Thus Eq. (24.31) becomes

$$E = m_0c^2 \cosh y \exp\left(-\frac{y}{\mathcal{V}/c}\right) \underset{\mathcal{V}/c\to 1}{\longrightarrow} m_0c^2\frac{1}{2}\left(1 + e^{-2y}\right)\,. \tag{24.32}$$

In the limit of $\mathcal{V}/c \to 1$ (that is for rocket exhaust having almost light velocity), the original observer sees the rocket split into two halves. She sees how, in a succession of many very small steps, 'half' of the rest energy is given to the expelled and combusted material. According to the action=reaction principle the recoiling rocket retains half of the original energy content for this observer, but with a vastly smaller fraction in its rest-frame, as is seen now dividing Eq. (24.32) by $\gamma = \cosh y$, which returns us to the main result Eq. (24.31).

Example 24.1

Mass-acceleration rocket relation

Consider a spaceship accelerating with a constant proper acceleration a_s and determine the manner in which the mass of the spaceship decreases as a function of proper spaceship time while the acceleration is maintained. Check your result by comparing with Eq. (24.31). Your derivation must apply to both, to the case of a relativistic propellant speed (rapidity $y_{\mathrm{s}\,\mathcal{V}}$) and to the usual nonrelativistic chemical combustion rocket. Discuss how one must choose the propellant rapidity to maximize the payload mass fraction. ◀

▶ **Solution** We begin with Eq. (24.24), which we cast into the format

$$\mathbf{1}\quad \frac{\Delta m}{m}\sinh y_{\mathrm{s}\,\mathcal{V}} = dy_{\mathrm{s}}\,.$$

When the magnitude of dy_{s} is small, given that $y_{\mathrm{s}} = 0$ for the comoving with spaceship observer we have, see also Eq. (23.2), for the effect of acceleration on the rapidity of the rocket

$$\mathbf{2}\quad dy_{\mathrm{s}} = \frac{dv_{\mathrm{s}}}{c} = \frac{a_{\mathrm{s}}}{c}dt_{\mathrm{s}}\,.$$

Combining these conditions we obtain

$$\mathbf{3}\quad \frac{a_{\mathrm{s}}}{c}dt_{\mathrm{s}} = \frac{\Delta m}{m}\sinh y_{\mathrm{s}\,\mathcal{V}}\,.$$

However, we cannot as yet integrate this equation as the ejected mass Δm is not the total mass loss dm, see Eq. (24.8): dm includes the mass δm consumed to give Δm its energy. We use Eq. (24.19) in Eq. 3 substituting the ejected mass Δm by the total change in mass dm and we obtain

$$\mathbf{4}\quad dt_{\mathrm{s}}\,a_{\mathrm{s}} = -c\frac{dm}{m}\tanh y_{\mathrm{s}\,\mathcal{V}}\,.$$

This equation can be integrated to yield

$$\mathbf{5}\quad m(t_{\mathrm{s}}) = m_0 \exp\left(-t_{\mathrm{s}}\,\frac{a_{\mathrm{s}}}{c\tanh y_{\mathrm{s}\,\mathcal{V}}}\right).$$

We now verify the consistency of this new result with the prior result, Eq. (24.31), where we saw how spaceship mass depends on spaceship speed, i.e. rapidity. Equating the arguments seen in the exponents with that in Eq. (24.31) we find

$$\mathbf{6}\quad y - y_0 = \frac{\mathcal{V}}{c}\,\frac{t_{\mathrm{s}}\,a_{\mathrm{s}}}{c\tanh y_{\mathrm{s}\,\mathcal{V}}} = t_{\mathrm{s}}\frac{a_{\mathrm{s}}}{c}\,,$$

where according to Eq. (24.13) we used $c\tanh y_{\mathrm{s}\,\mathcal{V}} = \mathcal{V}$. This is Eq. (23.10). This confirms equivalence with Eq. (24.31). Therefore we could have integrated Eq. 2 and substituted into Eq. (24.31) to obtain Eq. 6.

In Eq. 5 we note that $\tanh y_{s\mathcal{V}} \to 1$ for a relativistic exhaust matter beam, and in fact for $y_{s\mathcal{V}} = 1.5$ we already reach 90% of maximum effectiveness; that corresponds to $\gamma = \cosh y_{s\mathcal{V}} \to 2.35$. This exhaust rapidity is easily accessible with present day accelerator technology. The physical reason why one gains so little by increasing the particle beam rapidity further is the following: Once we achieve the relativistic domain where the kinetic energy dominates the rest energy, the mass consumed to produce the higher rapidity particle beam can be also used to make the beam more intense (large number of particles). High intensity relativistic particle beams are developed and employed today to produce secondary, typically pulsed, neutron beams – as opposed to continuous neutron beams obtainable from nuclear reactors.

Another example how such a 'stardrive' could be conceived is nucleon-anti-nucleon (e.g. antiproton $\bar{p}$ annihilating with proton p) i.e. antimatter pair annihilation emitting the material products of annihilation directly without further processing. This is so since the antiproton $m_N \simeq 940\,\text{MeV}/c^2$ annihilation at rest in the center of momentum frame of the rocket is accompanied by the production, on average, of little more than 5 pions, where $m_\pi \simeq 140\,\text{MeV}/c^2$. The energy set free in the annihilation process would yield on average a pion of energy $E_\pi \simeq 2 \times 940/5\,\text{MeV} = 375\,\text{MeV}$, which for $m_\pi c^2 \simeq 140\,\text{MeV}$ corresponds to $\gamma_\pi = 375\,\text{MeV}/m_\pi c^2 = 2.65$.

In the relativistic exhaust limit with $\tanh y_{s\mathcal{V}} \simeq 1$ we can look at the rocket we considered for the Milky Way explorers in Chap. 23, where we applied $a_s = g = 9.81\,\text{m/s}^2$. According to Eq. (23.1) we thus obtain

$$\textbf{7} \quad m(t) \simeq m_0 e^{-t_s/1\,\text{year}}\,, \qquad y_{s\mathcal{V}} > 2\,.$$

For non-relativistic fuels, such as a typical chemical rocket, we have $\mathcal{V} \simeq 3\,\text{km/s}$; thus $y_{s\mathcal{V}} \simeq 10^{-5}$. The decay time constant changes from one year to a fraction $10^{-5}\times$year; that is to 320 s = 5.3 min. Three times this is the typical operational time scale of chemical rockets before fuel is exhausted, leaving behind empty tanks. It is easy to trace the origin of the small factor to the small chemical energy mass defect fraction, $10^{-9} = (3 \times 10^{-5})^2$, which is as we discussed in Chap. 16. The mass defect defines the kinetic energy of produced fragments $\propto mc^2\beta^2$.

We conclude that rockets equipped with relativistic particle drives (stardrives) achieve an enormous gain by factor 10^5 in mass efficiency. They best operate in condition such that $\tanh y_{s\mathcal{V}} \to 1$ corresponding to $y_{s\mathcal{V}} > 2$ for maximum result – the quality function here is that we use the smallest fraction of rocket mass, maximizing the payload fraction. Such a stardrive-equipped rocket is still far in the future. We recall the interesting first step, the accelerator driven DAWN mission, see end of Sect. 24.1. ◀

Discussion 24.1 How realistic is space travel?

About the topic: This conversation is meant to please *Star Trek* and *Star Wars* aficionados.

- *Student:* I chose to study physics so I can one day travel to the stars.
- *Simplicius:* Even a trip to the moon will cost you millions; how can you afford this by doing physics?
- *Professor:* If you (speaking to the student) invent a stardrive, I am sure you will, in your lifespan, be able to travel to Mars.
- *Simplicius:* What exactly is a stardrive?
- *Professor:* For me it is a device that can produce exhaust propellant at speeds with rapidity two and above in the rocket rest-frame. This assures the best use of mass of propellant. As we saw this is a natural condition for particles emerging from the matter-antimatter annihilation process.
- *Simplicius:* I wonder how many thousands of years must pass before we can load a rocket with megatons of antimatter, keeping the payload module safe in the proximity of the large antimatter fuel tank.
- *Professor:* You draw here on the last example in the book. However, it is not necessary to set the rocket acceleration to be 1 g in a first effort. We can compromise and use a continuous accelerated rocket with, say, $10^{-5}\,g$. Such a rotating (to create habitat with one g) large spaceship equipped with a micro-stardrive burns just a tiny micro-fraction of spaceship mass each year.
- *Student:* Rather than 10 years, it would take a million years to reach the speed of light.
- *Professor:* Right, but such a spaceship could reach the edge of the solar system within a decade. Such spaceships are, in my opinion, on this century's horizon.
- *Student:* Particle accelerators easily provide such capability for exhaust rapidity. Why are they not yet used?
- *Professor:* We discussed the DAWN mission, see end of Sect. 24.1. I am sure that is not the end of this idea. However, I can see challenges related to both energy efficiency and delivered power. These challenges need to be overcome before our first trans-solar system spaceship is launched.
- *Student:* In other words, we need a lot of R&D. Who pays or what is the commercial return on such a huge investment?
- *Professor:* Our society produces every decade a handful of super rich who as we see it today have the same aspirations as do many students of physics. They pay for R&D because it is fun. I also think they will ultimately become even richer.
- *Simplicius:* Sending sports cars into space or mining for gold on Mars?

(continued)

- *Professor:* That too. Some gold meteorites must have hit in the past few billion years and their material is near the surface as Mars has been pretty quiet geologically.
- *Simplicius:* Why not do this on the Moon?
- *Professor:* Please ask a planetary mining expert.
- *Student:* We can right away agree we do not need to go far. If there are interesting impactors, they are also coming to near us. We just need these capable spaceships.
- *Simplicius:* I heard that one will use the stardrive ships to land on a near flyby meteor, and while mining, catch a free ride across the solar system.
- *Student:* Yes, this helps protect the traveler from space radiation; I now see the need to work on stardrive spaceship engines that deliver relativistic exhaust velocity with high energy efficiency and at high power. Just a small question, where do I begin?
- *Professor:* I believe that high intensity lasers will do the trick, but it takes some time, we are at the very beginning of this approach. The 2018 Nobel prize recognized the importance of this novel technology.[2] For this reason we have in this book discussed in Sect. 22 the physics of relativistic particles riding on EM waves. With such pulsed lasers it is possible to create high energy propulsion particle beam. However, a few decades will pass before we reach the required power.

Further reading: Jeremy S. Heyl[3] considers how the accelerated expansion of the Universe limits our ability to explore it. This topic continues to be investigated to this day.[4]

[2]Half of the 2018 Nobel prize in physics was awarded to Gérard Mourou and Donna Strickland "for their method of generating high-intensity, ultra-short optical pulses." See also Ref. 4 and Ref. [5].

[3]Jeremy S. Heyl, "The long-term future of space travel" *Phys. Rev.* **D 72**, 107302 (2005).

[4]Morteza Kerachian, "Uniformly accelerated traveler in an FLRW universe" *Phys. Rev.* **D 101**, 083536 (2020).

Index

J. Rafelski, *Modern Special Relativity*, https://doi.org/10.1007/978-3-030-54352-5